Lecture Notes in Civil Engineering

Volume 518

Series Editors

Marco di Prisco, Politecnico di Milano, Milano, Italy

Sheng-Hong Chen, School of Water Resources and Hydropower Engineering, Wuhan University, Wuhan, China

Ioannis Vayas, Institute of Steel Structures, National Technical University of Athens, Athens, Greece

Sanjay Kumar Shukla, School of Engineering, Edith Cowan University, Joondalup, Australia

Anuj Sharma, Iowa State University, Ames, USA

Nagesh Kumar, Department of Civil Engineering, Indian Institute of Science Bangalore, Bengaluru, India

Chien Ming Wang, School of Civil Engineering, The University of Queensland, Brisbane, Australia

Zhen-Dong Cui, China University of Mining and Technology, Xuzhou, China

Xinzheng Lu, Department of Civil Engineering, Tsinghua University, Beijing, China

Lecture Notes in Civil Engineering (LNCE) publishes the latest developments in Civil Engineering—quickly, informally and in top quality. Though original research reported in proceedings and post-proceedings represents the core of LNCE, edited volumes of exceptionally high quality and interest may also be considered for publication. Volumes published in LNCE embrace all aspects and subfields of, as well as new challenges in, Civil Engineering. Topics in the series include:

- Construction and Structural Mechanics
- Building Materials
- Concrete, Steel and Timber Structures
- Geotechnical Engineering
- Earthquake Engineering
- Coastal Engineering
- Ocean and Offshore Engineering; Ships and Floating Structures
- Hydraulics, Hydrology and Water Resources Engineering
- Environmental Engineering and Sustainability
- Structural Health and Monitoring
- Surveying and Geographical Information Systems
- Indoor Environments
- Transportation and Traffic
- Risk Analysis
- Safety and Security

To submit a proposal or request further information, please contact the appropriate Springer Editor:

- Pierpaolo Riva at pierpaolo.riva@springer.com (Europe and Americas);
- Swati Meherishi at swati.meherishi@springer.com (Asia—except China, Australia, and New Zealand);
- Wayne Hu at wayne.hu@springer.com (China).

All books in the series now indexed by Scopus and EI Compendex database!

Ping Xiang · Haifeng Yang · Jianwei Yan
Editors

Frontier Research on High Performance Concrete and Mechanical Properties

 Springer

Editors
Ping Xiang
Central South University
Changsha, Hunan, China

Haifeng Yang
Guangxi University
Nanning, China

Jianwei Yan
East China Jiaotong University
Nanchang, China

ISSN 2366-2557 ISSN 2366-2565 (electronic)
Lecture Notes in Civil Engineering
ISBN 978-981-97-4089-5 ISBN 978-981-97-4090-1 (eBook)
https://doi.org/10.1007/978-981-97-4090-1

© The Editor(s) (if applicable) and The Author(s) 2025. This book is an open access publication.

Open Access This book is licensed under the terms of the Creative Commons Attribution 4.0 International License (http://creativecommons.org/licenses/by/4.0/), which permits use, sharing, adaptation, distribution and reproduction in any medium or format, as long as you give appropriate credit to the original author(s) and the source, provide a link to the Creative Commons license and indicate if changes were made.
The images or other third party material in this book are included in the book's Creative Commons license, unless indicated otherwise in a credit line to the material. If material is not included in the book's Creative Commons license and your intended use is not permitted by statutory regulation or exceeds the permitted use, you will need to obtain permission directly from the copyright holder.
The use of general descriptive names, registered names, trademarks, service marks, etc. in this publication does not imply, even in the absence of a specific statement, that such names are exempt from the relevant protective laws and regulations and therefore free for general use.
The publisher, the authors and the editors are safe to assume that the advice and information in this book are believed to be true and accurate at the date of publication. Neither the publisher nor the authors or the editors give a warranty, expressed or implied, with respect to the material contained herein or for any errors or omissions that may have been made. The publisher remains neutral with regard to jurisdictional claims in published maps and institutional affiliations.

This Springer imprint is published by the registered company Springer Nature Singapore Pte Ltd.
The registered company address is: 152 Beach Road, #21-01/04 Gateway East, Singapore 189721, Singapore

If disposing of this product, please recycle the paper.

Preface

This book is based on a select collection of contributions from the academic conferences 2023 5th International Conference on Hydraulic, Civil and Construction Engineering (HCCE 2023) and 2023 9th International Conference on Hydraulic and Civil Engineering (ICHCE 2023), scientific research institutes, among others. This book aims to present the essence of the academic conferences and explore the importance and significance high-performance concrete in construction projects.

Concrete has become a main construction material all over the world in the past century. As the demand for construction rises, the need for concrete with stronger performance grows as well. This book brings together papers from leading experts and researchers from around the world who have made important breakthroughs and achievements in cutting-edge explorations in high-performance concrete and mechanical properties.

The content of this book analyzes study cases for the utilization of high-performance concrete in engineering practice, providing a comprehensive and systematic analysis of contributing factors to the performance of high-performance concrete. Through rigorous research methods and empirical analysis, these papers demonstrate the wide application of high-performance concrete in transportation, medical, industry, and other fields. They reveal the potential of high-performance concrete and its mechanical properties to increase productivity and improve the quality of construction projects in real world.

We sincerely hope that this book will stimulate readers' interest in the research and application of high-performance concrete and its mechanical properties, and promote exchanges and cooperation between academia and industry. We firmly believe that by sharing knowledge and experience, we can further advance the research and application in these domains.

Finally, we would like to thank all the authors and editorial teams who contributed to this book. Their efforts and enthusiasm made the publication of this book possible. We hope this book will inspire you and serve as a reference for your exploration in the fields of and related to high-performance concrete and its mechanical properties.

February 2024   The Editor in Chief

Contents

Mechanical Properties and Compressive Resistance Testing of High Performance Concrete

Study on the Influence of Basalt Fiber on Concrete Performance 3
Peipei Wei, Xiaofei Zhang, Chuan Yin, Jiulong He, Ming Kong, Xiaoxuan Chen, and Runxue Yang

Mechanical Behaviors of Encased Concrete of Composite Bridges with Corrugated Steel Web 15
Yaojun Wang, Bingbing Liu, and Qianqian Hao

Safety Analysis of Concrete Arch Dam Considering the Shear Action of Contraction Joints 27
Zhichao Miao and Guohua Liu

Experimental Study on Bearing Behaviour of Bored Piles Using End-side Association Post-grouting Technology 39
Shiding Su, Shuhui Lv, Jiaqi Wu, and Bo Zhang

A Procedure for Strength Parameters Inversion of Rock Slopes Based on Statistical Regression and Seepage-Stress Coupling: A Case Study 49
Rong Chen, Detan Liu, Haiku Zhang, Mao Liao, Zhangxin Huang, and Liqun Xu

Experimental Study on Creep Behavior of Prestressed Concrete 61
Tao Ge and Le Yan

Experimental Study on Mechanical Properties of Permeated Crystalline Concrete 71
Jingjing He, Haodan Lu, Haiting Wang, Wei Hu, Wenbo Wu, and Kunlong Zhao

Study on Seismic Performance of Waste Fiber Recycled Concrete Column Reinforced with CFRP 81
Jinghai Zhou, Shouyu Li, Jiagui Zhou, and Tianbei Kang

Sensitivity Analysis of Influencing Factors of Vertical Bearing Capacity of Rock-Socketed Piles Based on Orthogonal Test 93
Zongjun Sun, Yingfeng Han, Fei Liu, and Rui Min

The Influence of Hybrid Fiber Ratio on the Mechanical Properties of Concrete 107
Yabin He, Lei Yu, Sheng Chen, Chuankai Yan, Zhansheng Lin, and Chen Yu

Study on the New Preparation Method and Properties of Green Self-Compacting Transparent Concrete 117
Junliang Liu, Guangxiu Fang, Jiaxing Lu, and Han Jin

Superposed Element Method for the Temperature Field Simulation in Mass Concrete Structures Containing Cooling Pipes 129
Jianxin Ding and Qingzhou Yang

Effect of Machine-Made Sand Rate on the Compressive Strength, Workability, and Impermeability of Sleeper Concrete 141
Zhenchao Liu

Comparative Study of Tensile Capacity Testing Methods for Metal Connectors Used in Precast Concrete Insulated Sandwich Wall Panels 151
Puyan Wang, Feng Tu, Kai Shu, Yi Zhao, Weijun Zhong, and Jian Zhou

Basic Properties and Microstructure of Coal Gangue Pervious Concrete Under Acid Rain Environment 165
Junwu Xia, Linli Yu, Zhichun Zhu, Pengxu Li, Yuan He, and Jun Yu

Experimental Study on Axial Compressive Properties of Early Strength and High Ductility Cement-Based Composite Concrete 177
Weihong Jiang, Wenhong Duan, Jiaquan Yuan, Li Xiong, Huimei Li, Lin Mou, Xiaohua Yang, Xiaomin Huang, Weibing Xu, and Kun Yang

Numerical Simulation Analysis of Mechanical Properties of Semi-rigid Immersed Tube Tunnel Joints 193
Hai Ji, Yonggang Lv, and Qingfei Huang

Research on Anti-shear Performance of Steel-Mixed Combination Beam Bridge Single-Nail Shear Connector 209
Haiyuan Yang

Research on the Fracture Evolution Characteristics
and Mesoscopic Fracture Mechanism of Fissured Basalt Based
on PFC^{2D} 221
Jun Chen, Ning Liu, Chaoyi Wang, and Yaohui Gao

**Proportioning Design and Material Development of High
Performance Concrete**

Temperature Control Sensitivity Analysis and Research
on Thin-walled Hydraulic Tunnel Lining Concrete 239
Bu Zhang and Zhenhong Wang

Breakage Mechanism of Artificial Granular Materials 253
Longjiang Fan, Enlong Liu, Shijia Tang, and Yanlin Qin

Effect of Mesoscopic Heterogeneity of Concrete
on the Macro-mechanical Behavior 263
Qindong Lin, Chun Feng, Jianfei Yuan, Wenjun Jiao, and Yundan Gan

Experimental Study on Splitting Strength of Nano-active Powder
Concrete 271
Hailing Bao, Xiaofeng Ji, and Peibao Xu

Dynamic Characteristics Test and Microstructure Analysis of Silt
Soil Improved by Curing Agent 281
Qi Lu, Yongzhen Ma, Ganbin Liu, and Fuxin Ni

Numerical Analysis of Soil-Rock Mixture Subgrade Based on High
Density Resistivity Surveys 297
Jiangong Chen and Diming Lou

Experimental Study of the Fiber Improvement Test
on the Unsaturated Soil 311
Lina Guo, Yun Chen, and Minmin Luo

Experimental Study on Physical and Mechanical Characteristics
and Microstructure of Sandstone After High Temperature-Water
Cooling Treatment 323
Jinbin Lu, Lifeng Zheng, Feng Chen, Liang Yang, and Qiang Zhang

Investigating Effect of Bentonite Support Fluid on Soil-Pile
Interface Behavior by Ultra-Weak Fiber Bragg Gratings 339
Zhihong Li, Yu Gu, Xiaonan Jia, Xuehui Hu, Xuqun Zhang,
and Zhaofeng Li

Study on Improvement Measures of Hydraulic Engineered
Cementitious Composites Layer Bonding Performance 353
Yupu Wang, Jiazheng Li, and Yan Shi

Microscopic Investigation of Granular Materials in Filter Layer Based on LBM-DEM Method 363
Qirui Ma, Xing Peng, and Congpeng Zhang

Experimental Study on Strength of Luminous Concrete with Double Admixture of Fly Ash and Slag Powder 375
Meng Li, Guangxiu Fang, Haonan Wu, Chunming Wang, Huaiyu Li, and Zhoutong Li

Research on Design Method and Optimization of New Epoxy Resin Concrete Mix Ratio 389
Haoran Xu, Guangxiu Fang, and Baiyang Xue

Research on Service and Crack Control of Concrete in Ultra-High Altitude Environment 399
Zaifeng Yao, Lei Liu, Shuanye Han, Yaning Wang, and Xiang Lv

The Mix Proportion Optimization Design of Coal Gangue Pervious Concrete 411
Junwu Xia, Chao Luo, and Enlai Xu

Research on Axial Tensile Mechanical Properties of Early-Strength High Ductility Cementitious Composites 423
Wenhong Duan, Jiaquan Yuan, Li Xiong, Weihong Jiang, Huimei Li, Lin Mou, Xiaohua Yang, Xiaomin Huang, Weibing Xu, and Kun Yang

Development and Application of High Permeability and Low Shrinkage Synchronous Grouting Materials 441
Quanwei Liu, Zhijing Zhu, Weihao Li, Shoujie Ye, Rentai Liu, Mengjun Chen, and Linsheng Liu

Development and Field Application of Self-compacting and Highly Impermeable Backfill Materials 457
Peng Liu, Quanwei Liu, Shoujie Ye, Jia Yan, Rentai Liu, Mengjun Chen, Chao Zong, and Jinyan Jiang

Preparation and Blast Responses of Basalt Fiber-Reinforced Polymer (BFRP) Bar Reinforced Shield Tunnelling Segments 473
Ruiyi Jiang, Jiang Feng, and Min Hou

Seismic Behavior of Steel-Polypropylene Hybrid Fiber Reinforced Concrete Shear Wall 481
Luyang Zhang, Jitao Yao, and Yuting Tong

Investigation of Dynamic Response of Concrete Slab Under Air Blast Loading 493
Qindong Lin, Chun Feng, Yundan Gan, Jianfei Yuan, and Ying Yang

Research on Cumulative Damage of Quasi-Static Reinforced Concrete Short Columns with Low Cycle Fatigue 501
Hongyu Zhou, Juxin Guo, Qi Tang, and Haoda Wang

Study on the Damage Evolution Law of Railway High Pier of New Replaceable Components Under Near-Fault Ground Motion 513
Xudong Zhang, Xiushen Xia, and Heng Zhang

Damage Assessment of RPC Strengthened RC Columns Subjected to Blast Loading 531
Yu Fu, Siyuan Qiu, Zhifu Yu, Juan Su, and Xiaomeng Hou

Mechanical Properties and Compressive Resistance Testing of High Performance Concrete

Study on the Influence of Basalt Fiber on Concrete Performance

Peipei Wei, Xiaofei Zhang, Chuan Yin, Jiulong He, Ming Kong, Xiaoxuan Chen, and Runxue Yang

Abstract Basalt fiber has excellent quality, and adding appropriate amount of basalt fiber into concrete can enhance the mechanical properties and durability of concrete. Basalt fiber concrete has a broad development and application prospect. This paper summarizes the research progress of basalt fiber concrete in recent years and its application in water conservancy projects, analyzes the reinforcement mechanism of basalt fiber in concrete, and puts forward some suggestions for further research and development of basalt fiber concrete.

Keywords Basalt fiber · Concrete · Mechanical properties · Durability

1 Preface

As the most widely used material in the construction industry, concrete has the advantages of simple preparation process, relatively low cost, high compressive strength, good moldability and integrity, good durability and strong fire resistance [1]. According to statistics [2, 3], from 2010 to 2021, the output of commercial concrete in China increased from 1160 to 3293.3 million m^3, with an increase rate of 183.91%. With the construction of high-face rockfill dams, high arch dams, pumped storage power stations and other buildings subjected to high-speed water flow and sediment erosion in China's hydraulic construction, more and more hydraulic projects are built under dry, extremely cold and large temperature difference conditions, which means that the demand for concrete performance is more demanding. Improving and optimizing concrete performance is an urgent problem to be solved. Adding fibers into concrete to enhance concrete performance is a research hotspot at present [4].

P. Wei (✉) · C. Yin · J. He · M. Kong · X. Chen · R. Yang
Huaneng Lancang River Hydropower Inc Nuozhadu Power Plant, Yunnan, Pu'er 665000, China
e-mail: 624822930@qq.com

X. Zhang
State Key Laboratory Base of Eco-hydraulic Engineering in Arid Area, Xi'an University of Technology, Shaanxi, Xi'an 710048, China

© The Author(s) 2025
P. Xiang et al. (eds.), *Frontier Research on High Performance Concrete and Mechanical Properties*, Lecture Notes in Civil Engineering 518,
https://doi.org/10.1007/978-981-97-4090-1_1

Basalt fiber (BF), as a green environmental protection material in the twenty-first century, has excellent compatibility, simple manufacturing process, low cost and stable physical properties, and is a leader in the fiber industry. A large number of studies show that the addition of BF can improve the mechanical properties and durability of concrete. Through reading a large number of documents, this paper summarizes the research status of BF concrete, and provides ideas for the application of BF in water conservancy and civil engineering.

2 Study on Static Mechanical Properties of Basalt Fiber Reinforced Concrete

2.1 Effect of Basalt Fiber Content on Static Mechanical Properties of Concrete

As shown in Fig. 1, Long [5–17] and others explored the effect of BF content between 0.1 and 0.6% on the compressive strength of concrete. It can be seen that the addition of appropriate amount of BF can improve the strength of concrete specimens. The influence of BF content on the strength of concrete is basically that the compressive strength increases with the increase of fiber content. When the content exceeds a certain threshold, the fiber will have a negative effect, resulting in a decrease in the strength of the specimen. In the literature of this statistics, the fiber content between 0.1 and 0.3% reaches the threshold in the majority, and some literature reaches the threshold outside this interval. Among them, the largest increase in compressive strength is literature [16], the fiber content is 0.2%, and the strength increase rate is as high as 46%.

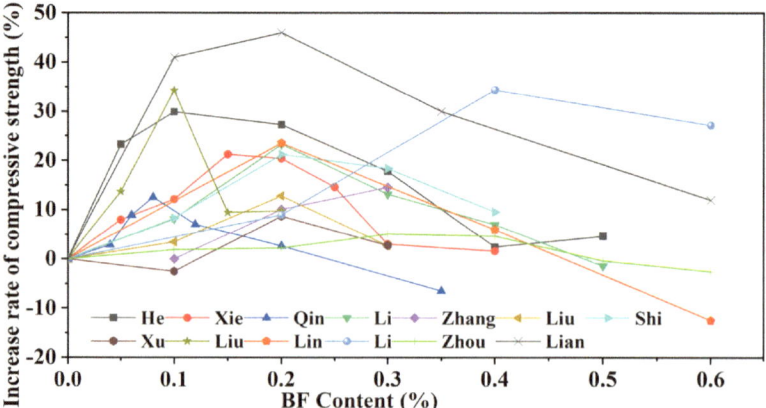

Fig. 1 Effect of BF content on compressive strength [5–17]

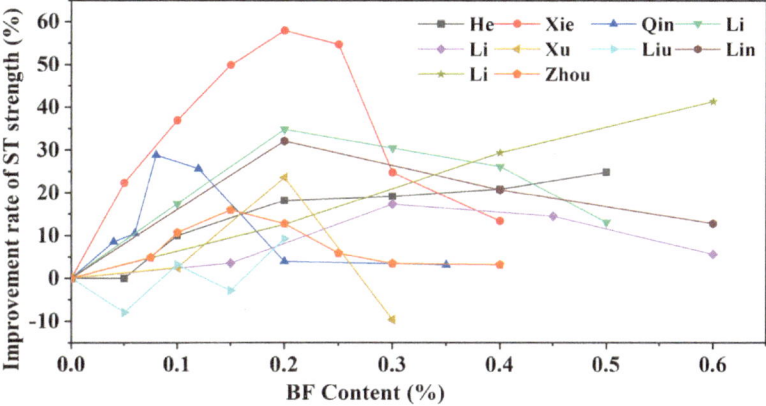

Fig. 2 Effect of BF content on splitting tensile strength [5–8, 12–15, 18, 19]

As shown in Fig. 2, Refs. [5–8, 12–15, 18, 19] explored the effect of BF content on the splitting tensile strength. It can also be found that the fiber content can improve the splitting tensile strength of the specimen. The law is that with the increase of fiber content, the strength of the specimen increases first and then decreases. When the fiber content exceeds the optimal content, the strength of the specimen decreases. As the BF content continues to increase, the strength of the specimen will even be lower than the strength of the ordinary specimen. In Refs. [5, 15], when the content is 0–0.6%, there is no inflection point in the splitting tensile strength curve, so it is impossible to judge whether 0.6% is the threshold of its content. In these statistical literatures, when the BF content is 0.2%, the number of literatures in which the splitting tensile strength of the specimen increases to the maximum value is the majority. The literature [6] has the largest increase in the splitting tensile strength of the specimen. The fiber content is 0.2%, the increase rate is 57.9%.

As shown in Fig. 3, Literature [7, 8, 11, 13, 16, 19] counts the influence of BF content on the flexural strength of the specimen, and its law is consistent with the compressive test and splitting tensile test. The flexural strength is improved the most in literature [15], with the fiber content of 0.4% and the improvement rate of 42.35%.

The mechanism of BF reinforced concrete mainly includes the following two points: ① At the initial stage of concrete pouring and mixing, the internal structure of concrete is improved by the grid structure formed by random distribution; ② When external force acts, the strength and toughness of concrete are enhanced mainly by pulling out and breaking.

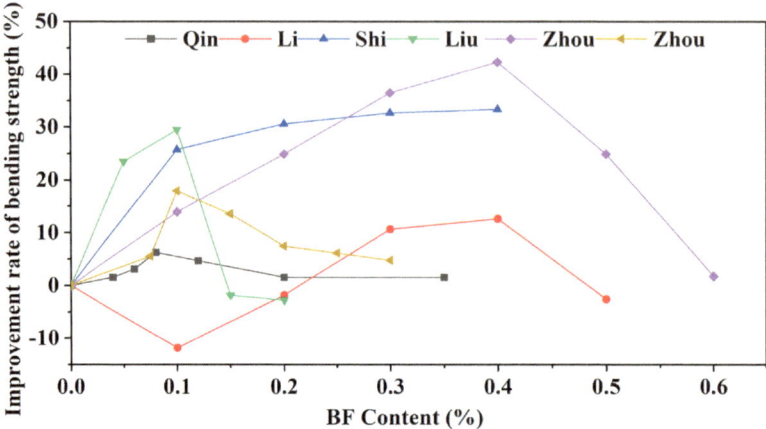

Fig. 3 Effect of BF content on flexural strength [7, 8, 11, 13, 16, 19]

2.2 Influence of Basalt Fiber Length on Static Mechanical Properties of Concrete

The research of Ruijun [20] shows that the strength enhancement effect of specimens with equal proportions of 6, 12 and 18 mm is better than that of single doping, because the fibers with three sizes are mixed to form a mesh lap with better stress, and the external force can be distributed more evenly throughout the interior. Shuhao [7] explored the BF concrete with 6, 18, 24, 30 and 48 mm, and found that the reinforcement effect was the best when the length was 24 mm; Zhijie [21] explored the influence of BF with 6, 16 and 50 mm length on the strength of specimens, and the reinforcement effect was 16 mm > 50 mm > 6 mm. The conclusion of literature [7, 21] is that too short fibers have no obvious enhancement of tensile strength and weak ability to inhibit cracks, while too long fibers are easy to reunite and intertwine, and the intertwined and unevenly dispersed BF will easily cause more weak surfaces and original defects in the specimen.

3 Study on Dynamic Mechanical Properties of Basalt Fiber Reinforced Concrete

Guangwei [22] compared the abrasion resistance of BF with the content of 0.8 kg/m^3 and polypropylene fiber concrete with the content of 0.9 kg/m^3, and pointed out that the former had better impact toughness than the latter. Changyu [23] pointed out that under high-speed impact load, BF concrete shows obvious strain rate effect: the dynamic compressive strength, ultimate strain and dynamic strength growth factor DIF of reactive powder concrete with the same BF volume fraction increase with the increase of strain rate, and DIF has a linear relationship with the logarithm of strain rate. Genshe [24] found that the appropriate amount of BF can significantly enhance the impact toughness of concrete, and the reasonable dosage range of BF is 5–10 kg/m^3. Jinyong [25] studied the impact compression toughness of BF concrete. In the range of 0–0.02 strain, the toughness of concrete with 0.2% BF content is the highest, especially in the range of 50/S–60/S strain rate, which is 12.7% higher than that of plain concrete. Jinyu [26] showed that 0.1% BF can significantly improve the impact compressive strength and toughness of the specimens, and the dynamic strength growth factor of BF concrete has a linear function relationship with the logarithm of strain rate, and the peak strain and the logarithm of strain rate have a quadratic polynomial function relationship. Study on durability of basalt fiber reinforced concrete.

3.1 Effect of Basalt Fiber on Specimen After Freezing and Thawing

Figure 4 [27–36] counts the influence of BF on the quality of specimens after freezing and thawing. From Fig. 4, it can be seen that BF can reduce the loss rate of the quality of the specimen after freezing and thawing, and the loss rate in some literatures has been in a downward trend with the increase of BF, and the improvement effect in some literatures first increases and then decreases with the increase of BF, and the loss rate corresponding to some BF contents even exceeds that of ordinary specimens, and the optimal dosage of the specimen in each literature is also different. The reason is that the water-cement ratio, concrete strength grade, fiber size and test conditions used in different documents are different, so the final optimal dosage is not consistent. For the literature whose loss rate has been declining, the reason is that the content set in the test has not reached the threshold of BF content, so the optimal content will inevitably appear with the gradual increase of BF. When the BF content exceeds the threshold under this ratio, it is well known that the extremes meet, too much BF will lead to the increase of internal defects in concrete, and the loss rate will inevitably rise.

3.2 Influence of BF on Permeability of Concrete

Figure 5 summarizes the influence of BF on the chloride ion penetration resistance of concrete in Refs. [18, 31, 32, 37–43]. There are two main rules: one is that the chloride ion electric flux decreases first and then increases with the gradual increase of BF; the second is that the chloride ion electric flux shows only a downward trend with the gradual increase of BF. The optimal dosage of each group of tests is also different. The reason is that the test conditions, mix ratio, material source

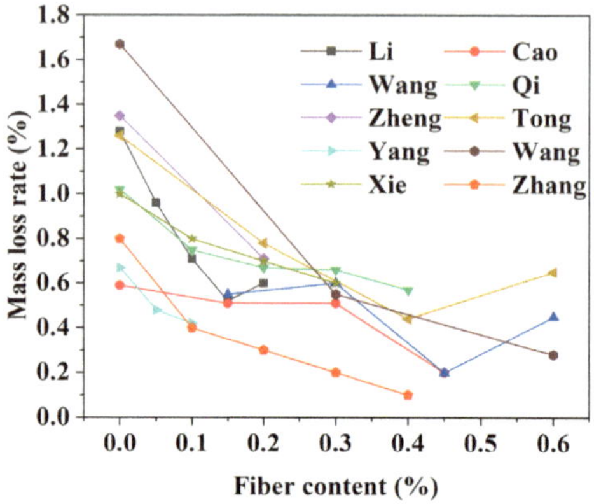

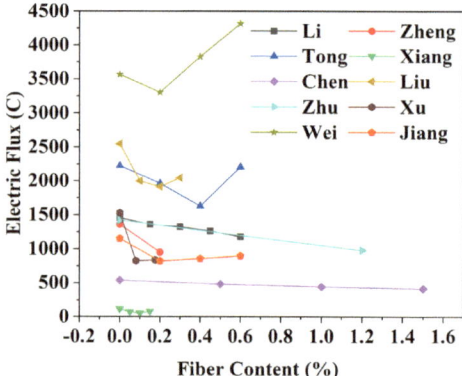

Fig. 4 Effect of BF dosage on mass loss under freeze–thaw action [27–36]

Fig. 5 Influence of BF content on electric flux [18, 31, 32, 37–43]

and number of test groups are different. Therefore, different conclusions have been drawn in various literatures. However, in general, the appropriate amount of BF can improve the chloride ion penetration resistance of concrete.

3.3 Influence of Basalt Fiber on Corrosion Resistance of Concrete

Figure 6 summarizes the influence of BF content in 8 Refs. [32, 44–50] on the strength corrosion coefficient of concrete. It can be found that the strength corrosion coefficient basically increases first and then decreases with the increase of fiber content, while there is a trend of only increasing in individual literatures. For the literature that only shows an upward trend, it is because its dosage has not reached the extreme value under this ratio. After BF is added to concrete, on the one hand, the internal pores of the specimen are reduced, that is, the diffusion channels of corrosive

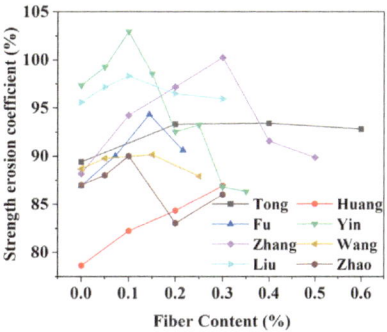

Fig. 6 Effect of BF content on strength corrosion coefficient [32, 44–50]

substances are reduced; In addition, the random distribution of BF can inhibit the generation and propagation of internal cracks in the specimen; However, too much BF will have the opposite effect, which will increase the internal defects of concrete, thus reducing the performance of the specimen. To sum up, proper BF can improve the corrosion resistance of concrete.

4 Application of Basalt Fiber in Water Conservancy Project

After many years of operation, the overflow structure of Xin 'anjiang Hydropower Station [51] has a large area of cracks and voids. The maintenance personnel have tried to repair it with various materials, but the final results are not satisfactory. In October 2010, the staff used mortar mixed with BF to repair it, and the BF content was 0.6%. The repaired building passed the subsequent flood discharge test, which opened a new chapter for the application of BF in water conservancy projects.

In 2011, Gongboxia Hydropower Station [51, 52] used BF concrete to repair the cracks and voids on the overflow floor. After the flood discharge in 2012, the repaired cracks were not damaged. In addition, the hydraulic structures repaired and strengthened with BF concrete include the spillway floor of Longyangxia Hydropower Station, the spillway retaining wall of Suiqingtang Reservoir Dam (BF content: 3 kg/m^3), the toe protection of Geshan Bridge (BF content: 2.5 kg/m^3), the channel project of Tongshanyuan Reservoir (BF content: 3 kg/m^3) and the restoration project of Xiushui River in Deyang City. The successful application of BF in these projects provides a very valuable reference for the future development of BF in water conservancy projects.

5 Conclusion

(1) Appropriate amount of BF can improve the static mechanical properties of concrete, such as compressive strength, splitting tensile strength and flexural strength. It can also improve the dynamic mechanical properties of concrete.
(2) Too short BF has a weak ability to suppress cracks and has little effect on splitting tensile strength, while too long fibers are prone to agglomeration and entanglement, which can easily cause more weak surfaces and original defects in the sample.
(3) Appropriate BF can improve the frost resistance, permeability and corrosion resistance of concrete, and improve the durability of concrete.
(4) There is an optimal BF content in concrete with different proportions. When the BF content is optimal, the reinforcement effect of concrete is maximized, which greatly improves the performance of concrete. However, when the BF content exceeds this threshold, the BF will have a negative effect, the reinforcement effect will be reduced, and its strength will even be worse than that of ordinary specimens.

Considering that BF has the advantages of green environmental protection, corrosion resistance, high temperature resistance, low cost, simple preparation process, light weight, good elasticity, etc., and the density and composition are similar to those of concrete, and the compatibility with concrete is good. These advantages make the application of basalt fiber in concrete have good competitiveness. In the future development of basalt fiber, its application in water conservancy projects needs to be considered as follows:

(1) The content, length, dispersion, concrete strength grade, ratio and test conditions of BF will all affect the strengthening effect of BF, so a lot of tests are needed to conduct a more comprehensive and systematic study on its performance.
(2) Under the impact of high-speed turbulence in water conservancy projects, a series of problems such as cavitation and wear often occur in overflow surfaces and spillway tunnels, and the dam body is subjected to cyclic and repeated loads in the seawater corrosion environment for a long time, so it is of great significance to study the performance of concrete members strengthened with basalt fiber under fatigue loads.
(3) At present, the research on BF concrete is mostly confined to the study of single influencing factor, but in practical engineering, concrete is often influenced by many factors, so it is essential to study the changes of mechanical properties of fiber reinforced concrete under the influence of many factors.

References

1. Liang Z (2019) Study on mechanical properties and microscopic mechanism of hybrid fiber modified concrete. Chang'an University
2. China Concrete Network (2020) Production and market analysis of commercial concrete in China in 2020 [EB/OL]. http://www.cnrmc.com/news/show.php?itemid=121658.htm
3. National Development and Reform Commission of the People's Republic of China (2021) Operation of building materials industry in 2021 [EB/OL]. https://www.ndrc.gov.cn/fgsj/tjsj/jjyx/mdyqy/202201/t20220130_1314182.html?code=&state=123
4. Peipei W (2022) Experimental study on properties of concrete mixed with basalt fiber and nano-SiO_2. Xi'an University of Technology
5. Long H (2022) Optimization design and mechanical properties of basalt fiber reinforced concrete. Shanxi Archit 48(22):114–117
6. Jindong X, Liang W, Zhihong L et al (2022) Experimental study on mechanical properties of chopped basalt fiber reinforced concrete. J Guizhou Univ (Nat Sci Ed) 39(04):105–109
7. Shuhao Q (2022) Experimental study on mechanical properties of new basalt fiber reinforced concrete. Guizhou University
8. Xitong L (2022) Study on durability of basalt fiber reinforced concrete under compound salt erosion. Tarim University
9. Yuting Z, Hong M, Shimao L (2021) Experimental study on mechanical properties of basalt fiber reinforced concrete. Bulk Cement 06:150–152
10. Yongwang L, Lei L, Xiang M (2022) Study on mechanical properties of basalt fiber reinforced concrete under different temperature and dosage conditions. Chem Miner Proces 51(06):26–31
11. Lei S (2021) Experimental study on mechanical properties of basalt fiber reinforced concrete based on acoustic emission method. Hebei University of Engineering
12. Xuefeng X (2021) Experimental study on mechanical properties of basalt fiber reinforced concrete with different wetting agents. Southwest Jiaotong University
13. Qian L (2021) Flexural performance test of chopped basalt fiber reinforced concrete beams. Southwest Jiaotong University
14. Jiajian L, Tianxia W, Guodong S et al (2020) Mechanical properties of quasi-static basalt fiber reinforced concrete and its BP neural network prediction model. Protect Eng 42(06):8–17
15. Dechao L, Chenxi Z (2020) Research on basic mechanical properties of basalt fiber reinforced concrete. Highway 65(06):237–241
16. Hao Z, Bin J, Hui H et al (2019) Experimental study on compressive and flexural mechanical properties of basalt fiber reinforced concrete. Ind Build 49(8):147–152
17. Jie L, Yongxin Y, Meng Y et al (2007) Experimental study on mechanical properties of chopped basalt fiber reinforced concrete. Mater Des 2007(06):8–10
18. Fuhai L, Hao G, Huiqi T et al (2022) Experimental study on basic properties of chopped basalt fiber reinforced concrete. J Railway Sci Eng 2022
19. Wei Z, Xinzhong W, Guoping H et al (2018) Experimental study on flexural strength and splitting strength of basalt fiber reinforced concrete. Municipal Technol 36(05):211–213
20. Ruijun L, Aimin G, Anyao L et al (2022) Effect of volume fraction and length of basalt fiber on mechanical properties of concrete. Jiangxi Build Mater 07:26–28
21. Zhijie W, Cheng X, Jiawei W et al (2020) Study on the influence of basalt fiber length on the mechanical properties of shotcrete. Tunnel Constr 40(S1):9–16
22. Guangwei L, Qiaoling M (2013) Effect of fiber on the abrasion resistance of hydraulic high performance concrete. Des Hydropower Station 29(02):87–91
23. Changyu L, Qirui W, Liyun Y et al (2022) Mechanical behavior and constitutive relationship of basalt fiber reactive powder concrete under impact load. Mater Guide 36(19):103–109
24. Genshe Y, Zongcai D, Jue W et al (2022) Experimental study on axial tension and impact toughness of basalt fiber high performance concrete. Concrete Cement Prod 2022(3):56–60
25. Jinyong Y, Jinyu X, Weimin L et al (2008) Impact compressive toughness of basalt fiber reinforced concrete. New Build Mater 06:69–72

26. Jinyu X, Feilin F, Erlei B et al (2010) Dynamic mechanical properties of basalt fiber reinforced concrete. J Undergr Space Eng 6(S2)
27. Zhen L (2020) Experimental study on salt frost resistance and mechanical properties of basalt fiber reinforced concrete. North China University of Water Resources and Electric Power
28. Lei C (2017) Basalt-polypropylene hybrid fiber concrete freeze-thaw cycle test research. Harbin Engineering University
29. Zituo W (2017) Experimental study on frost resistance of basalt fiber reinforced concrete. Harbin Engineering University
30. Guihua Q (2016) Experimental study on frost resistance of basalt fiber reinforced concrete. Jilin University
31. Peijin Z (2019) Experimental study on mechanical properties and durability of basalt fiber slag powder concrete. Liaoning University of Science and Technology
32. Huan T (2020) Experimental study on durability of basalt-PVA hybrid fiber reinforced concrete. Liaoning University of Technology
33. Jun Y (2018) Research on the effect of basalt fiber on the performance of high performance concrete. Southwest Jiaotong University
34. Liqiang W (2014) Experimental study on frost resistance of basalt fiber reinforced concrete. Inner Mongolia University of Technology
35. Xiaojie X, Yonggui W (2017) Research on freeze-thaw resistance of basalt fiber reinforced concrete. Concrete 09:108–111
36. Kaixiang Z (2018) Experimental study on frost resistance of basalt fiber reinforced sea sand concrete. Harbin Engineering University
37. Zhen X, Dehong W, Renjie S et al (2019) Experimental study on durability of basalt fiber reactive powder concrete. Concr Cement Prod 07:43–46
38. Feng C, Shenghao T, Jianhua R et al (2021) Study on impermeability of basalt fiber cement soil. J Shenzhen Univ (Sci Eng Ed) 38(02):157–162
39. Haozhe L (2017) Experimental study on chloride ion penetration resistance of basalt-polypropylene hybrid fiber reinforced concrete. Harbin Engineering University
40. Huajun Z (2009) Experimental study on durability of basalt fiber reinforced concrete. Wuhan University of Technology
41. Tiancheng X (2017) Experimental study on mechanical and durability properties of basalt fiber self-compacting concrete. Structural Engineering of Shenyang University
42. Kang W, Ben L, Qiao S (2022) Study on the improvement of chloride ion permeability of recycled concrete by basalt fiber
43. Zhongyou J (2021) Basalt fiber high strength and high titanium heavy slag concrete foundation performance and plate bending test research. Xihua University
44. Huang F (2021) Study on the effect of mixed fiber on the performance of recycled concrete. Xi'an University of Technology
45. Xiaorui F (2021) Experimental study on durability of basalt-cellulose hybrid fiber concrete. Liaoning University of Technology
46. Yulong Y (2015) Study on mechanical properties and durability of basalt fiber reinforced concrete. Chongqing Jiaotong University
47. Lanfang Z, Daofeng W (2018) Effect of basalt fiber content on sulfate corrosion resistance and impermeability of concrete. Silicate Bull 37(06):1946–1950
48. Kejian W (2020) Experimental study on mechanical properties and sulfate resistance of silica fume/basalt-polypropylene hybrid fiber reinforced concrete. Zhongyuan University of Technology
49. Yun L (2017) Erosion test and mechanism of basalt fiber reinforced concrete in acid environment. Xi'an University of Technology
50. Kai Z (2018) Experimental study on durability of basalt fiber reinforced concrete under sulfate attack environment. Xi'an University of Technology
51. Junzhi Z, Yanhong G, Xiaohua Z et al (2017) Basalt fiber hydraulic concrete and BFRP reinforcement. China Water Conservancy and Hydropower Press, Beijing

52. Zhiheng S, Yongfu Y, Deying L et al (2012) Spillway floor defect repair of Gongboxia Hydropower Station. In: Tenth annual meeting of Hydraulic Professional Committee of China Water Conservancy Society, Nanning

Open Access This chapter is licensed under the terms of the Creative Commons Attribution 4.0 International License (http://creativecommons.org/licenses/by/4.0/), which permits use, sharing, adaptation, distribution and reproduction in any medium or format, as long as you give appropriate credit to the original author(s) and the source, provide a link to the Creative Commons license and indicate if changes were made.

The images or other third party material in this chapter are included in the chapter's Creative Commons license, unless indicated otherwise in a credit line to the material. If material is not included in the chapter's Creative Commons license and your intended use is not permitted by statutory regulation or exceeds the permitted use, you will need to obtain permission directly from the copyright holder.

Mechanical Behaviors of Encased Concrete of Composite Bridges with Corrugated Steel Web

Yaojun Wang, Bingbing Liu, and Qianqian Hao

Abstract The encased concrete in a composite bridge with corrugated steel web (CSW) is a critical structure to improve the stability of the web and relieve the stiffness difference between the composite girder and intermediate crossbeam. This study aims to investigate the effects of the encased concrete on the mechanical behaviors of composite girder bridges with CSWs. The finite element model of a rigid frame bridge is established, the calculation method of the shear force sharing ratio of the encased concrete is proposed, and the effect of the length and thickness of the encased concrete is analyzed. The results show that the encased concrete was under compression in longitudinal direction, while vertical tensile stresses occur near the transition section between pure CSW and the composite web. The encased concrete shares 65–90% of the total shear force, and the shear force sharing ratio increases with the increase of the concrete thickness. Increasing the length of the encased concrete reduces the deflection of the girder and pre-compression stress of the top concrete slab. All the findings of present study may provide reference for the design of composite girder bridges with CSWs.

Keywords Composite girder bridges · Corrugated steel web · Encased concrete · Shear force shearing ratio

1 Introduction

The corrugated steel web (CSW) in a composite girder bridge can reduce the weight of the main girder, prevent web cracking, and improve the prestressing efficiency, which has been widely used in recent years [1–3].

Y. Wang (✉) · B. Liu
CCCC Infrastructure Maintenance Group Co. Ltd., Beijing, China
e-mail: adinamax@163.com

Q. Hao
Beijing Jingjiang International Engineering Consulting Co. Ltd., Beijing, China

For large-span composite girder bridges with CSWs, the shear stability of the web becomes a controlling factor in design due to the increased web height near intermediate supports. It is common to increase the shear stability of the CSW by pouring encased concrete. The encased concrete can also relieve the stress concentration caused by stiffness transition at the connection between the steel web and the crossbeam, effectively transferring the force of the web to the substructure. Nowadays, the encased concrete has become a necessary component for bridges with CSWs.

Some scholars have investigated the flexural or shear performances of the encased concrete in composite bridges. He et al. [4, 5] conducted bending tests on composite beams with CSW and encased concrete, and found that the encased concrete prevents buckling of the upper steel flange, and improving the bending strength and ductility of the composite beam. Jiang et al. [6] investigated the shear calculation method of the composite web with encased concrete, and compared the differences among three shear force distribution calculation methods. Deng et al. [7] studied the shear stress level of the web with encased concrete when it is arranged on one or both sides of a multi-cell box girder with CSWs in a cable-stayed bridge.

Although some researches have been conducted, few of them were based on real rigid-frame bridge with CSWs, and there is still no detail design method of the encased concrete in current specifications. Therefore, this paper takes a composite rigid frame bridge with span arrangement of 72 + 125 + 72 m [8] as the engineering background, and three-dimensional finite element model (FEM) of the bridge was established to obtain the stress distribution of the encased concrete. Then, a theoretical calculation method was proposed and verified to predict the shear force sharing ratio of the encased concrete. Finally, the effect of the encased concrete on the deflection of the girder, the shear force sharing ratio, and the pre-stressing efficiency was analyzed by parameter analysis. The results could provide reference for the design of composite girder bridges with CSWs.

2 Finite Element Model

The general finite element software ANSYS 16.0 was adopted to establish the 3D solid-shell finite element model of the rigid frame bridge, as shown in Fig. 1. The concrete was simulated using solid element SOLID65, the CSW and steel flange was simulated using shell element SHELL181, and the prestressing tendons were simulated by line element LINK10.

Coupling relationships were established between the CSWs and concrete top and bottom slabs, as well as between the encased concrete and the CSW, without considering the relative slip. Both the concrete and steel structures were simulated using elastic material, the elastic modulus was taken as 35,500 MPa [9] and 210 GPa [10], respectively. Fixed boundary conditions were applied at the base of the middle pier, and vertical (U_z) degrees of freedom were constrained at the side supports. The loads applied in the model included prestressing, second-stage permanent loads, and self-weight.

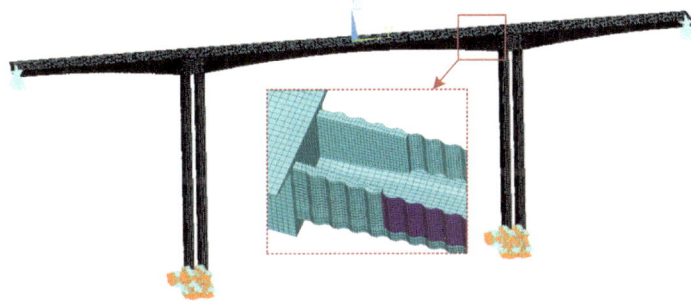

Fig. 1 Finite element model of the rigid frame bridge

Figure 2 shows the dimensions of the longitudinal and cross section of the FEM. The main girder adopts a single-cell box section, with a girder depth of 7.8 m at the middle support section and a depth of 3.5 m at the mid-span. The width of the top concrete slab is 12.5 m, with a thickness of 20 cm at the cantilever end and a thickness of 75 cm at the corner position. The length and thickness of the encased concrete of the basic model is 7.3 and 50 cm respectively. The CSW uses 1600-type corrugation and is made of Q345C steel.

Fig. 2 Dimensions of the FEM (cm)

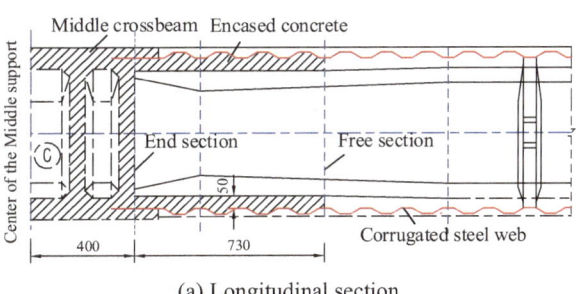

(a) Longitudinal section

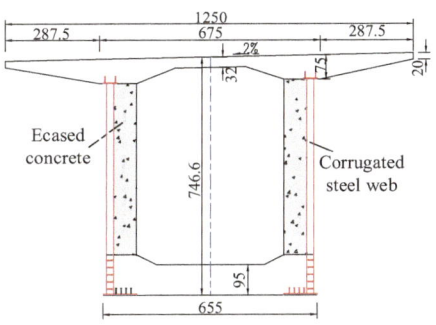

(b) Cross section

3 Stress Distribution of Encased Concrete

Figure 3 shows the longitudinal and vertical stress distribution of the encased concrete in the model. The origin of the horizontal axis is the transition section between the encased concrete and the middle crossbeam, with a positive direction towards the middle of the span. The vertical height h_c of the presented paths are 0.5, 1, 2.5, 4, and 4.5 m from the top surface of the encased concrete (lower surface of the top concrete slab).

The encased concrete is basically under compression in longitudinal direction, with stresses ranging from − 4.5 to 0 MPa. The longitudinal stress gradually decreases in a waved shape from the end to the free section of the encased concrete. This is mainly due to the thicker encased concrete at the concave side of the CSW, which leads to a smaller compressive stress, while the thinner encased concrete at the convex side of the CSW results in a relatively larger compressive stress. The longitudinal stress at the end of the encased concrete is evenly distributed along the

Fig. 3 Stress distribution of the encased concrete

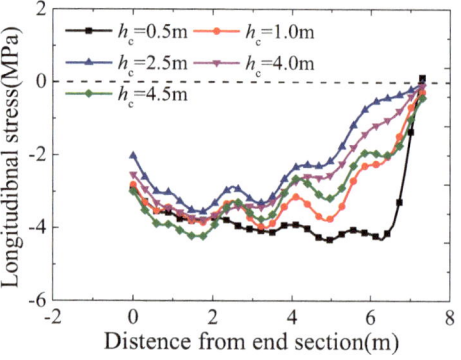

(a) Longitudinal stress distribution

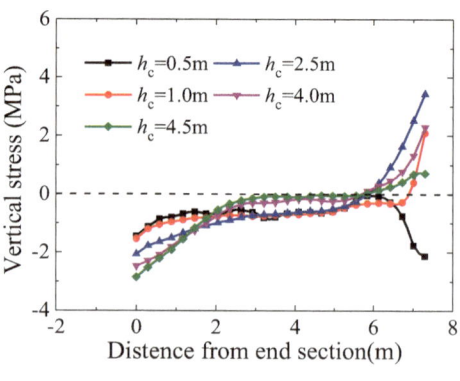

(b) Vertical stress distribution

vertical direction, while the stress at the free end tends to be larger in the upper and lower parts, and smaller in the middle part. The vertical stress of the encased concrete is mostly in a compressive state, and the compressive stress decreases from the end to the free section. Tensile stresses in vertical direction appear at the free end of the encased concrete, which is due to the transition between pure steel web and composite web at this section. Therefore, attention should be paid to setting reinforcement in the design of the transition section.

4 Shear Sharing Ratio of Encased Concrete

The encased concrete and CSW is usually connected by headed studs, and the slip between them could be ignored under serviceability limit state. Therefore, the shear stiffness of the composite web can be regarded as the superposition of the CSW and encased concrete.

As shown in Fig. 4, the unit shear force $V = 1$ at the free section of the composite web will lead to corresponding shear deformation δ, then:

$$V_s + V_c = 1 \tag{1}$$

where V_s and V_c represent the shear force sheared by the CSW and the encased concrete, respectively.

The shear stiffness of the encased concrete (G_c) and the shear stiffness of the CSW (G_s) can be calculated as follows:

$$G_c = \frac{E_c \cdot ht_{eq}}{2a(1 + v_c)} \tag{2}$$

$$G_s = \frac{E_s ht_s}{2a\eta(1 + v_s)} \tag{3}$$

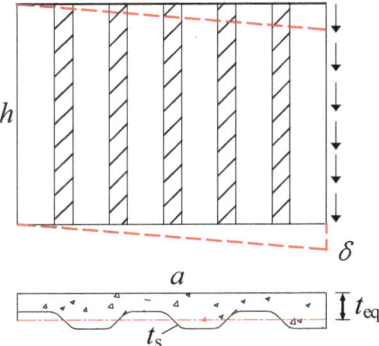

Fig. 4 Deformation of the composite web under shear force

where t_{eq} and t_s are the equivalent thickness of the encased concrete and the thickness of the CSW, respectively; E_c and E_s are the elastic modulus of concrete and steel, respectively; v_c and v_s are the Poisson's ratios of concrete and steel, respectively; η denotes the shape coefficient of the CSW, which is the ratio of the unfolding length to projection length of the CSW.

The shear deformation of the CSW (δ_s) and the shear deformation of the encased concrete (δ_c) under unit shear force are given as follows:

$$\delta_s = \frac{1}{k_s} V_s = \frac{2a\eta(1+v_s)}{E_s h t_s} V_s \tag{4}$$

$$\delta_c = \frac{1}{k_c} V_c = \frac{2a(1+v_c)}{E_c \cdot h t_e} V_c \tag{5}$$

According to the deformation compatibility condition, $\delta_s = \delta_c$, the shear force sharing ratio of the encased concrete can be obtained as:

$$\frac{V_c}{V} = \frac{(1+v_c)E_s t_s}{(1+v_c)E_s t_s + (1+v_s)\eta E_c t_{eq}} \tag{6}$$

Figure 5 shows the comparison of the shear sharing ratio of the encased concrete from the finite element analysis (FEA) and theoretical results. The thickness of the encased concrete (convex side of the CSW) is 50 cm, so its equivalent thickness (t_{eq}) is 61 cm. The thickness of the CSW is 20 mm.

It is seen that the shear sharing ratio of the encased concrete calculated from the FEA is 85.3% (taking the average result of three standard corrugations), and the theoretical result is 84.1%. The difference between the theoretical and FEA results is 1.2%. Therefore, the proposed formula can be used to predict the shear sharing ratio of the encased concrete in the composite web. The shear force sharing ratio of the encased concrete near the transition section between pure CSW and composite web obtained from the FEM is relatively small, and the theoretical formula gives conservative results.

Fig. 5 Comparison of FEA and theoretical results

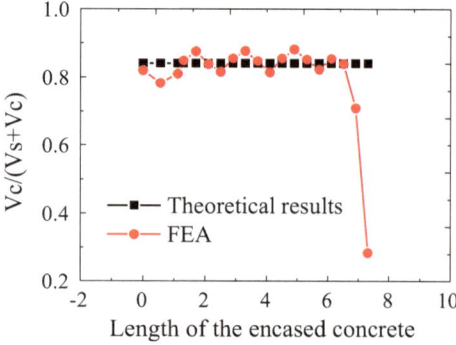

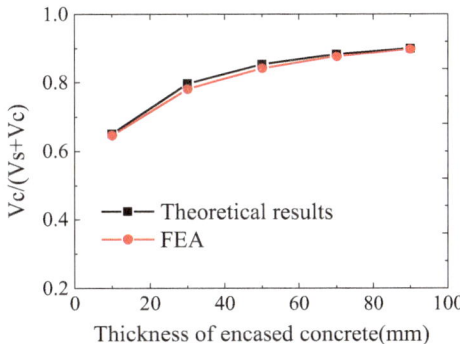

Fig. 6 Effect of the encased concrete thickness on shear sharing ratio

5 Parametric Analysis of Encased Concrete

5.1 Effect of Encased Concrete Thickness on Shear Sharing Ratio

The thickness of the encased concrete mainly affects its shear sharing ratio. CJJT 272–2017 [11] suggests that the thickness of the encased concrete should not be less than 20 cm. Therefore, the encased concrete thickness was varied to conduct parameter analysis. The thickness of the encased concrete was taken as 10, 30, 50, 70, and 90 cm in FEMs, respectively.

Figure 6 shows the effect of the encased concrete thickness on the shear sharing ratio. As the concrete thickness increases, the shear force shared by the encased concrete gradually increases. When the thickness of the encased concrete increases to a certain value, the shear sharing ratio tends to be stable. When the thickness of the encased concrete is 10 cm, the shear sharing ratio is 65%. When the concrete thickness increases to 90 cm, the shear sharing ratio increases to 90%. Therefore, the encased concrete shares a significant portion of the shear force, which should be taken into consideration in the design and a lot of steel can be saved.

5.2 Effect of Encased Concrete Length on Deflection

Figure 7 shows the effect of encased concrete length on bridge deflection. The deflection of the main span and side span of the bridge gradually decreases with the increase of the length of the encased concrete. Due to the significant reduction in shear stiffness of the CSW compared to concrete webs, it is necessary to consider the vertical deflection caused by shear deformation during design. After pouring the encased concrete on the inner side of the web, although the self-weight of the main beam is increased, the shear stiffness of the CSW is also increased, which reducing the vertical deflection caused by the shear deformation of the CSW.

Fig. 7 Effect of the encased concrete length on deflection

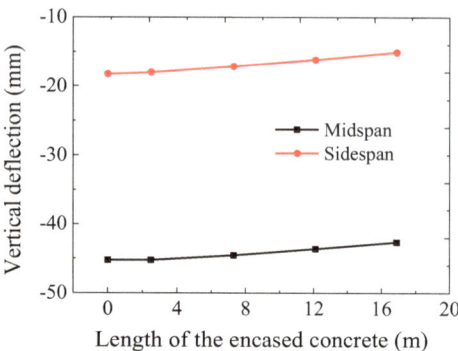

When there is no encased concrete on the inner side of the web, the deflection of the mid span is 45.2 mm. After pouring the encased concrete with lengths of 7.3 and 16.9 m respectively, the mid span deflection was reduced to 44.5 and 42.6 mm, with a reduction ratio of 1.55 and 5.75%, respectively. It is seen that the improvement of deflection by pouring encased concrete is slight, but it will significantly increase the self-weight of the main girder. Therefore, it is not suggested to use the method of pouring encased concrete to reduce the deflection of the main span caused by shear deformation of the CSW.

5.3 Effect of Encased Concrete Length on Prestressing Efficiency

To investigate the effect of the encased concrete length on the prestressing efficiency, the variation curves of longitudinal normal stresses of the top concrete slab were plotted, as shown in Fig. 8. The origin of the horizontal axis is the center of the middle support, and the load in the FEM only includes the prestressing effect. It is seen that with the increase of encased concrete length, the pr-compression stress of top concrete slab gradually decreases. When the length of the encased concrete is taken as 16.9 m, the maximum pre-compression stress decreases from 17.0 to 13.9 MPa.

Figure 9 shows the prestressing efficiency of different components of the girder at the end section of the encased concrete. The prestressing efficiency is defined as the ratio of the axial force of each component to the axial force of the whole cross-section. The negative ratio of bottom slab represents tensile force. It is seen that the top concrete slab shares 93.5% of the total prestress when there is no encased concrete. With increasing of the encased concrete length, the prestressing efficiency of the top concrete slab decreases, while the prestress shared by the encased concrete increases gradually. When the length of the encased concrete increases to 7.3 m, which is about 1.0 times the girder depth, the prestressing efficiency of the top concrete slab and

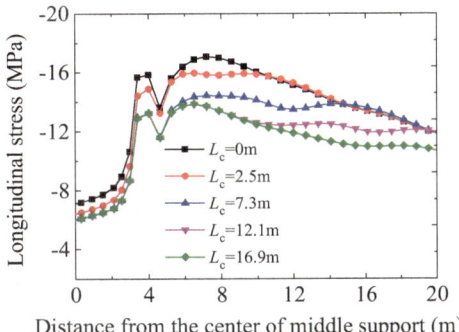

Fig. 8 Effect of the encased concrete length on longitudinal stresses

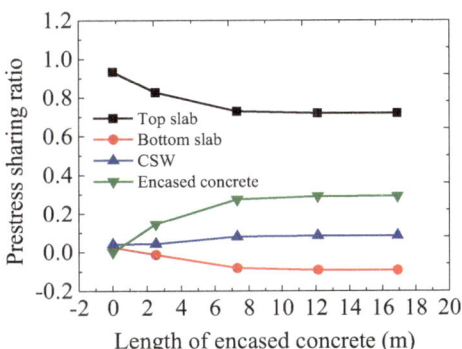

Fig. 9 Prestress sharing ratio of different components

encased concrete is about 72 and 27%, respectively. When the length of the encased concrete exceeds 7.3 m, the prestressing efficiency of different components keeps stable.

6 Conclusions

This paper investigates the mechanical behaviors of encased concrete in the composite bridge with CSWs, the shear force sharing ratio of the encased concrete was proposed, and the effects of the encased concrete length and thickness were analyzed. The following conclusions are obtained:

(1) The longitudinal stress of the encased concrete is basically in a compressive state, and decreases from the free to the end section. Tensile stress occurs near the transition section between pure CSW and the composite web due to stiffness variation. Therefore, attention should be paid to the arrangement of reinforcement at the transition position.

(2) The formula for predicting the shear sharing ratio of the encased concrete was derived, which is in good agreement with the finite element results. The encased concrete shares about 65–90% of the total shear force, and the sharing ratio gradually increases with the increase of concrete thickness.
(3) With the increase of the encased concrete length, the girder deflection gradually decreases. The encased concrete increases the shear stiffness of the web and reduces the vertical deflection caused by the shear deformation of the CSW.
(4) With the increase of the encased concrete length, the pre-compression stress and prestressing efficiency of the top concrete slab decreases. When the length of the encased concrete exceeds 1.0 times the girder depth, the encased concrete shares 27% of the total prestress and keeps stable.

References

1. Li M, Song J, Li W, Zhang J (2015) Application of prestressed concrete composite box-girder bridges with corrugated steel webs in bridge engineering in China. In: 5th international conference on civil engineering and transportation (ICCET 2015)
2. He J, Liu Y, Chen A, Yoda T (2012) Mechanical behavior and analysis of composite bridges with corrugated steel webs: state-of-the-art. Int J Steel Struct 12(3):321–338
3. Jiang R, Au F, Xiao Y (2015) Prestressed concrete girder bridges with corrugated steel webs: review. J Struct Eng ASCE 141(2):1–9
4. Wang D, He J, Chen A, Liu Y (2012) Experimental study on bending behavior of concrete-encased composite girder with corrugated steel web. J Tongji Univ (Sci) 40(9):1312–1317 (in Chinese)
5. He J, Liu Y, Chen A, Wang D, Yoda T (2014) Bending behavior of concrete-encased composite I-girder with corrugated steel web. Thin-Walled Struct 74:70–84
6. Jiang Y, Sun T, Wu Y (2007) Calculation of the shear resistance of composite web with encased concrete in PC box girder bridge with CSW. Highway 67(6):14–21 (in Chinese)
7. Deng W, Zhang J, Liu D, Hu J (2016) Study of lining concrete arrangement for multi-cell single boxcomposite girder with corrugated steel webs. World Bridg 44(2):77–81 (in Chinese)
8. He T, Wang Y (2020) Overall design of the main bridge of Molin expressway Diyoko No.1 Bridge. HighwTransp Technol (Appl Technol Ed) 1:235–237 (in Chinese)
9. GB 50010-2010 (2010) Code for design of concrete structures. China Architecture& Building Press, Beijing
10. JTG D64-2015 (2015) Specifications for design of highway steel bridge. China Communications Press Co. Ltd., Beijing
11. CJJT 272-2017 (2017) Technical standard for composite girder bridges with corrugated steel webs. China Architecture & Building Press, Beijing

Open Access This chapter is licensed under the terms of the Creative Commons Attribution 4.0 International License (http://creativecommons.org/licenses/by/4.0/), which permits use, sharing, adaptation, distribution and reproduction in any medium or format, as long as you give appropriate credit to the original author(s) and the source, provide a link to the Creative Commons license and indicate if changes were made.

The images or other third party material in this chapter are included in the chapter's Creative Commons license, unless indicated otherwise in a credit line to the material. If material is not included in the chapter's Creative Commons license and your intended use is not permitted by statutory regulation or exceeds the permitted use, you will need to obtain permission directly from the copyright holder.

Safety Analysis of Concrete Arch Dam Considering the Shear Action of Contraction Joints

Zhichao Miao and Guohua Liu

Abstract In this paper, taking an extra-high arch dam as an example, the simulation of the stress, deformation and overload failure of arch dam is carried out to investigate the effect of the shear strength of contraction joints on the safety of the arch dam based on the nonlinear trial load method. The numerical simulation results show that the overload safety factor of the arch dam decreases significantly with the decrease in shear strength of contraction joints. The maximum tensile stress of the arch dam is reduced when the shear strength of the contraction joints is reduced by 30% compared to the strength of the dam material. The results provide a reference for the design of the mechanical properties of the contraction joints.

Keywords Trial load method · Contraction joints · Shear strength · Overload safety factor

1 Introduction

To meet the temperature stress relief and construction needs of mass concrete, arch dams are usually cast in blocks, with contraction joints with keyways between adjacent dam sections [1]. The grout in the contraction joints has a certain tensile strength, while the keyway provides a certain shear strength. However, due to factors such as grout defects and different forms of keyways, the tensile strength and shear strength of the contraction joints are not exactly equivalent to the strength of the dam material [2]. The weakening of the strength of the contraction joints has a certain effect on the stress distribution, overall stability and safety of the dam. Therefore, it is important to study the influence of the variation of tensile strength and shear strength of the contraction joints on the safety of arch dams.

Z. Miao · G. Liu (✉)
College of Civil Engineering and Architecture, Zhejiang University, Hangzhou, Zhejiang, China
e-mail: zjuliu@163.com

Z. Miao
e-mail: 22112100@zju.edu.cn

At present, the numerical simulation method is the main method to simulate the action of contraction joints in arch dams. Based on the geometric and deformation characteristics of the contraction joints, scholars have conducted in-depth studies on the computational models of contraction joints, and the proposed models are two-dimensional two-node nonlinear spring model [3], three-dimensional contact unit model [4, 5], and contact mechanics model [6, 7]. In terms of the constitutive relation of the contact surface, the assumption that the contact force is the pressure is mostly adopted in the normal direction, and the Mohr Coulomb criterion or the assumption that the tangential constraint is infinite is often adopted in the tangential direction. The existing researches generally assume that the joint surface is flat, and most of them do not consider the influence of the initial normal tensile strength of the contraction joints. In the tangential direction, it is often assumed that the keyway has a strong shear capacity only when the contraction joints are closed, and its tangential restraint is no longer effective when the contraction joints are open. However, for the actual arch dam project, the grouting also has a certain tensile strength, which can withstand the tensile stress in the arch direction to a certain extent. It has a certain influence on the stress distribution of the dam blocks on both sides. Besides, the contraction joints can withstand a certain degree of shear capacity due to the restraint of the keyway after the contraction joints are opened. The calculation model used in this paper takes into account the initial tensile strength of the contraction joints and the shear strength after the contraction joints opening, which is more consistent with the real engineering situation.

This paper simulates the action of contraction joints by a generalized model based on the nonlinear trial load method. By analyzing the force, deformation and overload safety factor of the arch dam, the effect of the change in shear strength of the contraction joints on the stress distribution and overall safety of the arch dam is investigated.

2 Research Methods and Theories

2.1 Nonlinear Trial Load Method

The trial load method assumes that the material is elastic, builds the arch-beam grid, establishes equations based on the coordination conditions of the arch-beam intersection displacements, solves for the fundamental unknowns, and calculates the dam stresses and displacements. The available nonlinear trial load method [8] introduces nonlinear constitutive relations of the material under triaxial stress states and uses a damage model to simulate the stress–strain relationships of the material after cracking or crushing. The strength criterion uses the Hsieh -Ting-Chen model:

$$F = A\frac{J_2}{f_c^2} + B\frac{\sqrt{J_2}}{f_c} + C\frac{\sigma_1}{f_c} + D\frac{I_1}{f_c} - 1 = 0 \tag{1}$$

where: J_1, J_2, J_3 are stress invariants, I_1 is the first invariant of the strain tensor, $\sigma_1 = \frac{2}{\sqrt{3}}\sqrt{J_1}\cos\theta + \frac{1}{3}I_1$, f_c is the uniaxial ultimate compressive strength, A, B, C, D are the four parameters derived from unidirectional tensile, unidirectional compression, bidirectional compression, and lateral limit three-way compression.

The mechanical model for the nonlinear analysis of the dam material is as follows:

For a specific calculation point, the damage variables of the corresponding directions are defined for the three main stress directions, and the stress–strain relationship of concrete is constructed through the damage variables; the damage variables of a specific point are a cumulative change process that only increases but not decreases, and the current damage values in the new direction can be obtained by projection transformation when the main stress direction is changed.

For the stress–strain relationship in the direction of the principal compressive stress in concrete, the damage variable is introduced according to Saenz's formula as

$$D_a = 1 - \frac{1}{1 + X\left(\frac{\varepsilon}{\varepsilon_c}\right) + Y\left(\frac{\varepsilon}{\varepsilon_c}\right)^2 + Z\left(\frac{\varepsilon}{\varepsilon_c}\right)^3} \tag{2}$$

$$\sigma = E_u(1 - D_a)\varepsilon \tag{3}$$

where σ is the dam material stress, ε is the strain, ε_c is the strain value corresponding to the peak compressive strength, E_u is the initial modulus without damage, X, Y and Z are model parameters. The above equation can reflect the nonlinear full process curve of stress–strain of compressed concrete properly. The model parameters of X, Y and Z are determined by five material parameters: concrete initial modulus E_u, concrete peak compressive strength σ_c, and its corresponding strain value (peak strength strain value) ε_c, concrete crushing strength σ_u, and its corresponding strain value (ultimate compressive strain value) ε_u. The stress–strain nonlinear full process curves under compression is shown in Fig. 1.

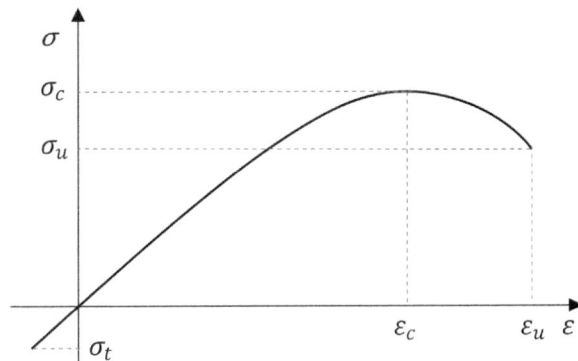

Fig. 1 Stress–strain curve of concrete

For the stress–strain relationship in the direction of the main compressive stress in concrete, when the main tensile stress does not exceed the permissible range, the damage variable $D_a = 0$, and the stress–strain is a linear elastic relationship. When the principal tensile stress exceeds the allowable tensile stress σ_t, the material cracks along the principal tensile stress surface, and the damage variable $D_a = 1$ in the direction of the principal tensile stress.

The nonlinear trial load method can simulate the failure of arch dam under water density overload conditions. On the basis of normal load, the water pressure at the upstream dam surface is increased by increasing the density of water. The ratio of the water density at the time of failure to the initial water density of the arch dam is defined as overload safety factor K [9].

2.2 Contraction Joints Model

The arch beam grid of A arch B beam of an arch dam project is used as the analytical calculation model. The (B-2) contraction joints are generalized to the sides of No.2–No.(B-1) beams in the arch beam grid.

Before loading, the contraction joints are closed and transmit tensile and shear stress. During the loading process, the opening and closing state of the joint surface can be changed repeatedly. When the joint surface is opened, the keyway of the contraction joints acts as a slip barrier according to the concrete shear strength. The shear strength uses the Mohr Coulomb criterion:

$$F = f'(\Sigma N + \Sigma U) + c'A \tag{4}$$

where F is the shear force of the joint face, ΣN is the sum of vertical forces at the seam joint face, ΣU is the sum of the lifting force, f' is the coefficient of friction, c' is the unit cohesive force, A is the area of the joint surface.

3 Calculation and Analysis

3.1 Project Description

A concrete arch dam is located in the southwest of China. It is a double-curvature arch dam with a parabolic horizontal arch ring. The maximum height of the dam body is 294.5 m, the total length of the dam roof is 901.8 m, the bottom thickness of the crown beam is 72.9 m, the top thickness is 12 m, and the width-to-height ratio of the river valley is 2.73.

3.2 Calculation Model and Material Parameters

According to the engineering data, the arch beam model of the concrete arch dam was established using the trial load method. The model mainly consists of 11 arches and 21 beams, as shown in Fig. 2.

The relevant parameters of the mechanical properties of the arch dam are shown in Table 1. The standard value of compressive and tensile strength of the dam material is obtained by converting the strength grade of the material according to the design code of hydraulic concrete. The dam shear parameters refer to the recommended values in the U.S. Bureau of Reclamation's Arch Dam [10], with the friction coefficient f′ taken as 1 and the cohesive force c′ taken as 10% of the material's compressive strength. The initial shear strength and tensile stress of the contraction joints are the same as that of the dam material. The shear strength is reduced by discounting f′ and c′ of the contraction joints.

The elasticity modulus of the dam foundation is in the range of 6.5–19 GPa and Poisson's ratio is in the range of 0.27–0.3, zoned according to the actual geological conditions.

The load includes upstream normal water pressure, downstream water pressure, sediment pressure, dead load of the dam, and temperature drop load. The relevant parameters of water and sediment pressure are shown in Table 2.

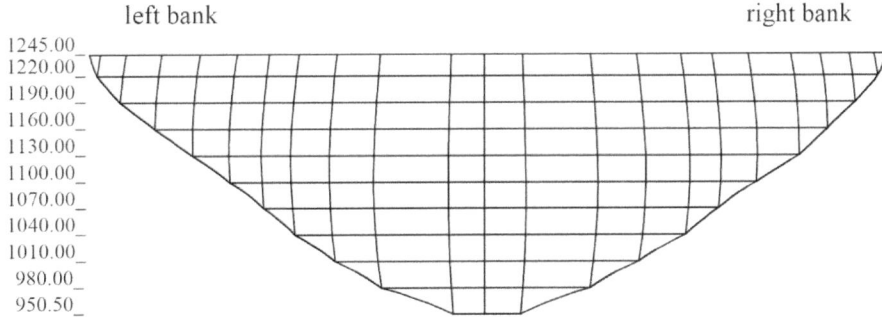

Fig. 2 Elevation view of arch beam grid

Table 1 Physical and mechanical parameters of the dam material

Concrete grade	Density (kN/m^3)	Elastic modulus (GPa)	Poisson's ratio	Coefficient of linear expansion (1/°C)	Thermal diffusivity (m^2/month)
C40	24	23	0.167	8.5×10^{-4}	2.336

Table 2 Parameters of water and sediment pressure

Reservoir level (m)	Downstream level (m)	Upstream silt sand elevation (m)	Silt sand friction angle (°)	Silt sand floating weight/(kN/m^3)
1242.51	1018.59	1097.00	24	9.6

3.3 Analysis of Calculation Results

Comparison of Four Models

Four typical calculation models are selected for stress analysis and overload simulation.

Model I: Dam material is linearly elastic. The effect of contraction joints is not considered.

Model II: Dam material is nonlinearly elastic. The initial tensile strength and shear strength of contraction joints are the same as the dam material.

Model III: Dam material is nonlinearly elastic. The Initial tensile strength of contraction joints is the same as the dam material and the shear strength is discounted by 50%.

Model IV: Dam material is nonlinearly elastic. The Initial tensile strength and shear strength of contraction joints are discounted by 50%.

It is calculated that the maximum tensile stress is located at the upstream surface of the dam and the maximum compressive stress is located at the downstream surface of the dam for all four models. Figure 3 shows the tensile stress distribution on the upstream surface of the 4 models. Figure 4 shows the compressive stress distribution on the downstream surface of the 4 models.

By comparing the stress results in Figs. 1, 2, 3 and 4, it is found that the shear strength reduction of contraction joints makes a difference to the stress distribution of the dam. The maximum compressive stress on the downstream surface of all four models is located at the left arch end of the middle part of the dam. The maximum tensile stresses in Model 1 and Model 2 are located at the left arch end of the middle and lower parts of the dam body. The maximum tensile stress in Model 3 is located at the left arch end of the middle of the dam body. The maximum tensile stress of Model 4 is located at the right arch end of the middle and upper part of the dam body. Both Model 3 and Model 4 have cracks and contraction joints openings at both ends of the bottom arch of the upstream surface.

The maximum stress, minimum stress, maximum radial displacement, and overload safety factor for the four models are shown in Table 3.

It can be seen from Table 3 that the maximum tensile stress of Model 2 is significantly reduced compared with Model 1. The nonlinear analysis is more in line with the actual situation because the appropriate constitutive model of concrete material is adopted. In Model 3, considering the weakening of the shear strength of contraction

Fig. 3 Tensile stress program on the upstream dam surface

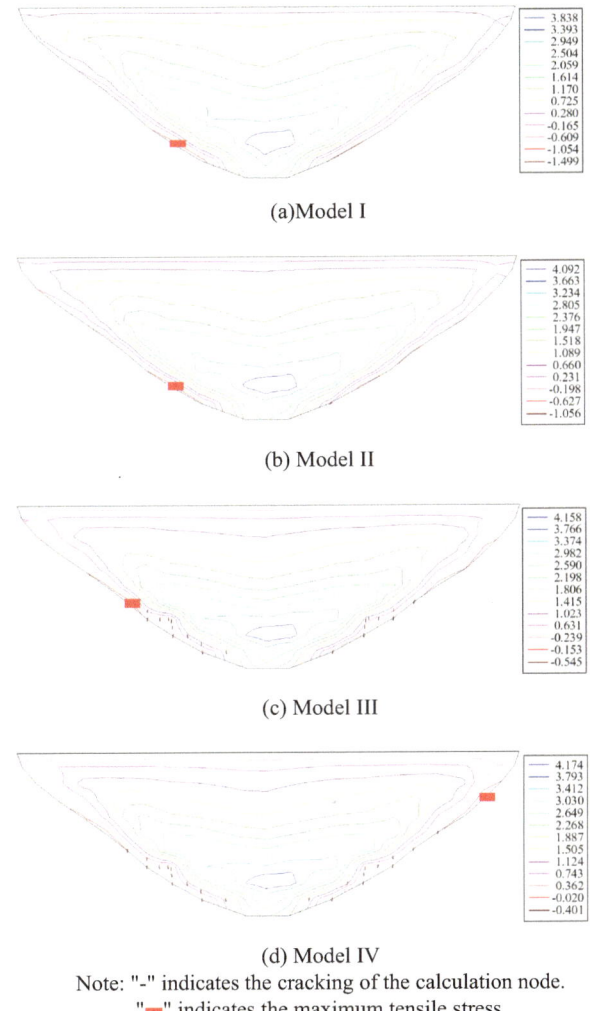

(a) Model I

(b) Model II

(c) Model III

(d) Model IV

Note: "-" indicates the cracking of the calculation node.
"■" indicates the maximum tensile stress.

joints, there is a significant decrease in the maximum tensile stress on the upstream surface and a slight increase in the maximum compressive stress on the downstream surface. Because of the opening contraction joints, the tensile stress of the arch ring is released, and part of the stress is transferred to the cantilever beam. The maximum radial displacement increases, which indicates that the arch dam tends to be unsafe. Meanwhile, the overload safety factor is reduced significantly. This is because the contraction joints weaken the shear strength of the arch, reducing the shear safety of the arch and transferring some of the shear stress to the beam, which in turn reduces the shear safety of the beam. This leads to a decrease in overall safety. Compared to Model 3, Model 4 takes into account the reduction in the initial tensile strength of contraction joints. In Model 4, the maximum tensile stress is slightly reduced, the

Fig. 4 Compressive stress program on the downstream dam surface

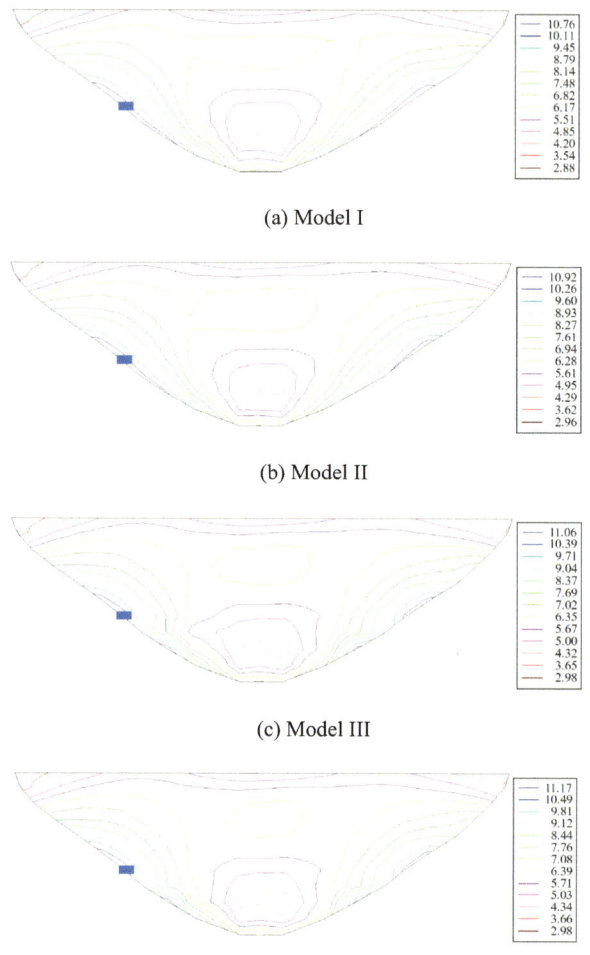

(a) Model I

(b) Model II

(c) Model III

(d) Model IV

Note: "■" indicates the maximum compressive stress

Table 3 Comparison of indicators of four model

	UT (MPa)	UC (MPa)	DT (MPa)	DC (MPa)	R (cm)	K
Model 1	1.5	7.01	0.23	10.76	20.23	
Model 2	1.06	6.93	0.12	10.92	19.48	2.8
Model 3	0.54	6.98	0.08	11.06	19.98	2.3
Model 4	0.4	6.99	0.03	11.17	19.98	2.2

Note "UT" represents the maximum tensile stress (MPa) at the upstream surface. "UC" represents the maximum compressive stress (MPa) at the upstream surface. "DT" represents the maximum tensile stress (MPa) at the downstream face. "DC" represents the maximum compressive stress (MPa) at the downstream surface. "R" represents the maximum radial displacement (cm). "K" represents the overload safety factor

maximum compressive stress is slightly increased, the maximum radial displacement is unchanged, and the K is slightly reduced.

Sensitivity Analysis of Contraction Joints Shear Strength

The shear strength of contraction joints is reduced by discounting the friction coefficient f' and the cohesive force c' of the contraction joints simultaneously. The shear strength sensitivity analysis of the contraction joints is done in two cases. One case is that the contraction-joints tensile strength is the same as the dam material. In the other case: the tensile strength of the contraction joints is reduced by 50%. The relationship between the shear strength of contraction joints and the maximum tensile stress of the dam is shown in Fig. 5. The relationship between the shear strength of contraction joints and the maximum compressive stress of the dam is shown in Fig. 6. The relationship between the shear strength of contraction joints and overload safety factor K is shown in Fig. 7.

As can be seen from Fig. 5, the maximum tensile stress in the dam body is constant until the shear strength of the contraction joints is reduced by 30%. After the 30% reduction, the maximum tensile stress gradually decreases until there is no tangential restraint in the contraction joints and the maximum tensile stress tends to zero. The maximum tensile stress is slightly lower when the initial tensile strength of the contraction joints is discounted by 50% compared to the tensile strength of the contraction joints without discounting. But the difference between the two maximum tensile stresses becomes smaller and smaller as the percentage reduction in shear strength increases.

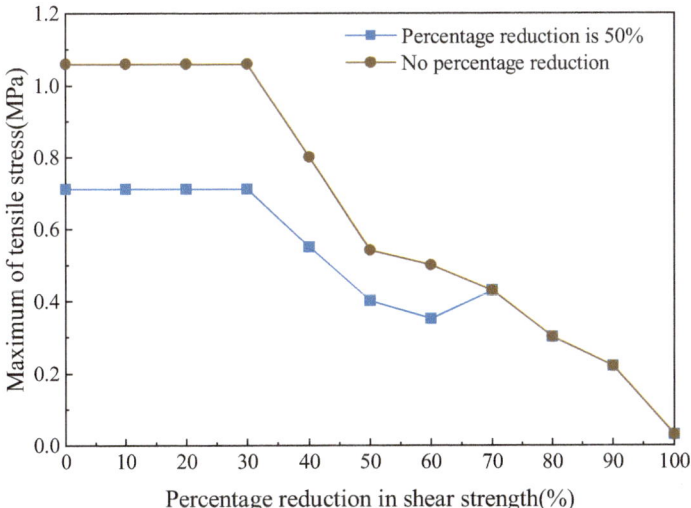

Fig. 5 Relationship between shear strength of contraction joints and maximum tensile stress of the dam

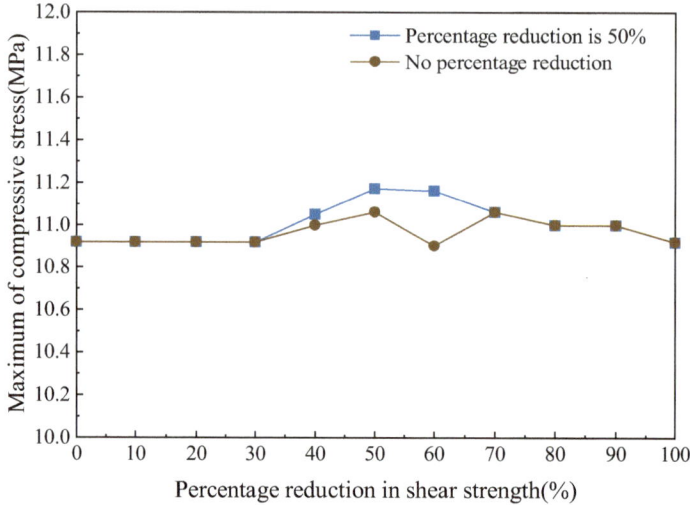

Fig. 6 Relationship between shear strength of contraction joints and maximum compressive stress of the dam

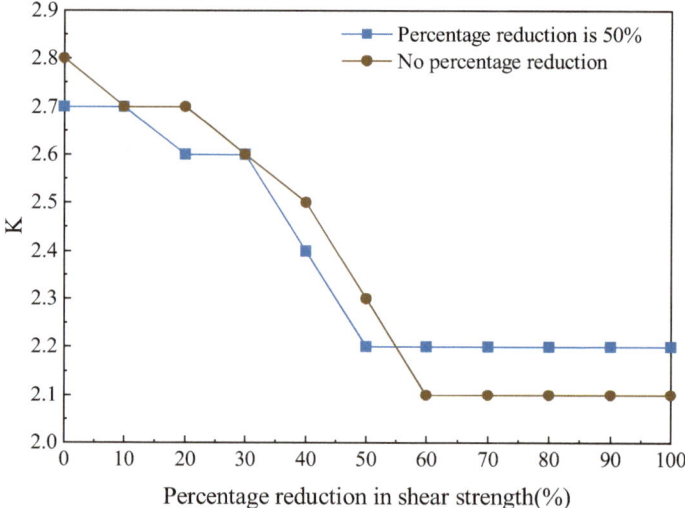

Fig. 7 Relationship between shear strength of contraction joints and overload safety factor K

It can be seen from Fig. 6 that the variation in contraction joints' shear strength has little effect on the maximum compressive stress in the dam body.

As can be seen from Fig. 7, as the percentage reduction in shear strength of contraction joints increases, the overload safety factor becomes lower and lower, and the dam becomes increasingly unsafe. After discounting to 60%, the arch dam safety remains at a stable lower level. The initial tensile strength of the contraction joints has no significant effect on the overload safety factor.

4 Conclusions

In this paper, the results of four different calculation models of contraction joints are compared and the shear strength sensitivity of contraction joints is analyzed. The following conclusions are obtained:

1. Contraction joints have a significant effect on the stress distribution and overload safety factor of the arch dam. The action of the contraction joints causes cracks in the lower and middle upstream face of the arch dam where tensile stress is released. The location of the maximum tensile stress in the dam body is changed. Because of the effect of the contraction joints, the overload safety factor is significantly reduced, the maximum compressive stress and the maximum radial displacement are slightly increased, and the arch dam tends to be unsafe.
2. With the decrease in the shear strength of the contraction joints, the maximum tensile stress in the dam body does not show significant changes in the early stage but significantly decreases in the later stage. There is no noticeable change in the maximum compressive stress, but the overload safety factor of the dam decreases. The reduction in tensile strength of contraction joints has a diminishing effect on the maximum tensile stress in the dam body, without significant impact on the overload safety factor. Therefore, when designing the keyway and determining the quality of the mortar, it is important to consider appropriate parameters such as friction coefficient, cohesion, and other equivalent factors.

Acknowledgements Thanks to the National Natural Science Foundation of China (No. 51979244) for supporting this project.

References

1. Lin JY (2009) Hydraulic buildings. China Water Conservancy and Hydropower Press, Beijing
2. Freitas M, Ben Ftima M, Leger P, Bouaanani N (2022) Three-dimensional failure envelope of concrete dam shear keys. Eng Struct 269:114766
3. Dowling ML, Hall JF (1989) Nonlinear seismic analysis of arch dams. J Eng Mechan 115(4):768–789
4. Xu YJ, Zhang CH, Wang GL et al (2001) Nonlinear seismic response analysis of arch dam with contraction joints. J Hydr Eng 04:68–74
5. Zhao LH, Li TC, Niu ZW (2007) The dynamic contact model of nonlinear seismic response of high arch dams with contraction joints. J Hydroelectr Eng 04:91–95
6. Lin G, Hu ZQ (2004) Studies of the effect of contraction joints and the effective earthquake-resistant measures for arch dams. World Earthq Eng 03:1–8

7. Niu ZY, Hu ZQ, Lin G (2017) Study on the contraction joints model of arch dam considering the change of normal initial tensile strength and tangential shear strength. Hydropower Pumped Storage 03:71–77
8. Liu GH, Wang SY, Bao ZR (1999) Study on nonlinear fully adjusted load sharing method for arch dams. J Zhejiang Univ Nat Sci Edn 33(1):7
9. Wang DW, Wang Y, Liu GH (2023) Study on shear safety analysis of typical arch dams in U-shaped wide river valleys. Hydroelectricity 49(04):47–54+80
10. U.S. Bureau of Reclamation, Arch Dam Design Translation Group (1984) Design of arch dams. Water Conservancy and Electric Power Press, Beijing

Open Access This chapter is licensed under the terms of the Creative Commons Attribution 4.0 International License (http://creativecommons.org/licenses/by/4.0/), which permits use, sharing, adaptation, distribution and reproduction in any medium or format, as long as you give appropriate credit to the original author(s) and the source, provide a link to the Creative Commons license and indicate if changes were made.

The images or other third party material in this chapter are included in the chapter's Creative Commons license, unless indicated otherwise in a credit line to the material. If material is not included in the chapter's Creative Commons license and your intended use is not permitted by statutory regulation or exceeds the permitted use, you will need to obtain permission directly from the copyright holder.

Experimental Study on Bearing Behaviour of Bored Piles Using End-side Association Post-grouting Technology

Shiding Su, Shuhui Lv, Jiaqi Wu, and Bo Zhang

Abstract In order to improve the bearing performances of bored cast-in-place piles in soft soil area, an end-side association post-grouting technology is proposed. Based on one highway bridge project, the composite post grouting and O-cell pile testing method on one cast-in-place concrete piles S1 are performed to study their bearing deformation characteristics and the improving effect of pile tip resistance and pile side frictional resistance. The research shows that pile side grouting has improved the mechanical characteristics of pile-soil interface and the pile tip grouting could strengthen pile end soil. The strengthening effects of pile shaft resistance in silty clay are in the range of 1.49–1.60 and the strengthening effects of pile tip resistance in gravel with clay is 1.71. End-side association post-grouting technology can effectively increase the axial bearing capacity of bored piles in soft soil and reduce the pile top settlement.

Keywords Bored cast-in-place pile · O-cell pile testing method · Post-grouting · Axial bearing capacity · Strengthening effects

---

S. Su (✉) · S. Lv · J. Wu · B. Zhang
Key Laboratory of Environment and Safety Technology of Transportation Infrastructure Engineering, CCCC Fourth Harbor Engineering Institute Co., Ltd., Guangzhou, China
e-mail: sushiding@cccltd.cn

S. Lv
e-mail: lvshuhui@cccltd.cn

J. Wu
e-mail: wujiaqi2@cccltd.cn

B. Zhang
e-mail: zhangbo95@cccltd.cn

1 Introduction

Bored cast-in-place piles are widely used in engineering construction, but accompanied by defects such as shaft mud cake and bottom sediment which would lower pile capacity. During construction process, optimization of pile foundation design including how to improve the bearing capacity of bored cast-in-place piles and decrease pile settlement is an important subject of scientific research. Lots of relevant studies show that post-grouting technology could not only enhance pile tip resistance, but also have an effect on the performance of pile side resistance, and raise the pile bearing capacity, diminish the settlement of piles [1–11]. However, the operation mechanism and effect are different for different soils. The main research of the current of the effect of end-side association post-grouting technology using in bored cast-in-place piles focus on sands but less in soft soils. Therefore, one cast-in-place concrete piles S1 with steel-string transducer, in deep soft soils, are selected to carry out the composite post grout and O-cell pile test before and after post grouting. The analysis results of test data are available for analyzing the vertical bearing performance of test pile in soft soils before and after post-grouting. The research is of benefit to the improvement of pile foundation defects and design optimization of bored cast-in-place piles in deep soft soils.

2 Basic Information of Geology and Test

The support project of major bridge is located in Yueqing, Wenzhou. The report of engineering geological exploration shows that the geologic conditions at Bridge site area are very complex. Test site adjacent to one borehole called ZK43 which is representative of geological conditions is selected. The soil layer is mainly composed of soft soils such as mud, clay and silty clay. Detailed physical and mechanical parameters of these soft soils are summarized in Table 1.

Bored cast-in-place pile with concrete strength of C40 is adopted as the foundation of this project. Test pile named S1 with diameters of 1200 mm are instrumented with steel-string transducer. The pile length of test pile S1 is 60 m with the pile top elevation of 3.28 m and the pile bottom elevation of − 56.72 m. The bearing stratum of test pile is gravel with clay. O-cell load test is performed on pile S1 before and after post-grouting. The end-side association post-grouting technology is applied to reach the purpose of research.

On the basis of the provisions of Code for Static Loading Test of Foundation Pile-self-balanced Method (JT/T738-2009) [12], the installation position of load cell shall be at the balance point position. According to the soil parameter and formulas stem from the Code for Technical Specifications for Building Pile Foundation (JGJ 94–2008) [13], the position can be determined. Test pile is divided into 10 sections, which are located at the interface of soil layers, and two transducers are installed at each section. In total of 20 steel-string transducers are used to measure the pile strain

Table 1 Soil physical and mechanical parameters

Type of soil	Top elevation/ m	Bottom elevation/ m	Water content (%)	Density ρ(g/cm^3)	Void ratio	Direct shear test	
						Internal friction angle φ_q (°)	Cohesion c_q (kPa)
①$_1$clay	3.28	1.64	41.7	1.80	1.159	9.1	18.5
②$_{12}$mud	1.64	− 11.36	67.8	1.57	1.970	2.9	7.1
②$_2$mud	− 11.36	− 21.49	63.5	1.59	1.840	3.3	8.2
③$_2$clay	− 21.49	− 37.44	42.5	1.74	1.256	7.8	16.6
④$_1$Silty clay	− 37.44	− 49.18	31.9	1.90	0.913	12.5	26.0
⑤$_1$Silty clay	− 49.18	− 51.46	34.7	1.85	0.996	10.5	20.5
⑥$_1$Silty clay	− 51.46	− 51.80	32.8	1.86	0.957	10.8	20.7
⑨$_2$Gravel with clay	− 51.80	− 57.70	–	–	–	–	–
⑩$_2$Strongly weathered ashstone	− 57.70	− 59.48	30.8	1.85	0.930	14.1	22.6
⑩$_3$Moderately weathered ashtone	− 58.63	− 68.00	–	–	–	–	–

of cast-in-place pile and pile side resistance and tip resistance can be calculated. Load-cell is installed at the position of − 62.92 m and the distance of 1.8 m above the pile bottom, seen in Fig. 1.

The mode of post-grouting is open grouting. Pile side grouting first and pile end grouting later. The interval time between pile side grouting and tip grouting shall be at least 2 h. Automatic control system is adopted to record grouting pressure, grouting amount and flow rate, detailed grouting parameters seen in Table 2. In situ grouting experiment can be seen in Fig. 2.

Fig. 1 Installation of load cell

Table 2 Grouting parameters

No.	Position	ID	Elevation/m	Cement content/t	Termination pressure /MPa
1	Pile shaft	SZ-1	− 12.72	0.968	1.74
2		SZ-2	− 24.72	0.970	2.10
3		SZ-3	− 32.72	0.975	1.30
4		SZ-4	− 40.72	0.980	1.00
5		SZ-5	− 48.72	0.979	1.62
6		SZ-6	− 56.72	1.048	1.50
7	Pile end	Pile end	− 64.72	3.135	1.21

Fig. 2 In situ grouting experiment

3 Test Results Analysis

3.1 Vertical Bearing Characteristic

As shown in the Fig. 3, the positive displacement indicates the upward movement of the upper pile and the negative displacement indicates the downward movement of the lower pile. With the load increasing, the displacement of top panel of pile S1 increases linearly, but not obviously. The maximum displacement of top panel is only 4.21 mm without pile post-grouting and 5.62 mm after post-grouting. Instead, the displacement of lower panel of test pile show that test pile happens with the sharp drop at the load of 9000 kN before grouting and the load of 13,500 kN after grouting. The maximum allowable test load of lower panel of pile S1 without grouting is 9000 kN and increases to 13,500 kN after grouting. Compared the curve data from the O-cell load test before and after post-grouting, the results indicate that the rate of displacement change slow down after grouting, and the displacement at the same load becomes small. Obviously, the ultimate bearing capacity of the cast-in-place pile can increase and displacement can reduce after post-grouting.

The Appendix B of Code for Static Loading Test of Foundation Pile-self-balanced Method (JT/T738-2009) give the method to transform the load–displacement curve

Fig. 3 Load-displacement curves of test pile

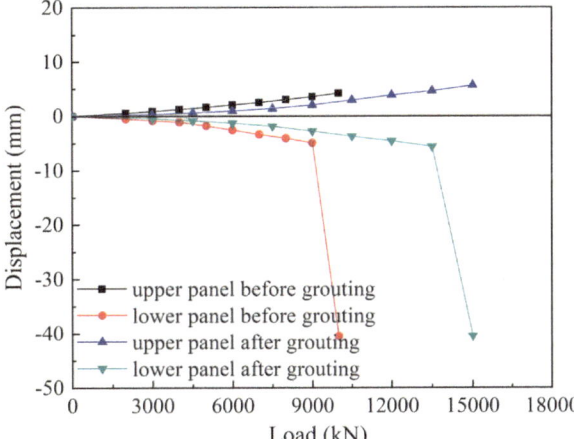

of O-cell load test into an equivalent curve of pile top loading which is similar to vertical compression static load test. The equivalent converted curves are shown in Fig. 4. The load–displacement curve reflects the pile failure, either before or after grouting.

The ultimate vertical bearing capacity of test pile S1 before and after grouting are 17,378.8 and 24,328.8 kN respectively. The ultimate bearing capacity of grouted pile increases by 40% compared to pile without post-grouting. For other piles using post grouting, design optimization of pile length can refer to the coefficient without another more field data.

The displacement corresponding to the former is 31.28 and 41.15 mm under the ultimate load applied on pile top. In addition, the variation of the displacement by applying multistage load has slowed affected by the strengthen effects of end-side association pile grouting.

Fig. 4 Equivalent pile top load–displacement curve

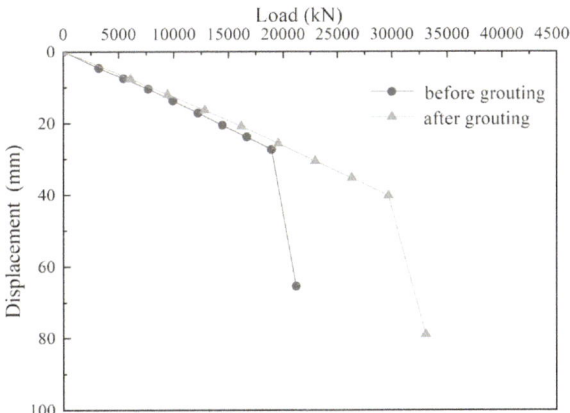

3.2 Pile Skin Resistance

During multi-staged loading of O-cell pile test, the measured pile strain of steel-string transducers embedded in the pile shaft can be converted to stress then transformed into pile axial force. On the basis of pile axial force results, skin resistance of test pile can be calculated and the distribution of the mobilized unit skin resistance at different soils are shown in Figs. 5 and 6. The positive skin resistance is in the upper pile and the negative skin resistance is in the lower pile.

The mobilized unit skin resistance of pile in different soils are shown in Table 3. The improvement in strengthen of skin resistance is obvious in different soils. There is little difference in improvement amplitude of different soil layers. The average percentage of improvement in skin resistance is 53.8%. The strongly weathered ashstone get the maximum increment of 72%. With respect to the soft soils such as

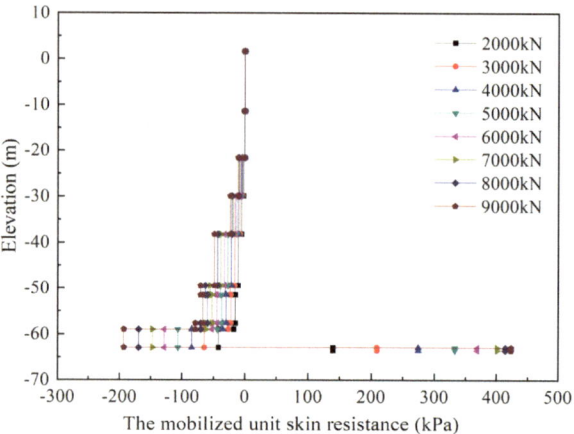

Fig. 5 The distribution of mobilized skin resistance of test pile before grouting

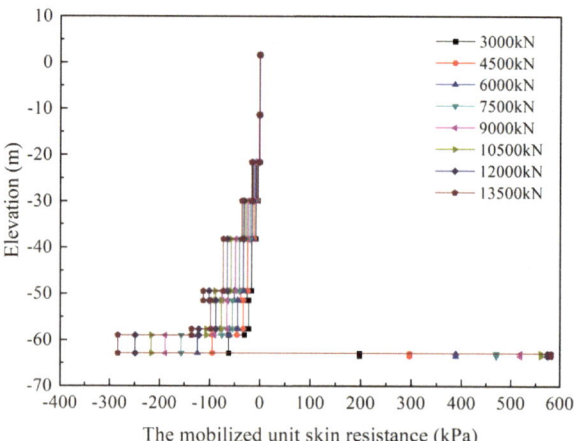

Fig. 6 The distribution of mobilized skin resistance of test pile after grouting

Table 3 Pile skin resistance before and after grouting

Section	Type of soil	The mobilized skin resistance (kPa)		Improve (%)
		Before grouting	After grouting	
G1-G2	②₁mud	− 0.5	− 0.8	60.00
G2-G3	②₂mud	− 1.0	− 1.49	49.00
G3-G4	③₂clay	− 11.0	− 16.5	50.00
G4-G5	③₂ clay	− 23.3	− 35.5	52.40
G5-G6	④₁silty clay	− 48.6	− 72.5	49.20
G6-G7	⑤₁silty clay	− 70.3	− 112.5	60.00
G7-G8	⑨₂gravel with clay	− 66.7	− 97.6	46.30
G8-G9	⑩₂strongly weathered ashstone	− 78.6	− 135.5	72.40
G9-G10	⑩₃moderately weathered ashstone	295.3	428.3	45.00

clay and silty clay, the strengthen coefficients are in the range of 1.49–1.60. Then these strengthen coefficients can be used to optimize the pile length which is contribute to decrease the construction cost under the premise of meeting design requirement.

3.3 Pile Tip Resistance

Load-cell is installed at the position of − 62.92 m and the distance of 1.8 m above the pile bottom. As shown in the above text, pile shaft has been divided into 10 sections. And each section has installed two transducers. The position of tenth section G10 of pile shaft is − 63.52 m, which is 1.2 m far away from pile bottom − 64.72 m. The axial force of the tenth section G10 can be used to estimate pile tip resistance. The unit skin resistance of pile shaft in the range of − 63.52 to − 64.72 m could be regarded as the same to G9-G10 skin resistance in moderately weathered ashstone. Take off the calculate skin resistance of soil in the range of − 63.52 to − 64.72 m, the modified pile axial force at the tenth section G10 can be thought to be pile tip resistance, as shown in Fig. 7. The maximum pile tip resistance is 8042.03 kN and the pile tip resistance corresponding to the ultimate bearing capacity is 7062.48 kN before end-side pile post-grouting. For the pile using post-grouting, the maximum pile tip resistance increased to 16,300.35 kN. For the ultimate bearing capacity condition, pile tip resistance is 14859.84 kN.

The tip resistance of test pile increases linearly with the increment of load. Considering the short distance between load cell and pile bottom, the load applied to lower panel are mostly undertaken by pile tip resistance. The increment of pile tip resistance is apparent which can be up to 50–100%. In addition, pile tip displacement also significantly decreases. The strengthen effects of pile tip resistance is induced by pile end grouting. Generally, soils beneath the pile end can be enhanced. According to

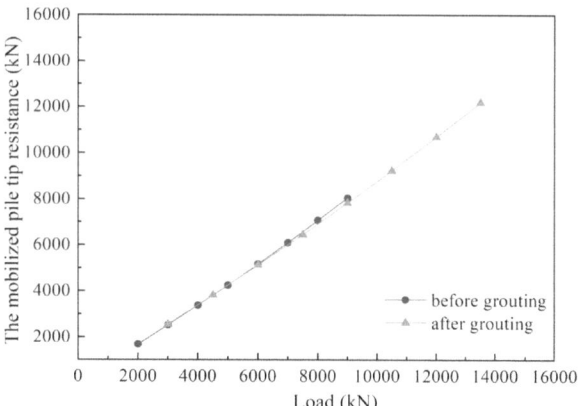

Fig. 7 Q-s curves of pile tip

the findings, pile end grouting can also regarded as an effect measures to improve the bearing capacity of pile bearing stratum.

4 Conclusions

Post-grouting technology has an obvious effect on eliminating the adverse impact on the bearing characteristic caused by the mud clamped in the pile shaft and sediment on pile end. For the reason of a lack of available field data, the effect of post-grouting on bored cast-in-place pile in the deep soft soils are not well recognised. Given that, this paper makes some study to strengthen effect of end-side association pile grouting on bored cast-in-place pile in soft soils. Research findings will help to the design optimization of bored cast-in-place pile in the deep soft soil. The conclusions are as follows.

All of the soft soils such as silty clay, clay and mud reflect the approximate strengthen coefficient of 1.54 which can be used to pile skin resistance calculation of grouted pile during design phase. And then pile design length can be optimized considering post grouting based on the bearing capacity increment.

With the contribution of pile end post grouting, pile tip resistance has an increment of 50–100% and the pile tip displacement becomes smaller during multistage load. The strengthen coefficient of pile tip resistance induced by post-grouting can be used to pile tip resistance calculation of grouted pile.

The axial bearing capacity of super-long cast-in-place piles can dramatically increase by end-side association pile grouting. And the axial displacement of pile also decreases. The end-side association pile post grouting used to optimum design of super-long bored cast-in-place pile in soft soils.

References

1. Safaqah O, Bittner R, Zhang X (2007) Post-grouting of drilled shaft tips on the Sutong Bridge: a case history. In: Geo-Denver, pp 1–10
2. Mullins G, Dapp SD, Lai P (2000) Pressure-grouting drilled shaft tips in sand. ASCE
3. Huang SG, Zhang XW, Cao H (2004) Mechanism study on bored cast-in-place piles with post-grouting technology. Rock Soil Mechan 25:251–254
4. O'Neil MW, Reese LC (2010) Drilled shafts: Construction procedures and design methods. In: Tunneling & underground space technology, pp 156–157
5. Yamato S, Karkee MB (2004) Reliability based load transfer characteristics of bored precast piles equipped with grouted bulb in the pile toe region. J Jpn Geotech Soc 44(3):57–68
6. GuoYC, Zhang JW, Dong XX (2014) Experimental study on bearing capacity of bored piles using pile tip and side post-grouting technology. J Highway Transp Res Dev
7. HuangSG, Shen JH, Meng LI (2019) Reliability analysis of bearing capacity of post-grouted bored piles. Rock and Soil Mechanics
8. Zhang ZM, Xin GF (2002) The application effect analysis of post grouting under bored piles with different bearing strata. J Build Struct
9. Dong J, Lin S, Dai Y (1994) The load transfer behavior of large diameter cast-in-situ pile in crushed pebble stratum. J Geotech Eng
10. Chen W (2013) Experimental study on the vertical bearing capacity of super-long bored pile in shanghai soft soil area. Chin J Undergr Space Eng 7(3):504–508
11. ChuCF, Li XC, Lu LH, Xi PS (2011) Load bearing behavior of pile tip post-grouting super-long large-diameter bored piles in Cohesionless soil. Chin J Geotech Eng
12. School of Civil Engineering of Southeast university (2009) JT/T738-2009 static loading test of foundation pile-self-balanced method. China Communications Press, Beijing
13. China Architecture Design and Research Group (2008) Technical code for building pile foundations: JGJ94-2008. China Architecture & Building Press, Beijing

Open Access This chapter is licensed under the terms of the Creative Commons Attribution 4.0 International License (http://creativecommons.org/licenses/by/4.0/), which permits use, sharing, adaptation, distribution and reproduction in any medium or format, as long as you give appropriate credit to the original author(s) and the source, provide a link to the Creative Commons license and indicate if changes were made.

The images or other third party material in this chapter are included in the chapter's Creative Commons license, unless indicated otherwise in a credit line to the material. If material is not included in the chapter's Creative Commons license and your intended use is not permitted by statutory regulation or exceeds the permitted use, you will need to obtain permission directly from the copyright holder.

A Procedure for Strength Parameters Inversion of Rock Slopes Based on Statistical Regression and Seepage-Stress Coupling: A Case Study

Rong Chen, Detan Liu, Haiku Zhang, Mao Liao, Zhangxin Huang, and Liqun Xu

Abstract Massive slope engineering exists in large water conservancy projects. The artificial and natural rock slopes are high and steep with diverse lithologies, and the stability and safety of high slopes are the guarantee for the overall project to be effective. This paper presents a parameter inversion procedure for rock slopes based on statistical regression models using monitoring data and coupled analysis of seepage and stress. It is applied to the intake slope of a hydropower station in Southwest China. The results show that based on the strength parameters from inversion, the relative error between the measured and calculated displacement of the monitoring points is less than 4.55% under the water level rise and fall inversion conditions, and less than 4.01% under the validation condition, meeting the requirements for engineering applications. The slope stability safety factor meets the requirements specified in the regulations. The inversion procedure proposed in this paper can provide reference for the determination of high slope strength parameters in this project and other projects.

Keywords Rock slopes · Parameter inversion · Statistical regression model · Seepage-stress coupling method

---

R. Chen
Datang Sichuan Power Generation Co., Ltd., Chengdu, China
e-mail: 184557348@qq.com

D. Liu · H. Zhang
Datang Hydropower Science & Technology Research Institute Co., Ltd., Chengdu, China
e-mail: zhk336@163.com

D. Liu · Z. Huang (✉) · L. Xu
College of Water Conservancy, Hydropower Engineering Hohai University, Nanjing, China
e-mail: zhangxin_huang@hhu.edu.cn

M. Liao
Sichuan Datang International Ganzi Hydroelectric Co., Ltd., Kangding, China
e-mail: 289838906@qq.com

1 Introduction

In the major hydropower projects in southwest China, there are numerous steep slopes, and slope stability issues are prominent [1]. The instability of high slopes in dam sites can directly endanger the safety of critical structures and personnel. It can also indirectly cause reservoir surges, leading to consequences such as obstruction of navigation and dam damage. Therefore, the stability of high slopes is crucial for the effective and beneficial operation of water conservancy projects. Determining the actual strength parameters of the slope during its operation is an important basis for assessing slope stability and is one of the key concerns of researchers.

Chu and Ji introduced a geological strength index (GSI), and calculated the shear strength parameters of dolomitic limestone and limestone for slope by regression analysis [2]. Yan et al. conducted a comparative analysis of the safety factors calculated using rigid body limit equilibrium and finite element methods for strength indicators such as cross-laminated timber and rapid shear and inverse methods and explored the reasons for the differences in calculated safety factors [3]. Liu et al. introduced the "wetting–drying" cyclic effect and cumulative damage ratio of uniaxial compressive strength of rock, made improvements to the Hoek–Brown strength criterion and determined the rock mechanics parameters for the reservoir bank slope [4]. Wan et al. analyzed fractured rock masses by assuming that other fractured rock masses have higher strength. They conducted slip zone searching calculations until the slip zone transferred and the safety factor met the actual requirements. Finally, they established a linear relationship between cohesion and internal friction angle under certain conditions [5]. Jiang and Wang adopted the BP neural network method to perform inverse analysis of the strength parameters of the natural slope at the intake of the Jinchuan Hydropower Station spillway. The limit equilibrium method was used to organize the training samples for the neural network. The safety factor was used as the input of the network, while cohesion and internal friction angle were the outputs of the network [6]. Walton discussed the process of back analysis using the cohesion-weakening-friction-strengthening (CWFS) [7]. Tawadrous et al. used artificial neural network (ANN) to constrain discontinuum model parameters [8].

In this study, statistical regression analysis is conducted on the deformation monitoring data of a slope at the intake of a hydropower station. A finite element model of the rock slope is established, and an analysis of seepage-stress coupling is performed. The inverse analysis is carried out by minimizing the sum of squared differences between the displacement values obtained from the statistical regression model and the finite element calculations. The aim is to provide a reference for the stability analysis of this slope and similar slopes in the future.

2 Methodology

2.1 Statistical Regression Model of Slope Displacement

The deformation monitoring data of observation points on the intake slope and the water level variation curve of the project reservoir is shown in Fig. 1.

Due to the presence of observation errors in the deformation monitoring values, it is difficult to analyze and describe the convergence of slope deformation using the measured values, especially when the slope deformation is relatively stable. To reduce the influence of observation errors, the water pressure component and time-dependent component are considered. The measured displacement values at each monitoring point are regressed and fitted using a trend displacement value. The water pressure component is linearly related to the upstream water depth H, H^2, and H^3, thus it is represented by a cubic polynomial of hydraulic head. The time-dependent component exhibits rapid initial changes, followed by a tendency towards stability, and gradually transitions from nonlinear to linear changes. To reflect these characteristics, the time-dependent component is represented by a combination of logarithmic and linear functions. The expression for the regression fitting model of the trend displacement values is as follows:

$$\begin{aligned}\delta &= \delta_H + \delta_T + \delta_\theta \\ &= a_0 + a_1 H + a_2 H^2 + a_3 H^3 + a_4 \cos(s) + a_5 \sin(s) \\ &\quad + a_6 \sin^2(s) + a_7 \sin(s)\cos(s) + a_8 t + a_9 \ln(t+1)\end{aligned} \quad (1)$$

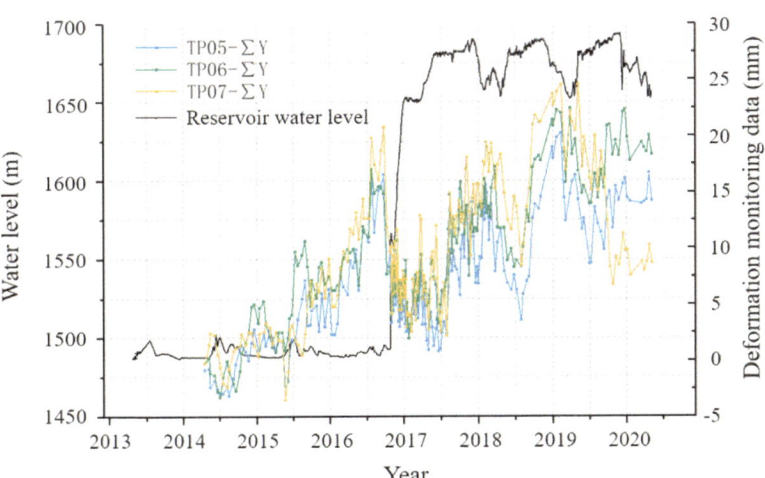

Fig. 1 Deformation monitoring data of observation points on the intake slope and the variation curve of reservoir water level

where: a_0 is a constant term; H is the reservoir water level; $s = \frac{2\pi t}{365.25}$, and t is the number of monitoring days starting from December 28, 2012.

2.2 Basic Theory of Seepage-Stress Coupling

Governing equation. For rock slopes, the rock mass is considered as a porous medium material and is treated as an ideal elastic body. Under this assumption, the governing equation for the seepage field can be expressed as:

$$\nabla[-\delta_K K_S/(\nabla p + \rho_w g \Delta z)] = \delta_Q Q_S \tag{2}$$

where: δ_K and δ_Q are flux and source proportionality coefficients, typically taken as 1.0; K_S is the permeability coefficient of the rock mass; Q_S is the source term of the fluid; H is the total hydraulic head; ∇ is the Laplacian operator.

For the rock mass considered as an ideal elastic body, the stress components σ_{ij} can be represented as follows:

$$\sigma_{ij} = 2G\varepsilon_{ij} + \lambda \delta_{ij} \delta_{kl} \varepsilon_{kl} - \alpha \delta_{ij} p \tag{3}$$

where: G is the shear modulus; λ is the Lamé parameter, given by $\lambda = E\nu/[(1+\nu)(1-2\nu)]$; ε_{ij} represents the strain components; α is the Biot coefficient; and δ_{ij} is the Kronecker delta.

According to the theory of elasticity and the equilibrium conditions of statics, the Navier–Stokes equation can be derived, and its expression is as follows:

$$Gu_{i,jj} + (G+\lambda)u_{j,ji} - \alpha p_i + F_i = 0 (i = x, y, z) \tag{4}$$

where: u_i is the displacement in the i-direction; F_i is the volume force in the i-direction.

By combining Eqs. (2) and (4), we can obtain the coupled seepage-stress equation for rock slopes. With appropriate initial and boundary conditions, the problem of seepage-stress coupling in rock slopes can be solved.

Coupling variable. For the seepage-stress problem, the permeability coefficient of the material is often selected as the coupling variable of the seepage field and the stress field of the rock mass. The permeability coefficient of porous media is a function of porosity, and the porosity of the material is related to its stress state, and the stress state is disturbed by the permeability coefficient due to the existence of pore water pressure. In this case, the following exponential formula is obtained:

$$k_f = k_0 e^{-\alpha' \sigma} \tag{5}$$

where: σ is the effective normal stress, $\sigma = \gamma H - p$, γ is the bulk density of rock mass, H is the thickness of rock mass; p is pore water pressure; $\alpha\prime$ is an empirical coefficient, which is related to the integrity of rock mass materials.

2.3 Inversion Procedure

The specific steps for the inversion of strength parameters in slope based on statistical modeling are as follows:

Step 1: Perform regression analysis on the displacement monitoring data to establish a statistical model for the displacements at various monitoring points on the slope.

Step 2: Build a finite element model of the slope and select a period with significant variation in reservoir water level. Considering the effect of pore water pressure, perform seepage-stress coupling calculations on the slope using the finite element method to obtain the calculated sequence of displacement changes for the monitoring points within that period.

Step 3: Compare the calculated displacement sequence from Step 2 with the displacement sequence obtained from the statistical model in Step 1. Calculate the sum of squared differences and select the displacement sequence with the minimum sum of squared differences as the strength parameters of the rock slope.

Step 4: After obtaining the inverted strength parameters, select another period and perform a recheck based on Step 2 to verify the rationality and accuracy.

3 Project Overview and Computational Model

3.1 Project Overview

A hydropower station is located in Kangding City, Garze Tibetan Autonomous Prefecture, Sichuan Province, China. It is one of the key projects in the recent development of the Dadu River mainstream hydropower cascade. The project includes a reservoir dam and an underground water diversion and power generation system. The water conveyance and power generation facilities are located on the left bank of the river and consist of the intake structure, pressure pipelines, main and auxiliary powerhouses, transformer chamber, switchyard, tailrace pressure regulation chamber, and tailrace tunnel.

Based on surface surveys and excavation, it has been revealed that there are no major regional faults at the intake structure site. The geological structure is characterized by secondary small faults, dense joint zones, and joint fractures. The longitudinal axis of the water diversion tunnel determines the layout of the excavated slopes at the inlet. The excavation will create a four-sided slope in the plan view, consisting of the tunnel face slope, inner slope, and outer slope. The elevation of the tunnel floor is 1628 m, and after excavation, the slopes will reach a height of approximately 70–190 m. The natural slopes in the area have a height of 800–1000 m.

3.2 Computation Model

The rock formation of the reservoir area slopes is the Jinning-Chengjiang period granite. After excavation, the intake slopes will reach a height of approximately 70–190 m. The design excavation slope ratio is 1:0.5–1:0.75, except for the lower 36 m section, which is a vertical slope. A primary bench is set at a height difference of 20–25 m with a width of 2 m. A two-dimensional finite element model of the intake slope is established, considering the rock mass as a linear elastic material. Anchor rods are simulated using rod elements.

Based on the topographical and geological conditions of the intake slope and the location of monitoring points, the boundary range of the model is defined as follows:

(1) The lower boundary is set at a depth of 55.50 m below the intake structure foundation;
(2) The upper boundary is set at a height of 27.50 m above the slope excavation.

The slope model is established based on design data, including cross-sectional diagrams and geological profiles. The model adopts the Cartesian coordinate system, with the coordinate origin at the lower left corner of the slope model. The X-axis points towards the water intake direction, with inward displacement towards the slope's inner side as positive and displacement towards the open face as negative. The Y-axis is vertical, with upward displacement as positive, consistent with the actual elevation. The two-dimensional finite element model is shown in Fig. 2.

The X-direction displacements of the left and right sides of the model are constrained, and X and Y direction constraints are applied at the bottom of the model, while the remaining boundaries are considered free. The left side of the model's slope is subjected to the water head boundary condition, with the applied water head corresponding to the upstream water level. The right side of the model's slope is set as a drainage boundary, while the remaining boundaries are considered free boundaries.

3.3 Calculated Work Conditions

After the deformation of the slope rock mass stabilizes due to unloading, the deformation of the rock mass is mainly caused by the changes in the upstream water level. Therefore, it is advisable to select time periods with significant changes in the upstream water level for the inversion analysis.

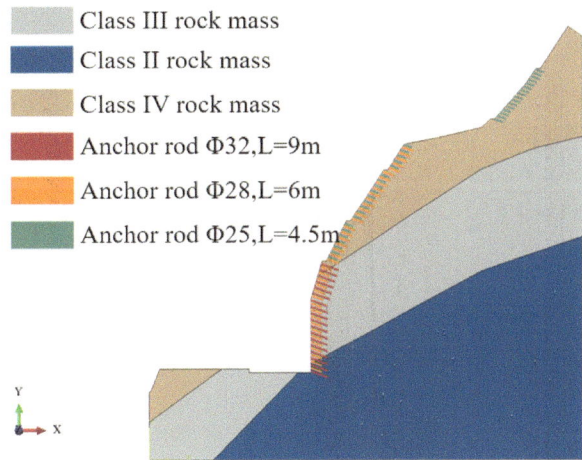

Fig. 2 The intake slope model

Based on the monitoring data of the reservoir water level and considering the dates of deformation monitoring at the slope displacement monitoring points, two time periods are identified for the inversion analysis as shown in Table 1. The first time period is from January 6, 2017, to July 5, 2017, during which the upstream water level increased from 1651.50 to 1679.10 m, with a significant water level change of 27.60 m. The second time period is from July 29, 2018, to February 9, 2019, during which the upstream water level decreased from 1680.10 to 1668.50 m, with a significant water level change of 11.60 m. Therefore, these two time periods will be selected for the inversion.

In addition, based on the recorded dates of reservoir water level monitoring data and deformation monitoring data, the time period from September 25, 2019, to March 28, 2020, (during which the upstream reservoir water level decreased from 1688.50 to 1666.05 m, with a water level variation of 22.45 m) is selected as the verification period for the inversion analysis results.

Table 1 Reservoir water level and slope groundwater level during different inversion periods (*Unit* m)

Inversion period	2017/1/6–2017/7/5		2018/7/29–2019/2/9	
	2017/1/6	2017/7/5	2018/7/29	2019/2/9
Reservoir water level	1651.50	1679.10	1680.10	1668.50
Groundwater level	1657.88	1657.88	1657.88	1657.88

Table 2 Calculation parameters of the material partitions

Material partition	Natural density (g/cm^3)	Deformation modulus (GPa)	Poisson's ratio	Internal friction Angle (°)	Cohesion (MPa)	Permeability coefficient (cm/s)
Class II rock mass	2.70	15–20	0.25	50.19–60.95	1.5–1.8	7.00×10^{-5}
Class III rock mass	2.62	8–10	0.30	45.00–50.19	1.0–1.5	1.00×10^{-4}
Class IV rock mass	2.60	1–3	0.35	16.70–26.57	0.3–0.5	5.00×10^{-4}
Anchor rod Φ32, L = 9 m	7.84	210	0.20	/	/	/
Anchor rod Φ28, L = 6 m	7.84	210	0.20	/	/	/
Anchor rod Φ25, L = 4.5 m	7.84	210	0.20	/	/	/

3.4 Calculation Parameters

The range for the inversion of the deformation modulus and shear strength of the slope is determined based on the results of field rock deformation tests. The calculated material parameters for the model are shown in Table 2.

4 Results and Discussion

4.1 Displacement Statistical Model

Based on Eq. (1), in combination with the displacement monitoring data in the Y-direction of the slope, the displacement statistical model for each monitoring point of the inlet slope is established, and the parameters of the statistical models for each monitoring point are obtained as shown in Table 3. Using the established displacement statistical models for each monitoring point, the displacement changes and water pressure components for the inversion period are obtained.

Table 3 Parameters of the displacement statistical model

Parameter	Monitoring point		
	TP05	TP06	TP07
a_0	8019.9826	7800.7032	− 14,558.1345
a_1	− 14.8600	− 14.2480	27.8903
a_2	0.0092	0.0087	− 0.0178
a_3	− 1.8913E−06	− 1.7594E−06	3.7897E−06
a_4	1.3861	1.0854	0.5844
a_5	− 0.5403	− 0.6583	0.3843
a_6	1.1978	0.5576	1.4819
a_7	0.6510	1.0094	1.3415
a_8	0.0121	0.0141	0.0052
a_9	− 1.1575	− 1.6439	4.6707

4.2 Inversion Results of Strength Parameters

The average values of rock deformation modulus and strength parameters obtained from the two inversion periods are calculated, resulting in the final values of slope rock material strength as shown in Table 4.

The relative errors between the inverted values and the measured values of displacement variation are calculated after obtaining the strength parameter from inversion, as shown in Table 5. It can be observed that the relative errors between the measured and inverted values of displacement changes are generally within 5%. The maximum relative error occurs at monitoring point TP07 in inversion period 1, which is 4.55%, while the minimum relative error occurs at monitoring point TP05 in inversion period 1, which is 0.58%. Overall, the relative errors of displacement changes meet the engineering requirements.

Table 4 Strength parameters of slope rock mass materials determined by inverse analysis

Material partition	Deformation modulus (GPa)	Internal friction angle (°)	Cohesion (MPa)
Class II rock mass	18.10	51.34	1.60
Class III rock mass	8.95	47.73	1.20
Class IV rock mass	2.20	38.66	0.50

Table 5 Comparison of inverted and measured values of displacement variations at each monitoring point of the slope during the inversion periods

Inversion period	Monitoring point	Measured displacement variation (mm)	Calculated displacement variation (mm)	Relative error (%)
2017/1/6–2017/7/5	TP05	− 0.1380	− 0.1372	0.58
	TP06	− 0.0817	− 0.0805	1.47
	TP07	− 0.0484	− 0.0462	4.55
2018/7/29–2019/2/9	TP05	0.1402	0.1385	1.21
	TP06	0.0660	0.0645	2.72
	TP07	0.0512	0.0501	2.15

Table 6 Comparison of inverted and measured values of displacement variations at each monitoring point of the slope during the verification periods

Inversion period	Monitoring point	Measured displacement variation (mm)	Calculated displacement variation (mm)	Relative error (%)
2019/9/25–2020/3/28	TP05	0.1477	0.1456	1.42
	TP06	0.0598	0.0574	4.01
	TP07	0.0542	0.0533	1.66

4.3 Check of Inversion Results

Table 6 provides the calculation results of the relative errors between the measured and calculated displacement variations for each monitoring point during the verification period. It can be observed from the table that the relative errors between the measured and calculated displacement changes for the monitoring points during the verification period are all within 5%. The maximum relative error in displacement change occurs at monitoring point TP06, with a value of 4.01%. The relative errors satisfy the engineering requirements, indicating that the inverted deformation modulus of the slope rock material is reasonably accurate and valid.

5 Conclusions

This study focuses on the high slope of the intake of a hydropower project on the Dadu River. It proposes an inversion procedure for the strength parameters of the rock slope. Firstly, considering the water pressure component and the time-dependent component, a regression analysis is performed on the measured displacement values at each monitoring point to reduce the influence of observation errors. Based on this, a coupled analysis of seepage and stress is conducted, and the inverse process is carried out by minimizing the squared differences between the regression model displacement and the calculated displacement. The relative errors between the measured and calculated values are within 4.55% under the inversion condition and within 4.01% under the verification condition, satisfying the requirements for engineering application. The stability safety factor of the slope meets the specifications under the inverted strength parameters. The proposed inversion procedure can serve as a reference for determining the strength parameters of high slopes in this project and other similar projects.

Acknowledgements The supports of the Scientific Research Project of China Datang Corporation Ltd. (CHB-FW170-2021) are much appreciated.

References

1. Yang J, Ma C, Cheng L, Lyu G, Li B (2019) Research advances in the deformation of high-steep slopes and its influence on dam safety. Rock Soil Mechan 6:2341–2353+2368
2. Chu C, Ji Y (2017) Calculation on high rock slope strength parameter based on GSI. J Liaon Tech Univ (Nat Sci) 5:488–493
3. Yan X, Wang J, Jiang X, Yang W (2020) Discuss on soil mechanics parameters for restriction stable calculation of high-slope on soft soil embankment. Water Resour Power 9:142–145
4. Liu X, Fu H, Qin Z, Feng B (2017) Stability analysis of high and steep rock slope under the action of reservoir water fluctuation. Sci Technol Eng 24:263–268
5. Wan L, Li J, Zhao Z, Xu Y, Yuan X (2012) Fractured rock strength parameter inversion of abutment slope of a hydropower station. Yangtze River 10:88–91
6. Jiang Z, Wang L (2014) BP neural network applied to strength parameters inverse analysis in unloading rock mass. Yellow River 1:109–110+114
7. Walton G (2019) Initial guidelines for the selection of input parameters for Cohesion-Weakening-Friction-Strengthening (CWFS) analysis of excavations in brittle rock. Tunnel Undergr Space Technol 84:189–200
8. Tawadrous AS, DeGagné D, Pierce M, MasIvars D (2009) Prediction of uniaxial compression PFC3D model micro-properties using artificial neural networks. Int J Numer Anal Methods Geomech 18:1953–1962

Open Access This chapter is licensed under the terms of the Creative Commons Attribution 4.0 International License (http://creativecommons.org/licenses/by/4.0/), which permits use, sharing, adaptation, distribution and reproduction in any medium or format, as long as you give appropriate credit to the original author(s) and the source, provide a link to the Creative Commons license and indicate if changes were made.

The images or other third party material in this chapter are included in the chapter's Creative Commons license, unless indicated otherwise in a credit line to the material. If material is not included in the chapter's Creative Commons license and your intended use is not permitted by statutory regulation or exceeds the permitted use, you will need to obtain permission directly from the copyright holder.

Experimental Study on Creep Behavior of Prestressed Concrete

Tao Ge and Le Yan

Abstract In order to study the influencing factors and variation patterns of the creep performance of prestressed concrete, the creep deformation of concrete specimens caused by long-term constant pressure and the instantaneous elastic deformation at the initial moment of applying this load were measured through experiments. The creep coefficient is mainly related to water reducing agents and additives, and the creep coefficient of concrete with single addition of fly ash at the same age is higher than that of concrete with double addition of fly ash and slag micro powder. Therefore, current research has not revealed the variation pattern of creep in prestressed concrete structures and the mechanism of its impact on prestressed relaxation. The measurement values of concrete creep and elastic modulus provided in this paper can be used as a reference for engineering construction control.

Keywords Prestress concrete · Water reducing agent · Creep

1 Introduction

The creep of concrete is an inherent characteristic of the material itself [1]. Since creep was observed in the early twentieth century, it has always been a research hotspot in the field of concrete research [2]. It directly affects a series of important impacts on the construction and use of concrete structures, such as cracks, durability, and stress redistribution between concrete and steel bars [3]. Especially for large-span structures using prestressed high-strength concrete, the impact is particularly significant, directly related to the loss of prestress, stress redistribution, and changes in secondary stress [4]. Extensive experimental and theoretical research has also been conducted on the creep performance of prestressed concrete both domestically and internationally [5, 6]. Due to the complexity of the materials, the results obtained have significant discreteness. This article combines the C50 prestressed concrete

T. Ge (✉) · L. Yan
Aeronautics Engineering College, Air Force Engineering University, Xi'an, China
e-mail: getaoge@163.com

© The Author(s) 2025
P. Xiang et al. (eds.), *Frontier Research on High Performance Concrete and Mechanical Properties*, Lecture Notes in Civil Engineering 518,
https://doi.org/10.1007/978-981-97-4090-1_6

used in a certain subway construction to conduct relevant creep performance tests and analyze them, attempting to draw some useful conclusions.

2 Test Preparation

2.1 Test Content and Purpose

Measure the creep deformation of concrete specimens caused by long-term constant pressure and the instantaneous elastic deformation at the initial moment of applying this load. The results will provide the creep degree and instantaneous elastic modulus of the concrete. The loading age of this experiment was 7 and 28 days, respectively.

Measure the self deformation of concrete and provide the self deformation (linear strain) of concrete.

2.2 Maintaining the Integrity of the Specifications

C50 pre-stressed pumping concrete with ratio numbers Y54 and Y57. This proportion of concrete is intended for use in the Nanjing Metro project, and the concrete proportion is listed in Table 1.

The raw materials for concrete are as follows:

The cement adopts Nanjing Jingyang 42.5 ordinary Portland cement; The fly ash adopts Class I ash from Jiangsu Nanjing Huaneng Power Plant; The slag powder is produced by Jiangnan Grinding Co., Ltd; The sand is made from Nanjing Dazhi Mixing Station's river sand; The crushed stone is from Nanjing Dazhi Mixing Station; The additive adopts NA-F2 from Ruidi New Materials Co., Ltd. of Nanjing Academy of Water Sciences.

Table 1 Concrete material mix ratio

Number	Water/binder ratio	Concrete material consumption (kg m^{-3})						
		Cement	Fly ash	Slag	Sand	Gravel	Water	NA-F2
Y54	0.32	440	50	–	616	1120	155	8.7
Y57	0.32	295	50	150	622	1106	156	8.8

3 Test Method

3.1 Creep Test of Prestressed Concrete

- Equipment and facilities. The constant pressure equipment of the test piece is equipped with a 300 kN three bar spring creep tester, a 300 kN hydraulic jack for loading and load reading, and a 400 kN steel ring dynamometer.
- Instrumentation. Strain sensor DI-25 differential resistance strain gauge, reading device QB-2 hydraulic proportional bridge.
- Temperature and humidity conditions. The laboratory temperature is $20 \pm 2\ °C$ and the humidity is $65 \pm 5\%$.
- Test piece. ɸ 20×60 cm cylinder, vertically formed, poured and vibrated in three layers; The strain gauge is placed in the central position of the specimen during molding; After the molding is completed, maintain the mold for 2 days, and then remove the mold for surface sealing treatment.
- Loading and testing. The maximum loading value is about 1/3 of the specimen strength, and the load application is divided into 3–4 levels, with an average loading rate of 100 kN/min; Lock the load after three loading and unloading cycles and completing the elastic model test; After the load is locked, the deformation of the loaded and verified specimens is measured at a certain interval, with a time interval of 1 time per day to 1 time per week.

3.2 Self Deformation of Prestressed Concrete

- Self deformed specimen. ɸ 20×60 cm cylinder is used in conjunction with the verification specimen for creep testing.
- Instrumentation. The DI-25 differential resistance strain gauge is placed in the center of the specimen during molding and parallel to the length direction of the specimen. The measuring instrument QB-2 hydraulic proportional bridge is used.

4 Results and Analysis

4.1 Measurement Calculation

Strain of the total deformation and self deformation of the loaded specimen ε

$$\varepsilon_j = f(z_1 - z_j) \tag{1}$$

In the equation ε_j is the strain (10^{-6}), with shortening being positive; Z_1 and Z_j is resistance ratio reference measurement value and tj time measurement value. Due to the minimal change in indoor temperature, the temperature compensation term is omitted here.

Creep and elastic modulus.

Creep degree $C(t, t_1)$:

$$C(t, t_1) = \frac{\varepsilon_j - \varepsilon_{0j}}{\sigma} \quad (2)$$

In the equation ε_j and ε_{0j} is the strain (increment, positive pressure) of the loaded specimen and the verified specimen from the reference time t_1 to the observation time t_j, respectively, σ It is the load stress (constant value).

Elastic modulus $E(t_1)$ at loading time t_1:

$$E(t_1) = \frac{\sigma}{f(z_0 - z_i)} \quad (3)$$

In the formula, Z_0 is the resistance ratio of zero load, and Z_j is the resistance ratio of the last level load, σ is the stress of the last level of load, and the above parameters are taken as the progress values of the third loading cycle.

4.2 Test Results

(1) The strength and elastic modulus of concrete are shown in Table 2.
(2) The creep degree and creep coefficient of concrete are shown in Tables 3, 4, 5 and 6. The creep curve of concrete are shown in Figs. 1, 2, 3 and 4.
(3) The self deformation results are shown in Tables 7 and 8. The deformation strain Curves are shown in Figs. 5 and 6.

Table 2 Concrete strength and elastic modulus

Number	7d strength (Mpa)	Elastic modulus (10^4 Mpa)	Relative error of elastic modulus (%)	28d strength (Mpa)	Elastic modulus (10^4 Mpa)	Relative error of elastic modulus (%)
Y54	42.8	4.08	3.2	58.4	4.34	2.1
Y57	40.0	3.73	2.2	60.2	4.41	0.7

Table 3 Creep degree and creep coefficient of Y54 concrete (t = 7 days)

Time (d)	Creep degree 10^{-6}/MPa	Creep coefficient	Time (d)	Creep degree 10^{-6}/MPa	Creep coefficient
0	0	0	56	21.0	0.86
1	6.0	0.24	65	21.0	0.86
3	9.6	0.39	75	22.2	0.90
5	10.8	0.44	85	22.2	0.90
8	12.6	0.51	95	22.5	0.92
11	14.4	0.59	105	22.8	0.93
14	15.0	0.61	135	23.7	0.97
21	16.8	0.68	155	23.7	0.97
28	18.6	0.76	185	24.0	0.98
35	19.5	0.79	205	24.3	0.99
42	20.1	0.82	225	24.3	0.99
49	20.7	0.84	245	24.3	0.99

Table 4 Creep degree and creep coefficient of Y54 concrete (t = 28 days)

Time (d)	Creep degree 10^{-6}/MPa	Creep coefficient	Time (d)	Creep degree 10^{-6}/MPa	Creep coefficient
0	0	0	56	11.0	0.48
1	3.5	0.15	65	11.0	0.48
3	5.4	0.23	75	11.8	0.51
5	5.9	0.26	85	11.8	0.51
8	6.7	0.29	95	12.1	0.52
11	7.3	0.32	110	12.6	0.55
14	8.1	0.35	125	12.9	0.56
21	8.9	0.39	145	13.2	0.57
28	9.7	0.42	155	13.4	0.58
35	10.2	0.44	175	13.7	0.59
42	10.5	0.46	205	14.0	0.61
49	10.5	0.46	225	14.0	0.61

4.3 Analysis of Test Results

The 7-day strength of both groups of concrete with different proportions is above 40.0 MPa, and the 28-day strength is above 58.4 MPa. The concrete strength of the specimens meets the expected requirements; The relative range of the measured elastic modulus values of the specimens at 7 and 28 days is between 0.7 and 3.2%, which is much lower than the control value of 15% proposed in the test regulations.

Table 5 Creep degree and creep coefficient of Y57 concrete (t = 7 days)

Time (d)	Creep degree 10^{-6}/MPa	Creep coefficient	Time (d)	Creep degree 10^{-6}/MPa	Creep coefficient
0	0	0	56	18.2	0.68
1	7.5	0.28	65	18.5	0.69
3	10.6	0.39	75	18.6	0.69
5	13.3	0.49	85	18.6	0.69
8	15.1	0.56	95	19.5	0.73
11	15.1	0.56	105	19.5	0.73
14	15.4	0.57	120	19.5	0.73
21	16.3	0.61	135	19.5	0.73
28	17.0	0.63	170	18.9	0.70
35	17.9	0.67	195	18.9	0.70
42	17.9	0.67	215	19.5	0.73
49	18.2	0.68	235	19.8	0.74

Table 6 Creep degree and creep coefficient of Y57 concrete (t = 28 days)

Time (d)	Creep degree 10^{-6}/MPa	Creep coefficient	Time (d)	Creep degree 10^{-6}/MPa	Creep coefficient
0	0	0	56	8.5	0.38
1	2.9	0.13	65	8.8	0.39
3	4.5	0.20	75	9.6	0.42
5	5.0	0.22	85	9.9	0.43
7	5.3	0.23	95	9.9	0.43
10	6.4	0.28	110	9.9	0.43
15	6.9	0.30	125	10.4	0.46
21	7.7	0.34	145	10.4	0.46
28	8.0	0.35	155	10.1	0.45
35	8.2	0.36	175	10.9	0.48
42	8.5	0.38	200	11.2	0.49
49	8.5	0.38	215	11.5	0.51

This indicates that the concrete specimens have relatively uniform molding and meet the requirements for molding and testing. The results can be used as a reference.

After all specimens are loaded, the constant pressure time is over 200 days, and the creep coefficient is between 0.45 and 1.0. The creep degree value is within the normal range, and the creep deformation has gradually stabilized.

The deformation value of Y54 concrete is within 60×10^{-6}, self shrinkage mainly occurs 40 days before the concrete is formed; The deformation value of Y57 concrete is within 100×10^{-6}, the self shrinkage still slowly increases after 150 days of

Experimental Study on Creep Behavior of Prestressed Concrete   67

Fig. 1 Creep curve of Y54 concrete for Nanjing metro project ($t = 7d$)

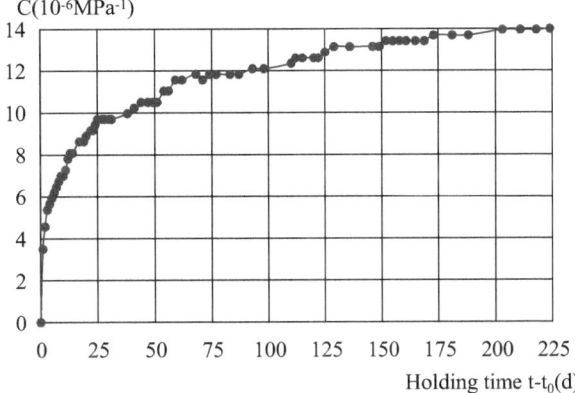

Fig. 2 Creep curve of Y54 concrete for Nanjing metro project ($t = 28d$)

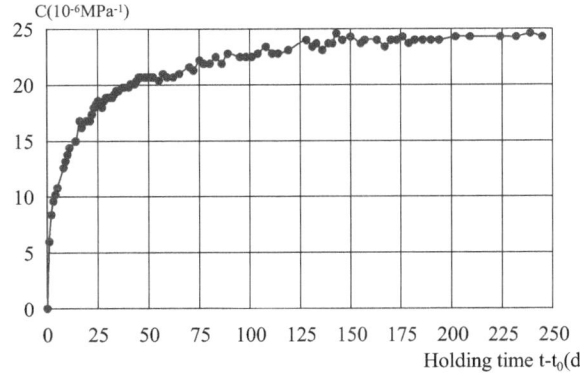

Fig. 3 Creep curve of Y57 concrete for Nanjing metro project ($t = 7d$)

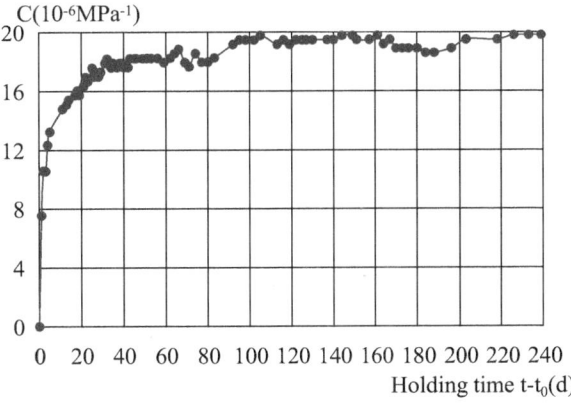

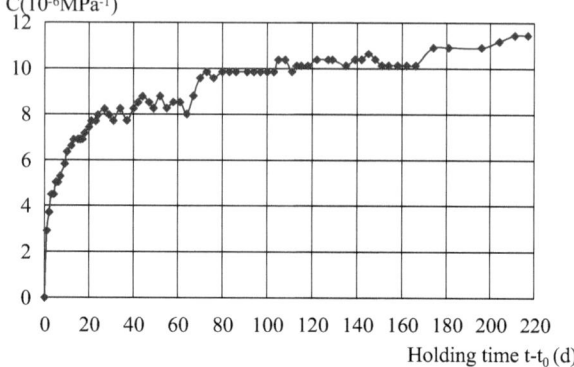

Fig. 4 Creep curve of Y57 concrete for Nanjing metro project (t = 28d)

Table 7 Self deformation results of Y54 concrete

Time (d)	Self deformation (10^{-6})	Time (d)	Self deformation (10^{-6})
1	0	65	54.74
3	3.42	75	54.74
5	10.26	85	54.74
7	15.4	95	51.32
10	18.82	105	49.62
14	27.38	135	46.2
21	37.62	155	47.92
28	46.18	185	47.9
35	51.32	205	47.9
45	56.44	225	47.9
55	53.04	245	47.9

Table 8 Self deformation results of Y57 concrete

Time (d)	Self deformation (10^{-6})	Time (d)	Self deformation (10^{-6})
1	0	65	81.6
3	3.4	75	91.8
5	3.4	85	95.2
7	6.8	95	88.4
10	10.2	105	91.8
14	27.2	120	91.8
21	44.2	135	95.2
28	57.8	170	102
35	64.6	195	102
45	71.4	215	102
55	78.2	235	105.4

Fig. 5 Deformation strain curve of Y54 concrete in Nanjing metro project (positive pressure)

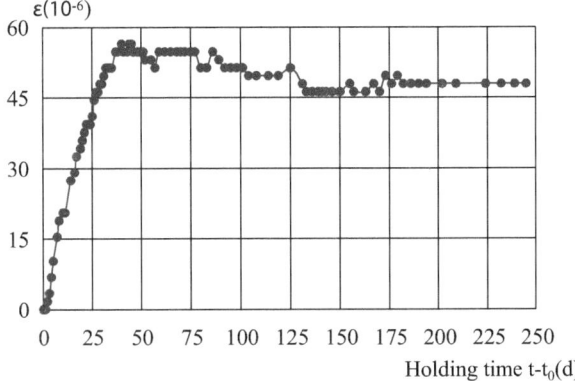

Fig. 6 Deformation strain curve of Y57 concrete in Nanjing metro project (positive pressure)

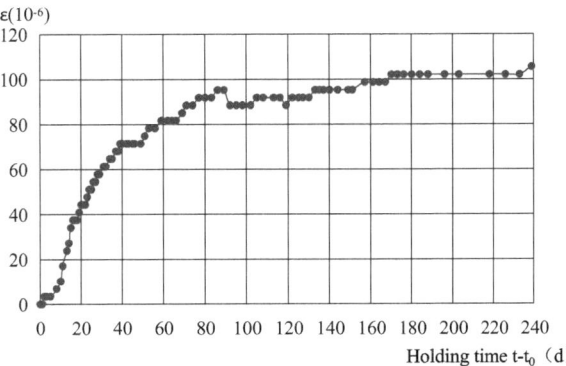

concrete formation. The difference in deformation between the two types of concrete may be related to the combination of materials (admixtures, admixtures).

The measurement values of concrete creep and elastic modulus provided in this paper can be used as a reference for engineering construction control. The recommended reference values are as follows:

① Y54 concrete:

Elastic modulus (10^4 Mpa): $E_7 = 4.08$ $E_{28} = 4.34$.

Creep Coefficient: $\varphi_7 = 0.99$ $\varphi_{28} = 0.61$.

② Y54 concrete:

Elastic modulus (10^4 Mpa): $E_7 = 3.73$ $E_{28} = 4.41$.

Creep Coefficient: $\varphi_7 = 0.74$ $\varphi_{28} = 0.51$.

5 Conclusions

The creep coefficient is mainly related to water reducing agents and additives, and the creep coefficient of Y54 concrete (with only fly ash added) at the same age is higher than that of Y57 concrete (with fly ash and slag micro powder added).

The measurement values of concrete creep and elastic modulus provided in this paper can be used as a reference for engineering construction control.

References

1. Ulm F-J, Acker P (2001) Creep and shrinkage of concrete; physical origins and practical measurements. Nucl Eng Des 2:15–26
2. Liu SH, Feng AY (2006) Summarization of research on concrete creep. Hongshui River 3(118–120):158
3. Wenchao H, Jing X, Wei W (2023) Research on strength and creep characteristics of concrete containing fly ash microbead. Inorg Chem Ind 1(124–128):158
4. Xue WC, Hu YM, Wang W (2008) Experiment on creep behaviors of prestressed concrete beams. China J Highway Transp 4:61–62
5. Zhou YL, Cao D, Wang K (2022) Analysis of creep effect of prestressed concrete continuous beam bridge in construction process. Ind Constr 9:101–107
6. Mokhtar A-SA, Ghali A (1988) Computer analysis and design of concrete beams and grids. J Struct Eng 114(12):2669–2691. https://doi.org/10.1061/(ASCE)0733-9445

Open Access This chapter is licensed under the terms of the Creative Commons Attribution 4.0 International License (http://creativecommons.org/licenses/by/4.0/), which permits use, sharing, adaptation, distribution and reproduction in any medium or format, as long as you give appropriate credit to the original author(s) and the source, provide a link to the Creative Commons license and indicate if changes were made.

The images or other third party material in this chapter are included in the chapter's Creative Commons license, unless indicated otherwise in a credit line to the material. If material is not included in the chapter's Creative Commons license and your intended use is not permitted by statutory regulation or exceeds the permitted use, you will need to obtain permission directly from the copyright holder.

Experimental Study on Mechanical Properties of Permeated Crystalline Concrete

Jingjing He, Haodan Lu, Haiting Wang, Wei Hu, Wenbo Wu, and Kunlong Zhao

Abstract Performance tests were conducted on concrete with different infiltration crystallization dosages. The influence of the dosage of permeable crystallization on the mechanical properties and durability of concrete is analyzed. The results show that infiltration crystallization has a significant improvement effect on the mechanical properties and durability of concrete. When the infiltration crystallization dosage is 1.2%, the increase in compressive strength, tensile strength, and compressive tensile elastic modulus of concrete reaches its maximum. When the infiltration crystallization dosage is 0.6%, the increase in the ultimate tensile strain value of concrete reaches its maximum. When the infiltration crystallization dosage increases within the range of 0–1.2%, the water seepage height and quality loss rate of concrete decrease to varying degrees, and there is a different degree of improvement compared to the dynamic modulus. The research results can provide reference for the mix design and engineering application of permeable crystalline concrete.

Keywords Permeated crystalline concrete · CCCW dosage · Mechanical properties · Durability properties

---

J. He (✉) · H. Lu · H. Wang · W. Hu · W. Wu · K. Zhao
PowerChina Northwest Engineering Co., Ltd., Xi'an, Shaanxi, China
e-mail: Hejing_86@126.com

H. Lu
e-mail: luhaod@nwh.cn

H. Wang
e-mail: wanghait@nwh.cn

W. Hu
e-mail: 824548188@qq.com

W. Wu
e-mail: wuwenb@nwh.cn

1 Introduction

Concrete materials are typical heterogeneous materials with numerous micro cracks and pores inside. These micro defects will not only affect the mechanical properties of concrete, but also penetrate, diffuse and develop into macro cracks in harmful environments. Ultimately, it will lead to a decrease in the durability of concrete [1, 2]. Therefore, how to improve the internal defects of concrete is the key to improving its durability, and it is also an important topic in concrete technology research.

Cementitious capillary crystalline waterproofing material (CCCW) is a type of waterproof material. It has the advantages of excellent waterproof performance, non-toxic and harmless properties, and is widely used in buildings, subways, dams and other projects [3, 4]. At present, research on permeable crystalline concrete mainly focuses on mechanical and durability properties and their mechanisms. The research shows that the permeable crystalline material can significantly improve the mechanical properties, impermeability and self-healing properties of concrete [5]. There are many engineering applications that use permeable crystalline materials as a waterproof coating, while there are relatively few studies and engineering applications that incorporate permeable crystalline materials into concrete. Permeable crystalline materials are an ideal admixture for repairing concrete cracks, and their mechanism of action is to fill the internal cracks and voids of concrete with crystals generated by their reaction with water. Therefore, the mechanical properties and durability of concrete with different dosages of permeable crystalline materials are tested. Analyze the influence of permeable crystalline materials on the performance of concrete. The research conclusion provides a basis for the application of permeable crystalline concrete engineering.

2 Test Overview

2.1 Raw Materials

Cement: Cement is one of the important components of concrete, and its strength directly affects the strength of concrete. This experiment selected P·O 42.5 ordinary Portland cement, and its performance indicators are shown in Tables 1 and 2.

Table 1 Physical performance indicators of cement

Project	Density /g/cm^3	BET surface area /m^2/kg	Standard consistency /%	Setting time/min	
				Initial	Final
GB 175-2007	–	≥ 300	–	≥ 45 min	≥ 45 min
P·O 42.5	3.05	389	27	145	145

Table 2 Mechanical performance indicators of cement

Project	Compressive strength/MPa			Flexural strength/MPa		
	3d	7d	28d	3d	7d	28d
GB 175-2007	≥ 17.0	–	≥ 42.5	≥ 3.5	–	≥ 6.5
P·O 42.5	28.9	39.4	55.7	5.7	7.7	9.5

Table 3 Physical and mechanical performance indicators of fibers

Fiber type	Tensile strength/MPa	Diameter/mm	Elastic modulus/GPa	Density/g/cm^3
Steel fibre	2620	1605	0.7	39

Table 4 Preliminary selection of concrete mix proportion

Project	Water	Cement	Fly ash	Sand	Small stone	Medium stone
JZ	115	224	56	731	593	725

Aggregate: Fine aggregate is made of natural river sand (Zone 2 sand). Artificial crushed stone is used for coarse aggregate.

CCCW: Cementitious calillary crystalline waterproofing material (CCCW) is produced by Beijing Chengrong Waterproof Materials Co., Ltd.

Fiber: Selecting rust resistant copper plated steel fibers for this experiment. The physical and mechanical indicators are shown in Table 3.

2.2 Concrete Proportioning

The design grade of concrete is C30W12F200. The proportion of the second graded aggregate is 55:45 for medium to small stones. The dosage of superplasticizer is 1.5%, and the dosage of air entraining agent is 1‰. The sand rate is 36%. The water cement ratio is 0.41. The volume addition rate of steel fiber is 1.0%. The CCCW dosage is 0, 0.6, 1.2, 1.8, 2.4, and 3.0% of the adhesive dosage, respectively. The preliminary mix proportion for concrete testing is shown in Table 4.

2.3 Preparation Process

The production and maintenance of the specimens shall strictly follow the provisions of the specifications GB/T 50,082-2009 and GB/T 50,081-2019. The concrete preparation process is shown in Fig. 1.

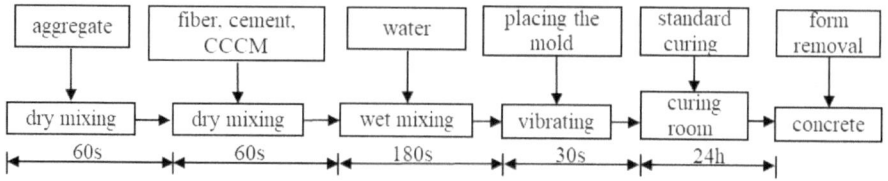

Fig. 1 Concrete preparation process

3 Experimental Study on Mechanical Properties

3.1 *Compressive and Splitting Tensile Strength*

Conduct concrete compressive strength test and splitting tensile strength test according to standard DL/T 5150-2017. The test pieces are all standard cubes with a side length of 150 mm. Cure the formed specimens under standard conditions until the age of 28 days for testing. The testing instrument is a microcomputer controlled pressure testing machine. The experimental results are shown in Figs. 2 and 3.

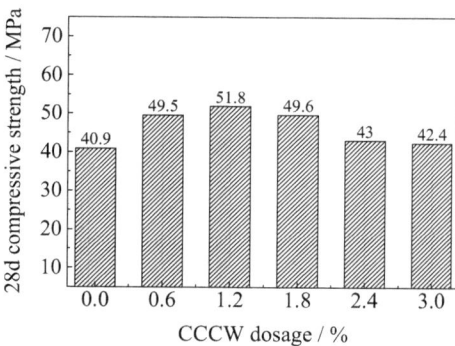

Fig. 2 The relationship between compressive strength and CCCW dosage

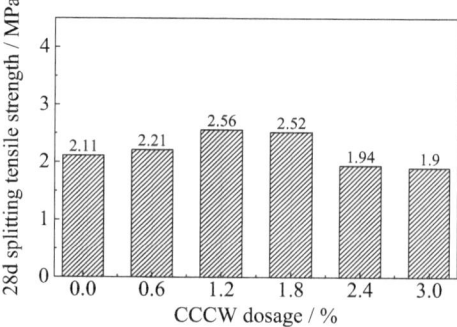

Fig. 3 The relationship between splitting tensile strength and CCCW dosage

According to Figs. 2 and 3, the compressive strength and splitting tensile strength of concrete without CCCW are 40.9 and 2.11 MPa, respectively. The compressive strength of concrete mixed with CCCW ranges from 42 to 52 MPa, and the splitting tensile strength ranges from 1.9 to 2.6 MPa. When the CCCM dosage is 1.2%, the compressive and splitting tensile strengths reach their maximum values, which are 51.8 and 2.6 MPa, respectively. Compared with the reference group without CCCM, it increased by 26.7 and 23.8%, respectively. Due to the fact that the maximum amount of CCCM only accounts for 3% of the cementitious material, and the cementitious material mainly provides strength in concrete. The precipitation crystallization strength generated by the reaction of active substances in CCCM is much lower than that of the cementitious material, the strength growth of concrete after adding CCCM is within a certain range.

The increase in compressive strength and splitting tensile strength of concrete shows a trend of first increasing and then decreasing with the increase of CCCM dosage. This is because the active chemical substances in the CCCM react with water during the hardening process of concrete, forming insoluble crystals that fill the internal pores of the concrete, thereby making the concrete more dense and improving its mechanical properties. However, when the CCCM dosage is higher than 2.4%, the improvement effect on the mechanical properties of concrete is not significant. This is because the number of pores and microcracks inside the concrete is limited. When the CCCM dosage is high, some active substances participate in the reaction to generate crystal precipitation, while the other part is dispersed inside the concrete in a dormant state, forming some pores, thereby slowing down the strength increase of the concrete.

3.2 Elastic Modulus and Ultimate Tensile Value

Static pressure elastic modulus test and ultimate Tensile testing of concrete shall be carried out according to DL/T 5150-2017. Concrete elastic modulus test specimen adopts $\Phi 150$ mm $\times$ 300 mm cylinder. Standard Fig. 8 specimen is used for ultimate Tensile testing. Cure the formed specimens under standard conditions until the age of 28 days for testing. The experimental results are shown in Figs. 4 and 5.

From Figs. 4 and 5, it can be seen that the compressive modulus of concrete with different infiltrating crystalline admixtures ranges from 30.9 to 34.1 GPa, while the tensile modulus is slightly higher than the compressive modulus, ranging from 31.2 to 34.6 GPa. The ultimate tensile value of concrete with different CCCM dosages is 1.19×10^{-4} to 1.31×10^{-4}. After adding CCCM, the compressive modulus, tensile modulus, and ultimate tensile value of concrete all increase to a certain extent, and the increase amplitude shows a trend of first increasing and then slowing down with the increase of CCCM dosage. When the CCCM dosage is 0.6%, the compressive modulus, tensile modulus, and ultimate tensile value of concrete are all increased by about 10%.

Fig. 4 The relationship between Elastic modulus and CCCW dosage

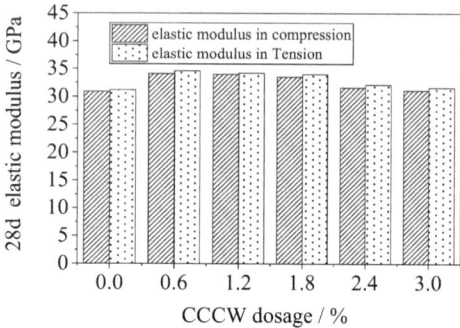

Fig. 5 The relationship between ultimate tensile value and CCCW dosage

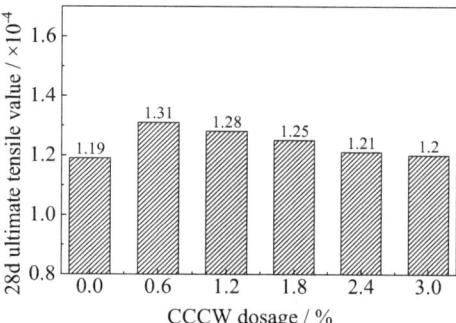

4 Experimental Study on Durability Properties

4.1 Impermeability Resistance

Conduct concrete permeability resistance test according to the DL/T 5150-2017 specification. During the test, after continuously pressurizing to the maximum water pressure of 1.3 MPa and maintaining a constant pressure for 8 h, the surface condition of the specimen was observed to be free from water seepage. According to the specifications, the impermeability level of each concrete mix is greater than or equal to W12. After the specimen is split, its average water penetration height is shown in Fig. 6.

From Fig. 6, it can be seen that the average water penetration height of the split specimen is between 26.1 and 66.3 mm, which is smaller than the height of the concrete impermeable specimen. The water seepage height of concrete decreases with the increase of infiltration crystallization dosage, indicating that infiltration crystallization has a significant improvement effect on the impermeability of concrete. On the one hand, due to the incorporation of permeable crystals into concrete, there are uniformly dispersed permeable crystals both inside and on the surface. The surface permeable crystals react with water during the curing process to form a layer of

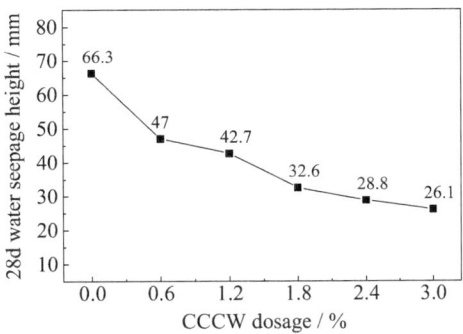

Fig. 6 The relationship between the height of concrete seepage and CCCW dosage

precipitation crystals, which is equivalent to coating the surface with a layer of waterproof material; On the other hand, when permeable water penetrates into the interior of concrete through capillaries, microcracks, etc., the active chemical substances in the permeable crystals undergo crystallization and precipitation reactions with the partially hydrated cement binder, generating stable crystals that fill the capillary pores, thereby preventing more permeable water from entering.

4.2 Frost Resistance

Conduct concrete frost resistance test according to the fast freezing method in DL/T 5150-2017. Evaluate the frost resistance level of concrete by reducing the relative dynamic modulus to the initial value ($\geq 60\%$) and the specimen mass loss rate ($\leq 5\%$). Perform 200 freeze–thaw cycles on each mix proportion of concrete. The frost resistance test results of concrete are shown in Table 5. The relationship between the loss rate of concrete quality and CCCW dosage are shown in Fig. 7. The relationship between the relative dynamic modulus and CCCW dosage are shown in Fig. 8.

Table 5 Results of concrete freeze–thaw cycle test

Number	CCCW dosage/%	Quality loss rate/%				Relative dynamic modulus/%			
		50	100	150	200	50	100	150	200
JZ	0.0	0.21	0.72	1.21	1.94	93	89	85	83
C0.6	0.6	0.20	0.56	1.20	1.76	97	95	93	90
C1.2	1.2	0.18	0.50	1.16	1.68	97	96	94	91
C1.8	1.8	0.18	0.42	1.15	1.66	97	96	95	92
C2.4	2.4	0.17	0.36	1.05	1.56	98	96	95	94
C3.0	3.0	0.16	0.38	0.95	1.45	98	97	96	94

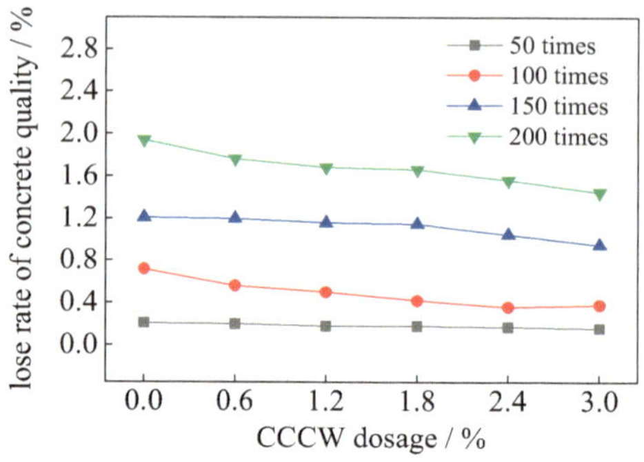

Fig. 7 The relationship between the loss rate of concrete quality and CCCW dosage

Fig. 8 The relationship between the relative dynamic modulus and CCCW dosage

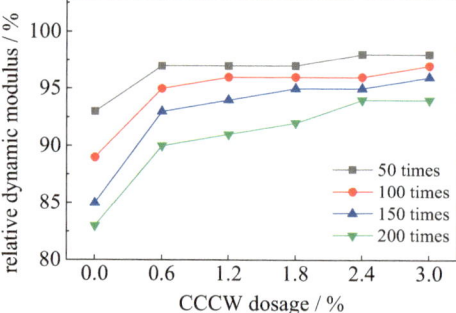

From Figs. 7 and 8, it can be seen that the addition of permeable crystals is beneficial for improving the frost resistance of concrete, and as the number of freeze–thaw cycles increases, the effect of permeable crystal dosage on the frost resistance of concrete becomes more significant. The mass loss rate of concrete decreases with the increase of infiltration crystallization dosage.

From Table 5, it can be seen that when the infiltration crystallization dosage is 3.0%, the mass loss rate of concrete after 200 freeze–thaw cycles is only 1.45%. The relative dynamic modulus of concrete increases with the increase of infiltration crystallization dosage. When the infiltration crystallization dosage is 3.0%, the relative dynamic modulus of concrete only decreases by 6% after 200 freeze–thaw cycles. Due to the significant improvement in the impermeability of concrete after the addition of permeable crystals, the precipitation and crystallization on the surface and inside of the concrete during the freeze–thaw process make it difficult for water to enter the interior of the concrete, resulting in a significant improvement in the frost resistance of the concrete.

In summary, when the infiltration crystallization dosage is 1.2%, the mechanical properties and durability performance of hydraulic concrete are better. The results of this study can provide reference for the mix design and engineering application of permeable crystalline concrete.

5 Conclusion

The increase in compressive and splitting tensile strength shows a trend of first increasing and then slowing down with the increase of CCCM dosage. Compared with the reference group without CCCM, the compressive and splitting tensile strength of concrete with CCCM dosage of 1.2% increased by 26.7 and 23.8%, respectively. However, when the CCCM dosage is higher than 2.4%, the improvement effect on the mechanical properties of concrete is not significant.

After adding CCCM, the compressive modulus, tensile modulus, and ultimate tensile value of concrete all increase to a certain extent, and all show a trend of first increasing and then decreasing with the increase of CCCM dosage. When the CCCM dosage is 0.6%, the compressive modulus, tensile modulus, and ultimate tensile value of concrete are all increased by about 10%.

The addition of permeable crystals is beneficial for improving the durability of concrete. The water permeability height of concrete decreases with the increase of permeable crystal dosage, the mass loss rate of concrete decreases with the increase of permeable crystal dosage, and the relative dynamic modulus increases with the increase of permeable crystal dosage.

When the infiltration crystallization dosage is 1.2%, the mechanical properties and durability performance of hydraulic concrete are better.

Acknowledgements The authors gratefully acknowledge the support of the Natural Science Basic Research Program of Shaanxi China (Program No. 2021JQ-983).

References

1. Wang L (2021) Experimental research on the influence of permeable crystalline materials on concrete compressive strength. Railway Constr Technol 8:51–55 (in Chinese)
2. Yi D, Yang GJ (2022) Low temperature microbial induced calcium carbonate precipitation and its application in concrete crack repair. Bull Chin Ceram Soc 41(3):959–968 (in Chinese)
3. Liu ML, Liu P, Yu ZW et al (2022) Review on the action mechanism and performance evaluation of cementitious capillary crystalline waterproofing materials. New Build Mater 49(4):135–140
4. Yu F, Zhang J, Cheng B (2018) Research on spraying crystal infiltration activity of parent material. New Build Mater 45(6):133–136
5. Ren GY, Wang ZJ, Liu ZK et al (2022) Research on mechanical properties of concrete with permeable cementitious capillary crystalline waterproof material. Concrete 2:159–161+166

Open Access This chapter is licensed under the terms of the Creative Commons Attribution 4.0 International License (http://creativecommons.org/licenses/by/4.0/), which permits use, sharing, adaptation, distribution and reproduction in any medium or format, as long as you give appropriate credit to the original author(s) and the source, provide a link to the Creative Commons license and indicate if changes were made.

The images or other third party material in this chapter are included in the chapter's Creative Commons license, unless indicated otherwise in a credit line to the material. If material is not included in the chapter's Creative Commons license and your intended use is not permitted by statutory regulation or exceeds the permitted use, you will need to obtain permission directly from the copyright holder.

Study on Seismic Performance of Waste Fiber Recycled Concrete Column Reinforced with CFRP

Jinghai Zhou, Shouyu Li, Jiagui Zhou, and Tianbei Kang

Abstract In this paper, the effect of CFRP reinforced in different ways on the mechanical properties of WFRC columns is investigated by using numerical simulations. By studying the stress cloud, hysteresis curve, skeleton curve and energy dissipation performance, the development of seismic performance of WFRC columns reinforced by CFRP was investigated. The results show that the damage patterns of CFRP-reinforced specimens are basically the same as those of unreinforced specimens, but the seismic performance is slightly improved.

Keyword CFRP reinforced column · Waste fiber recycled concrete · Numerical analysis

1 Introduction

With accelerated urbanization, the recycling of construction waste has become a new research hotspot nowadays. Recycled concrete can make full use of solid construction waste and reduce pollution to the environment. However, the performance of recycled concrete is inferior to that of natural concrete [1, 2]. Adding waste fibers to recycled concrete can effectively improve the performance of recycled concrete [3], and can also make use of the waste fabric to reduce pollution. CFRP has the advantages of

---

J. Zhou
Green and Livable Rural Construction Institute, Shenyang Jianzhu University, Shenyang, China
e-mail: zhoujinghai@sjzu.edu.cn

S. Li (✉) · J. Zhou · T. Kang
School of Civil Engineering, Shenyang Jianzhu University, Shenyang, China
e-mail: Krasnyi20@stu.sjzu.edu.cn

J. Zhou
e-mail: zhoujiagui@stu.sjzu.edu.cn

T. Kang
e-mail: kangtianbei@sjzu.edu.cn

high stiffness-to-mass ratio, good corrosion resistance, good fatigue resistance and environmental friendly [4, 5]. CFRP has been used to repair and strengthen building structures, however, it is time consuming and tedious to reinforce the structure after it has been damaged. If the structure can be reinforced at the time of erection, it can save trouble. In this paper, the seismic performance of CFRP-reinforced WFRC columns and ordinary WFRC columns are compared and studied by numerical simulation.

2 Models and Methods

In order to investigate the seismic behavior of CFRP-reinforced specimens and normal WFRC columns, the finite element program CAE 2019 was used to numerically simulate the specimens. The principal structure model of WFRC in this paper is chosen from the equation proposed by Wang et al. [6].

$$\varepsilon_{crf} = 1.16(1.471V_f + 0.926)(0.006l_f + 0.904)(0.12r + 0.905)\varepsilon_{cf} \quad (1)$$

$$\sigma = (1 - d_{cf})E_{cf}\varepsilon \quad (2)$$

In the equations, ε_{crf} is the peak strain of WFRC, d_{cf} is the uniaxial compression damage of WFRC, E_{cf} is the elastic modulus of WFRC.

$$d_{cf} = \begin{cases} 1 - \frac{\rho_{cf}^n}{n-1+x^n} & x \leq 1 \\ 1 - \frac{\rho_{cf}}{\alpha_{cf}(x-1)^2 + x} & x > 1 \end{cases} \quad (3)$$

$$\rho_{cf} = \frac{f_{crf}}{E_{cf}\varepsilon_{crf} - f_{crf}} \quad (4)$$

$$n = \frac{E_{cf}\varepsilon_{crf}}{E_{cf}\varepsilon_{crf} - f_{crf}} \quad (5)$$

$$x = \frac{\varepsilon}{\varepsilon_{crf}} \quad (6)$$

$$\alpha_{crf} = 0.157f_{crf}^{0.785} - 0.905 \quad (7)$$

In the equations, ρ_{cf} is the intermediate parameter and f_{crf} is the uniaxial compressive strength representative value of WFRC.

At present, there are three main types of reinforcement principal structure models in FEM analysis, which are no obvious flow amplitude reinforcement principal structure model, three-fold line reinforcement principal structure model and double-fold line principal structure model. In this paper, the double-folded principal structure model is obtained by ignoring the elastic–plastic phase of the reinforcement and considering the reinforcement as the ideal elastic–plastic material.

Fig. 1 Loading scheme

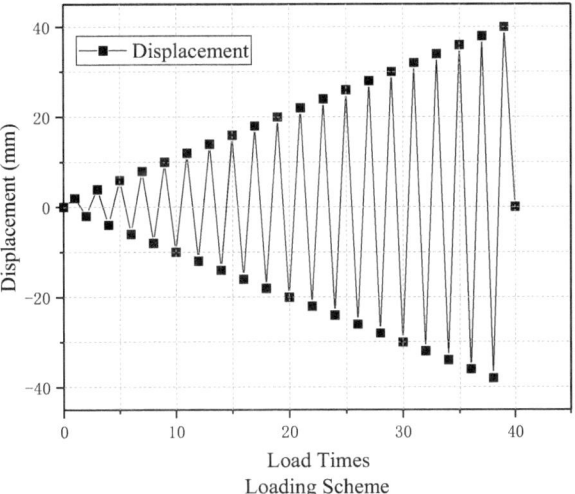

The C3D8R element were used to simulate the core concrete, and the S4R shell element were also used to simulate the CFRP. The boundary condition of the bottom of the column is set as a fixed constraint which the translation and rotation in three directions are restricted. The upper end is unconstrained, and the vertical load is applied to the top of the column according to the code. The analysis step is set to 3, where the first step is to apply the axial load analysis step and transfer it to the second step; the second step is to apply the low reversed cyclic load. For this numerical simulation, the incremental step is set to 5000, the initial incremental step is set to 0.1, the minimum incremental step is set to 1e−10, and the maximum incremental step is set to 1. Axial load and horizontal displacement are applied to the reference point at the top of the column to simulate the seismic action, the loading scheme is shown in Fig. 1.

3 Results and Discussions

3.1 Results

Figure 2 show the distribution of stress cloud for each specimen. The figs show that the damage of the specimens is concentrated in the plastic hinge region at the bottom of the column.

The concrete equivalent plastic stress cloud shows that the damage of the specimen concrete is mainly concentrated in the range of plastic hinge area at the bottom of the column, and the damage pattern of the specimen is basically the same. From the stress cloud of the reinforcement, it can be found that the stress of the longitudinal

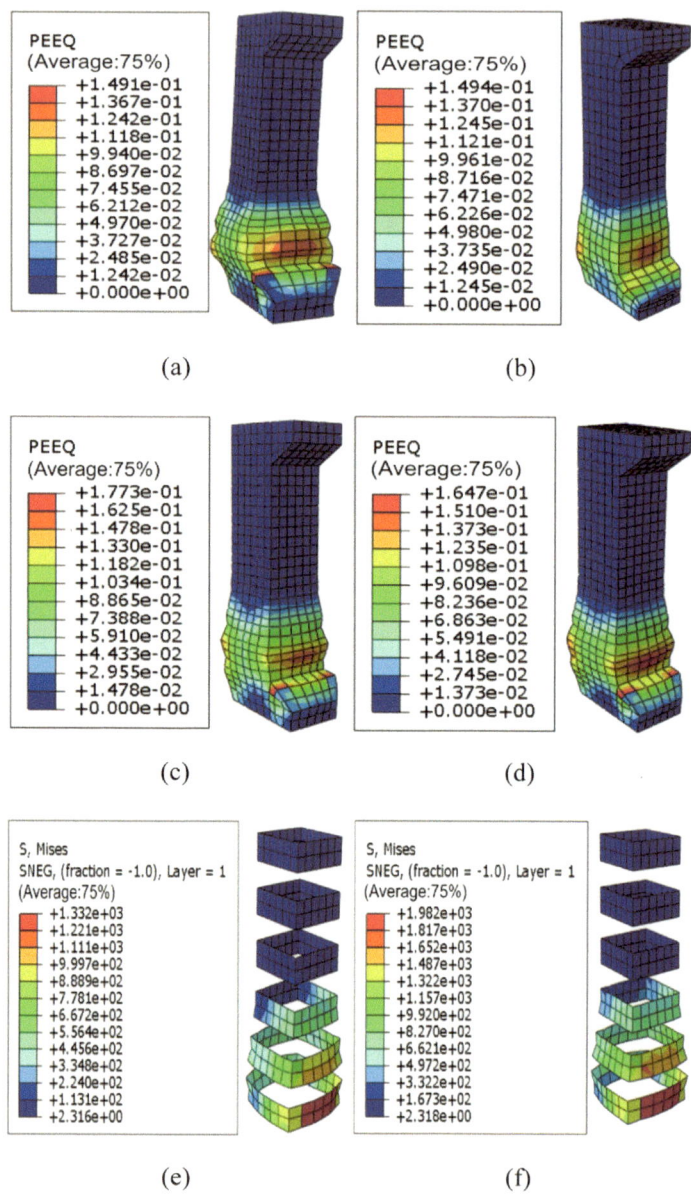

Fig. 2 CFRP plastic damage cloud map of specimens

reinforcement at the bottom is larger, the deformation of the stirrups at the bottom indicates that the concrete at the bottom of the specimen is crushed and dislodged. It can also be verified from the stress cloud of CFRP that the stress is mainly concentrated in the wrapped area at the bottom of the column. When the concrete at the bottom of the column expands and deforms, the stress borne by CFRP is greater than its tensile strength, and then it is cracked.

The damage pattern of the CFRP-reinforced specimen is basically the same, but there are still minor differences. It can be seen from Fig. 2a, b that the compression damaged area at the bottom of the CFRP reinforced specimen is smaller, and the load borne by the column is reduced. While the equivalent plastic damage of the reinforced specimen is greater compared with the unreinforced specimen, and the ultimate displacement is greater when the specimen is damaged. CFRP effectively improved the ductility of the specimens and limited the expansion of cracks, thus reducing the load carried by the reinforcement. As shown in Fig. 2c, d, the damage area of concrete with 4 layers of CFRP is obviously reduced, and the plastic strain of concrete at the position of bracket at the bottom of column is also less than that of 2 layers of CFRP. The stress areas of CFRP under different number of layers are the same when the specimen is damaged, but the stress of 4-layer CFRP is relatively smaller, which may be due to the increase of the number of layers makes the CFRP stiffness increase, thus reducing the effective stress of CFRP.

3.2 Hysteresis Loop

Comparing the hysteresis curves of the specimens, the specimen is in the elastic zone and the displacement is small in the initial loading stage, the slope of the hysteresis curve basically remains the same and presents a line. As the displacement gradually increases, the unreinforced specimen yields and the hysteresis curve gradually tends to the inverse S-shape and the area of the hysteresis curve gradually expands. The specimen starts to deviate from the vertical load axis and turns to the displacement axis, the specimen enters the elastic–plastic region. As the displacement increases, the bearing capacity of the specimen increase slightly until the ultimate bearing capacity is reached, the specimen is damaged and the simulation is ended.

As shown in Fig. 3a, b, after reinforced by CFRP, the ultimate bearing capacity and ultimate displacement of the specimen are increased, and the hysteresis curve is fuller than that of the unreinforced specimen. The ultimate displacement of the reinforced specimen reaches 36 mm, which is significantly higher than that of the unreinforced specimen at 30 mm. When the specimen reaches the ultimate load capacity, the load capacity of the unreinforced specimen decreases rapidly, while the load capacity of the CFRP-reinforced specimen decreases relatively smoothly.

When other conditions are the same, the increase in the number of CFRP layers makes the hysteresis curve of the specimens fuller, the ultimate load of the specimen does not show a significant improvement, but the ultimate displacement increases significantly. This indicates that the increase in the number of layers has some effect

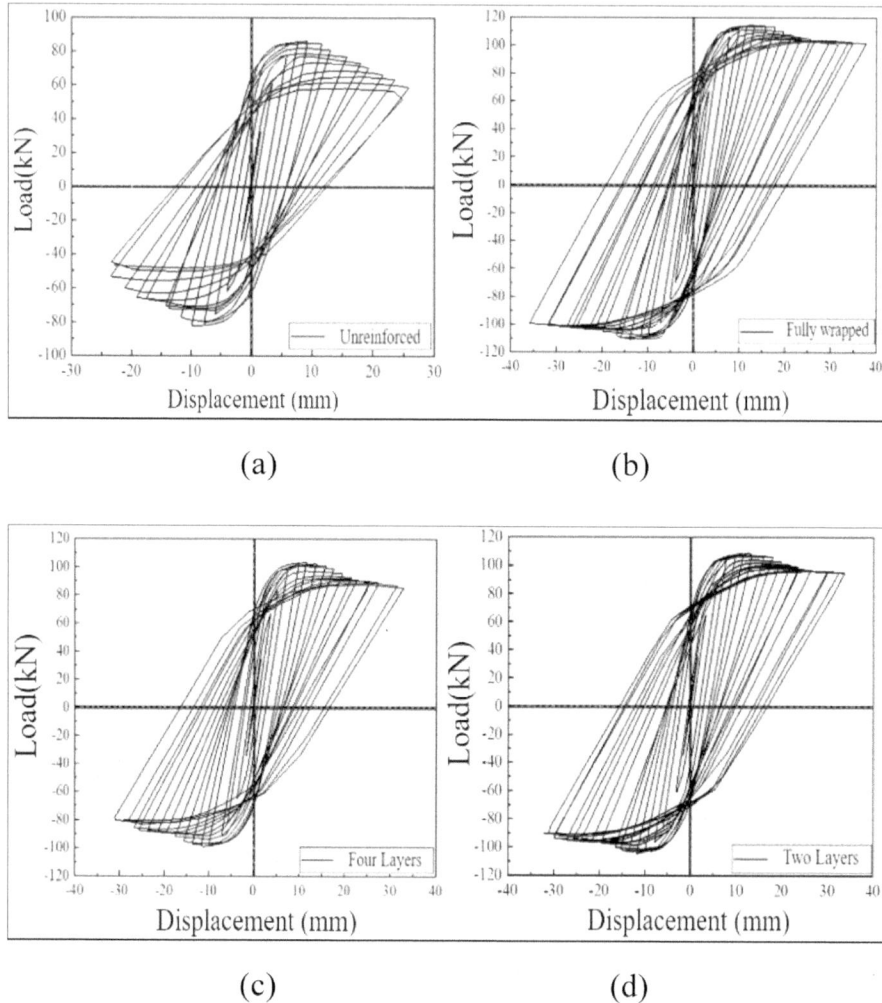

Fig. 3 Hysteresis loop of different specimens

on the seismic performance of CFRP-reinforced WFRC columns. Figure 3c, d shows that the increase of layers allows CFRP to restrict the concrete more effectively to avoid its premature cracking, thus ensuring the bond between concrete and reinforcement which in turn leads to an increase in the ultimate displacement of the specimen.

3.3 Bearing Capacity

The skeleton curve can reflect the overall force characteristics and equilibrium of the structure, determine the support points, location and force path of the structure, etc. It can also classify the building structure according to the force situation. The skeleton curves of the specimens are shown in Fig. 4.

All specimens showed similar skeleton curve trends, going through obvious elastic, strengthening and strength degradation stages. The rising phase of the curves basically overlap, indicating that the annularly covered CFRP has less influence on the initial stiffness of the specimens. While the declining section of CFRP reinforced specimen is longer and the declining trend is relatively gentle, which shows better performance than the unreinforced specimen. This indicates that the CFRP reinforcement can effectively improve the deformation capacity and ductility performance of the specimen and provide better seismic resistance. After the unreinforced specimen reaches the ultimate bearing capacity, the bearing capacity decreases rapidly as the horizontal displacement loading increases, and there is almost no obvious horizontal section. The bearing capacity of the reinforced specimen shows a slower decreasing trend, which indicates that CFRP can effectively suppress the deformation of concrete and improve its bearing capacity. The same trend was shown for the different number of layers of reinforcement in Fig. 4b. The increase in the number of layers led to an increase in the peak load and ultimate displacement of the specimens. The falling section after the peak load became smoother, and was approximately horizontal when the specimens were close to damage. This indicates that the increase in the number of layers effectively improves the ductility performance of the WFRC columns and makes the specimens have better seismic resistance. The comparison in the figure shows that CFRP reinforcement does not significantly increase the ultimate bearing

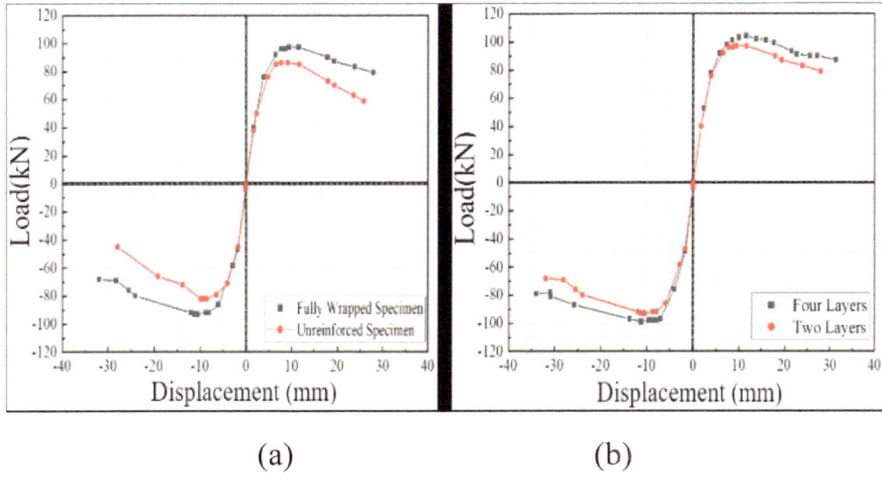

Fig. 4 Skeleton curves of different specimens

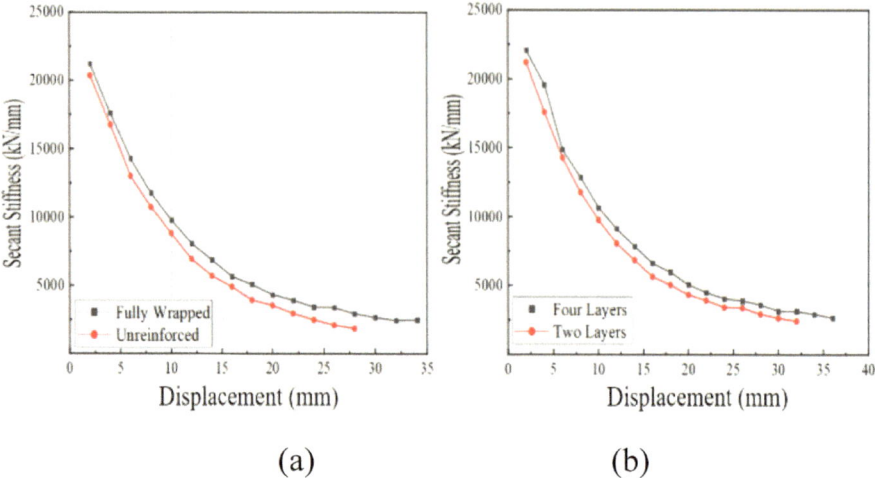

Fig. 5 Stiffness degradation curves of different specimens

capacity of the specimens. The load-bearing force is mainly controlled by the longitudinal reinforcement for the transversely wrapped CFRP does not have a strong restraining effect on the longitudinal load. Therefore, CFRP does not significantly improve the bearing capacity of the members.

3.4 Stiffness Analysis

It is found in Fig. 5 that the stiffness degradation curves of all specimens show an exponential decrease with the increase of the number of cycles, and the basic shape is the same, this indicates that the damage form and mechanism of the reinforced specimen do not change but the stiffness degradation becomes more slowly. This indicates that CFRP can effectively improve the ductility performance of the specimens, and has good load-bearing capacity even after the members reach the ultimate load, making the stiffness degradation relatively flat. Also, the full wrapping method and increasing layers of CFRP can limit the development of cracks in the WFRC columns better and slow down the stiffness degradation of the specimens.

3.5 Energy Dissipation Capacity

From Fig. 6, it can be seen that the energy dissipation capacity of CFRP-reinforced specimens is significantly stronger than that of unreinforced specimens, and the

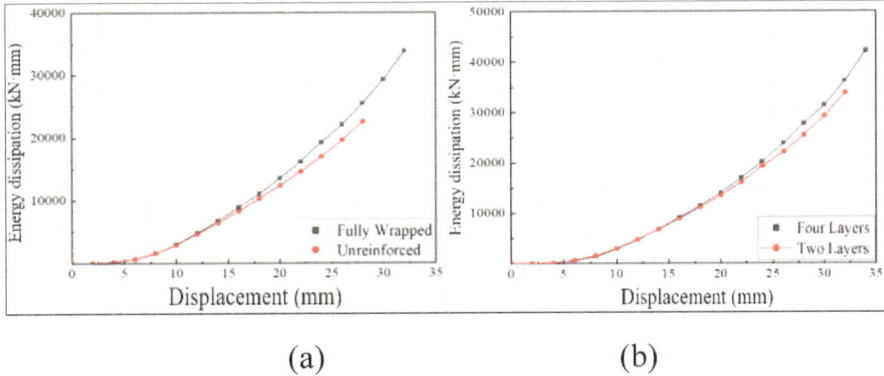

Fig. 6 Energy consumption curves of different specimens

energy dissipation capacity of fully wrapped specimen is stronger than that of strip-reinforced specimens. Increasing the number of CFRP layers can also significantly increase the cumulative energy dissipation of specimens. This is because the stiffness of the specimen is improved after CFRP reinforcement, and the seismic energy dissipation of the specimen increases with the increase of stiffness under the seismic load. In addition, the restraint effect of CFRP is better than the stress state of concrete at the time of damage, which makes the core concrete three-way stressed. Thus delaying the cracking of concrete and yielding of longitudinal stressing bars and improves the energy dissipation capacity of the specimen.

3.6 Ductility Analysis

From the simulation results, the seismic performance indexes are analyzed and the load, displacement, and ductility coefficients are obtained, where the ductility coefficients are taken as average values. The specific values are shown in Table 1.

The ductility coefficient of the CFRP-reinforced WFRC column increased from 3.9 to 5.9, which is a significant improvement. CFRP does not enhance the load significantly, the load capacity of the reinforced specimens is enhanced within 20 kN, among which the reverse peak load enhancement is the best with 15 kN, and the

Table 1 Equivalent viscous damping coefficient and Ductility factor of specimens

Specimen	Equivalent viscous damping coefficient h_c	Ductility factor
All wrapped	0.343	10.37
Unreinforced	0.309	3.83
Two layers	0.322	5.11
Four layers	0.329	5.81

forward yield load enhancement is the weakest with only 8 kN. The displacements of the specimens after CFRP reinforcement were all improved, among which the forward ultimate displacement was improved the most, reaching 12 mm, and the reverse peak displacement was improved the weakest, only 0.7 mm.

4 Conclusion

By comparing the seismic performance of CFRP-reinforced WFRC columns under low cyclic loading, the damage morphology of the specimens was analyzed by stress cloud diagrams, as well as the hysteresis curve, bearing capacity, energy dissipation capacity, and ductility performance. The following conclusions were obtained:

(1) The damage pattern of the CFRP-reinforced specimen is similar to that of the unreinforced specimen. However, the damage area of the reinforced specimen is smaller, and the concrete force and steel bar deformation are lower than those of the unreinforced specimen.
(2) The hysteresis curves of both specimens showed a shuttle shape, but the hysteresis curves of the reinforced specimen are fuller, and the bearing capacity and ultimate displacement were both improved. Among them, the ultimate displacement improvement was relatively obvious, but the improvement of ultimate bearing capacity was not.
(3) The CFRP reinforcement effectively improved the energy dissipation capacity and ductility properties of the specimen and has a relatively gentle decrease.

Acknowledgements This research was supported by the National Natural Science Foundation of China (Nos. 52108235), and Liaoning Provincial Department of Education Fund (LT2019011). The financial support is gratefully acknowledged.

References

1. Etxeberria M, Vázquez E, Marí A et al (2007) Influence of amount of recycled coarse aggregates and production process on properties of recycled aggregate concrete. Cem Concr Res 37(5):735–742
2. Akbarnezhad A, Ong KCG, Zhang MH et al (2011) Microwave-assisted beneficiation of recycled concrete aggregates. Constr Build Mater 25(8):3469–3479
3. Suda VBR, Sutradhar R (2021) Strength characteristics of micronized silica concrete with polyester fibres. Mater Today Proc 38:3392–3396
4. Dong PAV, Azzaro-Pantel C, Cadene AL (2018) Economic and environmental assessment of recovery and disposal pathways for CFRP waste management. Resour Conserv Recy 133:63–75
5. Xie HB, Wang YF, Gucunski N et al (2019) Environmental efficiency of strengthening schemes for concrete beams with externally bonded fibre reinforced plastic composites. Struct Infrastruct E 15(1):127–135

6. Zhang JC, Yang WT, Zhou JH (2013) Study on mechanical properties and compressive constitutive relationship of waste fiber recycled concrete. China Concr Cem Prod 2:54–58

Open Access This chapter is licensed under the terms of the Creative Commons Attribution 4.0 International License (http://creativecommons.org/licenses/by/4.0/), which permits use, sharing, adaptation, distribution and reproduction in any medium or format, as long as you give appropriate credit to the original author(s) and the source, provide a link to the Creative Commons license and indicate if changes were made.

The images or other third party material in this chapter are included in the chapter's Creative Commons license, unless indicated otherwise in a credit line to the material. If material is not included in the chapter's Creative Commons license and your intended use is not permitted by statutory regulation or exceeds the permitted use, you will need to obtain permission directly from the copyright holder.

Sensitivity Analysis of Influencing Factors of Vertical Bearing Capacity of Rock-Socketed Piles Based on Orthogonal Test

Zongjun Sun, Yingfeng Han, Fei Liu, and Rui Min

Abstract In order to investigate the sensitivity of influencing factors of rock-socketed piles' vertical bearing capability. The field static load test of embedded rock piles was performed in the article, and the pile-soil interaction model was built using ABAQUS software. The orthogonal test is properly designed to examine the sensitivity of the influencing elements of embedded rock piles by varying the relevant pile, pile side soil and bedrock parameters. The conclusions are as follows: (A) From large to small, the following factors have varying degrees of influence on the bearing capacity of rock-socketed piles: rock internal friction angle, pile diameter, rock elastic modulus, pile elastic modulus, rock poisson's ratio, soil density, soil thickness, rock-socketed depth, rock friction coefficient, rock cohesion, pile density, soil poisson's ratio, soil friction coefficient, soil elastic modulus, rock density, soil cohesion, pile poisson's ratio and soil internal friction angle; (B) Rock internal friction angle, pile diameter, rock elastic modulus, pile elastic modulus, rock poisson's ratio, soil density, soil thickness, and depth of the rock-socketed piles all have a greatly significant impact on the vertical bearing capacity of these piles. The bearing capacity of rock-socketed piles is significantly impacted by the rock friction coefficient and rock cohesiveness. The research results can provide a basis for the design of rock-socketed piles and the prediction of bearing capacity of rock-socketed piles.

Keywords Static load test · Orthogonal test · Rock-socketed pile · Sensitivity analysis

---

Z. Sun · Y. Han (✉) · F. Liu
College of Civil Engineering and Architecture, Shandong University of Science and Technology, Qingdao, China

F. Liu
e-mail: 455928221@qq.com

Z. Sun
Qingdao Ruihan Technology Group Co., Ltd., Qingdao, China

R. Min
Qingdao West Coast Traffic Investment & Comstruction Co., Ltd., Qingdao, China

© The Author(s) 2025
P. Xiang et al. (eds.), *Frontier Research on High Performance Concrete and Mechanical Properties*, Lecture Notes in Civil Engineering 518,
https://doi.org/10.1007/978-981-97-4090-1_9

1 Introduction

Rock-socketed pile has a high bearing capacity, a minimal amount of settlement, and superior seismic resistance, which are all reasons why it is frequently employed in many engineering fields. The current design of rock-socketed piles is still based on semi-human experience, which makes the design too cautious [1]. This is due to the complicated geological context and several unknown human interference effects. Therefore, in order to improve the engineering design, a reasonable quantitative study of the parameters that affect the vertical bearing capacity of rock-socketed piles is required. Qiu [2] found that the elastic modulus of pile and the elastic modulus of embedded rock are the main factors affecting the bearing capacity and settlement of single pile. Zhang [3] focuses on the impact of pile size on the bearing properties of rock-socketed piles using the nonlinear finite element method. It has been discovered that the bearing capacity of rock-socketed piles can be increased by reducing the depth of the rock-socketed and increasing the pile diameter.

There are numerous academics who study the sensitivity of the influencing elements of the pile's bearing capacity, but the majority of them stick to single factor analyses. There exist interference factors, and their experimental examination is straightforward. The multi-factor analysis method, which includes range analysis, variance analysis, and grey correlation analysis [4–7], corresponds to this. Yang [8] analyzed the influencing factors of rock-socketed pile side resistance by designing orthogonal test; Xin [9] analyzed the influencing factors of bearing capacity of large diameter belled pile by grey correlation method. In light of this, this research uses ABAQUS software to simulate the pile-soil interaction model and uses actual engineering to confirm the model's viability. The orthogonal test is logically intended to examine how different variables affect the strength of rock-socketed piles.

2 Finite Element Simulation of Rock-Socketed Pile

2.1 Project Overview

This project is situated on Qingdao west coast new area. The foundation is constructed using piles. Percussive cast-in-place piles made of reinforced concrete make up the type of foundation pile. The No. 35 pile is the subject of this essay's research. The pile measures 17.21 m in length, has a concrete strength of C45, is 800 mm in diameter, and has 6.15 m of embedded weathered tuff. Table 1 displays the details of the pile and the surrounding soil.

Table 1 Pile and soil parameter table

Soil and pile	h:m	ρ: kg/m³	E:MPa	υ	φ :°	c:kPa
①-Plain fill	9.56	1800	10	0.35	25	6
②-Strongly weathered tuff	1.5	2100	60	0.25	30	50
③-Mid-weathered tuff	Not debunked	2400	1500	0.23	42	125
Pile	–	2450	33,500	0.2	–	–

h, thickness; ρ, density; E, elastic modulus; υ, poisson's ratio; φ, angle of internal friction; c, force of cohesion. The following similar parameter units are consistent

2.2 Finite Element Modeling

The pile-soil solid model is created in this research using the ABAQUS program. The soil's horizontal perimeter measures 10 times the diameter of the pile, while its vertical barrier measures 1.5 times the pile.

The pile is elastic material, and the Mohr–Coulomb constitutive model is selected for the soil. The pile-soil contact is frictional, hard contact is present in the normal direction of the contact surface, and Coulomb friction is present in the tangential direction.

2.3 Validation of Finite Element Fitting Results

Field Bearing Test

According to the Technical Code for Testing of Building Foundation Piles (JGJ106-2014), a single pile is subjected to a static load test, Fig. 1 depicts the field compressive static load test. The method of slow maintenance loading is used. 12,000 kN is the maximum vertical loading. Each stage's loading quantity is one tenth of the maximum loading amount, with the first stage requiring one-fifth of the maximum load value. Table 2 displays the test pile load and settlement.

Comparison of Simulation Results

The load settling data from the simulation are displayed in Table 3 and combined with genuine engineering data to create the Q–S curve in Fig. 2. The simulated curve's axial trend matches that of the real engineering curve. The final settlement calculated using finite elements is 11.27 mm, with an 8.4% inaccuracy when compared to the actual settlement. The simulation effect is within the permitted range, with an average inaccuracy of 12.5%.

Fig. 1 In-situ static load test diagram

Table 2 Pile load test data

Load: kN	The settlement of this level: mm	Cumulative settlement: mm
0	0.00	0.00
2400	1.81	1.81
3600	0.62	2.43
4800	0.69	3.12
6000	0.90	4.02
7200	1.20	5.22
8400	1.61	6.83
9600	1.69	8.52
10,800	1.83	10.35
12,000	1.96	12.31

Table 3 Finite element simulation data table

Load: MN	0	2.4	3.6	4.8	6.0	7.2	8.4	9.6	1.08	1.20
Settlement: mm	0	1.90	2.84	3.92	5.03	6.17	7.36	8.61	9.92	11.28

3 Orthogonal Design

3.1 Influencing Factors

The soil parameter data from the test pile side are homogenized and processed in accordance with formula (1) to make the calculation simpler.

$$\alpha = \sum_{i=1}^{n} \alpha_i h_i / h, i = 1, 2 \dots n \tag{1}$$

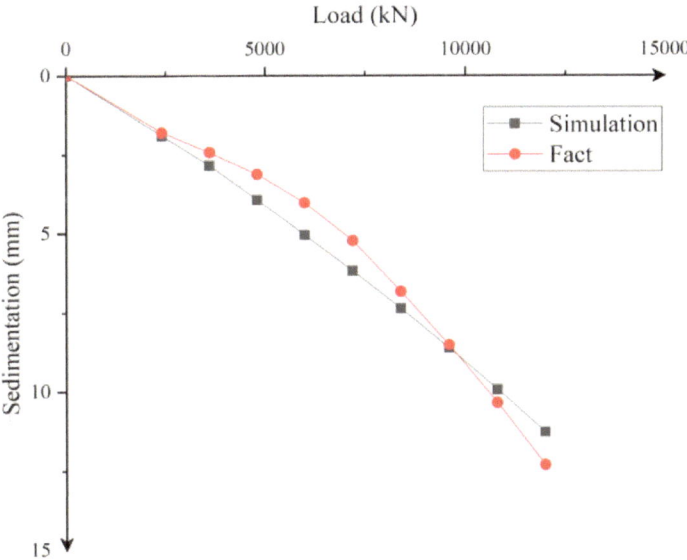

Fig. 2 Q–S curve comparison diagram

α is the pile side's final parameter, α_i is the i-layer side's soil parameter, h_i is the i-layer soil's thickness, and h is the thickness of the layer. Table 4 displays the treated standard soil properties. Table 5 displays the pile's standard parameters.

By reading the relevant literature [10], most of the factors affecting the bearing capacity of rock-socketed piles are analyzed from the perspectives of pile, pile side soil and bedrock. The pile diameter, pile density, pile elastic modulus, pile poisson's ratio, soil thickness, soil density, soil elastic modulus, soil poisson's ratio, soil internal friction angle, soil cohesion, rock-socketed depth, rock density, rock elastic modulus,

Table 4 Soil's standard parameter table

Soil horizon	ρ: kg/m³	E: MPa	υ	φ:°	c: kPa	μ
Pile side soil	1842	17	0.34	25	12	0.3
Bed rock	2400	1500	0.23	42	125	0.53

μ, coefficient of friction

Table 5 Pile's standard parameter table

Factor	d: m	ρ_P: kg/m³	E_P: MPa	υ_P	l_S: m	l_R: m
Value	0.9	2450	33,500	0.34	10	3d

ρ_P, pile density; E_P, pile elastic modulus; υ_P, pile poisson's ratio; l_S, soil thickness; l_R, rock-socketed depth

rock poisson's ratio, rock internal friction angle, rock cohesion, soil friction coefficient and rock friction coefficient are selected as the influencing factors of the bearing capacity of rock-socketed piles, and the standard parameters of the same amplitude are scaled. Each factor value is divided into three levels, and each factor and its level value are shown in Table 6.

3.2 Orthogonal Test Table Design

The proper orthogonal test table is chosen, and ABAQUS software is used to simulate the pile-soil interaction under vertical load. The Q-S curves for several tests are discovered. The bearing capacity of a rock-socketed pile is chosen as the load corresponding to a settlement of 40 mm. Table 7 displays the estimated orthogonal test table and bearing capacity.

4 Orthogonal Test Data Processing and Analysis

4.1 Range Analysis

The range analysis method is the most common method to deal with the results of orthogonal test. It can swiftly determine how much an independent variable has an impact on a dependent variable. Tables 8 display the range analysis findings for the test data in Table 7.

The range represents how frequently the data change. The influence of this element on the bearing capacity of rock-socketed piles increases with the size of the range. As shown in Table 8, the influence degree of each factor on the bearing capacity of rock-socketed piles is from large to small: internal friction angle of rock, pile diameter, rock elastic modulus, pile elastic modulus, rock poisson's ratio, soil density, soil thickness, rock-socketed depth, rock friction coefficient, rock cohesion, pile density, soil poisson's ratio, soil friction coefficient, soil elastic modulus, rock density, soil cohesion, pile poisson's ratio, soil internal friction angle.

4.2 Analysis of Variance

In order to further investigate the significant level of the influence of various factors on the vertical bearing capacity of rock-socketed piles, the test data were analyzed by variance analysis. The analysis results are shown in Table 9, where p is the adjoint probability calculated by F. When p is less than 0.01, it indicates that this factor has a greatly significant effect on the bearing capacity of rock-socketed piles. When p is

Table 6 Influencing factors and levels

Level factor	d	ρ_P	E_P	υ_P	l_S	ρ_S	E_S	υ_S	φ_S	c_S	l_R	ρ_R	E_R	υ_R	φ_R	c_R	μ_S	μ_R
1	0.8	2181	29,815	0.18	10	1639	15	0.30	22	11	2.5d	2136	1335	0.20	37	111	0.27	0.47
2	0.9	2450	33,500	0.20	11	1842	17	0.34	25	12	3.0d	2400	1500	0.23	42	125	0.3	0.53
3	1.0	2720	37,185	0.22	12	2045	19	0.38	28	13	3.5d	2664	1665	0.26	47	139	0.33	0.59

ρ_S, soil density; E_S, soil elastic modulussoil; υ_S, soil poisson's ratio; φ_S, soil internal friction angle; c_S, soil cohesion; ρ_R, rock density; E_R, rock elastic modulus; υ_R, rock poisson's ratio; φ_R, rock internal friction angle; c_R, rock cohesion; μ_S, soil friction coefficient; μ_R, rock friction coefficient

Table 7 Orthogonal test scheme and simulation results

Test number	d	l_S	l_R	ρ_P	E_P	υ_P	ρ_S	E_S	υ_S	φ_S	c_S	ρ_R	E_R	υ_R	φ_R	c_R	μ_S	μ_R	P
1	1	1	1	1	1	1	1	1	1	1	1	1	1	1	1	1	1	1	30.10
2	1	1	1	1	1	2	2	2	2	2	2	2	2	2	2	2	2	2	36.86
3	1	1	1	1	1	3	3	3	3	3	3	3	3	3	3	3	3	3	43.55
4	1	1	3	2	3	1	1	3	2	3	2	2	1	3	1	3	3	2	34.07
5	1	1	3	2	3	2	2	1	3	1	3	3	2	1	2	1	1	1	37.16
6	1	1	3	2	3	3	3	2	1	2	1	1	3	2	3	2	2	1	42.50
7	1	2	2	3	3	1	2	2	3	3	1	2	2	3	3	1	2	1	40.97
8	1	2	2	3	3	2	3	3	1	1	2	3	3	1	1	2	3	1	36.60
9	1	2	2	3	3	3	1	1	2	2	3	1	1	2	2	3	1	1	34.51
10	1	2	3	1	2	1	2	3	1	2	3	1	2	3	1	2	3	1	35.27
11	1	2	3	1	2	2	3	1	2	3	1	2	3	1	2	3	1	2	38.43
⋮	⋮	⋮	⋮	⋮	⋮	⋮	⋮	⋮	⋮	⋮	⋮	⋮	⋮	⋮	⋮	⋮	⋮	⋮	⋮
44	3	2	1	1	3	2	1	3	3	2	3	2	1	1	3	1	3	2	32.27
45	3	2	1	1	3	3	2	1	1	3	1	3	2	2	1	2	1	3	31.66
46	3	2	3	2	1	1	3	1	3	2	1	3	1	3	2	1	3	1	30.39
47	3	2	3	2	1	2	1	2	1	3	2	1	2	1	3	2	1	2	31.78
48	3	2	3	2	1	3	2	3	2	1	3	2	3	2	1	3	2	3	31.41
49	3	3	2	1	2	1	1	3	2	3	1	1	3	2	3	1	1	3	35.74
50	3	3	2	1	2	2	2	1	3	1	2	2	1	3	1	2	2	1	29.03
51	3	3	2	1	2	3	3	2	1	2	3	3	2	1	2	3	3	2	31.72
52	3	3	3	3	3	1	1	1	1	1	3	3	3	3	3	2	2	2	37.93
53	3	3	3	3	3	2	2	2	2	2	1	1	1	1	1	3	3	3	28.35
54	3	3	3	3	3	3	3	3	3	3	2	2	2	2	2	1	1	1	32.84

P(MN), bearing capacity of rock-socketed pile

greater than 0.05, it means that this factor has no obvious influence on the bearing capacity of rock-socketed piles. When it is between 0.01 and 0.05, this factor has obvious influence on the bearing capacity of rock-socketed piles.

Table 9 shows how the sensitivity of each factor can be arranged by the size of the p. From large to small, the following factors have the influence on the vertical bearing capacity of rock-socketed piles: rock internal friction angle, pile diameter, rock elastic modulus, pile elastic modulus, rock poisson's ratio, soil density, soil thickness, rock-socketed depth, rock friction coefficient, rock cohesion, soil poisson's ratio, pile density, soil friction coefficient, soil elastic modulus, rock density, soil cohesion, pile poisson's ratio and soil friction angle. The outcomes largely agree with the range analysis, and the results of the test analysis are also basically consistent with the results of Qiu [2] and Zhang [3], which confirms the reliability of the analysis results.

Table 8 Range analysis results

Level	d	l_S	l_R	ρ_P	E_P	υ_P	ρ_S	E_S	υ_S	φ_S	c_S	ρ_R	E_R	υ_R	φ_R	c_R	μ_S	μ_R
1	661	627	628	624	600	620	613	620	620	621	622	620	583	602	580	617	620	617
2	620	621	618	621	623	622	623	622	619	622	620	621	622	620	622	621	623	620
3	582	615	618	618	640	621	627	622	624	621	621	623	658	641	661	625	621	626
Range	26.32	4.13	3.44	1.90	13.32	0.43	4.71	0.82	1.70	0.42	0.56	0.79	25.19	12.89	26.68	2.80	1.15	2.83

Table 9 Results of analysis of variance

Factor	Type III sum of squares	Degree of freedom	Mean square	F	p	Significance
Pile diameter	173,217,547.73	2	86,608,773.86	490.88	9.19e−16	**
Soil thickness	4,256,364.56	2	2,128,182.28	12.06	5.48e−4	**
Rock-socketed depth	3,733,616.12	2	1,866,808.06	10.58	1.04e−3	**
Pile density	911,810.49	2	455,905.25	2.58	0.105	
Elastic modulus of pile	44,698,863.47	2	22,349,431.74	126.67	6.13e−11	**
Pile poisson's ratio	48,047.20	2	24,023.60	0.14	0.874	
Soil density	5,737,855.42	2	2,868,927.71	16.26	1.13e−4	**
Soil elastic modulus	198,783.92	2	99,391.96	0.56	0.580	
Soil poisson's ratio	928,747.58	2	464,373.79	2.63	0.101	
Internal friction angle of soil	44,636.21	2	22,318.10	0.13	0.882	
Soil clay cohesion	84,682.17	2	42,341.08	0.24	0.789	
Rock density	181,843.94	2	90,921.97	0.52	0.606	
Rock elastic modulus	158,756,322.27	2	79,378,161.13	449.90	1.90e−15	**
Rock poisson's ratio	41,683,168.40	2	20,841,584.20	118.13	1.07e−10	**
Internal friction angle of rock	178,074,292.55	2	89,037,146.27	504.64	7.30e−16	**
Rock cohesion	1,963,224.70	2	981,612.35	5.56	0.014	*
Soil friction coefficient	335,316.87	2	167,658.44	0.95	0.406	
Rock friction coefficient	2,066,369.85	2	1,033,184.93	5.86	0.012	*
Error	2,999,418.30	17	176,436.37			

Greatly significant effect: **; Significant effect: *

From Table 9, it can be seen that the internal friction angle of rock, pile diameter, rock elastic modulus, pile elastic modulus, rock poisson's ratio, soil density, soil thickness and rock-socketed depth have a very significant impact on the bearing capacity of rock-socketed piles. The influence of rock friction coefficient and rock cohesion on the vertical bearing capacity of rock-socketed piles is significant, and the influence of other factors on the vertical bearing capacity of rock-socketed piles is not obvious. Therefore, characteristics that significantly affect the bearing capacity of rock-socketed piles can be chosen as input variables when predicting the ultimate bearing capacity of rock-socketed piles by machine learning [11–13].

4.3 Application of Experimental Results

The pile-soil parameters are updated by the sensitivity analysis of the elements that affect the bearing capacity of rock-socketed piles, and the Q-S curves before and after the modification are drawn as shown in Fig. 3. Tables 10 display the modifications to the parameters and bearing capacity before and after the adjustment. The improved rock-socketed piles' bearing capacity is discovered to increase by 34%, from 34.75 MN to 46.48 MN.

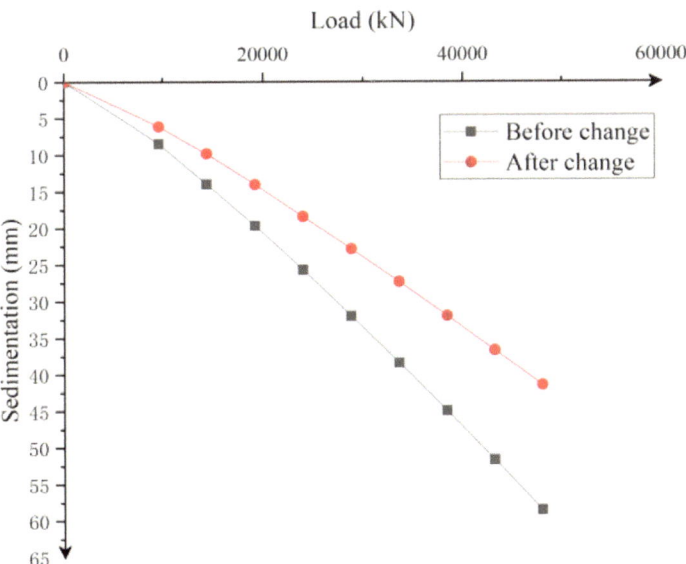

Fig. 3 Q-S curve before and after parameter changes

Table 10 Comparison table of parameters

State factor	d	ρ_P	E_P	v_P	l_S	ρ_S	E_S	v_S	φ_S	c_S	l_R	ρ_R	E_R	v_R	φ_R	c_R	μ_S	μ_R	P
Before change	0.9	2450	33,500	0.2	11	1842	17	0.34	25	12	3d	2400	1500	0.23	42	125	0.3	0.53	34.75
After change	0.8	2181	37,185	0.2	10	2045	17	0.38	25	11	2.5d	2664	1665	0.26	47	139	0.27	0.59	46.48

5 Conclusion

By comparing the settlement of rock-socketed piles under the same load between the simulation and the actual project, the reliability of the finite element model is proved.

After a logically designed orthogonal test table, the test data are subjected to range and variance analyses. The outcomes are essentially constant. From large to small, the sensitivity of each factor to the vertical bearing capacity of rock-socketed piles is listed: rock internal friction angle, pile diameter, rock elastic modulus, pile elastic modulus, rock poisson's ratio, soil density, soil thickness, rock-socketed depth, rock friction coefficient, rock cohesion, pile density, soil poisson's ratio, soil friction coefficient, rock density, soil cohesion, pile poisson's ratio, soil friction angle.

The vertical bearing capacity of rock-socketed piles is discovered to be significantly influenced by the internal friction angle, pile diameter, rock elastic modulus, pile elastic modulus, rock poisson's ratio, soil density, soil thickness, and rock-socketed depth. The bearing capacity of rock-socketed piles is primarily influenced by the rock friction coefficient and rock cohesiveness, with little or no influence from other parameters.

Acknowledgements The High-level Talent Team Project of Qingdao West Coast New Area (Number: RCTD-JC-2019-06) provides funding for the research project of the article.

References

1. Zhang JX, Ye HD, Du HJ (2003) Discussion on several problems in the design of rock-socketed piles. J Rock Mechan Eng 7:1222–1225
2. Qiu Y, Zhou L, Liu SY (2003) Finite element analysis of bearing behavior of single deep long large diameter rock-socketed pile. J Civil Eng 10:95–101
3. Zhang JX, Wu DY, Zhang SC (2007) Finite element analysis of size effect of rock-socketed piles. Geotech Mechan 137:1221–1224
4. Das PP, Chakraborty S (2022) Application of grey correlation-based EDAS method for parametric optimization of non-traditional machining processes. Scientia Iranica 29:864–882
5. Sabarish KV, Parvati TS (2021) An experimental investigation on L9 orthogonal array with various concrete materials. Mater Today Proc 37:3045–3050
6. Zhai MY, Sheng JL, Dong S (2017) Coupling analysis of slope sensitivity based on orthogonal design and grey relational analysis. Chem Miner Proc 2:69–72
7. Cheng JL, Zheng M, Lou JQ (2012) Comparison of common experimental optimization design methods. Lab Res Explor 31:7–11
8. Yang XL, Tao GL, Wang CL, Qiao ZY (2016) Sensitivity analysis of influencing factors of side resistance of rock-socketed piles based on orthogonal design. Waterway port 37:622–625
9. Xin XY, He KQ, Wang ST, Guo YY, Zhang LG (2018) Analysis of influencing factors of ultimate bearing capacity of single pile of large diameter belled pile based on grey correlation theory. Eng Constr 50:1–5
10. Wang YG (2010) Finite element analysis of bearing behavior of rock-socketed piles. J Yangtze River Acad Sci 4:44–48

11. Ranajeet M, Shakti S, Sarat KD (2018) Prediction of vertical pile capacity of driven pile in cohesionless soil using artificial intelligence techniques. Int J Geotech Eng 12:209–216
12. Pijush S (2011) Prediction of pile bearing capacity using support vector machine. Int J Geotech Eng 5:95–102
13. Wei JB, Chen HX, Huang DY (2005) Prediction of vertical bearing capacity of rock-socketed piles using BP neural network. Earth Environ S1:91–94

Open Access This chapter is licensed under the terms of the Creative Commons Attribution 4.0 International License (http://creativecommons.org/licenses/by/4.0/), which permits use, sharing, adaptation, distribution and reproduction in any medium or format, as long as you give appropriate credit to the original author(s) and the source, provide a link to the Creative Commons license and indicate if changes were made.

The images or other third party material in this chapter are included in the chapter's Creative Commons license, unless indicated otherwise in a credit line to the material. If material is not included in the chapter's Creative Commons license and your intended use is not permitted by statutory regulation or exceeds the permitted use, you will need to obtain permission directly from the copyright holder.

The Influence of Hybrid Fiber Ratio on the Mechanical Properties of Concrete

Yabin He, Lei Yu, Sheng Chen, Chuankai Yan, Zhansheng Lin, and Chen Yu

Abstract The concrete with different fibre and ratio can lead to different performance. This paper studies the mechanical properties of concrete by changing the ratio of PVA fibre and steel fibre. Three kinds of ratio of the PVA fibre and steel fibre are studied. They are 0% PVA fibre and 0% steel fibre, 0.25% PVA fibre and 0.75% steel fibre, 0.5% PVA fibre and 0.5% steel fibre. The compressive strength, flexural strength, bending toughness are tested. The test result shows that the compressive strength of the P0.25S0.75 increases. While the group of P0.5S0.5 shows a little range of decreasing. The 28d flexural strength of P0.25S0.75 and P0.5S0.5 increases 2.3 and 30.2% respectively. The loading-deflection curves of P0.25S0.75 and P0.5S0.5 all appear strain harden property. By the mixing amount of the PVA fibre increasing, the cracking strength and limitation bearing ability can all be enhanced. The areas under the curve enlarge and the toughness of the concrete are enhanced. The bending toughness of P0.5S0.5 is much better than the P0.25S0.75 but the difference is not very large.

Keywords Hybrid fibre · Concrete · Mixing ratio · Mechanical properties

---

Y. He · L. Yu (✉)
Beijing Xinqiao Technology Development Co., Ltd., Beijing 100088, China
e-mail: yuleimabel@163.com

L. Yu
Research Institute of Highway Ministry of Transport, Beijing 100088, China

S. Chen · C. Yan
China Railway 10th Bureau Group Second Engineering Co., Ltd., Zhengzhou 450003, Henan, China

Z. Lin · C. Yu
Shandong Road and Bridge Group Co., Ltd., Jinan 250014, Shandong, China

1 Introduction

Mixing hybrid of steel fibre and polymeric fibre can enhance the mechanical properties of concrete such as tensile strength, flexural strength, bending toughness and fatigue property. While the ratio of different fibre plays an important role in enhancing theses performances. Jun Huang puts PVA fibre and steel fibre into concrete and found that the tensile strength of the concrete is not affected obviously while the toughness of the concrete is enhanced obviously [1]. Zhongwen Ou found that the PVA fibre concrete has higher impact energy absorption capacity than steel fibre [2]. Nemkumar, Zongcai Deng put steel fibre and PVA fibre into concrete and the test result shows that the performance of concrete is better than single kind fibre while the PVA fibre will decrease the flexural impact resistance [3]. Shuzhao Shi put different ratio of PVA fibre and steel fibre into concrete and found that the PVA fibre mixing can enhance the workability of the concrete than mixing single kind of steel fibre [4, 5]. Fenglei Li put PVA fibre and steel fibre into C30 concrete and the results shows that the 8:1 ratio of steel fibre and PVA fibre and the total volume is 0.9% put out the best bending roughness performance [6]. Some other researcher such as Cong Zhang put other reactive powder into the concrete and the result shows that bending toughness of concrete with the PVA and steel fibre group are all well [7–10].

The research status shows that the amount of the mechanical properties of the hybrid fibre concrete with PVA and steel fibre is not very perfect by now. The optimal ratio of the two kinds of fibre still needs to be studied. This paper studies the mechanical properties of the hybrid fibre concrete with three kinds of ratio of the PVA fibre and steel fibre and the relative optimal ratio is also studied.

2 Materials and Mix Proportion

2.1 *Materials*

The raw materials include cement, sand, crushed stone, expansion agent, silica fume, organosilicon waterproofing agent, polycarboxylate superplasticizer, tap water, polyvinyl alcohol fibre (PVA) and steel fibre. The cement is P.O 52.5 ordinary Portland cement and the properties are shown in Table 1. The sand is river one with a fineness modulus of 2.8. The crushed stone is 5–20 mm continuous graded. The water reducing rate of the poly carboxylate superplasticizer is 20%. The content of the expansion agent in Table 2. The Properties of the fibres are shown in Table 3.

Table 1 Properties of cement

Fineness (80 μm square hole sieve)/(%)	Water requirement of normal consistency/%	Setting time (h: min)		Compressive strength (MPa)		Splitting strength (MPa)		Stability (boiling method)
		Initial set	Final set	3d	28d	3d	28d	
3.1	27.6	2:59	3:59	45.8	59.9	9.7	12.3	Conformity

Table 2 Content of the expansion agent

Component	SiO_2	Al_2O_3	Fe_2O_3	CaO	MgO	SO_3
Content/(%)	3.67	10.65	1.35	43.87	1.84	29.8

Table 3 Properties of the fibres

Type	Diameter/mm	Length/mm	Density/(g/cm^3)	Tensile strength/MPa	Elasticity modulus/GPa
PVA fibre	0.039	12	1.2	1600	40
Steel fibre	0.75	35	7.8	1100	210

2.2 Mix Proportion

The mix ratios of 3 groups of ordinary concrete are designed, as shown in Table 4. The group of P0S0 is the control group, the P0.25S0.75 and the P0.5S0.5 are all test group. When mixing with concrete, the concrete slump is 150 ± 30 m.

Table 4 Mix proportion (kg/m^3)

ID	Cement	Water	Sand	Silica fume	Crush stone
P0S0	399.4	174.8	690.4	49.9	1035.6
P0.25S0.75	399.4	174.8	690.4	49.9	1035.6
P0.5S0.5	399.4	174.8	690.4	49.9	1035.6
ID	Superplasticizer	Expansion agent	Waterproofing agent	PVA fibre	Steel fibre
P0S0	6.4	49.9	5	0	0
P0.25S0.75	6.4	49.9	5	3.13	58.88
P0.5S0.5	6.4	49.9	5	6.25	39.25

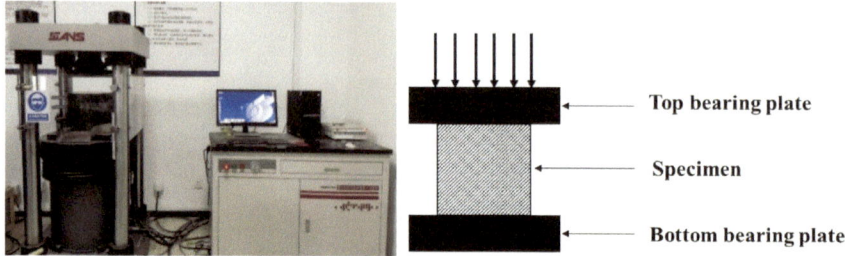

a) Pressure testing machine picture   b) Cube specimen Compressed test model

Fig. 1 Compressed strength test device

3 Experimental Method

The compressive strength, flexural strength, bending toughness are test. The experimental methods are shown as follows.

3.1 Compressive Strength

According to ISO 4012-1978 to test the compressive strength. The size of the specimen is 150 mm × 150 mm × 150 mm. The rate of the loading is 0.60 MPa/s. when the shape of the specimen changes quickly, the test can stop. The peaking load is recorded. The test picture is shown in Fig. 1.

3.2 Flexural-Tensile Strength

According to ISO 4013-1978 to test the flexural-tensile strength. The size of the specimen is 100 mm × 100 mm × 400 mm. The rate of the loading is 0.10 mm/min. The test picture and model are shown in Fig. 2.

3.3 Bending Toughness

According to ASTM C1018 to test the bending toughness. The device is the same as the flexural-tensile strength one. The bending toughness index is calculated as shown as Fig. 3.

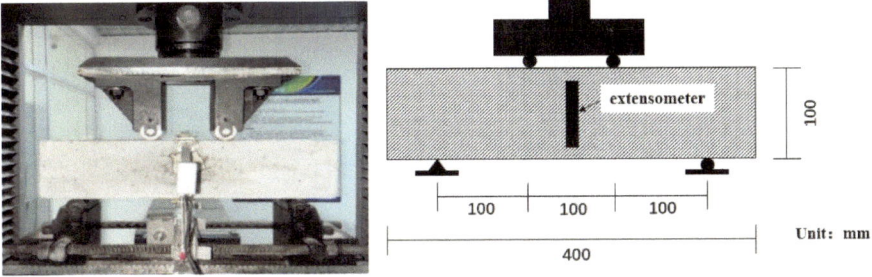

a) Flexural-tensile strength test picture   b) Flexural-tensile strength test model

Fig. 2 Flexural-tensile strength test device

Fig. 3 Schematic diagram of the bending toughness calculated

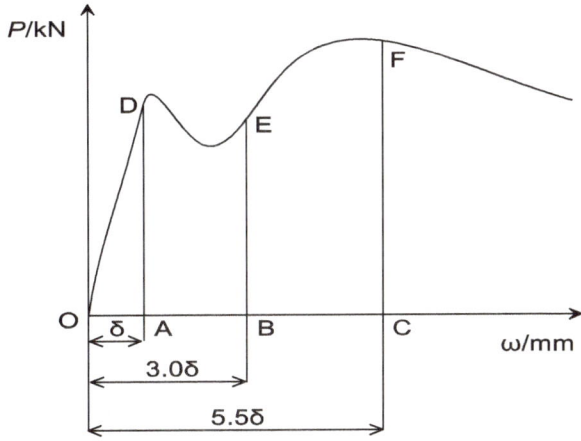

4 Analysis of the Test Results

4.1 Compressive Strength

Figure 4 shows the compressive strength of the concrete with different fibre ratio at 7d and 28d curing time. As can be seen from the picture, The compressive strength of the concrete with 7d curing time can rise to 68% of the concrete with 28d curing time. The compressive strength can satisfy the design value of c50 concrete. Comparing to the group of P0S0, the compressive strength of the P0.25S0.75 increases. While the group of P0.5S0.5 shows a little range of decreasing.

The steel fibre and the PVA fibre can disperse in the concrete and form a kind of supporting structure like 3D net. The fibre concrete become more homogeneity and continuity. At the same time, the PVA fibre was deal with hydrophilia so that the hydration products can adhere to the surface of the PVA fibre to enhance the

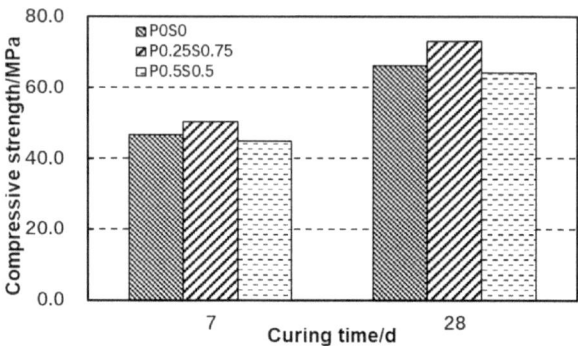

Fig. 4 Compressive strength of different fibre ratio concrete

binding performance with the cement base. By this method, the steel fibre can't sink to the bottom of the fresh concrete. While, the fibre can increase the porosity of the concrete and decrease the area of the bonding surface between the fibre and the cement base. So, when the interface of the cement-fibre is the main supporting mechanical property, the compressive strength of the concrete will decrease.

4.2 Flexural Strength

Figure 5 shows the flexural strength of the concrete with different fibre ratio at 7d and 28d curing time. Comparing to the P0S0 group, the 28d flexural strength of P0.25S0.75 and P0.5S0.5 increases 2.3 and 30.2% respectively. It verifies that the flexural strength of the fibre concrete will increase by the increasing amount of PVA fibre admixture. The reason why the P0.25S0.75 group making a less increasing is that the PVA mixing ratio is less. The steel fibre will make a remarkable role when the width of the cracking achieves to some extent. This phenomenon can be seen in Fig. 3.

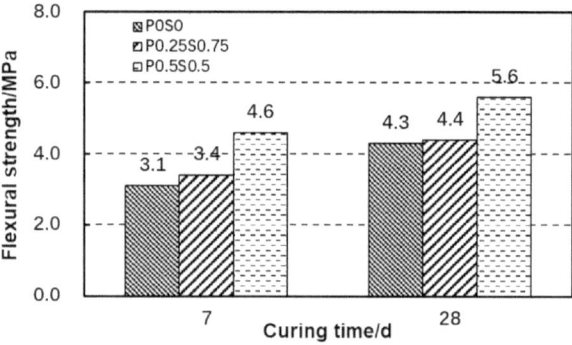

Fig. 5 Flexural strength of different fibre ratio concrete

Fig. 6 Load VS deflection curve

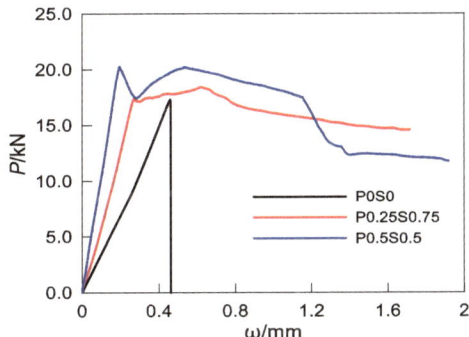

Figure 6 shows the load changing law by the midspan deflection increasing at the 28d curing time. The P0S0 curve shows an obvious brittle failure. The stress drops to 0 after the specimen broken. The P0.25S0.75 and P0.5S0.5 all appear strain harden property. The curve can be divided into 3 stages. They are elastic stage, strain hardening stage and strain softening stage. By the mixing amount of the PVA fibre increasing, the cracking strength and limitation bearing ability can all be enhanced. The areas under the curve enlarge and the toughness of the concrete are enhanced.

4.3 Bending Toughness

Figure 7 shows the index of bending toughness with different fibre ratio. The trends of I_5 and I_{10} are almost the same. The indexes of bending toughness are all enhanced when mixing the PVA and steel fibres into the concrete. And the real values are all larger than the ideal elastic plastic value of 5 and 10. This verifies that the PVA and steel fibre can make positive hybrid effect. From the data of P0.25S0.75 and P0.5S0.5, the bending toughness of P0.5S0.5 is much better than the P0.25S0.75 but the difference is not very large.

Fig. 7 Bending toughness of different fibre ratio concrete

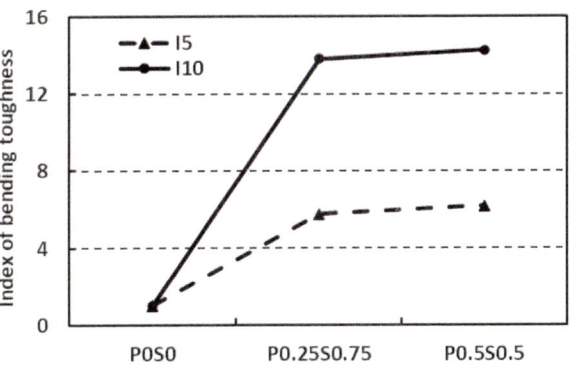

5 Conclusion

This paper study the mechanical properties of the fibre concrete with different fibre ratio. The fibres are hybrid with PVA fibre and the steel ones. The ratios are P0.25S0.25 and P0.5S0.5. The compressive strength, flexural strength, bending toughness are tested. The test result shows as follow as:

(1) the compressive strength of the P0.25S0.75 increases. While the group of P0.5S0.5 shows a little range of decreasing. When the interface of the cement-fibre is the main supporting mechanical property, the compressive strength of the concrete will decrease.
(2) The 28d flexural strength of P0.25S0.75 and P0.5S0.5 increases 2.3 and 30.2% respectively. It verifies that the flexural strength of the fibre concrete will increase by the increasing amount of PVA fibre admixture.
(3) The P0.25S0.75 and P0.5S0.5 all appear strain harden property. By the mixing amount of the PVA fibre increasing, the cracking strength and limitation bearing ability can all be enhanced. The areas under the curve enlarge and the toughness of the concrete are enhanced.
(4) The bending toughness of P0.5S0.5 is much better than the P0.25S0.75 but the difference is not very large.

Acknowledgements This work has been supported by Special Fund Project for the Research and Development Centre of Automation Operation Technology in the Transportation Industry.

References

1. Huang J (2006) Axial tensile test of short fibre and hybrid fibre concrete mortar. Concrete 12:04
2. Wang J, Ou Z, Liu J et al (2018) Review of research on PVA fiber reinforced concrete. Fly Ash Compreh Util 2018(4):97–103
3. Deng Z, Zeng H, Wang L (2009) Experimental study on flexural toughness of layered steel and PVA mixed fiber reinforced concrete beams. China Concrete Cement Prod 1:44–47
4. Shi S, Qi Y, Yan Y, Pan S (2009) Research on mechanical performance and microstructure of fiber reinforced active powder concrete at high temperature. Mater Introduction 10:65–68
5. Chen Z, Ling P, Chanjuan Y, Yang Q (2016) Design of hybrid fibre reinforced self-consolidating concrete based on fire resistance property-mechanical performance at room temperature. Concrete 3:86–88
6. Li F, Sun M (2017) Test and research on bending toughness of steel—PVA hybrid fibers concrete. J Suzhou Univ Sci Technol (Eng Technol) 1:19–25
7. Li Q, Zhao X, Shilang X (2017) Impact compression properties of nano-SiO_2 modified ultra-high toughness cementitious composites using a split Hopkinson pressure bar. Eng Mechan 2:85–93
8. Zhang C, Cao M (2014) Mechanical property test of a multi-scale fibre reinforced cementitious composites. Acta Materiae Compositae Sinica 3:661–668
9. Bai R, Liu S, Yan C et al (2020) Flexural cracking performance of strain-hardening cementitious composites with polyvinyl alcohol: experimental and analytical study. Constr Build Mater 247:118110

10. Yu J, Meng L, Leung CKY (2016) Pull-out response of single steel fiber embedded in PVA fiber reinforced cementitious matrix. In: International conference on fracture mechanics of concrete & concrete structures

Open Access This chapter is licensed under the terms of the Creative Commons Attribution 4.0 International License (http://creativecommons.org/licenses/by/4.0/), which permits use, sharing, adaptation, distribution and reproduction in any medium or format, as long as you give appropriate credit to the original author(s) and the source, provide a link to the Creative Commons license and indicate if changes were made.

The images or other third party material in this chapter are included in the chapter's Creative Commons license, unless indicated otherwise in a credit line to the material. If material is not included in the chapter's Creative Commons license and your intended use is not permitted by statutory regulation or exceeds the permitted use, you will need to obtain permission directly from the copyright holder.

Study on the New Preparation Method and Properties of Green Self-Compacting Transparent Concrete

Junliang Liu, Guangxiu Fang, Jiaxing Lu, and Han Jin

Abstract Through reading and integrating a large amount of literature, this paper innovatively proposed a new method of making transparent concrete. On this basis, we added fly ash and slag powder to replace cement according to different volume ratios, and developed four new types of green self-compacting transparent concrete with different mix ratios. By testing the compressive strength, tensile strength, elastic modulus and thermal conductivity of these concrete, found that according to (20%, 20%) added fly ash and slag powder cement instead of new green self-dense light transparent concrete in all aspects of the best performance, test results show that the 28 days compressive strength of 36.2 Mpa, 28 days of tensile strength 4 Mpa, 28 days of elastic modulus of 27.1 Gpa. This new type of green self-dense transparent concrete not only has high compressive strength, tensile strength and elastic modulus, but also has excellent transparent thermal conductivity. The experimental results show that the average illumination of four self-dense transparent concrete of 25W bulb is 152XL, and the average illumination of 40W bulb is 246XL and the average temperature is 32.125 °C. In addition, the photothermal conductivity of the four different mix ratios showed good consistency, indicating that the addition of fly ash and slag powder has no negative impact on the light transmittance of the transparent concrete. This new type of transparent concrete not only has high mechanical properties, but also has good transparent thermal conductivity, which has wide application prospects and provides reference for subsequent research and application.

Keywords Light permeable concrete · Fly ash · Slag powder · Compressive strength · Split tensile strength

---

J. Liu · G. Fang (✉) · J. Lu · H. Jin
Civil Engineering, School of Engineering, Yanbian University, Yanji, Jilin, China
e-mail: gxfang@ybu.edu.cn

1 Introduction

Transparent concrete is a new type of building material that combines photoelectric technology with concrete to enable concrete to penetrate light. This material has a wide range of applications, including landscape design, architectural design, interior decoration, etc. [1]. However, there are still some problems in transparent concrete, such as its complex production process, high production cost, transparent thermal conductivity and strength to be improved. Self-compacting concrete is a high performance concrete due to its high fluidity, high filling and high separation resistance [2]. This kind of concrete does not need to vibrate in the production process, so it can solve the complex production process and the dislocation of light guide material, and at the same time can improve the integrity of transparent concrete to improve its quality and durability. Fly ash and slag powder are two common wastes, their application in concrete can reduce environmental pollution, in line with the concept of sustainable development. At the same time, a lot of studies show that the proper incorporation of fly ash and slag powder can improve the mechanical properties of concrete [3–5]. The purpose of this study is to use self-compacting concrete as the base of light-permeable concrete and add different proportions of fly ash and slag powder to develop four new types of green self-compacting light-permeable concrete with different mix ratios. By analyzing the properties of these four kinds of concrete, including compressive strength, split tensile strength, elastic modulus, light transmittance and thermal conductivity, etc., the best mix ratio can be selected, so as to improve the performance [6] of transparent concrete and reduce environmental pollution. The test results show that the mixture ratio of 20% fly ash and 20% slag powder has the best performance. This new green self-compacting transparent concrete has high strength, high elastic modulus, good transparent thermal conductivity and low production cost. In addition, it can also reduce environmental pollution and meet the requirements of green buildings.

2 Raw Materials and Test Methods

2.1 Raw Materials

Portland cement P O 42.5 grade ordinary Portland cement; fly ash is grade 1 fly ash; slag is S95, grade slag; the chemical components of cement, fly ash and slag powder are shown in Table 1. The fine aggregate is Yanji River sand with fineness modulus of 2.18; the coarse aggregate is 1 ~ 10 mm; the admixture is polycarboxylic acid water efficient agent produced by Xiamen Kezhijie Building Materials Co., Ltd., with water reduction rate of 45%; the optical fiber is plastic with light diameter of 5 mm and light transmittance of 80% [7–9]. According to the Design Code for Self-compacting Concrete (JGJ / T283-2012), C40 concrete is configured with four different volume ratios of (0%, 0%) (10%, 30%) (20%, 20%) (30%, 10%) with fly ash and slag powder to replace cement, insert optical fiber by drilling 4 × 4 matrix arrangement, fixed with boards on all sides, make 150 × 150 × 150 test mold, and the bottom board is shown in Fig. 1. The mix ratio of new green self-compacting transparent concrete and the fluidity of new green self-compacting transparent concrete are shown in Table 2.

Table 1 Oxide composition of cement and fly ash and slag

Raw material	w (CaO)	w (SiO_2)	w (Al_2O_3)	w (Fe_2O_3)	w (SO_3)	w (MgO)	w (K_2O)	w (other)
Cement	65.65	18.24	4.26	3.68	2.44	0.15	0.96	4.11
Pulverized fuel ash	61.70	21.30	7.00	4.50	1.99	0.48	0.50	0.35
Ground slag	42.96	27.60	13.51	9.31	1.38	0.54	0.46	0.39

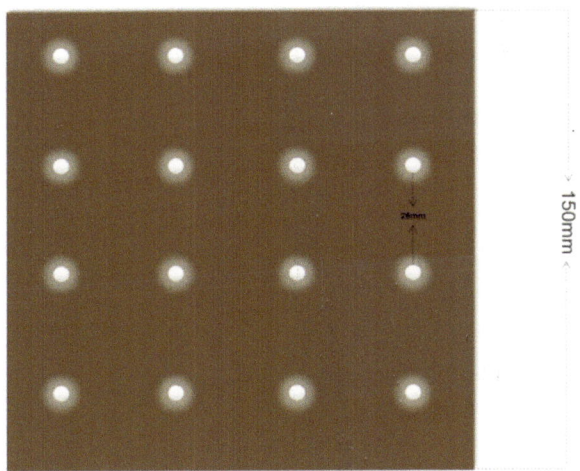

Fig. 1 Fiber arrangement and grinding tool model

Table 2 Mix proportions and fresh properties of SCC

Number	Volume fraction of each component/(kg.m^{-3})							Slumps (mm)	Divergence (mm)
	Cement	Pulverized fuel ash	Ground slag	Sand	Spall	Water	Water reducer		
SCC1	343.75	0	0	942.19	942.19	171.875	4.125	260	690
SCC2	224.18	37.37	112.09	927.42	927.42	171.875	4.483	260	670
SCC3	229.17	76.39	76.39	923.09	923.09	171.875	4.583	265	680
SCC4	229.17	114.58	38.19	923.09	923.09	171.875	4.583	265	640

2.2 Test Protocol

Compression and tensile strength test

According to the Standard for Test Methods for Mechanical Property of Ordinary Concrete(GB/T 50081-2019), The mechanical properties of the compressive strength of SCC at 7d, 14d and 28 d and the split tensile strength ("split strength"), Prepare 150×150×150 mm cube block according to standard requirements, After curing for 7d, 14d, and 28d under standard conditions, Place the test block on the test machine, Keeping its axis coincides with the loading axis of the experimental machine, And then gradually loaded at a strain rate of 1–2%, Record the deformation and loading force of the test block, And draw the stress–strain curves [9]. The maximum stress during the test block failure is read on the curve and used as the compressive strength or tensile strength of the test block. The 2000 kN concrete universal pressure tester is shown in Fig. 2.

Elastic modulus test

According to the Standard for Test Methods of General Concrete (GB/T50081-2019), the test block of $150 \times 150 \times 600$ mm is used. Test the compressive strength with a universal testing machine and measure the strain values with a strain gauge. Before the experiment, the test block is placed on the test machine, so that its axis coincides with the loading axis of the test machine, and then loads at a certain speed, record the deformation and loading force of the test block, and draw the stress–strain curve. The elastic modulus of the test block is calculated according to the elastic phase of the curve [10, 11]. For 2000 kN concrete universal pressure test machine, see Fig. 2.

Optical transmission and thermal conductivity test

The test block is placed on the test instrument and surrounded by a black screen to prevent the interference of the external light source. The transparent concrete test

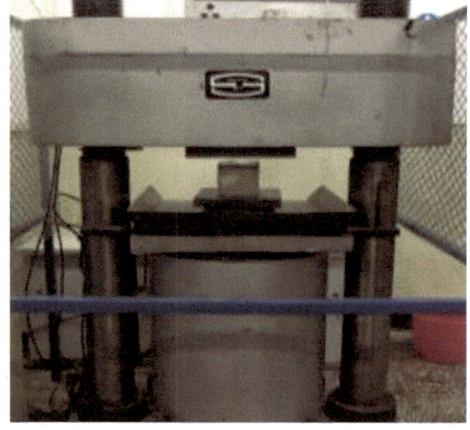

Fig. 2 A 2000 kN concrete universal pressure test machine

Fig. 3 Optical transmission and thermal conductivity tester

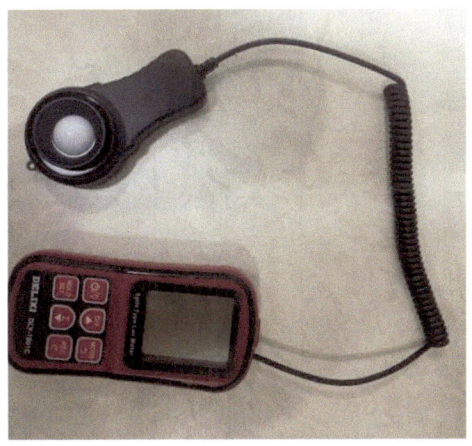

block used is a standard test block of 150 × 150 × 150 mm. Make the surface of the test block in contact with the test instrument, set an incandescent light source on the lower surface of the test block, choose 25 and 40 W. 25 W incandescent 2 h surface temperature of 45 °C and illumination of 220 Lx. The surface temperature of the 40 W incandescent lamp is 60 °C and the illumination is 350 Lx, and then the thermal conductivity of the test block is measured through the test instrument. The transheat imeter is shown in Fig. 3.

3 Test Results and Analysis

3.1 Mechanical Properties Analysis

Results and analysis of compressive strength and tensile test

The compressive strength experimental data are shown in Table 3, the relationship between compressive strength (Mpa) and age of 7d, 14d and 28d is shown in Fig. 4, the tensile strength (Mpa) of 7d, 14d and 28d is shown in Table 4, and the relationship between tensile strength (Mpa) of 7d, 14d and 28d and age is shown in Fig. 5.

Table 3 7d, 14d, 28d compression strength (Mpa)

Number	Compressive strength (MPa)		
	7d	14d	28d
SCC1	25.7	27.4	32.2
SCC2	26.1	29.2	33.6
SCC3	27.4	31.3	36.2
SCC4	26.4	28.7	34.2

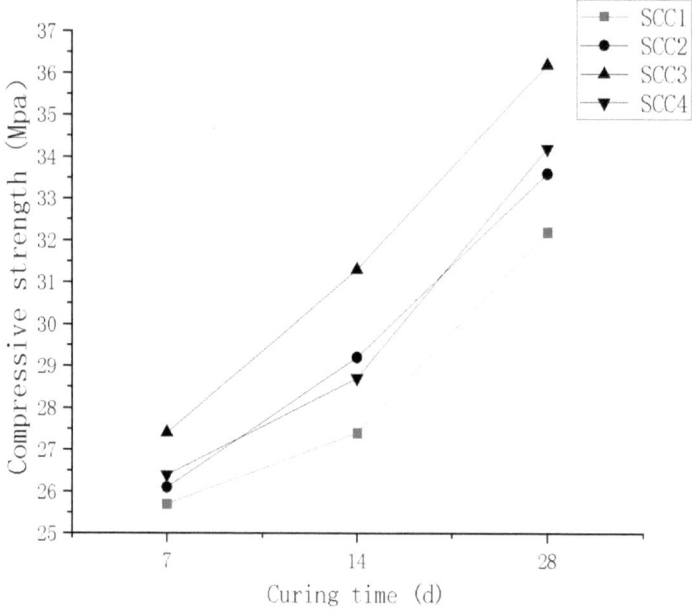

Fig. 4 7d, 14d, 28d Compressive Strength (Mpa)

Table 4 7d, 14d, 28d split-resistant tensile strength (Mpa)

Number	Compressive strength (MPa)		
	7d	14d	28d
SCC1	0.8	1.9	2.7
SCC2	1.1	2.3	3.1
SCC3	1.5	2.4	4.0
SCC4	1.3	2.1	3.4

According to the line chart of compressive strength for each group at 7d, 14d, and 28d, the transparent concrete with 20% fly ash and 20% slag powder showed the best performance in terms of compressive strength and splitting tensile strength, depending on the age. For example, at an age of 7 days, the compressive strength and splitting tensile strength of the transparent concrete in this ratio may be 27.4 MPa and 1.5 MPa, respectively. At an age of 14 days and 28 days, the corresponding strength values may be 31.3 MPa, 2.4 MPa and 36.3 MPa, 3.4 MPa, respectively. The compressive strength and splitting tensile strength of transparent concrete with other ratios may be lower than this value.

Analysis of the reasons for the test results

There may be the following reasons for the experimental results:

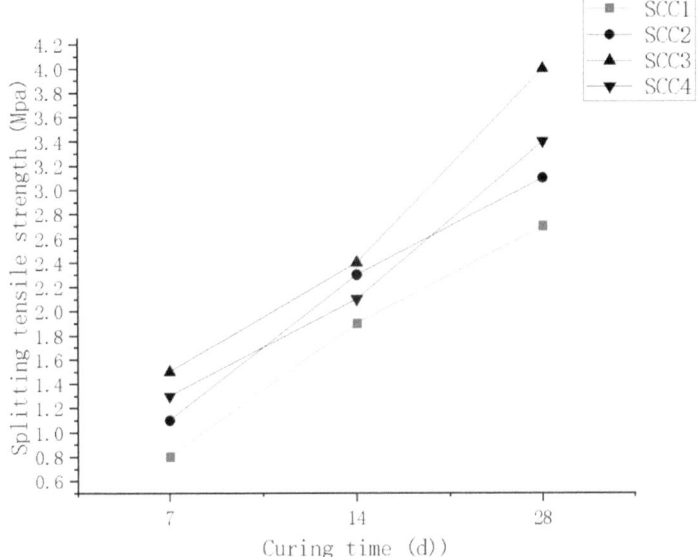

Fig. 5 7d, 14d, 28d split-resistant tensile strength (Mpa)

(1) The addition of fly ash and slag powder improves the microstructure of transparent concrete, enhances its compactness, and thus enhances its compressive strength and splitting tensile strength.
(2) Fly ash and slag powder have good pozzolanic activity and can undergo hydration reactions with the calcium components in concrete, generating more hydration products and further improving the strength of concrete.
(3) The addition of fly ash and slag powder may also reduce the porosity and water absorption of transparent concrete, reduce internal defects, and improve its compressive strength and splitting tensile strength.
(4) Self compacting transparent concrete has good flowability and can better fill the formwork, reduce pores and defects, thereby improving its compressive strength and splitting tensile strength.
(5) Due to the presence of hard optical fibers, it has a certain positive effect on the compressive strength and splitting tensile strength of transparent self compacting concrete

In summary, transparent concrete with a ratio of 20% to 20% fly ash and slag powder exhibits the best performance in terms of compressive strength and splitting tensile strength, which may be related to the improvement of the microstructure, porosity, and water absorption of transparent concrete by the addition of fly ash and slag powder, the flowability of self compacting concrete, and hard straight optical fiber.

3.2 Results and Analysis of the Elastic Modulus Test

The experimental data are shown in Fig. 6.

The results show that the elastic modulus of transparent self-compacting concrete is 7.2, 15.7 and 27.1 Gpa at 7d and slag powder (20%, 20%), 14d and 28d. The elastic modulus is slightly lower than that of ordinary concrete, but the difference is not large.

For the test results, there may be the following reasons:

The elastic modulus of transparent self-compacting concrete double mixed with fly ash and slag powder (20%, 20%) is 27.1 Gpa at 28d, which is slightly lower than that of ordinary concrete. This may be due to the presence of optical fiber in the transparent concrete, which has some influence on the elastic modulus. As a heterogeneous material, the presence of optical fiber will change the stress distribution inside concrete, which will affect the elastic mode value. In addition, the addition of fly ash and slag powder in transparent concrete may have some influence on the elastic mode value. Fly ash and slag powder have good volcanic ash activity, which can hydrate with the calcium composition in concrete to generate more hydration products, so as to improve the strength and elastic mode value of concrete.

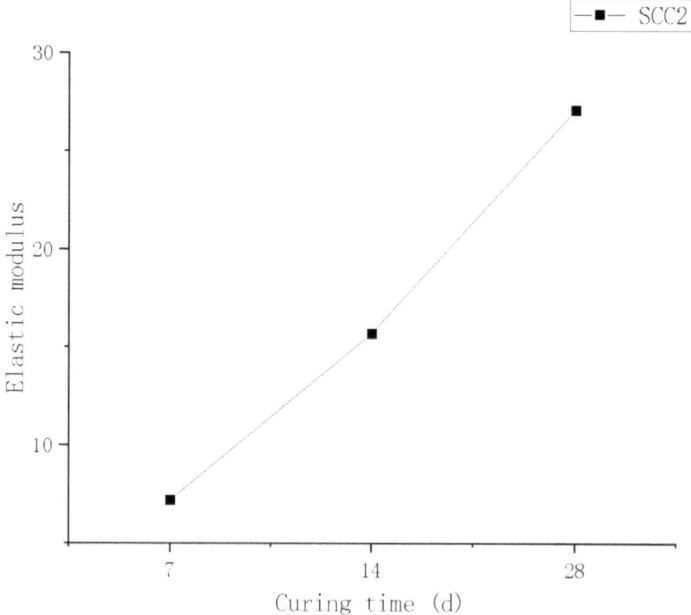

Fig. 6 Hows the elastic modulus of 7d, 14d, 28d of scc 3

Study on the New Preparation Method and Properties of Green …

Table 5 Photothermal conductivity of four different mix proportion

	Light source power consumption (W)	Illuminance	Temperature/ 2 h	Light source power consumption (W)	Illuminance	Temperature/ 2 h
SCC1	25	152Lx	24.4 °C	40	243LX	32.0 °C
SCC2	25	151Lx	24.1 °C	40	246LX	32.3 °C
SCC3	25	153Lx	24.2 °C	40	247LX	32.2 °C
SCC4	25	152Lx	24.0 °C	40	248LX	32.0 °C

3.3 Analysis of Photothermal Conductivity of New Green Self-Compacting Light Permeable Concrete

According to the experimental data, the average illumination of the self-compact concrete of four different mixing ratios was 152XL and the average temperature of 25 W bulb for two hours, the average temperature was 24.175 °C, the average illumination was 246XL and the average temperature was 32.125 °C, and the thermal conductivity of the concrete of four different mixing ratios showed good consistency, indicating that the addition of fly ash and slag powder had no negative effect on the transparent performance of the concrete. In fact, because these two materials have a small particle size, they can fill the gap in concrete, reduce the reflection and scattering of light, thus improving the light transmittance of concrete. From the point of view of optical fiber and light propagation, optical fiber plays a role in capturing and guiding the light in the transparent concrete, making the light pass through the concrete more evenly, thus enhancing its light transmission performance. At the same time, because the distribution of the optical fiber in the transparent concrete is more uniform, so the transparent concrete with different mix ratios shows a good consistency in the thermal conductivity of the light transmittance. Secondly, the thermal conductivity of the transparent concrete increases with the illumination and surface temperature in the test. When the intensity of the light source increases, more light can penetrate the concrete, making its transparent thermal conductivity improved. In addition, the thermal conductivity of concrete is also affected by its thickness, optical fiber arrangement and irradiation time (Table 5).

4 Conclusion

1. (20%20%) of fly ash and slag powder added ratio of transparent concrete, its 28 days compressive strength of 36.2 Mpa, the 28 days of tensile strength 4 Mpa, showed excellent mechanical properties, although because of transparent concrete internal optical fiber, but added in the transparent concrete of fly ash and slag powder can optimize the compressive and tensile strength, the reason is that

the hybrid material can produce synergistic effect, can improve the mechanical properties of concrete.
2. (20%20%) of fly ash and slag powder added ratio of transparent concrete, the elastic modulus of 28 days is 27.1 Gpa. This is because of the existence of internal optical fiber, there will be a certain gap between optical fiber and concrete, so that its 28 day elastic modulus is slightly lower than the elastic modulus of ordinary concrete, but the difference is not large. This is because fly ash and slag powder have good volcanic ash activity, which can hydrate with the calcium composition in concrete to generate more hydration products, so as to improve the strength and elastic mode value of concrete.
3. The new type of green self compacting transparent concrete studied in this experiment has good transparency and thermal conductivity. Four different proportions of self compacting transparent concrete have an average illuminance of 152XL and an average temperature of 24.175 °Cunder two hours of exposure to a 25 W light bulb, and an average illuminance of 246XL and an average temperature of 32.125 °Cunder two hours of exposure to a 40 W light bulb, Moreover, the transparency and thermal conductivity of transparent concrete with four different mix ratios showed good consistency, indicating that the addition of fly ash and slag powder did not have a negative impact on the transparency of transparent concrete. Its transparency and thermal conductivity increase with the intensity of the external light source. It has good environmental benefits.

References

1. Sun H, Zhang Q, Cai Z, et al (2023) Study on the influence of fly ash and slag powder on concrete properties. World of Concr (07):53–56
2. Yu X, Guan S, Xie G, et al (2022) Study on the preparation and properties of light-permeable concrete. Concr World (09):8–11
3. Deng Y, Cheng Z, Su J, et al (2019) Progress in the development and application of transparent concrete. New Build Mater 46(07):1–7
4. Jiang Z, Feng Y, Zhang J, et al (2018) Preparation methods and research progress of resin photopermeable concrete. New Chem Mater 46(03):235–238
5. Li M, Lu Y, Chen Z, et al (2021) Research on the mechanical properties and optical properties of plastic fiber optic concrete. Concrete (10):6–9
6. Danial N, Zahra A, Alireza R et al (2023) Developing light transmitting concrete for energy saving in buildings. Case Stud Constr Mater 18
7. Shahmir GN, Bhat M (2020) Structural and luminance properties of light transmitting concrete. Ann Chimie Sci Mat 44(3)
8. Wang S, Wu Y, Wu X et al (2016) Overview of the preparation, properties, and application of light-permeable concrete. Mater Guide (S1) 30:467–469+477
9. Shengxian W, Xuefang W, Shaofei J (2022) The effect of fly ash and slag on the early age cracking resistance of self compacting concrete. J Shenyang Jianzhu Univ (Nat Sci Ed) 38(06):1104–1113
10. Qifeng L, Pengfei D, Anguo C (2024) Mechanical strengths and optical properties of translucent concrete manufactured by mortar-extrusion 3D printing with polymethyl methacrylate (PMMA) fibers. Compos Part B 268

11. Jianzhou Y, Shunchao C, Chunyan D et al (2023) Comparative experimental study on the strength and elastic modulus of long-term self compacting concrete and ordinary concrete. Silic Bull 42(10):3530–3537. https://doi.org/10.16552/j.cnki.issn1001-1625.2023.10.10

Open Access This chapter is licensed under the terms of the Creative Commons Attribution 4.0 International License (http://creativecommons.org/licenses/by/4.0/), which permits use, sharing, adaptation, distribution and reproduction in any medium or format, as long as you give appropriate credit to the original author(s) and the source, provide a link to the Creative Commons license and indicate if changes were made.

The images or other third party material in this chapter are included in the chapter's Creative Commons license, unless indicated otherwise in a credit line to the material. If material is not included in the chapter's Creative Commons license and your intended use is not permitted by statutory regulation or exceeds the permitted use, you will need to obtain permission directly from the copyright holder.

Superposed Element Method for the Temperature Field Simulation in Mass Concrete Structures Containing Cooling Pipes

Jianxin Ding and Qingzhou Yang

Abstract Pipe cooling is an important measure for controlling the temperature in mass concrete structures, so it is one of the key problems to specifically simulate the effect of the cooling pipes in temperature field analysis. Because the size of cooling pipe is quite small from that of mass concrete structure, its exact simulation is the focus of simulation calculation. In this paper, a new method called SEM (superposed element method) is proposed to analyze the temperature field of mass concrete with cooling pipes in it. First, the structure is divided into two independent meshes. One is the global mesh without cooling pipes, and the other is the local mesh of the cooling pipes together with its adjacent region. Then the elements within the local mesh is treated based on FEM, and the two independent meshes are coupled together through the coupling surface by the coordination of heat conduction. Finally, using the SEM proposed in this paper, the numerical model with a single pipe was analyzed and the calculation precision was validated. The convenient process of grid discretization and good performance in both precision and efficiency show the feasibility of the new method in engineering application.

Keywords Hydraulic structure · Mass concrete · Temperature field · SEM (superposed element method) · Cooling pipe

---

J. Ding (✉)
Changjiang Survey, Planning, Design and Research Co., Ltd, Wuhan 430010, China
e-mail: whudjx@126.com

Q. Yang
Power China Hubei Electric Engineering Corporation Limited, Wuhan 430040, China

1 Introduction

The pipe cooling is an important cooling method for the mass concrete structures, so how to accurately simulate the cooling pipe is one of the key problems in temperature field analysis. Due to the large number of cooling pipes, small pipe diameter, serpentine layout, constant change of water temperature along the way, and the phased cooling in the actual process, the temperature field of the mass concrete structures containing cooling pipes is a complex three-dimensional transient temperature field, and it is difficult to simulate accurately.

At present, the numerical calculation methods of pipe cooling effect generally include equivalent method and discrete method. At present, the finite element method [1–3] is often used, which regards the cooling pipe as a heat sink and considers the cooling effect of the water pipe in an average sense, and cannot simulate the local temperature gradient near the cooling pipe. The discrete simulation method places dense elements around each cooling water pipe [4], and can calculate the temperature distribution near the water pipe more accurately. However, the finite element model of this method requires a lot of work, and it is not realistic when applied to practical engineering problems. Recently, some scholars proposed the composite element method [5], which can discretely simulate the role of cooling pipes without the need to discretely disperse cooling pipes in the model, and the pre-processing is greatly reduced, but its accuracy still depends on the overall finite element mesh size, and the calculation speed needs to be improved.

In this paper, the Superposed Element Method (SEM) for simulating cooling pipes in temperature field analysis of mass concrete is proposed by referring to the ideas of superposed element method in stress-strain analysis [6] and seepage analysis [7]. This method is simple and easy to use, and the local mesh containing the cooling pipe can participate in the calculation dynamically, which not only ensures the high precision near the cooling pipe, but also reduces the workload in the simulation. It is expected to be applied to the engineering practice.

2 Superposed Element Method

2.1 Mesh Discretization

For the structure containing cooling pipes, it is discrete into a global mesh without cooling pipe and a local mesh of the cooling pipe together with its adjacent zone, and the meshes are independent of each other. Because the existence of cooling water pipe is not considered in the global mesh, the element size can be larger, so the global mesh is simple and the element shape is easy to control. The local mesh only considers the smaller area of the cooling pipe, and the cooling pipe can be simulated in fine detail with a smaller element size. In actual engineering, usually a pouring layer is 1–3 m, and the mesh size is generally 0.5–1 m. Referring to the results in reference [6, 7], the local mesh of the cooling pipe here can be taken as 1.2–1.5 m.

2.2 The Interpolation Model

After the global mesh and the local mesh of the cooling pipe are divided independently, they need to be superimposed in order to describe the temperature field uniformly. Figure 1 shows the schematic diagram of mesh superposition. The parts covered by local mesh in the global mesh (including complete and incomplete elements) are regarded as invalid elements, and the remaining effective elements are divided into the following three categories:

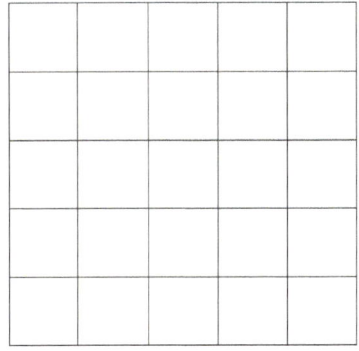
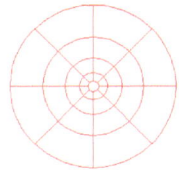

(a) The global mesh without cooling pipe

(b) The local mesh of the cooling pipe together with its adjacent zone

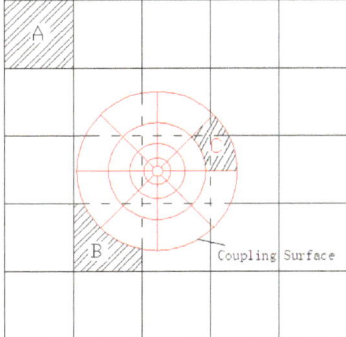

(c) The local mesh of the drainage hole together with its adjacent zone

Fig. 1 The sketch map of the mesh superposition

(1) Type A: the complete elements which are not covered at all by the local mesh in the global mesh.
(2) Type B: the incomplete elements which are covered partly by the local mesh in the global mesh.
(3) Type C: the complete elements in the local mesh.

For the complete elements of type A and type C, the conventional finite element format can be used to discrete the temperature function. For the incomplete elements of type B, because of its irregular shape, the numerical manifold format based on the finite element is used to discrete the temperature function by treating the original nodes of the complete elements as their virtual nodes [8, 9].

2.3 The Assumption of Coupling Surface

Because the global mesh and the local mesh of the cooling pipe are independent of each other, the temperature field function is discontinuous at the outer interface of the local mesh of the cooling pipe. In order to realize the coupling between the global mesh and the local mesh of the cooling pipe, it is necessary to assume that there is a coupling surface on the outer boundary surface of the local mesh of the cooling pipe, as follows:

(1) The interface is assumed to have a small virtual thickness t;
(2) There is no adiabatic temperature rise at the interface, and the normal heat flux remains unchanged in the thickness direction, that is, the temperature at the interface changes linearly in the thickness direction;
(3) The normal temperature coefficient a_n on the interface is very large, and the tangent temperature coefficient $a_s = 0$.

2.4 Heat Conduction Equation

For any point in the effective element of the global mesh and the local mesh of the cooling pipe, the temperature function can be expressed as:

$$\begin{aligned} T_G &= [N]_G \{\overline{T}\}_G \\ T_L &= [N]_L \{\overline{T}\}_L \end{aligned} \quad (1)$$

In the formula, the subscripts G and L represent the global and local mesh respectively, $[N]$ is the shape function matrix, and $\{\overline{T}\}$ is the node temperature. The temperature equation can be expressed as:

$$\begin{aligned} [K]_G \{\overline{T}\}_G &= \{F\}_G \\ [K]_L \{\overline{T}\}_L &= \{F\}_L \end{aligned} \quad (2)$$

where, $[K]$ is the heat conduction matrix, $\{F\}$ is the heat vector.

Since the heat generation of the interface is not considered, the functional on the interface can be written as:

$$I(T)_{GL} = \iiint_{\Omega_{GL}} \left(\frac{1}{2}(\{S\}T)^T[a]_{GL}(\{S\}T)\right) d\Omega_{GL} \tag{3}$$

Since the thickness of the interface t is relatively small, its volume element can be written as:

$$d\Omega_{GL} = dx_j dy_j \cdot t \tag{4}$$

The functional of Eq. (3) can then be rewritten as:

$$I(T)_{GL} = t \iint_{\Gamma_{GL}} \frac{1}{2}(\{S_j\}T)^T[a]_{GL}(\{S_j\}T) \, dx_j dy_j \tag{5}$$

If we regard $\{\overline{T}\}_G$, $\{\overline{T}\}_L$ as the basic unknowns, then the interface functional is a function of the two. In order to obtain the minimum value of the functional, the first partial derivative of $\{\overline{T}\}_G$, $\{\overline{T}\}_L$, respectively, is set to zero. At the same time, assuming $a_s = 0$, we can obtain:

$$\begin{cases} \dfrac{\partial I(T)_{GL}}{\partial \{\overline{T}\}_G} = \dfrac{t}{2} \iint \left(\dfrac{2a_n}{t^2}[N]^T[N])\{\overline{T}\}_G - \dfrac{2a_n}{t^2}[N]^T[N])\{\overline{T}\}_L\right) dx_j dy_j = 0 \\ \dfrac{\partial I(T)_{GL}}{\partial \{\overline{T}\}_L} = \dfrac{t}{2} \iint \left(-\dfrac{2a_n}{t^2}[N]^T[N])\{\overline{T}\}_G + \dfrac{2a_n}{t^2}[N]^T[N])\{\overline{\varphi}\}_L\right) dx_j dy_j = 0 \end{cases} \tag{6}$$

This binary is denoted as:

$$\begin{cases} [H]_{GG}\{\overline{T}\}_G + [H]_{GL}\{\overline{T}\}_L = 0 \\ [H]_{LG}\{\overline{T}\}_G + [H]_{LL}\{\overline{T}\}_L = 0 \end{cases} \tag{7}$$

where $[H]_{GG}$, $[H]_{GL}$, $[H]_{LG}$, $[H]_{LL}$ is the heat conduction matrix of the interface, which reflects the influence of the interface on the global mesh and local mesh and its calculation formula is as follows:

$$\begin{cases} [H]_{GG} = [H]_{LL} = \dfrac{t}{2} \iint_{\Gamma_{GL}} \left(\dfrac{2a_n}{t^2}[N]^T[N]\right) dx_j dy_j \\ [H]_{GL} = [H]_{LG} = \dfrac{t}{2} \iint_{\Gamma_{GL}} \left(-\dfrac{2a_n}{t^2}[N]^T[N]\right) dx_j dy_j \end{cases} \tag{8}$$

It can also be seen from the above formula that the normal temperature coefficient a_n and the virtual thickness t of the interface always appear in the form of $\frac{a_n}{t}$. For the convenience of narration, it is defined $\frac{a_n}{t}$ as the temperature conductance coefficient of the coupling surface, and its value will be discussed later combined with a concrete example analysis.

By superimposing Eqs. (7 and 2), the overall heat conduction equation can be obtained:

$$\begin{bmatrix} [H]_G + [H]_{GG} & [H]_{GL} \\ [H]_{LG} & [H]_L + [H]_{LL} \end{bmatrix} \begin{Bmatrix} \{\bar{T}\}_G \\ \{\bar{T}\}_L \end{Bmatrix} + \begin{Bmatrix} [C]_G \{\dot{T}\}_G \\ [C]_L \{\dot{T}\}_L \end{Bmatrix} = \begin{Bmatrix} \{P\}_G \\ \{P\}_L \end{Bmatrix} \quad (9)$$

It can be seen that the heat transfer matrix of the superposed element method is similar to that of the ordinary finite element method in form.

3 Key Technologies

3.1 The Time Discrete

The discretization of the temperature field in the time domain is the same as that of the finite element method, and the basic equation of the heat conduction problem can be discretized as follows:

$$\{[H] + \frac{1}{\Delta t}[C]\}\{T_{n+1}\} + \{-\frac{1}{\Delta t}[C]\}\{T_n\} - \{R\} = 0 \quad (10)$$

3.2 The Numerical Integration

(1) For the class A and class C elements (Fig. 1), as with the conventional finite element principle, the conventional second-order Gaussian integral can meet the accuracy requirements;
(2) For the class B effective elements, because they are irregularly shaped incomplete elements, the numerical integration results can be approximated by using the encrypted Gaussian point method. First, all class B elements are encrypted with Gaussian points; Then, determine whether each Gaussian point within a class B element is in a local element. If not, the contributions of these Gauss points are added up when calculating the heat transfer matrix. If so, these points are omitted, so that the contribution of global elements cut by local elements to the heat conduction matrix is obtained.

(3) For the integration of the coupling surface, considering that the size of the local mesh is generally smaller than that of the global mesh, the conventional second-order Gaussian integral can be directly used on the coupling surface. It should be noted that for a Gaussian point on the coupling surface, its local coordinates in the global mesh need to be calculated according to its global coordinates in order to calculate the shape function matrix of the global mesh.

4 Numerical Example

Based on the theory of the superposed element method in temperature field, a program for the three-dimensional transient temperature field is written.

The Fig. 2 shows a 50 m long, 5 m wide and 5 m high example. There is a cooling pipe with a diameter of 0.032 m in the center of the specimen. The initial temperature of concrete is 22 C, the temperature of cooling water is 10 °C, and the flow rate of water is 0.4 m/s. The conventional finite element method and the superposed element method proposed in this paper were respectively used to analyze the temperature field of the specimen, and the mesh was shown in Fig. 3. In the finite element method, the mesh near the cooling pipe is the same as the local mesh of the superposed element method, and the rest of the mesh is obtained by free subdivision, the total number of elements is 3200, and the number of nodes is 4131. In the superposed element method, the local mesh of the cooling pipe is a cylinder with a diameter of 1.5 m, the number of elements is 1200, and the number of nodes is 1632. The whole mesh was evenly divided, with 625 elements and 936 nodes.

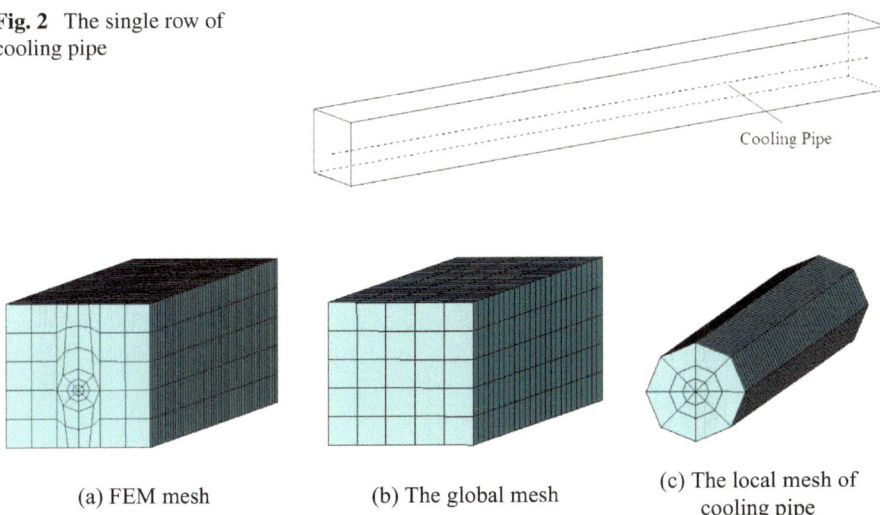

Fig. 2 The single row of cooling pipe

(a) FEM mesh (b) The global mesh (c) The local mesh of cooling pipe

Fig. 3 The mesh of a single row of cooling pipe

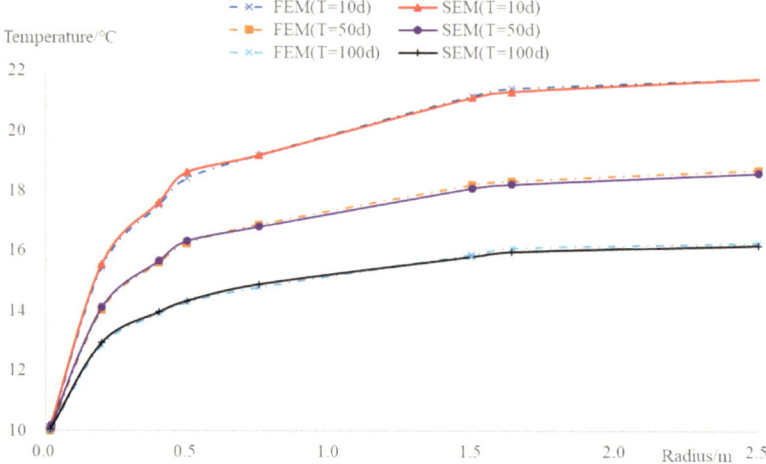

Fig. 4 The temperature distribution diagram of typical section at different times (cooling pipe intermediate interface)

Compared and analyzed the calculation results of the two methods without considering the heat of hydration, the temperature distribution diagram of the middle section of the cooling pipe at different times was calculated and sorted out respectively. Due to the symmetry of the model, only the temperature distribution of points along the cooling pipe circle center as the origin and from the right to the outer boundary of the specimen was taken here, as shown in Fig. 4.

It can be seen that the results of the superposed element method are in good agreement with those of the finite element method, and there is no obvious discontinuity at the outer boundary of the local mesh.

In the previous calculation and analysis, the temperature conductivity coefficient of the outer interface of the cooling pipe is the given value. In order to reveal the influence of this parameter on the calculation accuracy of the superposed element method, the sensitivity analysis is carried out below. In order to facilitate the description of the results, two feature points A and B in the middle section of the model (Fig. 5) are selected, where point A is 0.184 m away from the surface of the cooling pipe, and point B is located on the outer boundary surface of the local mesh of the cooling pipe.

The point A is used to investigate the error of the results calculated by the superposed element method compared with the results calculated by the finite element method. The formula for calculating the relative error is as follows:

$$\varepsilon_A = \frac{|T_{SEM} - T_{FEM}|}{T_{FEM}} \times 100\% \quad (11)$$

where T_{SEM} and T_{FEM} represent the temperature at point A calculated by the superposed element method and the finite element method respectively. The point B is

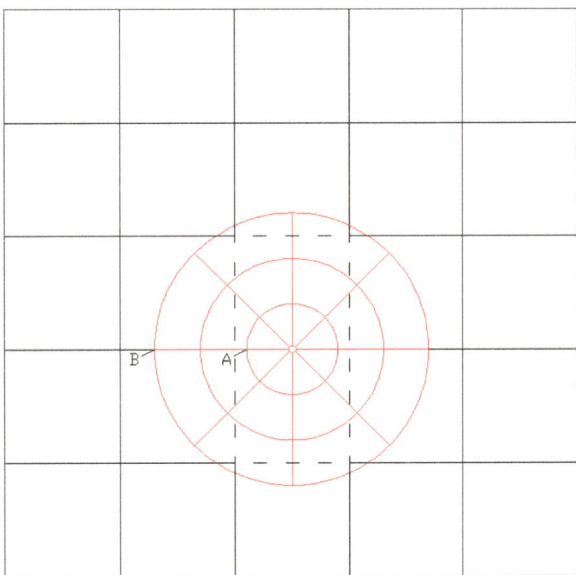

Fig. 5 The sketch map of feature point

to analyze the temperature continuity between the outer interface of the local mesh of the cooling pipe and the global mesh in the superposed element method, and its relative error is calculated as follows:

$$\varepsilon_B = \frac{|T_{SEM-G} - T_{SEM-L}|}{T_{FEM}} \times 100\% \quad (12)$$

Figure 6 shows the sensitivity analysis results of the temperature conductance coefficient of the outer interface of the local mesh of the cooling pipe. Logarithm lg(a) of the interface temperature conductivity coefficient a is used as the horizontal coordinate in the figure. The left ordinate is the relative error of the point A and B, ε_A, ε_B representing the calculation accuracy; The vertical coordinate on the right is the number of iterative steps in solving the temperature equations, representing the calculation speed.

It can be seen that with the gradual increase of the value of the interface temperature conductance coefficient a, the number of iterative steps in solving the temperature equations gradually increases, while the temperature accuracy of the feature points near the cooling pipe increases first and then decreases. Obviously, if the value of the interface temperature conductance coefficient a is too small, the temperature on both sides of the interface is obviously discontinuous, and the calculation accuracy is low. If the value of the interfacial temperature conductance coefficient a is too large, the properties of the temperature equations are poor, the calculation speed is slow, and the calculation accuracy is not high. Based on the above analysis, it is suggested that the interface temperature conductivity coefficient a should be 10–100 m^2/d.

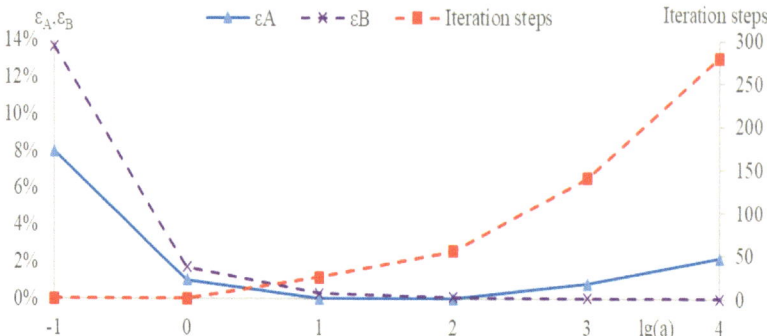

Fig. 6 The influence of interfacial temperature conductance coefficient a on the calculation result

5 Conclusion

Aiming at the problem of cooling pipe simulation in the temperature field simulation analysis of mass concrete, a new method, the superposed element method, is proposed in this paper. The method is simple and easy to use. By using the dynamic participation of the local mesh of the cooling pipe, the calculation accuracy can be ensured, and the calculation amount can be controlled reasonably. It makes it possible to accurately simulate the function of the cooling pipe in practical engineering, and has a wide application prospect.

An example of single row cooling pipe specimen shows that the calculation results of the superposed element method are in good agreement with those of the finite element method, which shows that the method in this paper can effectively simulate the cooling pipe in practical engineering.

References

1. Bofang Z (2014) Thermal stresses and temperature control of mass concrete. Beijing: Tsinghua University press
2. ZHU BF (2003) The equivalent heat conduction equation of pipe cooling in mass concrete considering influence of external temperature. J Hydraul Eng 34(3):49–54
3. Bofang Z (2020) The finite element method fundamentals and applications in civil, hydraulic, mechanical and aeronautical engineering, Beijing. Tsinghua University press
4. ZHU YM, XU ZQ, HE JR, et al (2003) A calculation method for solving temperature field of mass concrete with cooling pipes. J Yangtze River Sci Res Inst 20(2):19
5. Ding J, Chen S (2013) Simulation and feedback analysis of the temperature field in massive concrete structures containing cooling pipes. Appl Therm Eng 61(2):554–562
6. Ding J, Yang Q (2023) Superposed element method for the numerical simulation of structure crack. Adv Eng Technol Res 5(1):53
7. Jianxin D, Zhengtao X, Qingzhou Y (2023) Superposed element method for the numerical simulation of seepage problems containing drainage holes. The 9th international conference on hydraulic and civil engineering and smart water conservancy and safety engineering forum

8. Chen S (2023) Advanced computational methods and geomechanics, Germany. Springer press
9. Hong Z (2022) Numerical manifold method. Science press, Beijing

Open Access This chapter is licensed under the terms of the Creative Commons Attribution 4.0 International License (http://creativecommons.org/licenses/by/4.0/), which permits use, sharing, adaptation, distribution and reproduction in any medium or format, as long as you give appropriate credit to the original author(s) and the source, provide a link to the Creative Commons license and indicate if changes were made.

The images or other third party material in this chapter are included in the chapter's Creative Commons license, unless indicated otherwise in a credit line to the material. If material is not included in the chapter's Creative Commons license and your intended use is not permitted by statutory regulation or exceeds the permitted use, you will need to obtain permission directly from the copyright holder.

Effect of Machine-Made Sand Rate on the Compressive Strength, Workability, and Impermeability of Sleeper Concrete

Zhenchao Liu

Abstract The exploitation and utilization of river sand along the SGR Phase II A Project in Kenya was rare. This work aims to explore the effect of the machine-made sand rate on the compressive strength, workability, and impermeability of sleeper concrete. The research demonstrated that the 28d compressive strength (46.30 MPa) of sleeper concrete (KD-45) were the optimal when the machine-made sand rate in the sleeper concrete was 45%. These results indicated that the compactness of the KD-45 was improved, leading to an increase in compressive strength. The slump (175 mm) and extensibility (455 mm) of the sleeper concrete KD-45 were the highest as the machine-made sand rate was 45% in the sleeper concrete. This result proved that the workability and flowability of KD-45 pre mixed sleeper concrete were excellent. The electrical flux (840 °C) of KD-45 was lowest when the machine-made sand rate in the sleeper concrete was 45%. This phenomenon demonstrated that the ability of sleeper concrete KD-45 to resist chloride ion penetration was higher compared to other sleeper concrete. These properties meet the requirements of C40 sleeper concrete. Therefore, this work presents novel insight on the performance improvement of sleeper concrete based on the machine-made sand rate.

Keywords Machine-made sand · Sleeper concrete · Compressive strength · Workability · Impermeability

1 Introduction

Kenya was the first stop of the Belt and Road Initiative in East Africa [1]. The Mombasa-Nairobi Railway has been completed and opened to traffic, and the SGR Phase II A Project was along the Mombasa-Nairobi Railway [2]. The SGR Phase II A Project is a flagship project planned for Kenya in 2030. The transportation situation in

---

Z. Liu (✉)
China Road and Bridge Corporation, Beijing 100010, China
e-mail: liuzc@crbc.com

the northern region of Kenya could be improved by the SGR Phase II A Project, while cross-border logistics transportation costs was reduced [3]. Employment and regional economic development in the country also were promoted [4]. The integration process of transportation in the East African region has been promoted by the integration of the SGR Phase II A Project and the Mombasa-Nairobi Railway [5].

Most areas of Kenya were year-round dry, with few rivers and a shortage of river sand [6]. Environmental protection was valued by the Republic of Kenya, and laws prohibiting the exploitation of river sand have been enacted [7]. There was very little river sand mined and utilized along the SGR Phase II A Project, and the use of machine-made sand was an inevitable choice. The application research of machine-made sand in railway concrete mainly focuses on culverts and bridges [8]. However, there were relatively few applications for sleeper concrete. The effect of machine-made sand rate on the compressive strength, workability, and impermeability of sleeper concrete has been investigated to fill the research gap mentioned above.

In this work, sleeper concrete was prepared based on different machine-made sand rate from multistage mechanical crusher. This research aims to determine the optimal performance of the sleeper concrete corresponding to the machine-made sand rate. The effect of machine-made sand rate on the compressive strength, workability, and impermeability of sleeper concrete has also been considered. Therefore, this article could provide a theoretical basis for the production of sleeper concrete.

2 Material and Methods

2.1 Raw Materials

The sleeper concrete was mainly prepared with the following raw materials:

(1) Cement: Kenya Bamburi CEM I 52.5 cement was used. It's the initial and final setting was 155 min and 255 min, respectively. The 3 days (d) and 28 d compressive strength of 52.5 cement were 27.2 MPa and 58.6 MPa, respectively. The specific surface area of cement was 320 m^2/kg.
(2) Fly ash: Grade 1 fly ash was used in this experiment. The fineness of fly ash was 45 microns (10.1%). The water demand ratio of fly ash was 85%, and the combustion loss was 1.45%.
(3) Fine aggregate: The basalt was ground and processed by a multi-stage crusher in this work. Basalt sand was used as the fine aggregate for this work, which sourced from the SGR Phase II A Project. The apparent density of basalt fine aggregate was 2840 kg/m^3, with a water absorption rate of 1.2% and a crushing value of 6%. The fineness modulus of fine aggregate was 2.7, the content of stone powder was 7.3%, and the MB value was 0.8.

(4) Coarse aggregate: 5–20 mm continuous graded gravel was produced from the SGR Phase II A Project, and the lithology was basalt. The base material strength of coarse aggregate was 160 MPa. The crushing value of coarse aggregate was 6%, and the needle content was 3%.
(5) Water reducing agent: Polycarboxylic acid water reducing agent was used, with a solid content of 30% and a water reducing rate of 29.3%.

2.2 Test Method

The preparation and performance experiments of sleeper concrete were executed based on GB/T37330-2019 (Sleeper for ballasted track-Concrete sleeper) [9]. Compressive strength of sleeper concrete was detected by a model Sye-3000 compression testing machine based on the prescribed process of GB/T37330-2019 (Sleeper for ballasted track-Concrete sleeper). Then, the strength of sleeper concrete was calculated from the average of three concretes. Slump and extensibility of sleeper concrete were tested by a standard slump tube and extensibility tester according to GB/T37330-2019 (Sleeper for ballasted track-Concrete sleeper). The electrical flux of the sleeper concrete was detected by an electrical flux meter.

Figure 1 shows the process flow of wet processing for machine-made sand. Firstly, the raw ore after soil removal was roughly crushed from a jaw crusher. Secondly, the crushed ore was transported to the secondary crushing equipment for intermediate crushing. Crushed ore was crushed using a conical crusher. Thirdly, the crushed ore was crushed by an impact crusher. The crushed ore was screened to produce crushed stone products (4.75 ~ 9.5 mm, 9.5 ~ 19 mm, and 19 ~ 26.5 mm) that meet the specifications. Machine-made sand (Particle size < 4.75 mm) was obtained by finely crushing crushed ore using a vertical axis impact crusher (sand making machinery equipment). However, particles larger than 26.5 mm were returned to the secondary crushing equipment.

2.3 Mix Proportion of Sleeper Concrete

Table 1 shows the mix proportions of four types about sleeper concrete. Sleeper concrete was composed of cement, fly ash, sand, gravel, water, and water reducing agents. The machine-made sand rates corresponding to four types of sleeper concrete KD-41, KD-43, KD-45, and KD-47 were 41%, 43%, 45%, and 47%, respectively. Next, the slump, extensibility, electrical flux, and compressive strength of sleeper concrete are investigated in the following chapters.

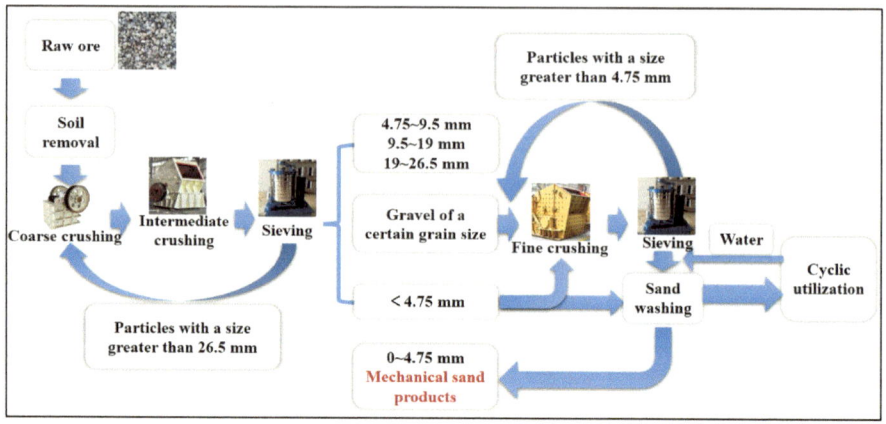

Fig. 1 Process flow diagram of wet processing of machine-made sand

Table 1 Mix proportion of sleeper concrete

No.	Machine-made sand rate (%)	Cement (kg/m³)	Fly ash (kg/m³)	Sand (kg/m³)	Gravel (kg/m³)	Water (kg/m³)	Water reducing agent (kg/m³)
KD-41	41	340	60	752	1083	160	5.20
KD-43	43	300	100	789	1046	160	5.20
KD-45	45	260	140	826	1009	160	5.20
KD-47	47	220	180	862	973	160	5.20

2.4 Experimental Process

Figure 2 shows the experimental procedure and scenarios of this work. First, the machine-made sand was obtained by a jaw crusher. Coarse and fine aggregates are collected. Cement, fly ash, and water reducing agents are used as cementitious materials. Second, the cementitious materials, aggregates, and water were weighed according to Table 1. Third, the cementitious materials, aggregates, and cement were mixed and stirred. Fourth, the slump and expansion of pre mixed concrete were measured. Fifth, pre mixed concrete was formed and cured, which was used to test compressive strength and electrical flux after hardening.

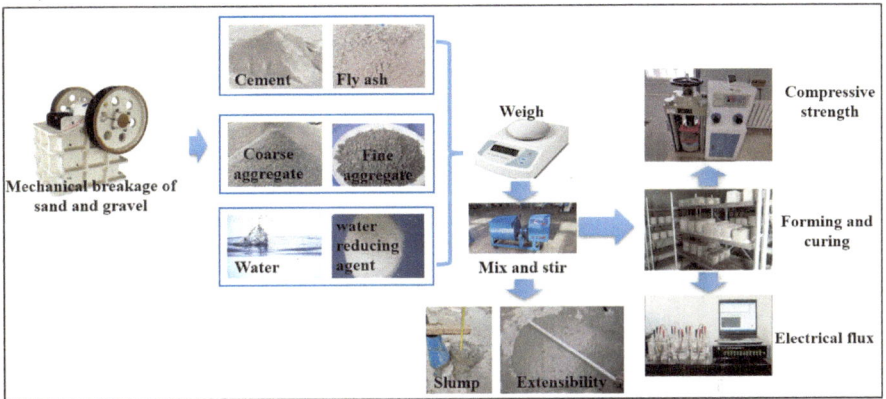

Fig. 2 Experimental procedure and scenarios

3 Results and Discussion

3.1 Slump and Extensibility

The slump of sleeper concrete mainly depends on its plasticizing performance and pumpability. The diffusivity of concrete was the flow ability of flowing concrete in a naturally accumulated state. Figure 3 shows the results of slump and extensibility for four types (KD-41, KD-43, KD-45, and KD-47) of sleeper concrete samples. It can be clearly seen that the slump and extensibility values of the sleeper concrete first increased and then decreased as the machine-made sand rate in the sleeper concrete from 41 to 47%. Among them, the slump (175 mm) and extensibility values (455 mm) of KD-45 sleeper concrete were the highest when the machine-made sand rate was 45% in the sleeper concrete. This result indicated that the workability and flowability of KD-45 pre mixed sleeper concrete were optimal. The reason for this phenomenon was that the total specific surface area of the aggregate was suitable as the machine-made sand rate was 45% in the sleeper concrete. Cement slurry on the aggregate has a certain thickness, which plays a good lubrication role.

The permeability value of sleeper concrete could be evaluated by measuring the voltage value and cumulative electrical flux value of the sleeper concrete. The evaluation method for the resistance of sleeper concrete to chloride ion penetration, abbreviated as the electrical quantity method. Figure 4 showed the electrical flux results of four types (KD-41, KD-43, KD-45, and KD-47) of sleeper concrete. Overall, it could be seen that the electrical flux of the sleeper concrete first decreased and then increased as the machine-made sand rate in the sleeper concrete increased from 41 to 47%. Among them, the electrical flux of the KD-45 sleeper concrete was the lowest as the machine-made sand rate was 45% in the sleeper concrete. This phenomenon proved that the ability of sleeper concrete KD-45 to resist chloride ion penetration was higher when the machine-made sand rate in sleeper concrete was 45%, compared

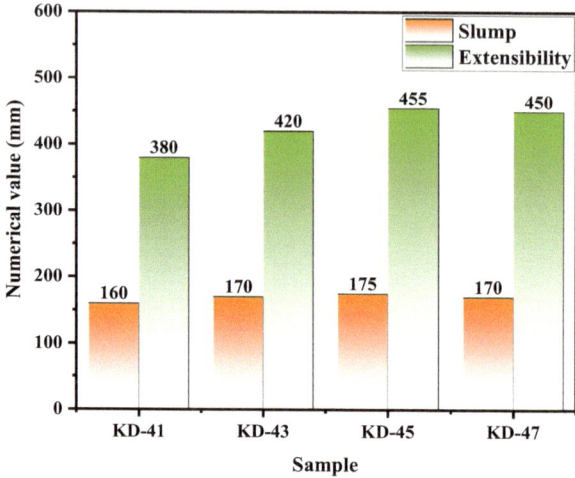

Fig. 3 Slump and extensibility of sleeper concrete

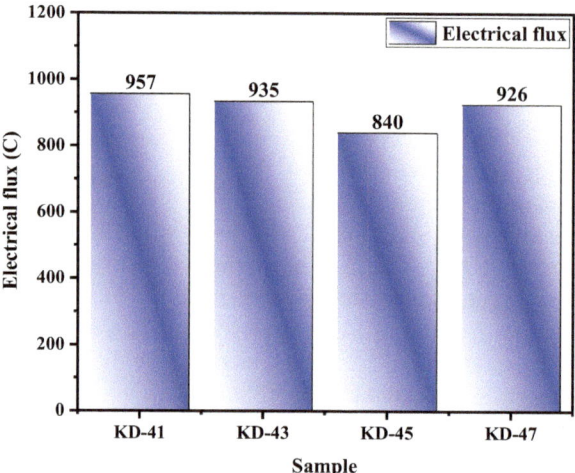

Fig. 4 Electrical flux of sleeper concrete

to other sleeper concrete. The reason for this result was that the porosity of the sleeper concrete KD-45 was reduced by an appropriate machine-made sand rate (45%), and the compactness and resistance to chloride ion penetration of the sleeper concrete were improved.

3.2 Compressive Strength

The compressive strength of sleeper concrete was one of the important indicators to measure its qualification [10]. Figure 5 shows the compressive strength of sleeper

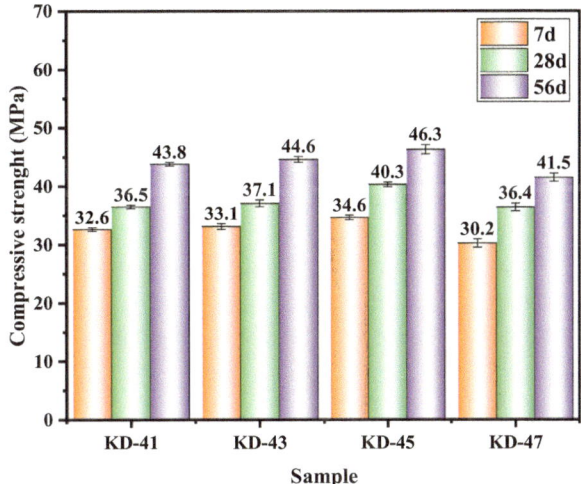

Fig. 5 Compressive strength of sleeper concrete based on different machine-made sand rates

concrete with different machine-made sand rates at 3 days (d), 7d, and 28d. From Fig. 5, it could be clearly seen that the compressive strength of sleeper concrete first increased and then decreased with the machine-made sand rate increased from 41 to 47% in the sleeper concrete. Thereinto, the compressive strength of sleeper concrete KD-45 was the highest as the machine-made sand rate was 45% in the sleeper concrete, and its compressive strengths was 34.6 MPa, 40.3 MPa, and 46.3 MPa at 3d, 7d, and 28d, respectively. The reason for this phenomenon was that the ability of sleeper concrete KD-45 to resist deformation was improved by the synergistic effect of cement, fly ash, machine-made sand rate, crushed stone, water, and water reducing agents.

4 Conclusions

In this work, the basalt ore was gradually crushed into less than 4.75 mm by a mechanical crusher, which promoted the density improvement of the sleeper concrete. The slump, extensibility, electrical flux, and compressive strength of sleeper concrete were investigated in different machine-made sand rates of sleeper concrete. The relevant conclusions were as follows:

(1) The slump (175 mm) and expansion values (455 mm) of KD-45 sleeper concrete were the highest as the machine-made sand rate was 45% in the sleeper concrete. This result indicated that the workability and flowability of KD-45 pre mixed sleeper concrete were optimal. This performance was beneficial for the construction of sleeper concrete on the production line.

(2) The electrical flux of the sleeper concrete KD-45 was the lowest as the machine-made sand rate was 45% in the sleeper concrete. This phenomenon proved that the ability of sleeper concrete KD-45 to resist chloride ion penetration was higher compared to other sleeper concrete. The impermeability of KD-45 sleeper concrete was strengthened in strong wind and fog environments.

(3) The compressive strength of sleeper concrete KD-45 was the highest with the machine-made sand rate was 45% in the sleeper concrete, and its compressive strengths was 34.6 MPa, 40.6 MPa, and 46.3 MPa at 3d, 7d, and 28d, respectively. The reason for this phenomenon was that the ability of sleeper concrete KD-45 to resist deformation was improved by the synergistic effect of cement, fly ash, machine-made sand rate, crushed stone, water, and water reducing agents.

References

1. Cannon BJ, Mogaka S (2022) Rivalry in East Africa: the case of the Uganda-Kenya crude oil pipeline and the East Africa crude oil pipeline. Extr Ind Soc 11:101102. https://doi.org/10.1016/j.exis.2022.101102
2. Zhang W, Gu J, Zhou X, et al (2021) Circulating fluidized bed fly ash based multi-solid wastes road base materials: hydration characteristics and utilization of SO_3 and f-CaO. J Cleaner Prod 316:128355. https://doi.org/10.1016/j.jclepro.2021.128355
3. Li S, Cao X, Liao W, et al (2020) Factors in the sea ports-of-entry and road ports-of-entry cross-border logistics route choice. J Transp Geogr 84:102689. https://doi.org/10.1016/j.jtrangeo.2020.102689
4. FokamDNDT, Kamga BF, Nchofoung TN (2023) Information and communication technologies and employment in developing countries: effects and transmission channels. Telecommun Policy 102597. https://doi.org/10.1016/j.telpol.2023.102597
5. Muhammad S, Ahmad M, Toufic M et al (2023) Policy and economic challenges towards scalable green-H2 transition in the middle east and north Africa region. Int J Hydrogen Energy. https://doi.org/10.1016/j.ijhydene.2023.05.083
6. Crystele L, Stéphanie D, Olivier H et al (2013) Floods and livelihoods: the impact of changing water resources on wetland agro-ecological production systems in the Tana River Delta, Kenya. Glob Environ Chang 23(1):252–263. https://doi.org/10.1016/j.gloenvcha.2012.09.003
7. Wei Z, Xiaoming L, Yaguang W et al (2021) Binary reaction behaviors of red mud based cementitious material: hydration characteristics and Na^+ utilization. J Hazard Mater 410:124592. https://doi.org/10.1016/j.jhazmat.2020.124592
8. Marko H, Luca B, Sami K (2022) Receiver sand mitigation measures along railways: CWE-based conceptual design and preliminary performance assessment. J Wind Eng Ind Aerodyn 228:105109. https://doi.org/10.1016/j.jweia.2022.105109
9. GB/T 37330–2019 (2019) Ballast track sleeper concrete. https://www.doc88.com/p-6864785742728.html
10. Zhang W (2023) Stabilization and synergistic utilization of unstable components in circulating fluidized bed fly ash-blast furnace slag-red mud based cementitious materials. https://doi.org/10.26945/d.cnki.gbjku.2023.000270

Open Access This chapter is licensed under the terms of the Creative Commons Attribution 4.0 International License (http://creativecommons.org/licenses/by/4.0/), which permits use, sharing, adaptation, distribution and reproduction in any medium or format, as long as you give appropriate credit to the original author(s) and the source, provide a link to the Creative Commons license and indicate if changes were made.

The images or other third party material in this chapter are included in the chapter's Creative Commons license, unless indicated otherwise in a credit line to the material. If material is not included in the chapter's Creative Commons license and your intended use is not permitted by statutory regulation or exceeds the permitted use, you will need to obtain permission directly from the copyright holder.

Comparative Study of Tensile Capacity Testing Methods for Metal Connectors Used in Precast Concrete Insulated Sandwich Wall Panels

Puyan Wang, Feng Tu, Kai Shu, Yi Zhao, Weijun Zhong, and Jian Zhou

Abstract Metal connectors are important load-bearing components in precast concrete insulated sandwich wall panels, and their tensile capacity need to be determined through testing. Due to the variety of connector styles and different anchoring structures, this article focuses on the commonly used single-pull and double-pull test methods, using typical specifications of pin connectors and flat connectors for tensile testing, and considering the influence of different constraints on the single-pull test method. By comparing the failure modes, tensile strains and tensile capacity of the connectors, the study shows that using the double-pull test method to test metal connectors results in more stable and safer test results, and it is recommended to prioritize this method.

Keyword Precast concrete insulated sandwich wall panel · Metal connector · Tensile capacity · Experimental method

1 Introduction

Precast concrete insulated sandwich wall panels are important elements in prefabricated buildings, and the metal connectors that connect the inner and outer wythes of the insulated sandwich wall panels are important accessories [1]. Metal connectors mainly made of stainless steel are widely used due to their good bearing performance and durability. Several typical specifications of metal connectors and anchoring structures are shown in Fig. 1. It is evident that there are significant differences in size and anchoring structure among different metal connectors. Some connectors even have different anchoring structures on the inner and outer wythes of the panel.

P. Wang · F. Tu · K. Shu · W. Zhong
Ningbo Power Supply Company, State Grid Zhejiang Electric Power Co., Ltd., Ningbo 315020, China

Y. Zhao · J. Zhou (✉)
China Academy of Building Research, Beijing 100013, China
e-mail: zhoujian@cabrtech.com

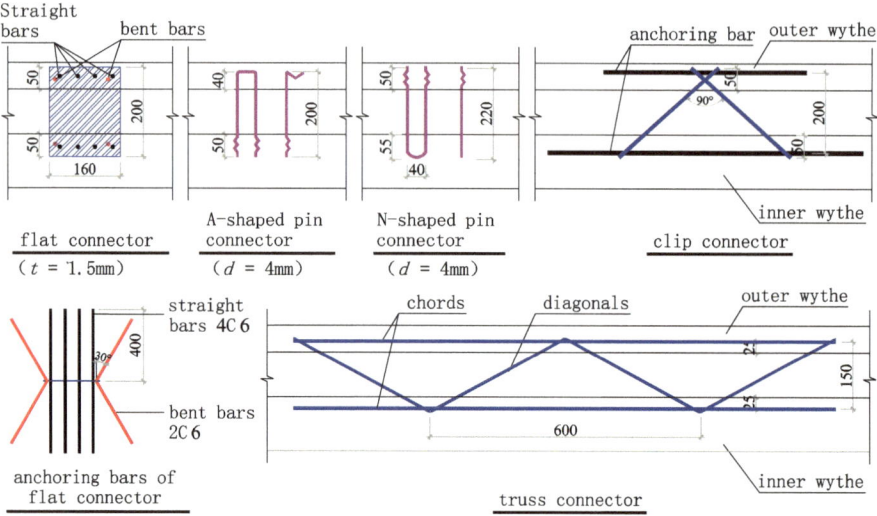

Fig. 1 Several typical specifications of metal connectors

Under wind load and out-of-plane seismic actions, the connectors will bear tensile forces, oh which the characteristic value of the tensile capacity of the connectors should be determined through tests, with a safety factor of not less than 95% [2]. Common testing methods for connectors include single-pull testing and double-tensile testing, which differ based on the test specimen shape and testing apparatus used [3]. The specimen used in the single-pull test is anchored to the concrete slab at one end of the connector, and the other end is a clamping end. The test device generally consists of a jack loading system and a reaction support frame, as shown in Fig. 2. The single-pull test method has the advantages of simple specimen form and easy operation, and is widely used for pull-out testing of anchor bolts and other anchors [4]. Meng et al. [5] and Jiang et al. [6] also used a single pull test method in the tensile testing of U-shaped connectors and rod-like connectors. However, the setting of the reaction support frame will inevitably impose different degrees of constraints on the concrete slab, which may affect the test results. In order to make the concrete in an unconstrained state, ASTM E488/E488M-18 *Standard Test Methods for Strength of Anchors in Concrete Elements* [7] requires that the net distance between the reaction support frames should not be less than four times the effective embedment depth of the anchor, which is difficult to achieve for connectors with a large anchoring range, and requires the preparation of multiple support frame specifications. On the other hand, in order to achieve non-constrained conditions, the distance between supports is generally large, and the concrete slab is equivalent to simply supported. The maximum bending moment and deflection occur at the point of concentrated force, and the concrete may crack under tension, which weakens the constraint. In addition, when the anchoring structures at both ends of the connector are different, tensile tests need to be conducted on both sides separately.

Fig. 2 Single-pull testing apparatus

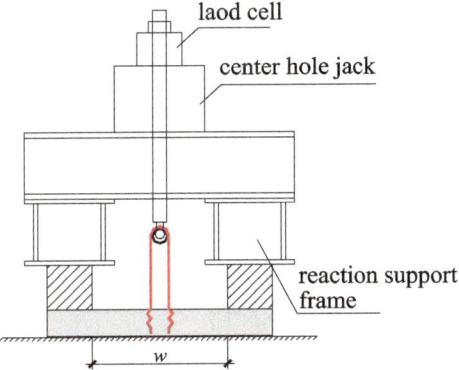

In the specimen used in the double-pull test method, both ends of the connector are anchored in the concrete slab, and clamping fixtures are embedded outside the concrete slab for loading, as shown in Fig. 3, which avoids the use of reaction support frames and ensures that the concrete is in an unconstrained state. This method has been widely used in tensile testing of rod connectors [8, 9], clip connectors [10], hook-type connectors [11], and truss connectors [12]. The double-pull test method may have the influence of introducing additional moments or even shear forces due to the misalignment between the clamped reinforcement and the axis of the connector. In addition, during transportation and storage, the specimen may apply premature pressure and shear forces to the connectors, which may affect the test results.

Currently, there is no consensus on the preferred method of tensile testing for connectors. Therefore, this article takes the typical specification A-shaped pin

Fig. 3 Form of double-pull test specimens

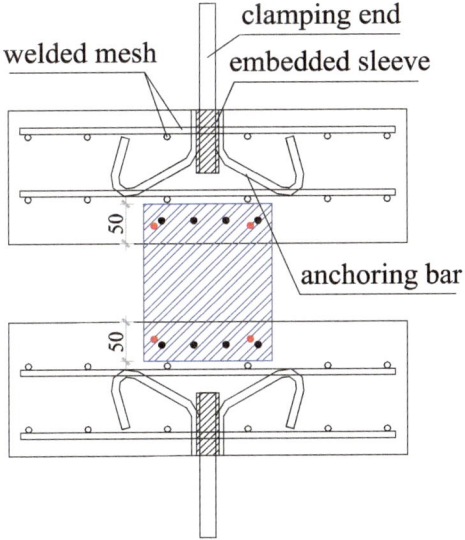

connector, N-shaped pin connector, and flat connector shown in Fig. 1 as the objects, and conducts tensile tests with different restraint methods and testing methods to verify the applicability of the two testing methods and provide a basis for standardizing the testing methods.

2 Experimental Plan

In this test, three types of connectors with specifications shown in Fig. 1, including A-shaped pin connector, N-shaped pin connector, and flat connector, were selected, and the minimum value of the anchoring depth was taken as reference from the product technical manual [13].

2.1 Single-Pull Test

The specimen parameters are shown in Table 1. The concrete slab is available in three forms: outer wythe, inner wythe, and widened and thickened outer wythe, all of which have a square cross-section. The connectors are centrally arranged. The typical cross-sectional dimensions of the specimens are shown in Fig. 4. The concrete strength grade is C30, and single-layer bidirectional reinforcement is provided in the outer wythe, and double-layer bidirectional reinforcement is provided in the inner wythe. All adopt HRB400 steel bars with a diameter of 6 mm. Two specimens of each type were made, and different reaction restraint methods were used during loading, resulting in a total of 12 specimens.

The test device is arranged according to Fig. 2. The reaction support frame may differ depending on the specific method of reaction restraint used. A displacement gauge is set up to measure the vertical relative displacement between the anchor and concrete slab. Strain gauges are placed along the direction of tension at a distance of 20 mm from the surface of the concrete for both the rod type and plate type anchors. The loading speed is controlled at 0.1 kN/s, and the load is applied until the specimen fails.

2.2 Double-Pull Test

The specimen parameters are shown in Table 2, and the specimen construction is shown in Fig. 3. Two specimens of each type were made, resulting in a total of six specimens. The concrete strength grade of all specimens was C30.

Install a displacement sensor in the center of each side of the specimen to measure the relative displacement between the two concrete slabs. Attach strain gauges along the direction of the tensile force at the exposed position of the tension member to

Table 1 Single-pull test specimen parameters

Specimen No.	Connector type	Anchorage	Concrete slab dimensions (mm)		Reaction restraint	w^b (mm)
			Side length	Thickness		
ZAT-E-1	A-shaped pin connector	Outer wythe	500	60	Fixed at four corners	500
ZAT-E-2	A-shaped pin connector	Outer wythe	500	60	Fixed by loading ring	330
ZAT-I-1	A-shaped pin connector	Inner wythe	500	100	Fixed at four corners	500
ZAT-I-2	A-shaped pin connector	Inner wythe	500	100	Fixed by loading ring	330
ZNT-E-1	N-shaped pin connector	Outer wythe	500	60	Fixed at four corners	500
ZNT-E-2	N-shaped pin connector	Outer wythe	500	60	Fixed by loading ring	330
ZNT-I-1	N-shaped pin connector	Inner wythe	500	100	Fixed at four corners	500
ZNT-I-2	N-shaped pin connector	Inner wythe	500	100	Fixed by loading ring	330
BT-E-1	Flat connector	Outer wythe	500	60	Fixed at opposite sides	250
BT-E-2	flat connector	Outer wythe	500	60	Fixed at four corners	300
BT-E-3	flat connector	Outer wythe (wider and thicker)[a]	500 + 200 × 2	60 + 100	Fixed at four corners	550
BT-E-4	flat connector	Outer wythe (wider and thicker)[a]	500 + 200 × 2	60 + 100	Fixed by loading ring	560

[a] This set of specimens added 200 mm width and 160 mm thickness of concrete around the ordinary-sized concrete slab, as shown in Fig. 1
[b] w refers to the net distance at the bottom of the reaction support frame, as shown in Fig. 2

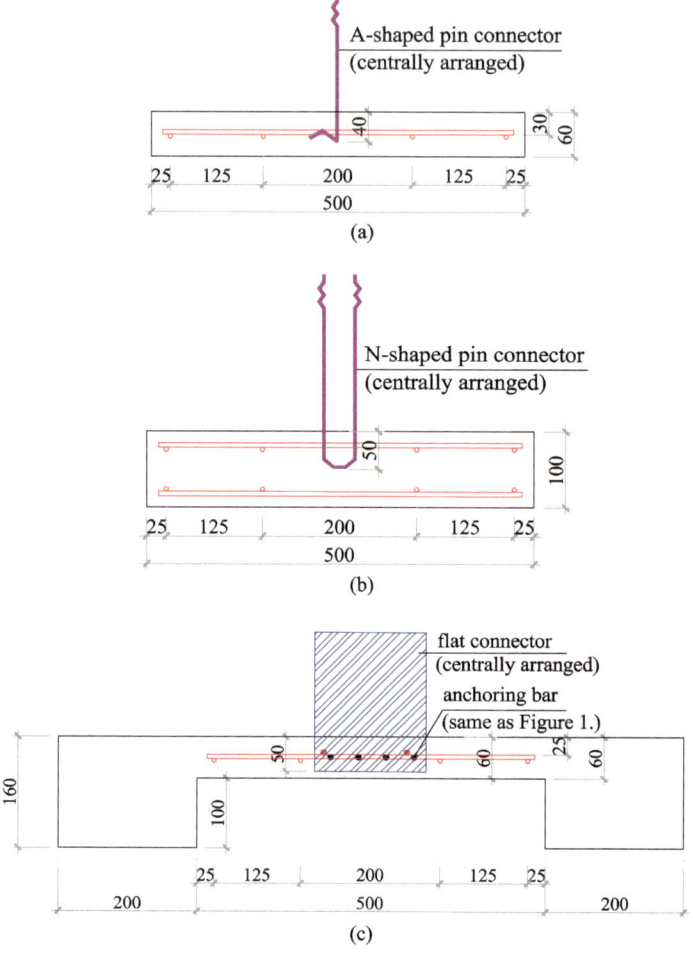

Fig. 4 Cross-sectional view of a single-pull test specimen with connectors

Table 2 Parameters of double-pull test specimens

Specimen No.	Connector type	Concrete slab dimensions (mm)	
		Side length	Thickness
ZAT-B-1	A-shaped pin connector	300	170
ZAT-B-2			
ZAT-I-1	N-shaped pin connector	300	170
ZAT-I-2			
ZNT-B-1	Flat connector	500	170
ZNT-B-2			

Fig. 5 Double-pull testing apparatus for connectors

measure the strain during the tensile process of the tension member. The arrangement of the double-tension test device and the displacement meter is shown in Fig. 5. The loading speed of the test is controlled at 0.1kN/s, and loaded until the specimen is destroyed.

2.3 Material Properties

According to the GB/T 228.1-2021 *Metallic materials–Tensile testing–Part 1: Method of test at room temperature* [14], mechanical property tests were carried out on the stainless steel tension members used, and it was found that the needle-type tension member had a specified plastic elongation strength $R_{p0.2}$ of 692 MPa, tensile strength R_m of 861 MPa, and elongation A of 45% after fracture; the plate-type tension member had a specified plastic elongation strength $R_{p0.2}$ of 576 MPa, tensile strength R_m of 841 MPa, and elongation A of 41% after fracture. All of the results met the requirements of the product standard.

3 Test Results

3.1 Failure Mode

The statistical results of the failure modes of each specimen are shown in Table 3, and typical photos of the failure modes are shown in Fig. 6. According to Table 3 and Fig. 6:

(1) The anchoring structure of the A-shaped needle-type tension member on the inner and outer wythes differ significantly, and the failure mode is also different. The tension member failed in all inner wythe specimens, but the concrete at the root of the tension member was basically undamaged. The failure mode of the

Table 3 Failure modes and peak loads and displacements of specimens

Connector type	Specimen No.	Failure mode	Peak point Load (kN)	Peak point Displacement (mm)	Average peak load (kN)
A-shaped pin connectors	ZAT-E-1	Concrete surface layer overall pullout	16.8	7.8	16.8
	ZAT-E-2	Rod weld fracture	16.8	9.3	
	ZAT-I-1	Rod weld fracture	19.7	10.1	19.4
	ZAT-I-2	Rod weld fracture	19.0	5.4	
	ZAT-B-1	Outer wythe concrete failure	12.4	6.0	13.7
	ZAT-B-2	Outer wythe concrete failure	15.0	4.2	
N-shaped pin connectors	ZNT-E-1	Concrete plug local pullout	14.6	6.4	17.2
	ZNT-E-2	Rod pullout	19.8	5.5	
	ZNT-I-1	Rod weld fracture	20.7	6.8	20.5
	ZNT-I-2	Rod weld fracture	20.2	11.0	
	ZNT-B-1	Inner wythe concrete failure	21.2	7.6	21.7
	ZNT-B-2	Rod fracture	22.2	26.5	
Flat connectors	BT-E-1	Concrete failure	41.2	7.4	40.0
	BT-E-2	Concrete failure	38.7	9.1	
	BT-E-3	Concrete failure	50.4	10.1	55.0
	BT-E-4	Concrete failure	59.6	3.2	
	BT-B-1	Concrete failure	43.4	1.1	44.4
	BT-B-2	Concrete failure	45.4	1.9	

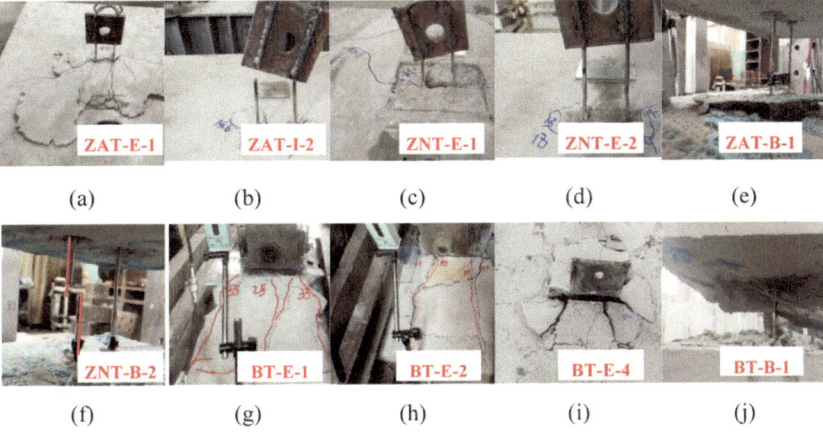

Fig. 6 Typical failure modes of specimens

outer wythe specimens was affected by the constraint method. When the support net distance was large, the surface concrete was pulled out integrally because there was a transverse bend hook at the anchorage end of the A-shaped needle-type tension member on the outer wythe, but when the support net distance was small, the concrete was not damaged by the constraining force, and the tension member was broken. The double-tension specimens also had concrete anchorage failure on the outer wythe.

(2) The tension member failed in all N-shaped needle-type tension member inner wythe specimens, but the concrete at the root of the tension member was undamaged. The failure mode of the outer wythe specimens was affected by the constraint method. When the support net distance was large, the concrete truncated cone around the tension member was partially pulled out, and when the support net distance was small, the bent section of the tension member was pulled out from the concrete. The double-tension specimens experienced either concrete failure or tension member breakage, and there was no concrete failure on the relatively weakly anchored outer wythe. This is partly because the concrete slab was subjected to bending moment during single-tension testing, which made it easier to crack, and also because of the discrete nature of the concrete strength.

(3) The failure mode of the single-tension specimen for the plate-type tension member was concrete failure, and the range of concrete failure and cracking path were directly related to the constraint boundary. Both double-tension specimens of the plate-type tension member experienced concrete failure, and the anchoring bars of the tension member lifted the concrete as a whole.

3.2 Load-Strain Curves

Except for the strain gauge damage on specimen ZAT-E-1, the load-strain curves before the failure of the remaining specimens are shown in Fig. 7. According to the figure:

(1) The strain curve of Type A pin fastener is basically identical to the strain curve of metal tension without yielding point. The plastic strain before the failure of the specimen can reach 2000με, and the strain curve of the inner wythe single tensile specimen can enter the platform section, while the strain curve of the other specimens is basically in the elastic stage.

(2) The strain curve of Type N pin fastener is basically identical to the strain curve of metal tension without yielding point. The plastic strain before the failure of the specimen can reach 2000 με, the strain curve of the outer plate single

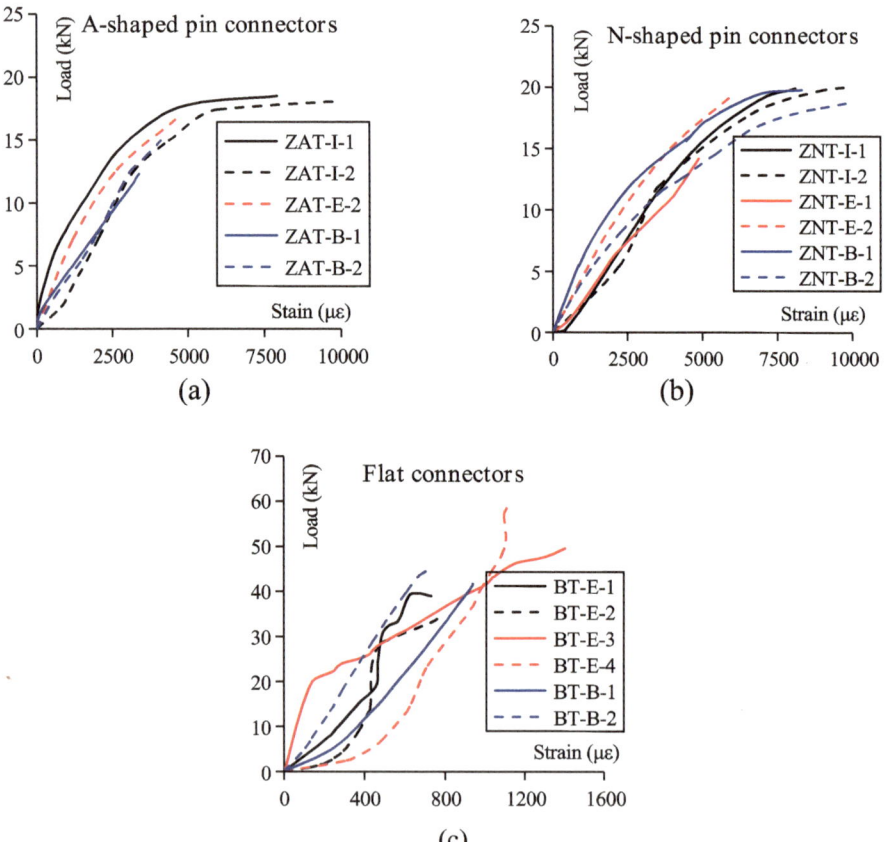

Fig. 7 Load-strain curves for each specimen

tensile specimen is basically in the elastic stage, and the strain curve of the other specimens can enter the platform section.
(3) Due to the influence of the installation deviation of the fastener and strain gauge, the strain curve of the plate-type fastener is relatively discrete, but it is obviously visible that all plate-type fasteners are in the elastic stage before the failure of the specimens.

3.3 Load–displacement Curves and Tensile Capacity

The load–displacement curves of each specimen before failure are shown in Fig. 8 and the peak load and displacement of each specimen are shown in Table 3. Based on Fig. 8 and Table 3:

(1) In the A-shaped pin-shaped tensile connector specimens, the peak load of the double tensile test is significantly lower than that of the single tensile test, indicating that the bearing capacity of the A-shaped pin-shaped tensile connector obtained by using the single tensile test method may be unsafe.
(2) In the N-shaped pin-shaped tensile connector specimens, except for the ZNT-E-1 specimen where the outer wythe failed, the measured bearing capacity of all other specimens was relatively close, approximately equivalent to 0.9 times the tensile capacity of the connector parent material.
(3) In the plate-type tensile connector specimens, the bearing capacity measured from the widened and thickened outer wythe specimens was significantly higher than that of other specimens, and the constraint method had a significant impact on the test results, which is not recommended. The bearing capacity obtained from the double tensile specimens was similar to that of the ordinary single tensile specimens of the outer plate, with an average increase of 11%, as the concrete slab in the single tensile test was more prone to failure due to additional bending moment.
(4) Overall, the results of the double tensile tests were relatively stable, and the measured bearing capacity tended to be safe.

4 Result Analysis

Based on the test results and discussions presented above, the conclusions are obtained as below:

(1) The test results of Type A pin pull connectors are greatly affected by the method of constraint and the test device. The double-pull test method can reflect its real failure mode, and the measured bearing capacity is biased towards safety.
(2) The test results of Type N pin pull connectors are less affected by the method of constraint and the test device, and the bearing capacity measured by the double-pull test is relatively stable.

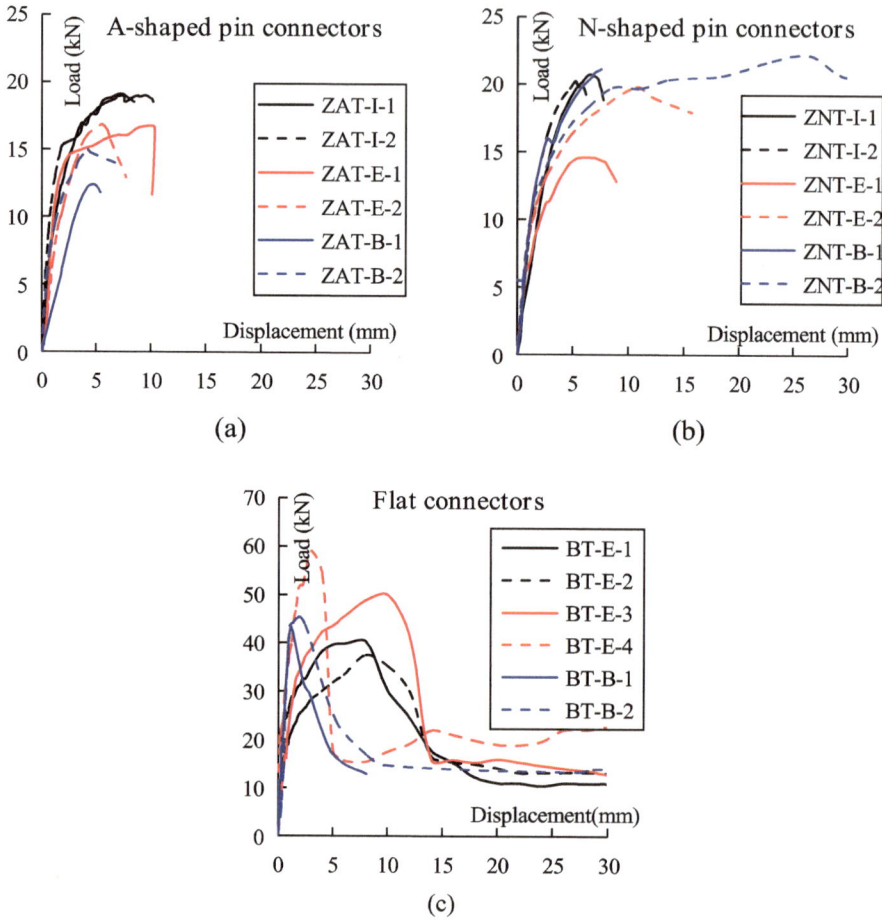

Fig. 8 Load–displacement curves for each specimen

(3) The failure modes of the plate-type pull connectors are consistent under both test methods, and the test results are also relatively close. The bearing capacity measured by widening and thickening the specimen is biased towards unsafety, so it is not recommended to use this method.

(4) There are various types of metal pull connectors, and their anchoring structures are also different. Based on the test results in this article, it is recommended to use the double-pull test method as a priority when testing the tensile capacity of metal connectors.

Acknowledgements This work was financially supported by the Science and Technology Program of State Grid Zhejiang Electric Power Co., Ltd (5211NB21N005).

References

1. Wu XC, Li JJ, Wang PS et al (2021) Review of research and application of precast concrete sandwich panels. Build Struct 51(S2):1197–1202
2. Beijing municipal commission of housing and urban-rural development (2023) Technical specification for application of precast concrete insulated sandwich wall panels: DB11/T 2128-2023, Beijing
3. Li Y, Hu X, Gu S et al (2018) Review of testing methods for FRP connectors in precast concrete sandwich insulation wall panels. Constr Technol 47(12):87–91
4. Wu SN, Li C (2014) Tensile test study of screw anchors in concrete. Struct Eng 30(1):149–154
5. Meng XH, Zhou AL, Liu HC et al (2014) (2014) Experiments of mechanical properties on the connectors of sandwich insulation wallboard. J Shenyang Jianzhu Univ (Naural Sci) 30(02):227–234
6. Jiang QJ, Wu HJ, Qi CC et al (2018) Study on mechanical properties of non metallic connector for non combination sandwich facade panel. Archit Technol 49(01):72–76
7. ASTM international(2018) Standard test methods for strength of anchors in concrete elements: ASTM E488/E488M-18, West Conshohocken
8. Xue WC, Zhang S, Su RJ et al (2018) Test study on pull-out behavior of rod-like stainless steel connectors in precast sandwich insulation wall panels. Constr Technol 47(12):92–94
9. Yang JL, Xue WC, Shu X (2013) Mechanical properties test of FRP connectors in precast sandwich insulation wall panels. J Jiangsu Univ (Nat Sci Edition) 34(06):723–729
10. Lyu AA, Zhang SW, Zhao ZG et al. (2020) Analysis on mechanical performance of metal clamp connectors on precast concrete sandwich wall with ultra-low energy consumption. Chin Concr Cem Prod (05):61–65+91
11. Jiang HZ, Guo ZX, Shen F et al (2017) Pull-out tests for evaluations of anti pulling behavior of hook-typesteel core composite connector in precast concrete sandwich wall panels. Constr Technol 46(02):124–128
12. Xue WC, Jiang WQ, Song JZ et al (2018) Experimental research on pull-out and shearing capacity of stainless stee truss connectors in precast concrete sandwich insulation facade wall panels. Constr Technol 47(12):95–99
13. Deutsches Institut für Bautechnik (2019) Halfen FA flachanker, Berlin
14. State administration for market regulation (2021) Metallic materials–Tensile testing–art 1: Method of test at room temperature: GB/T 228.1—2021. China Standard press, Beijing

Open Access This chapter is licensed under the terms of the Creative Commons Attribution 4.0 International License (http://creativecommons.org/licenses/by/4.0/), which permits use, sharing, adaptation, distribution and reproduction in any medium or format, as long as you give appropriate credit to the original author(s) and the source, provide a link to the Creative Commons license and indicate if changes were made.

The images or other third party material in this chapter are included in the chapter's Creative Commons license, unless indicated otherwise in a credit line to the material. If material is not included in the chapter's Creative Commons license and your intended use is not permitted by statutory regulation or exceeds the permitted use, you will need to obtain permission directly from the copyright holder.

Basic Properties and Microstructure of Coal Gangue Pervious Concrete Under Acid Rain Environment

Junwu Xia, Linli Yu, Zhichun Zhu, Pengxu Li, Yuan He, and Jun Yu

Abstract To systematically investigate the durability performance of coal gangue pervious concrete (CGPC) under acid rain environment, various factors are taken into account, including designed porosity, acid rain pH value, and erosion time. These factors affect the mass, compressive strength, and permeability coefficient of CGPC. Moreover, the research delves into the erosion mechanism of acid rain. The results indicate that the compressive strength of CGPC shows an initial increase followed by a subsequent decrease under acid rain environment. The permeability coefficient slightly decreases in the early stage of acid rain erosion, followed by an increase in the later stage. In the acid rain, the reaction between SO_4^{2-} and Ca^{2+} generates expansive products such as gypsum. Initially, these products fill the pores, increasing the structural density. However, as erosion progresses, the continuous accumulation of gypsum leads to the formation of expansive microcracks, resulting in a reduction in strength.

Keywords Coal gangue pervious concrete · Acid rain · Compressive strength · Permeability · Microstructure

---

J. Xia
Jiangsu Design Institute of Geology for Mineral Resources (Testing Center, CNACG), Xuzhou 221006, China

CNACG Key Laboratory of Mineral Resources in Coal Measures, Xuzhou 221006, China

J. Xia · L. Yu (✉) · Z. Zhu · P. Li · Y. He · J. Yu
China University of Mining and Technology, Xuzhou 221116, China
e-mail: cumt_yulinli@cumt.edu.cn

L. Yu
Jiangsu Collaborative Innovation Center of Building Energy-Saving and Construction Technology, Jiangsu Vocational Institute of Architectural Technology, Xuzhou 221116, China

1 Introduction

Coal gangue is a solid waste generated during the process of coal mining and processing [1]. It has been pilled up to approximately 7 billion tons, causing land occupation and environmental pollution [2]. Thus, the utilization of coal gangue in situ has become an urgent issue [3, 4]. Concurrently, based on the concept of sponge city, the use of pervious concrete for road paving presents an opportunity for the recycling of water resources and alleviation of urban waterlogging [5]. Therefore, employing coal gangue as an aggregate to prepare pervious concrete and applying it to permeable road in mining areas has significant social and economic benefits. In recent years, the concern about acid rain has grown. Atmospheric precipitation contains anions such as SO_4^{2-}, NO_3^-, and Cl^-, as well as cations like NH_4^+, Ca^{2+}, Na^+, K^+, and Mg^{2+}, which can impact the performance of pervious concrete road surface [6]. Xie et al. [7] compared the cyclic immersion method with the spray test method, found out cyclic immersion method is deemed more suitable for laboratory experiments. Gao et al. [8] simulated acid rain solution's erosion on pervious concrete through dry–wet cycles. After 12 cycles, the compressive and flexural strength of the specimens decreased by 30.7% and 40.8%, respectively. The addition of fly ash and silica fume enhanced the acid rain erosion resistance of pervious concrete [9, 10]. Zhou et al. [11] suggested that under acid rain environment, concrete experiences both H^+ corrosive erosion and SO_4^{2-} expansive erosion simultaneously, causing damage to pervious concrete. Unlike ordinary sand and gravel, the durability performance of coal gangue pervious concrete (CGPC) under acid rain environment requires systematic investigation. Therefore, this study investigates the variations in mass, compressive strength, and permeability coefficient of CGPC under acid rain environment. Additionally, the research delves into the microscopic erosion mechanisms of acid rain on CGPC.

2 Test Materials and Methods

2.1 Raw Materials

CGPC is prepared by coal gangue coarse aggregate, cement, water, and permeable agent. The cement used is P.O 52.5 ordinary Portland cement with a density of 3.1 g/cm^3, meeting the Chinese standard GB175-2020, produced by Yangchun Cement Co., Ltd. in Zhucheng City, Shandong Province. The coal gangue used is from Zhangshuanglou Coal Mine in Xuzhou City, Jiangsu Province. After washing, crushing, and sieving, coal gangue coarse aggregate with a particle size of 9.5 ~ 16 mm is obtained. The coal gangue is mainly composed of SiO_2 and Al_2O_3, as shown in Table 1, tested by X-ray fluorescence. Basic properties of the coal gangue are determined according to the standard GB/T14685-2011, including a bulk density of 1405 kg/m^3, an apparent density of 2818 kg/m^3, and a water absorption rate of

Table 1 Chemical components of coal gangue coarse aggregate

Chemical component	SiO₂	Al₂O₃	CO₃	Fe₂O₃	K₂O	CaO	Na₂O	MgO	Others
Content (%)	61.69	19.11	5.83	4.16	3.04	2.35	2.28	0.64	0.90

Table 2 Physical and mechanical performance parameters of permeable agent

Performance index	Moisture content (%)	Fineness (%)	pH	Total alkali content (%)	SiO₂ (%)	Cl⁻ (%)
	0.45	2.32	9.50	4.77	70.03	0.25

2.68%. The permeable agent is from Jiangsu Jiajing Ecological Engineering Technology Co., Ltd., serving as a reinforcement agent. The physical and mechanical properties of the agent are presented in Table 2. The water used in the experiment is laboratory tap water.

The acid rain erosion solution is prepared by NaCl, $(NH_4)_2SO_4$ and HNO_3. $(NH_4)_2SO_4$ and NaCl are produced by Xilong Scientific Co., Ltd., and HNO_3 used is a standard dilute nitric acid solution.

2.2 Mix Proportion and Preparation of Specimens

The basic proportion of the CGPC include a coarse aggregate particle size of 9.5–16 mm, a w/c ratio of 0.29, and permeable agent dosage of 3.2%. The experiment focuses on three key factors: acid rain erosion pH values (3 and 4), designed porosity (18%, 20%, and 22%), and erosion duration (0 days, 30 days, 60 days, and 90 days). The study aims to analyze how these factors collectively influence the fundamental properties of CGPC under acid rain environment. A comprehensive experimental design involves 24 groups in a full factorial experiment. The nomenclature employs "A" to denote the pH value of the acid rain erosion solution and "P" for the designed porosity. For example, "A3P18-90d" signifies the erosion of CGPC with a designed porosity of 18% in an acid rain solution with a pH value of 3 over a duration of 90 days.

Given the designed porosity, w/c ratio, and coarse aggregate density, the mix proportion of CGPC can be determined by substituting these values into the appropriate formula:

$$\frac{M_G}{\rho_g} + \frac{M_C}{\rho_c} + \frac{M_W}{\rho_w} + P = 1 \tag{1}$$

$$M_G = \alpha \cdot \rho_G \tag{2}$$

where M_G, M_C, M_W is the mass of coal gangue coarse aggregate, cement and water per unit volume (kg/m³), respectively; ρ_g, ρ_c, ρ_w is apparent density of coal gangue coarse aggregate, cement, and water (kg/m³), respectively; P is the designed porosity(%); α is the correction factor for coarse aggregate, a dimensionless parameter, typically taken as 0.98; ρ_G is the compacted density of coal gangue coarse aggregate (kg/m³).

After calculating the mix proportions for different designed porosity, CGPC is prepared using the cement-wrapping method, and compacted to form 100 × 100 × 100 mm molds. Three identical specimens are cast for each group. After curing for 28 days, the specimens are exposed to acid rain solution. Basic properties are tested after reaching the corresponding erosion age.

2.3 Acid Rain Corrosive Environment

The simulated acid rain solution is prepared by mixing HNO_3, $(NH_4)_2SO_4$, and H_2O. The pH values for the simulating acid rain are set at 3 and 4, with an SO_4^{2-} concentration of 0.05 mol/L. HNO_3 is used to adjust the pH of the acid rain solution, while $(NH_4)_2SO_4$ is employed to regulate the concentration of SO_4^{2-} in the solution. The acid rain erosion test is carried out by immersing cyclically within the dynamic testing framework. Specimens are immersed in the acid rain solution for 5 days, left to air-dry for 1 day in the natural environment, and the cycle is repeated. The pH value is adjusted daily using HNO_3 to maintain the stability of the erosion solution's pH.

2.4 Test Methods

Mass change. After each cycle, the mass of the specimens is measured by a balance with a precision of 0.01 g. The mass change rate of CGPC is calculated using the following formula:

$$\omega_t = \frac{m_t - m_0}{m_0} \qquad (3)$$

where m_0 and m_t is the mass of CGPC before erosion and after t hours of acid rain erosion, respectively (g). The results are obtained by averaging the values from three identical specimens.

Compressive strength. According to the GB/T 50081-2019, compressive strength tests are conducted on the CGPC specimens after acid rain erosion using the YAW-3000 hydraulic press, with a loading rate of 0.3 MPa/s, as illustrated in Fig. 1. The compressive strength of the specimens after acid rain erosion is calculated using the

Fig. 1 Compressive test of CGPC

following formula:

$$f_{cu} = 0.95 \frac{F}{A} \qquad (4)$$

where f_{cu} is the compressive strength of CGPC (MPa), F is the maximum load value (kN), and A is the bearing area (mm^2). The results are obtained by averaging the values from three specimens.

Permeability coefficient. In the study, the permeability coefficient of CGPC is assessed with a fixed water level, as depicted in Fig. 2. Four CGPC specimens, with side surfaces sealed, are positioned within a prefabricated steel mold. The transparent containers and steel mold are securely affixed using AB glue and rubber mud, creating a unified structure. The assembly is then submerged into a water tank, and the inflow is carefully regulated to maintain a stable water level difference.

Measure the water level difference between the transparent container and the water tank, denoted as H. Record the time of the water outflowing, denoted as t. After three repetitions of the measurement, the average of the outflow is calculated, denoted as Q. The permeability coefficient of CGPC can be calculated by the following formula:

$$K_T = \frac{QL}{AHt} \cdot \frac{\eta_T}{\eta_{15}} \qquad (5)$$

where K_T is permeability coefficient of the specimen at a water temperature of T °C (mm/s), Q is the amount of water permeating within t seconds (m^3), L is specimen thickness (mm), A is upper surface area of the specimen (mm^2), H is water level difference (mm), t is the during time (s), η_T and η_{15} is the relative viscosity of water at the temperature of T °C and 15 °C respectively, dimensionless parameters.

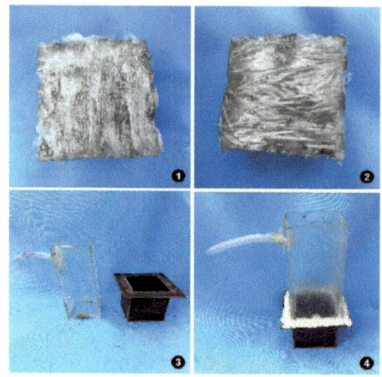

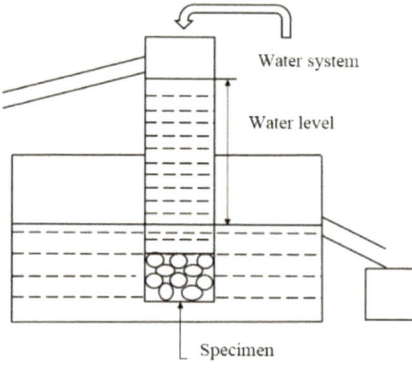

Fig. 2 Test procedure of permeability coefficient

3 Results and Discussion

3.1 Mass Change

Figure 3 illustrates the variations in mass and mass loss rate of CGPC after acid rain erosion. It is evident shown in the figure that with continuous acid rain erosion, the mass of the CGPC specimens initially increases and then decreases. In the early stage of erosion, there is a slight rise in the mass of the specimens, with the highest growth rate observed for A3P22, reaching approximately 0.5%. However, With the increase of erosion time, the mass gradually decreases, and part of specimens experience the detachment of corner aggregates. The trends in mass change for CGPC show variations under distinct acid rain erosion environments. Compared to the erosion environment with a pH of 4, it is evident that the mass loss rate of CGPC is higher in the pH 3 environment, which indicates that as the acidity of the simulated acid rain erosion solution increases, the erosion effect on CGPC intensifies. In the erosion solution with a pH of 3, as the designed porosity increases, the mass loss rate gradually

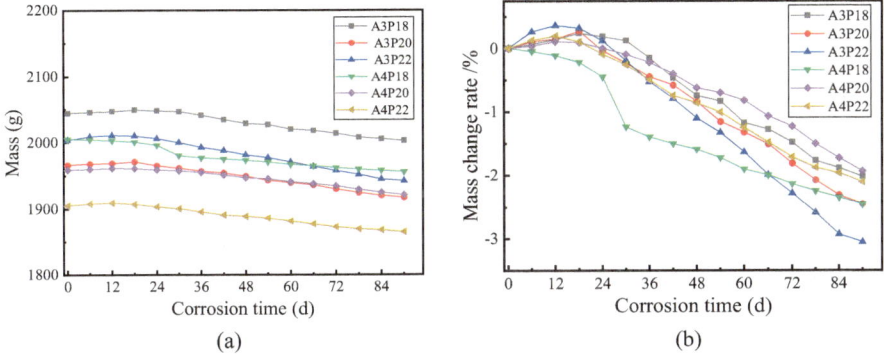

Fig. 3 Mass change and change rate of CGPC after acid rain corrosion

rises. After 90 days of erosion, the mass loss rate of specimens in the A3P18 group is around 2%, while the A3P22 group specimens have a mass loss rate around 3%. This suggests that CGPC specimens with higher designed porosity experience more severe erosion in acid rain.

3.2 Compressive Strength

The development pattern of compressive strength in CGPC during acid rain erosion is illustrated in Fig. 4. According to the figure, after 28 days of curing, the compressive strength of CGPC with designed porosity of 18% and 20% exceeds 20 MPa, meeting the requirements for pervious concrete compressive strength. However as the designed porosity increases, the compressive strength gradually decreases. This can be attributed to the increased internal voids, which reduce the amount of coarse aggregate and cementitious material, resulting in a decrease in compressive strength. Similar to the mass change pattern, the compressive strength of CGPC slightly increases in the early stage of acid rain erosion, but with erosion time increasing, the compressive strength gradually decreases. This is primarily attributed to the initial reaction between ions in the acid rain solution and the hydration products of cement, leading to the formation of gypsum that fills voids and makes the structure denser, causing a slight increase in strength. As erosion time increases, the accumulated gypsum induces expanding effects, resulting in the development of microcracks and a subsequent decrease in strength.

After 90 days of acid rain erosion, the compressive strength loss rates for A3P18, A3P20, and A3P22 are 13.8%, 14.7%, and 20.7%, respectively. This indicates that under lower designed porosity, the compressive strength loss rate reduces, suggesting a stronger resistance to acid erosion. At the erosion time of 90d, the compressive strength loss rates for A3P18 and A4P18 are 13.8% and 9.5%, respectively. This

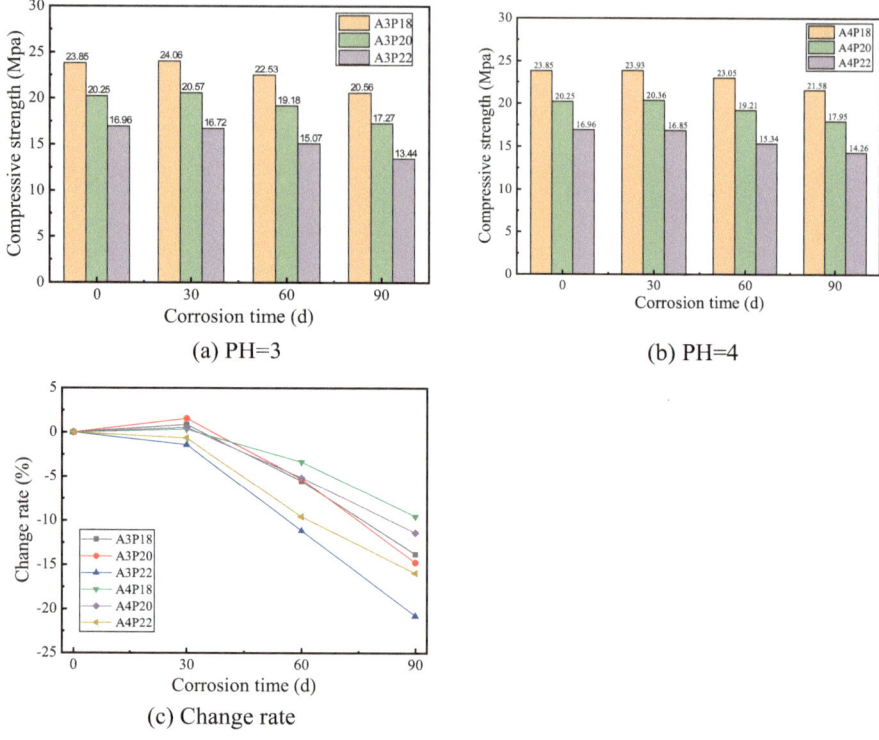

Fig. 4 Compressive strength and change rate of CGPC after acid rain corrosion

suggests that as the acidity of the erosion solution increases, the degradation of CGPC becomes more severe.

4 Permeability Coefficient

The permeability coefficient is a crucial indicator for assessing the permeability of pervious concrete. Figure 5 illustrates the variation pattern of the permeability coefficient in CGPC after acid rain erosion. As observed from the figure, initial permeability coefficients for CGPC with designed porosity groups of 18%, 20%, and 22% are recorded as 1.37 mm/s, 1.56 mm/s, and 2.28 mm/s, respectively, meeting the requirements for pervious concrete permeability coefficient.

In the initial stage of erosion, there is no significant change in the permeability coefficient, and it may even slightly decrease. The permeability coefficient gradually increases over time. This is because in the early stage of erosion, the expansive nature of gypsum increases the structural density, leading to a slight decrease in the permeability coefficient. However, with the accumulation of expansive substances over

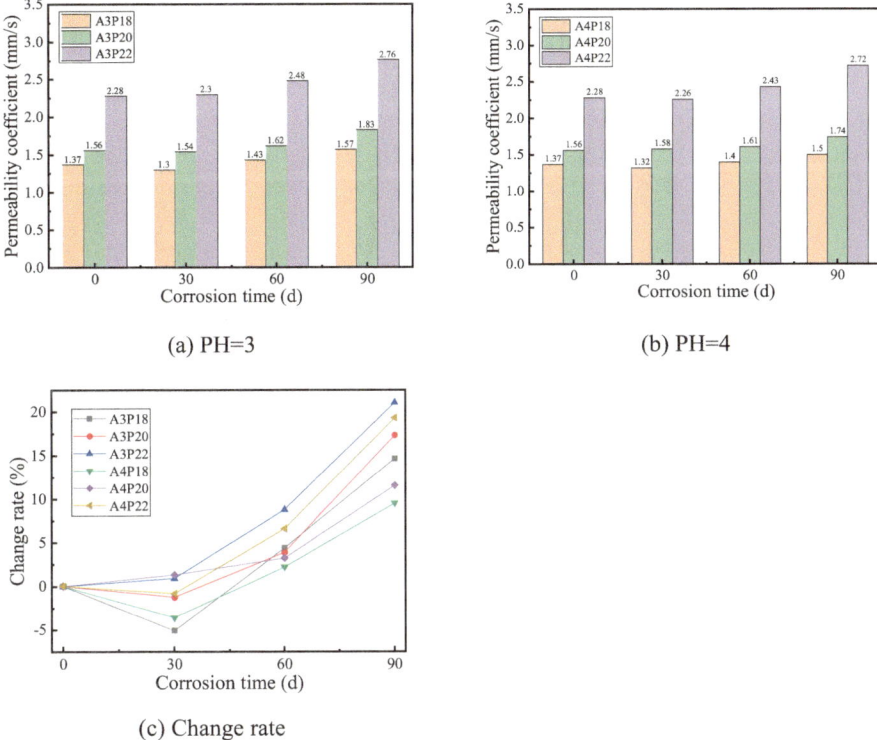

Fig. 5 Permeability coefficient and change rate of CGPC after acid rain corrosion

time, microcracks are induced, resulting in an increase in the permeability coefficient. Additionally, with an increase in designed porosity, the permeability coefficient of CGPC specimens continues to rise. This is primarily because the permeability coefficient of pervious concrete is positively correlated with the connected porosity, and specimens with higher designed porosity have more connected pores. After 90 days of acid rain erosion, the permeability coefficient change rates for A3P18 and A4P18 are 14.6% and 9.4%, respectively. The higher the acidity of the erosion solution, the more severe the erosion, and the more pronounced the change in the permeability coefficient of the specimens.

5 Micro Mechanism of CGPC Subjected to Acid Rain Environment

The fundamental mechanical properties of concrete are closely related to its microscopic structure. To delve into the erosion mechanism of acid rain on CGPC, SEM and EDS tests were carried out on specimens before and after acid rain erosion, with the microscopic morphology illustrated in Fig. 6.

In Fig. 6a, it can be observed that in the CGPC that has not undergone sulfate erosion, there is a significant amount of amorphous hydrated calcium silicate gel (C-S-H) and needle-like ettringite in the hydration products. These components are interconnected to form a mesh structure, which is conducive to the development of strength in CGPC specimens. EDS scanning of the needle-like regions reveals main elements such as Ca, Al, O, Mg, and C. Coal gangue aggregate contains C and Mg elements, confirming that the needle-like crystals are ettringite ($3CaO\ Al_2O_3\ 3CaSO_4\ 32H_2O$). Furthermore, after experiencing acid rain erosion, the hydration products in CGPC show the presence of flaky gypsum ($CaSO_4 \cdot 2H_2O$), as depicted in Fig. 6b. Several hexagonal block-shaped gypsum crystals are stacked together, and the surfaces and vicinity of the aggregate and gypsum crystals are still filled with powdery gel, while it is challenging to find needle-like or columnar ettringite

 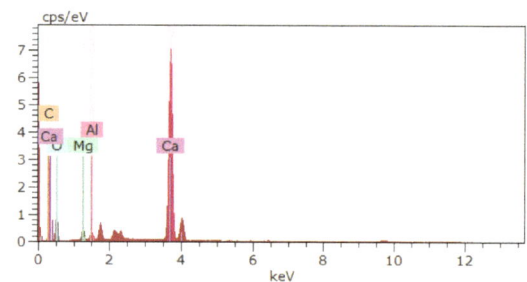

(a) without acid rain corrosion

 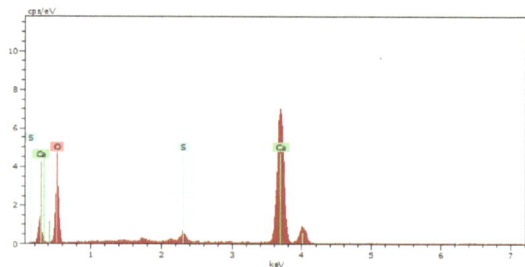

(b) after acid rain corrosion

Fig. 6 Microstructure of CGPC before and after acid rain corrosion

crystals. This leads to a reduction in the strength of CGPC. The EDS results show that the main products of acid rain erosion consist of elements such as Ca, S, and O, indicating the presence of gypsum. Therefore, combining the microscopic results, the reaction in CGPC under acid rain conditions can be inferred as follows:

$$Ca(OH)_2 + 2H^+ \rightarrow Ca^{2+} + 2H_2O \tag{6}$$

$$C-S-H + 2H^+ \rightarrow Ca^{2+} + SiO_2 \cdot nH_2O \tag{7}$$

$$C-A-H + 2H^+ \rightarrow Ca^{2+} + Al_2O_3 \cdot nH_2O \tag{8}$$

$$C_3A \cdot 3CaSO_4 \cdot 32H_2O + 2H^+ \rightarrow Ca^{2+} + CaSO_4 \cdot 2H_2O + Al_2O_3 \cdot nH_2O \tag{9}$$

$$Ca^{2+} + SO_4^{2-} + 2H_2O \rightarrow CaSO_4 \cdot 2H_2O \tag{10}$$

6 Conclusion

(1) Under acid rain environment, the compressive strength of CGPC shows an initial increase followed by a subsequent decrease. The stronger the acidity of the acid rain solution and the larger the designed porosity, the greater the compressive strength loss rate in the later stage of erosion.
(2) The permeability coefficient of CGPC slightly decreases in the initial stage of acid rain erosion. As acid rain erosion continues, the permeability coefficient gradually increases.
(3) Acid rain erosion is primarily characterized by reactions with ions such as Ca^{2+} and SO_4^{2-}, leading to the formation of expansive products like gypsum. In the early stage, the filling of pores by gypsum increases the structural density. In the later stage, the continuous accumulation of gypsum results in the appearance of expansive microcracks in the specimens, leading to a reduction in strength.

Acknowledgements This study was supported by the CNACG Key Laboratory of Mineral Resource in Coal Measures (No. KFKT-2020-4), Jiangsu Collaborative Innovation Center for Building Energy Saving and Construction Technology (No. SJXTBS2121) and the National Natural Science Foundation of China (No. 52074270).

References

1. Peng B, Guo D, Qiao H et al (2018) Bibliometric and visualized analysis of China's coal research 2000–2015. J Clean Prod 197:1177–1189
2. Li Y, Liu S, Guan X (2021) Multi-technique investigation of concrete with coal gangue. Constr Build Mater 301:124114
3. Dong Z, Xia J, Fan C, Cao J (2015) Activity of calcined coal gangue fine aggregate and its effect on the mechanical behavior of cement mortar. Constr Build Mater 100:63–69
4. Sun J, Zhou C et al (2022) Green synthesis of ceramsite from industrial wastes and its application in selective adsorption: performance and mechanism. Environ Res 214:113786
5. Zhou J, Zheng M, Zhan Q et al (2023) Study on mesostructure and stress–strain characteristics of pervious concrete with different aggregate sizes. Constr Build Mater 397:132322
6. Zhou C, Zhu Z, Wang Z et al (2018) Deterioration of concrete fracture toughness and elastic modulus under simulated acid-sulfate environment. Constr Build Mater 176:490–499
7. Xie S, Qi L, Zhou D (2014) Investigation of the effects of acid rain on the deterioration of cement concrete using accelerated tests established in laboratory. Atmos Environ 38(27):4457–4466
8. Gao L, Lai Y, Islam Pramanic MR et al (2021) Deterioration of portland cement pervious concrete in Sponge Cities subjected to acid rain. Materials 14(10):2670
9. Adil G, Kevern JT, Mann D (2020) Influence of silica fume on mechanical and durability of pervious concrete. Constr Build Mater 247:118453
10. Liu H, Luo G, Wang L et al (2019) Strength time-varying and freeze-thaw durability of sustainable pervious concrete pavement material containing waste fly ash. Sustain Basel 11(1):176
11. Zhou C, Zhu Z, Zhu A et al (2019) Deterioration of mode II fracture toughness, compressive strength and elastic modulus of concrete under the environment of acid rain and cyclic wetting-drying. Constr Build Mater 228:116809

Open Access This chapter is licensed under the terms of the Creative Commons Attribution 4.0 International License (http://creativecommons.org/licenses/by/4.0/), which permits use, sharing, adaptation, distribution and reproduction in any medium or format, as long as you give appropriate credit to the original author(s) and the source, provide a link to the Creative Commons license and indicate if changes were made.

The images or other third party material in this chapter are included in the chapter's Creative Commons license, unless indicated otherwise in a credit line to the material. If material is not included in the chapter's Creative Commons license and your intended use is not permitted by statutory regulation or exceeds the permitted use, you will need to obtain permission directly from the copyright holder.

Experimental Study on Axial Compressive Properties of Early Strength and High Ductility Cement-Based Composite Concrete

Weihong Jiang, Wenhong Duan, Jiaquan Yuan, Li Xiong, Huimei Li, Lin Mou, Xiaohua Yang, Xiaomin Huang, Weibing Xu, and Kun Yang

Abstract Cement concrete is widely used in various levels of highways, and it is prone to diseases such as cracking, potholes, and local collapse during operation. The maintenance and repair of existing cement concrete has become an important task in the daily operation and maintenance of roads and bridges. However, traditional cement-based repair materials have problems such as long maintenance periods and low bonding strength with existing concrete. In view of this, this article proposes an early strength and high ductility cement composite material, aiming to achieve good curing age, compressive performance, and relatively low price. The compressive performance tests of a new type of composite material were conducted to study the effects of PVA fiber content, fly ash (FA) content, and sand cement ratio on the axial compression performance. The results show that the FA content is 50wt.% When the PVA fiber content is around 2% and the sand cement ratio is 0.36, the compressive performance of early strength and high ductility cement composite materials at different ages is better. Under optimal mix ratio conditions, the axial compressive strength of early strength and high ductility cement-based composite materials at 2 h, 1 day, 3 days, and 28 days reaches 18.1, 27.5, 34.4, and 37.6 MPa; The peak strain at each age is greater than 0.5%.

Keywords Cement based repair materials · Curing age · PVA fiber · Axial compressive properties · High ductility

---

W. Jiang · W. Duan · J. Yuan · L. Xiong · H. Li · L. Mou · X. Yang
Dali Danan Expressway Co., Ltd, Dali 671005, China

X. Huang (✉)
Kunming University of Science and Technology, Kunming 650031, China
e-mail: Hjs.sy@163.com

W. Xu · K. Yang
Beijing University of Technology, Beijing 100124, China

1 Introduction

Cement concrete has a wide range of applications in bridges and roads. At present, cement concrete is prone to diseases such as cracking, potholes, and local collapse during operation. The maintenance and repair of related diseases have become an important part of daily operation and maintenance of road bridges. Traditional cement-based repair materials have problems such as long maintenance period, low bond strength with existing concrete, insufficient tensile and compressive bearing capacity, and low ductility [1–5].

In view of this, scholars at home and abroad have carried out a lot of research on new cement-based composite materials for road and bridge repair. The related improvement measures include the use of special cement and the addition of fibers. In terms of modification research of repair materials based on sulfoaluminate cement (SAC), Fu pointed out that through the combination of early strength agent and SAC, the 4 h compressive strength of early strength SAC reached 40 MPa, and the flexural strength reached 7 MPa, but the flexural strength appeared to shrink after 28 d [6]. In order to overcome the shrinkage of flexural strength and reduce the cost, Zhang used Ordinary Portland Cement (OPC) to improve the early performance of SAC, studied the early compressive strength of concrete with different OPC and SAC mixing ratios, and found that adding 10–20 wt.% OPC can make OPC and SAC mixed (OPC-SAC) concrete quickly set, the final setting time of OPC-SAC concrete is shortened to 11 min, and the 2 h age compressive strength reaches 32 MPa [7]. The setting time of OPC-SAC concrete is less than the setting time of any single cement concrete, which provides a basis for the joint use of SAC and OPC [8]. OPC-SAC concrete not only has good early mechanical properties, but also has better resistance to sulfuric acid and seawater erosion than OPC concrete [9, 10]. In addition, the hydration of SAC has the characteristic of micro-expansion [11], which is not easy to shrink and creep, and helps to improve the service life of repair materials.

In terms of research on high-ductility cement-based composite materials based on PVA fibers, Victor C. Li studied the compressive performance of high-ductility cement-based composite materials and summarized the characteristics of PVA-ECC's compressive performance and elastic modulus [12]. Xu et al. developed an ultra-high toughness cement-based composite material (UHTCC) based on cement mortar and polyvinyl alcohol fiber, and carried out mechanical performance tests. The results show that this material has good toughness and can maintain good integrity in the limit state [13–15]. Deng and Li pointed out that the elastic modulus of high-ductility concrete (HDC, FRCC) is lower, but its axial compression peak strain is higher [16, 17]. It should be pointed out that the addition of PVA fibers can improve the toughness of OPC, but it also leads to a deterioration of its rheological properties and an increase in matrix defects [18]. In view of this, Ghosh developed a cement-based material with low cost, early strength, self-compacting, and good fluidity by mixing SAC with other cements and adding different types of fibers [19]. Yoo Doo-Yeol used polyethylene fiber to enhance the SAC-OPC system, and obtained an early-strength high-ductility cement-based composite material with 4 h and 28 d

Table 1 Material parameters of PVA fiber

Name	ρ (g/cm^3)	L (mm)	D (mm)	$f_{t\text{-}PVA}$/MPa	E (GPa)
PVA fiber	1.30	12	0.04	1200	4.5

compressive strengths of 38 and 60 MPa, and tensile strengths greater than 5 and 9 MPa [20], but the introduction of polyethylene fibers also significantly increased the cost of the matrix [21], which is not conducive to promotion and use. Existing research results show that existing repair materials still have difficulty in having characteristics such as early strength, small shrinkage deformation, high ductility, and low cost; the mechanical performance research of early-strength high-ductility cement-based composite materials is still not perfect, especially the mechanical performance research results of such materials at different maintenance ages are lacking.

In view of this, this article proposes an early-strength high-ductility cement-based composite material, and carries out uniaxial compressive tests on this type of composite material at 2 h, 1 d, 3 d, and 28 d maintenance ages. The effects of PVA fiber content, FA content, sand-binder ratio, etc. on the peak stress, peak strain and other performance indicators of the material were analyzed.

2 Experimental Program

2.1 Materials

The raw materials for the experiment include 42.5 grade ordinary Portland cement, 42.5 grade rapid hardening sulfoaluminate cement, polycarboxylic acid water reducer (PCE), polymer defoamer, quartz sand, first-grade FA, and domestic PVA fiber. Tap water is used as the mixing water, the particle size of quartz sand is 80–120 mesh, and the parameters of PVA fiber are shown in Table 1. Based on the relevant research results of fiber-reinforced concrete, the mix proportions of various materials are determined as shown in Table 2.

2.2 Preparation of Specimens

Mechanical stirring is used to add the weighed cement, quartz sand, FA, defoamer, and water reducer to the mixer for slow stirring for 60 s to fully mix the materials; then water is added for stirring for 60 s, and finally PVA fiber is slowly dispersed into the mixer, the time for adding PVA fiber should be less than 60 s; after adding PVA fiber, stir for another 60 s. After the specimens are vibrated and cast, they are all covered with plastic wrap; the specimens are demolded 2 h after being placed at room temperature, and after demolding, they are cured at a constant temperature of

Table 2 The fit ratio of test specimens

No.	Name	PVA fiber (%)	FA (wt.%)	S/C	Water-binder ratio	PCE (wt.%)	Defoam agent content (wt.%)	Calcium formate content (wt.%)
1	PVA0	0	30	0.64	0.25	0.5	0.1	1.0
2	PVA0.5	0.5	30	0.64				
3	PVA1.0	1	30	0.64				
4	PVA1.5	1.5	30	0.64				
5	FA 30	2	30	0.64				
6	FA 40	2	40	0.64				
7	FA 50	2	50	0.64				
8	FA 60	2	60	0.64				
9	SA0.50	2	50	0.50				
10	SA0.36	2	50	0.36				

20 ± 3 °C, with a relative humidity of 90 ± 10%, until the specified age. It should be pointed out that the 2 h age specimens are put into the curing box immediately after final setting, and taken out 10 min before the test. The specimens are prismatic bodies of 100 × 100 × 300 mm.

2.3 Test Method

The axial mechanical performance test is carried out according to "Standard Test Method for fiber reinforced concrete" (CECS13:2009). There are 3 specimens in each group, using a mixed loading mode of load and displacement, the early loading speed is 0.5 MPa/s, and after reaching the peak load, it is changed to displacement loading, with a loading rate of 0.1 mm/min. The loading of the specimen is shown in Fig. 1.

3 Test Results and Discussion

3.1 Failure Modes

As can be seen from Fig. 2, the specimens with PVA fibers show obvious toughness failure characteristics during the compression failure process. The specific failure process is as follows: at the beginning of loading, there is no obvious change in the specimen; when the load reaches 60–70% of the peak load, the first small crack

Fig. 1 Full curve test for axial compressive stress–strain

Fig. 2 Failure modes of specimen

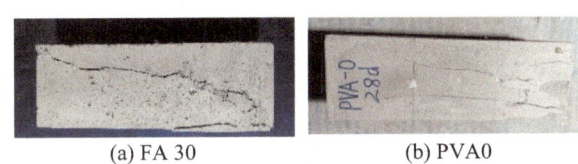

(a) FA 30   (b) PVA0

appears on the surface of the specimen; as the load continues to increase, new small cracks continuously appear on the surface of the specimen, and the width of the original small cracks increases and extends along the surface of the specimen; when the limit load is reached, the specimen forms multiple oblique through cracks, and the original small cracks also close; then the bearing capacity of the specimen begins to decrease, when the load on the specimen drops to 20–40% of the peak load, the width of the surface crack of the specimen reaches about 2 mm, and the test is terminated. Correspondingly, the specimens without PVA fibers show obvious brittle characteristics during the compression failure process. The specific failure process is as follows: there is no obvious cracking phenomenon from the beginning of the test to before the peak load; after reaching the limit load, a small amount of longitudinal through cracks appear, the bearing capacity drops rapidly, and the test is terminated.

3.2 Test Results

In order to conveniently compare the stress–strain full curves of the same factors at different levels, the following figure uses normalization treatment, that is, stress/peak stress (f/f_c) as the vertical axis, strain/peak strain ($\varepsilon/\varepsilon_c$) as the horizontal axis, and the peak point coordinates are (1, 1).

Effect of fiber

The normalized stress–strain curves of different PVA fiber contents at each age are shown in Fig. 3. It can be seen that with different PVA fiber contents, the stress–strain curves of the specimens change to different degrees. As the fiber content increases, the peak strain of the specimen increases, and the rising section of the normalized curve is also fuller. The influence curves of fiber content on the axial compressive strength, peak strain, and elastic modulus of the specimens are shown in Fig. 4.

(1) Compressive strength

As shown in Fig. 4a, at the curing age of 2 h, the axial compressive strength of the prismatic specimens has reached 14.5–17.5 MPa. With the increase of fiber content, the compressive strength of the prismatic specimens first decreases and then increases. When the fiber content is 1%, the axial compressive strength is the smallest; the addition of fibers will reduce the 2 h axial compressive strength of the specimens (between −5.3 and −16.9%). The reason for this is that the addition of PVA fibers will introduce air bubbles, and the bond strength between the fiber and the matrix is still small at 2 h, so the axial compressive strength of the specimens is reduced. It should be pointed out that when the PVA fiber content is greater than 1%,

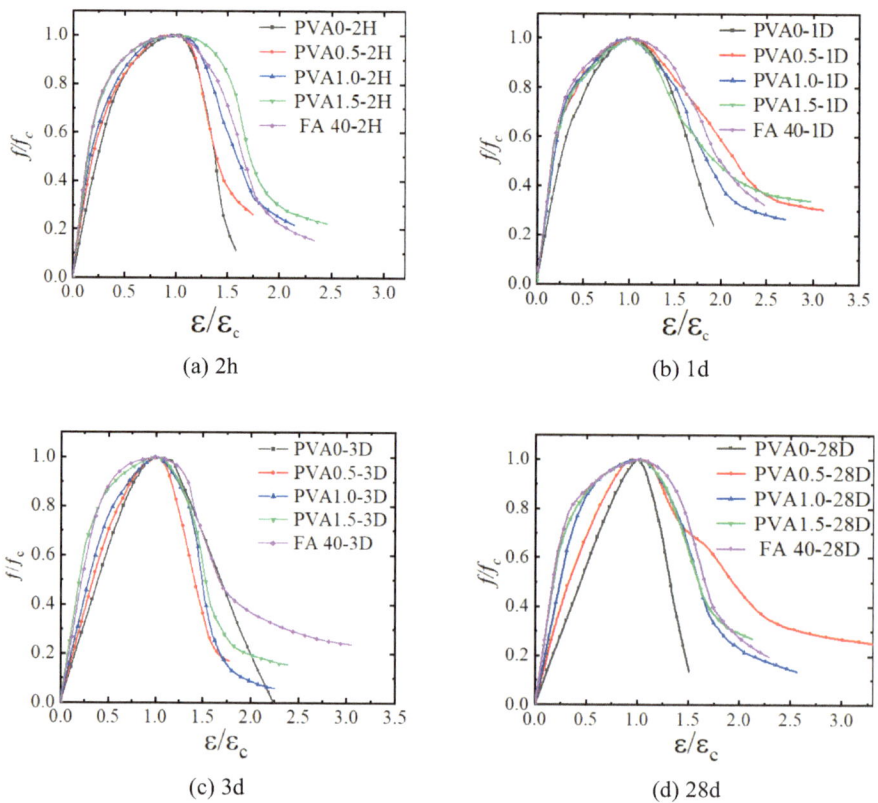

Fig. 3 Normalized stress–strain curves of PVA fiber content at different ages

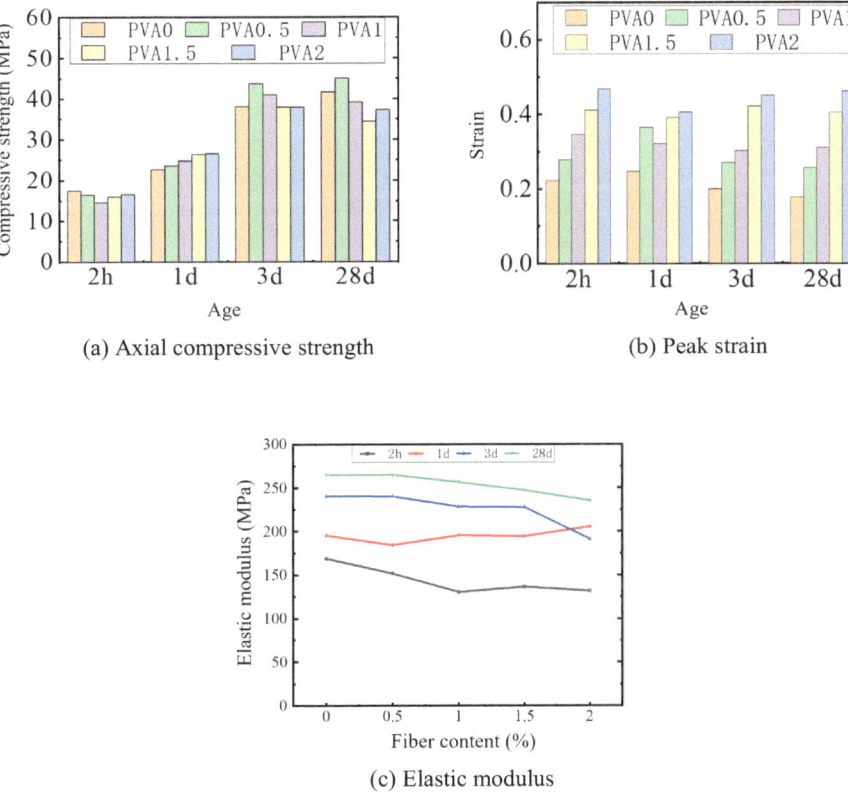

Fig. 4 Effect of PVA fiber content

the compressive strength of the specimens gradually increases, indicating that the addition of fibers can enhance the early age compressive strength to a certain extent.

At the curing age of 1 d, the axial compressive strength of the prismatic specimens has reached 22.6–26.5 MPa. With the increase of PVA fiber content, the axial compressive strength of the specimens continues to increase. Compared with the condition without fiber, the axial compressive strength of the specimen with 2% fiber content increased by 14.6%. The addition of PVA fibers can significantly improve the axial compressive strength of the 1 d curing age. In addition, the increases in axial compressive strength of specimens with PVA fiber contents of 0.5, 1, 1.5, and 2% are 7.1, 10.3, 10.4, and 10.0 MPa, respectively, indicating that when the PVA fiber content is greater than 1%, it can greatly improve the axial compressive strength of the 1 d age prismatic specimens.

At the curing age of 3 d, the axial compressive strength of the prismatic specimens has reached 37.9–43.6 MPa. With the increase of PVA fiber content, the axial compressive strength of the specimens shows a trend of first increasing and then decreasing. Compared with the condition without fiber, the axial compressive strength of the specimen with 0.5% fiber content increased by 14.7%. The addition of a small amount of fiber can significantly improve the axial compressive strength of the specimens at the 3 d curing age. In addition, compared with the 1 d age specimens, the addition of fibers has a greater effect on the increase in axial compressive strength of the 3 d age specimens, with an increase range between 11.4 and 20.0 MPa.

At the curing age of 28 d, the axial compressive strength of the prismatic specimens is in the range of 34.5–45.1 MPa. With the increase of PVA fiber content, the axial compressive strength of the specimens shows a trend of first increasing and then decreasing. Compared with the specimens without fiber, the axial compressive strength of the specimens with 0.5% fiber content increased by 8.6%; however, when the fiber content is greater than or equal to 1%, the axial compressive strength of the specimens decreases to varying degrees (-5 0.7–17.0%).

In summary, the axial compressive strength of the prismatic specimens mainly forms within the 0–3 d age period, and the difference between the axial compressive strength of the 3 d age specimens and the 28 d age specimens is between 1.6 and 9.1%. When the curing age is 2 h, the axial compressive strength of the specimens reaches 36.5 and 46.4% of 28 days; when the curing age is 1 d, the axial compressive strength of the specimens reaches 52.3–76.5% of 28 days. The addition of a small amount of fiber (less than 0.5%) has a significant improvement effect on the early age of the specimens, without affecting the later strength of the specimens.

(2) Peak strain

As shown in Fig. 4b, at the curing age of 2 h, the peak strain of the specimens is in the range of 0.22–0.47%. With the increase of PVA fiber content, the peak strain of the specimens increases. When the fiber content reaches 2%, the peak strain of the specimen is 2.1 times that of the specimen without fiber. At the curing age of 1 d, the peak strain of the specimens also shows the same trend with the increase of PVA, but when the fiber content is greater than 0.5%, the peak strain of the specimens is not significantly affected by the change in fiber content. The reason for this is that the bridging effect between the fiber and the matrix is weak at an early age. At the curing age of 3 d, the specimens with fiber content in the range of 0–1% show similar peak strains, indicating that it is difficult for specimens with low fiber content to greatly improve the peak strain of the specimens; specimens with fiber content between 1.5 and 2% show similar effects, and the peak strain of the specimens has increased by 110.9–120.6%, the ductility of the specimens at the age of 3 d and above is greatly affected by the fiber content, and the bridging effect between the fiber and the matrix is significantly enhanced at this time.

(3) Elastic modulus

As shown in Fig. 4c, at the curing age of 2 h, the elastic modulus of the specimens has reached 13.17–16.87 GPa. With the increase of fiber content, the elastic modulus of the specimens continues to decrease. At the curing age of 1 d, the elastic modulus of the specimens has reached 18.42–20.52 GPa. With the increase of fiber content, the elastic modulus of the specimens first decreases and then increases, and when the fiber content is 0.5%, the elastic modulus of the specimens is the smallest. Compared with the 2 h curing age specimens, the elastic modulus of the 1 d curing age specimens has significantly increased, with an increase range of 2.63–7.34 GPa, and the greater the fiber content, the more obvious the increase effect. At the curing age of 3 d, the elastic modulus of the specimens has reached 19.12–24.01 GPa. With the increase of fiber content, the elastic modulus of the specimens continues to decrease; and compared with the 1 d curing age specimens, the elastic modulus of the 3 d curing age specimens has significantly increased. At the curing age of 28 d, the elastic modulus of the specimens has reached 23.47–26.49 GPa. With the increase of fiber content, the elastic modulus of the specimens continues to decrease; but the addition of a small amount of fiber (0.5%) will not reduce the elastic modulus of the specimens.

In summary, the elastic modulus of the specimens mainly forms before the 2 h age period, and the elastic modulus of the 2 h specimens reaches 53.9–63.7% of the peak elastic modulus. With the increase of fiber content, the elastic modulus of the specimens at each age basically shows a downward trend, but when a small amount of fiber is added (less than 1%), the elastic modulus of the specimens decreases less.

Effect of FA

The normalized stress–strain curves of different FA content at different ages are shown in Fig. 5. Different FA content leads to differences in the axial compressive strength and elastic modulus of the specimens. The influence of FA content on the axial compressive strength and elastic modulus is consistent, and the overall trend is that the larger the FA content, the fuller the curve. The influence of FA content on the axial compressive strength, peak strain and elastic modulus of the specimen is shown in Fig. 6.

(1) Compressive strength

As shown in Fig. 6a, at the curing age of 2 h, the axial compressive strength of the specimens has reached 9.8–18.2 MPa, and with the increase of FA content, the axial compressive strength of the specimens continuously decreases. However, it is worth noting that when the FA content increases from 30 to 50 wt.%, the strength only decreases by 13.4%. For specimens at the 2 h curing age, the FA content should not exceed 50 wt.%. The reason for this is that on the one hand, the increase in FA content leads to a decrease in the proportion of cement, resulting in fewer hydration products; on the other hand, the reaction of FA with water reduces the contact between cement and water, further reducing the hydration reaction rate. At the curing age of 1 d, the axial compressive strength of the specimens has reached 17.0–31.8 MPa. With the increase of FA content, the axial compressive strength of the specimens continuously decreases, and for every 10 wt.% increase in FA content, the axial compressive strength decreases by 4.4–5.3 MPa. Compared with the specimens at the 2 h curing age, the axial compressive strength of the specimens at the 1 d curing age has significantly increased, with an increase range between 40.3 and 75.1%. At the curing age of 3 d, the axial compressive strength of the specimens has reached 22.2–40.4 MPa. With the increase of FA content, the axial compressive strength of the specimens also continuously decreases. When the FA content is less than 50 wt.%, for every 10 wt.% increase in FA content, the axial compressive strength

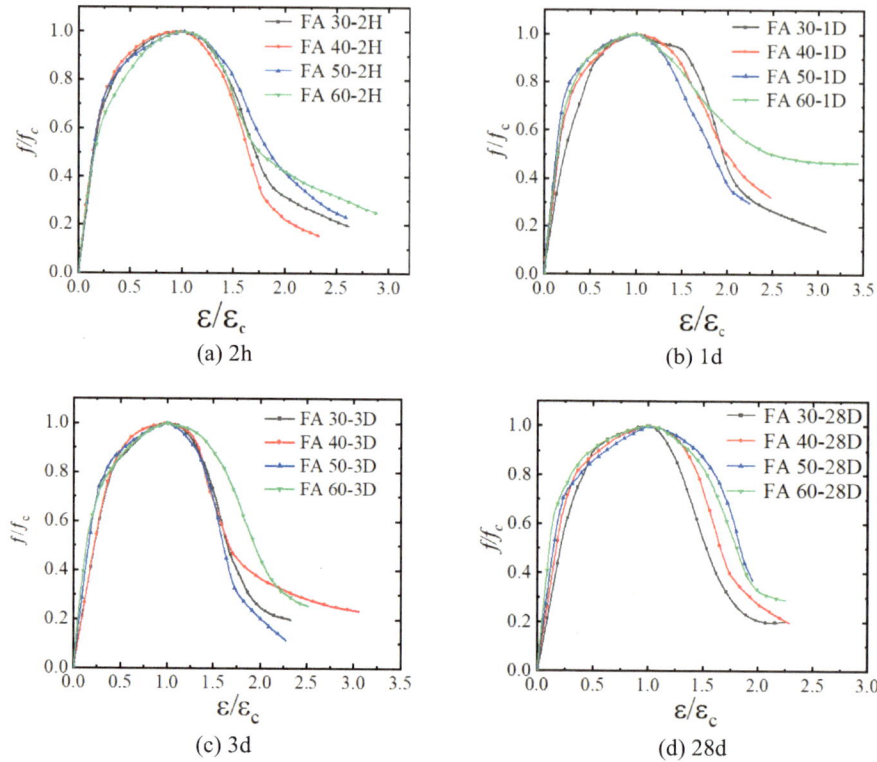

Fig. 5 Normalized stress–strain curves with different FA content at different ages

decreases by 2.4–3.8 MPa. At the curing age of 28 d, the axial compressive strength of the specimens has reached 23.5–39.6 MPa, and with the increase of FA content, the trend of change in the axial compressive strength of the specimens remains consistent with other ages. And when the FA content is less than 50 wt.%, for every 10 wt.% increase in FA content, the axial compressive strength decreases by 2.3–5.6 MPa.

On the whole, with the increase of FA content, the axial compressive strength of each age specimen decreases continuously. When the FA content is less than 50 wt.%, the axial compressive strength of the prismatic specimens at 2 h and 3 d curing ages reaches 41.6–49.6 and 69.7–80.5% of the axial compressive strength of the specimens at 28 days, and it reaches its peak at the curing age of 3 d; for FA content greater than 50 wt.%, the axial compressive strength of the specimens increases with the increase of curing age, but its strength is still mainly formed before the age of 3 d (94.5%).

(2) Peak strain

As shown in Fig. 6b, at the curing age of 2 h, the peak strain of the prismatic specimens is in the range of 4400–5000 με; with the increase of FA content, the peak strain of the specimens also increases, and when the FA content increases from

Experimental Study on Axial Compressive Properties of Early Strength ...

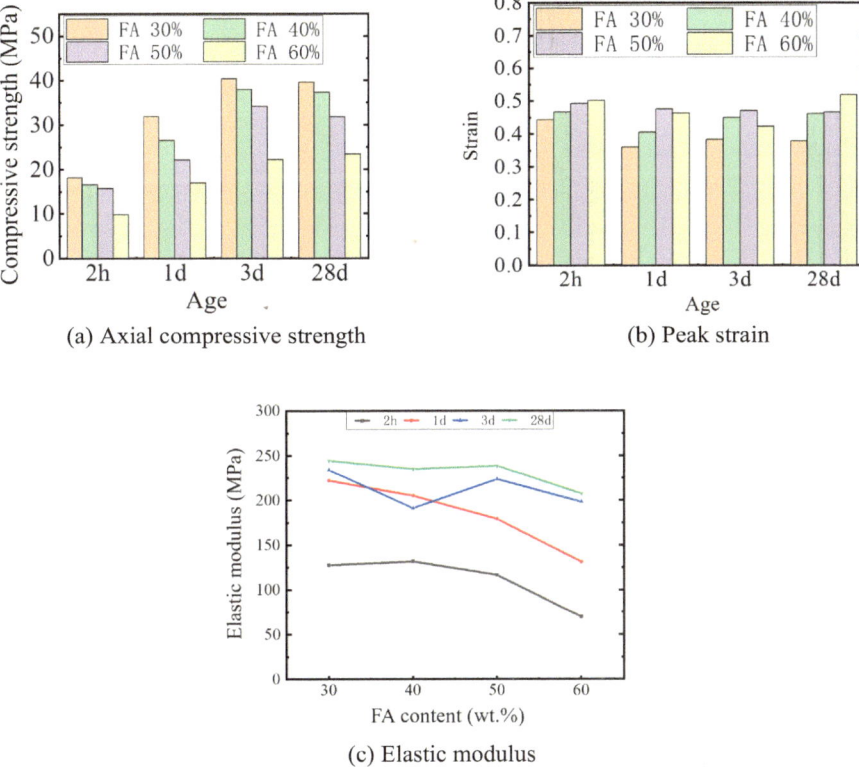

Fig. 6 Effect of FA content

30 to 60 wt.%, the peak strain increases by 13.1%. At the curing age of 1 d, with the increase of FA content, the peak strain of the specimens first increases and then decreases, and the peak strain of the specimens is the largest when the FA content is 50 wt.%. At the curing age of 3 d, the peak strain of the specimens also shows the same trend as at the age of 1 d. At the curing age of 28 d, the peak strain of the specimens increases with the increase of FA content, and except for the peak strains of the specimens with 40 and 50 wt.% FA content being close, the others are quite different, with a difference of 1400 με between the peak strains of the specimens with 60 wt.% and 30wt.% FA content. Overall, the effect of curing age on the peak strain of the specimens is less than the effect of FA content.

(3) Elastic modules

As shown in Fig. 6c, at the curing age of 2 h, the elastic modulus of the specimens has already reached 6.97–13.17 GPa, and with the increase of FA content, the elastic modulus of the specimens first slightly increases and then decreases, and the elastic modulus of the specimens is the largest when the FA content is 40 wt.%, and when the FA content is greater than 50 wt.%, it will have a greater impact on the elastic modulus of the 2 h age specimens. At the curing ages of 1, 3, and 28 d, the elastic modulus of the specimens approximately decreases with the increase of FA content, and the impact of FA content on the elastic modulus decreases with the increase of curing age.

Effect of sand cement ratio

The normalized stress–strain curves of different sand-cement ratios at various ages are shown in Fig. 7. The sand cement ratio (S/C) has a small impact on the stress–strain curve before the peak of the specimens, but has a larger impact on the mechanical properties in the later stage of early age. The impact curves of S/C on the axial compressive strength, peak strain, and elastic modulus of the specimens are shown in Fig. 8.

(1) Compressive strength

As shown in Fig. 8a, with the decrease of the sand-cement ratio, the axial compressive strength of the specimens continuously increases. At the curing age of 2 h, the axial compressive strength of the specimens reaches 15.7–18.1 MPa, and the axial

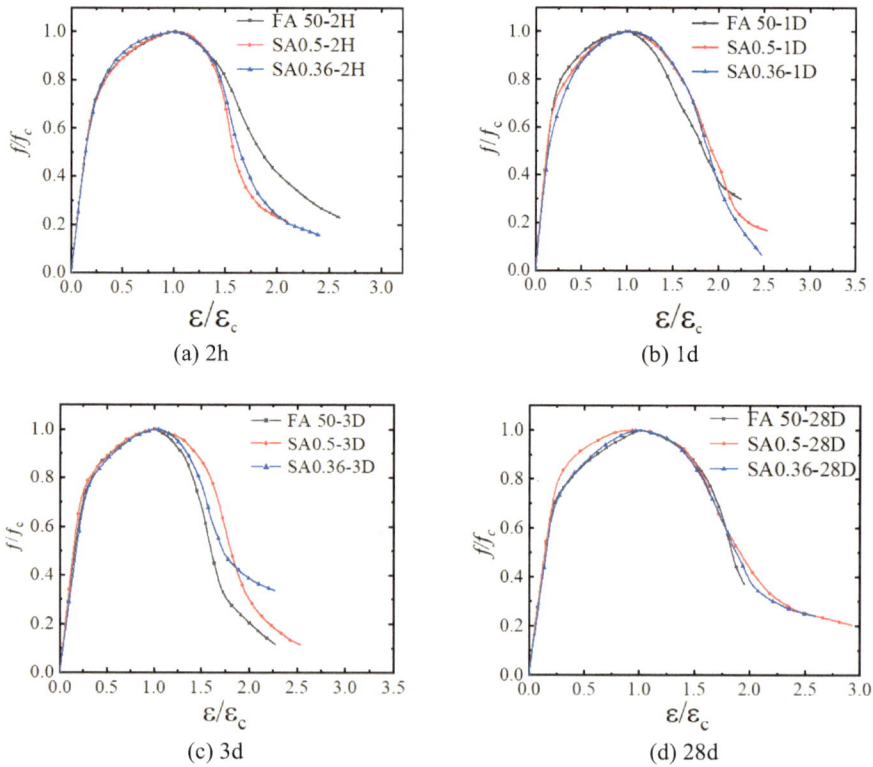

Fig. 7 Different S/C at different ages normalized stress–strain curves

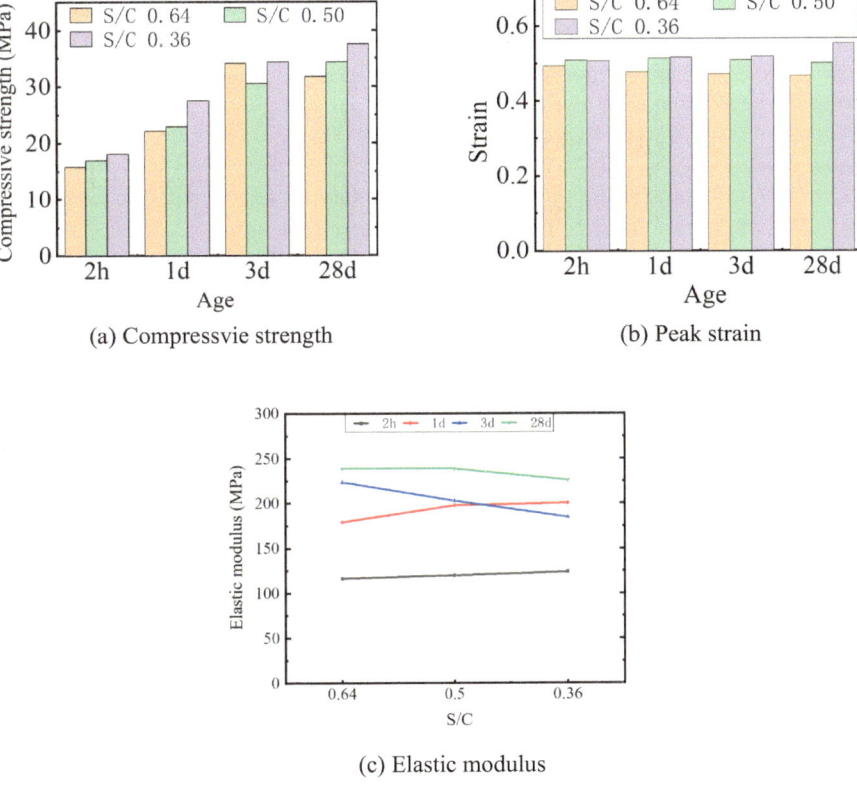

Fig. 8 Effect of S/C

compressive strength of the 2 h age specimens reaches 48.1–49.6% of the axial compressive strength of the 28 d age specimens; the axial compressive strength of the specimens with different sand-cement ratios shows a trend of increasing with the increase of curing age, and the impact of the sand-cement ratio on the axial compressive strength is less than the factor of curing age.

(2) Peak strain

As shown in Fig. 8b, the sand-cement ratio has a significant impact on the peak strain of the specimens, and the smaller the sand-cement ratio, the larger the peak strain of the specimens; the curing age has a small impact on the peak strain of the specimens with different sand-cement ratios, and the peak strain of the specimens at each age does not differ much.

(3) Elastic modulus

As shown in Fig. 8c, at the curing age of 2 h, the elastic modulus of the specimens has already reached 11.64–12.38 GPa, and with the decrease of the sand-cement ratio, the elastic modulus of the specimens increases. The reason for this is that the decrease in the sand-cement ratio increases the amount of cementitious material. At the curing age of 1 d, the elastic modulus of the specimens reaches 17.89–20.01 GPa, and with the decrease of the sand-cement ratio, the elastic modulus of the specimens increases. Compared with the specimens at the 2 h age, the elastic modulus of the specimens at the 1 d age has significantly increased (6.25–7.78 GPa), and the smaller the sand-cement ratio, the more obvious the increase effect. At the curing ages of 3 d and 28 d, the elastic modulus of the specimens decreases with the decrease of the sand-cement ratio; but compared with the specimens at the 3 d age, the elastic modulus of the specimens at the 28 d age is higher. Overall, the sand-cement ratio has a small impact on the elastic modulus, and the elastic modulus of the specimens is mainly affected by the curing age, and the elastic modulus of the 2 h age specimens is only 26.2–33.8% of the elastic modulus of the 28 d age specimens.

4 Conclusions

This paper studies the impact of FA content, sand-cement ratio, and PVA volume ratio on the axial compressive performance (including the complete stress–strain curve) of early-strength high-ductility cement-based composite prismatic specimens through axial compression mechanical performance tests. The specific conclusions are as follows:

(1) With the increase of FA content, the axial compressive strength and elastic modulus of the early-strength high-ductility cement-based composite prismatic specimens decrease, but the peak strain corresponding to the peak strength increases. To achieve a higher early axial compressive strength, the FA content should not exceed 50 wt.%.
(2) With the increase of PVA fiber content, the axial compressive strength of the early-strength high-ductility cement-based composite prismatic specimens shows different trends at each age, but the peak strain corresponding to the peak strength gradually increases with the increase of fiber content. At the 2 h age, the axial compressive strength of the specimens decreases after adding PVA fibers; at the 1 d age, the axial compressive strength of the specimens increases with the increase of PVA fiber content; at the 3 and 28 d ages, adding 0.5% PVA fibers helps to increase the axial compressive strength. When the PVA fiber content is 2%, the maximum peak strain of the specimens at each age reaches 5000 $\mu\varepsilon$.
(3) With the decrease of the sand-cement ratio, the axial compressive strength of the early-strength high-ductility cement-based composite prismatic specimens gradually increases. When the sand-cement ratio is 0.36, the axial compressive strengths at 2 h, 1, 3, and 28 d reach 18.1, 27.5, 34.4, and 37.6 MPa, respectively.

Acknowledgements This study was partly supported by the Yunjiaoke Jiaobian (Grant No. [2021] 91) and Beijing Natural Science Foundation (Grant No. 8232001). Their support is gratefully acknowledged.

References

1. Jianping F (2015) Preparation and properties of super-early-strength cementitious materials. Southeast University (In Chinese)
2. Chong WH, Yaoling Luo Z, Zhou L (2015) Preparation and mechanism analysis of fast-setting and fast-hardening high-strength concrete. J Huazhong Univ Sci Technol (Natural Science Edition) 43(2):89–93 (In Chinese)
3. Mingming ZY, Liang Z, Wang Z (2011) Research on the compounding of inorganic salt early-strengthening agent and polycarboxylic acid high-efficiency water reducing agent. Concrete 4:91–94 (In Chinese)
4. Zhenguo GY, Cahngrui W, Wang Z (2009) Research on the formulation and action mechanism of alkali-free concrete early-strengthening agent. J Wuhan Univ Technol 31(7):81–83 (In Chinese)
5. Wang Y, Sun L, Liu S, Li S, Guan X, Luo S (2022) Development of a novel double-sulfate composite early strength agent to improve the hydration hardening properties of portland cement paste. Coatings 12(10):1485
6. Tingyang FB, Yuping G, Zhang J (2018) Application research of super-early-strength sulfoaluminate cement concrete repair material. Concrete 2:140–144 (In Chinese)
7. Zhang J, Li G, Ye W, Chang Y, Liu Q, Song Z (2018) Effects of ordinary Portland cement on the early properties and hydration of calcium sulfoaluminate cement. Constr Build Mater 186:1144–1153
8. Qing YX, Zhang D, Xiang L, Zhang YX (2018) Research on hydration performance and mechanism of ordinary silicate-alumina sulfate composite cementitious system. Mater Rep 32(A02):517–521 (In Chinese)
9. Cao R, Yang J, Li G, Liu F, Niu M, Wang W (2022) Resistance of the composite cementitious system of ordinary Portland/calcium sulfoaluminate cement to sulfuric acid attack. Constr Build Mater 329:127171
10. Bertola F, Gastaldi D, Irico S, Paul G, Canonico F (2020) Behavior of blends of CSA and Portland cements in high chloride environment. Constr Build Mater 262:120852
11. Kothari A, Tole I, Hedlund H, Ellison T, Cwirzen A (2022) Partial replacement of OPC with CSA cements–effects on hydration, fresh and hardened properties. Adv Cem Res 35(5):207–224
12. Kanda T, Li VC (1999) New micromechanics design theory for pseudostrain hardening cementitious composite. J Eng Mech 125(4):373–381
13. Shilang XX, Cai Y, Zhang Y (2009) Experimental determination and analysis of uniaxial compressive stress-strain full curve of ultra-high toughness cementitious composites. Chin Civil Eng J 42(11):79–85 (In Chinese)
14. Shilang XX (2009) Basic mechanical properties of ultra-high toughness fiber reinforced cementitious composites. J Hydraul Eng 9:1055–1063 (In Chinese)
15. Shiyong JH, Yao GW, Tao ST (2018) Progress in the study of mechanical properties and intrinsic relationship of ECC materials. Mater Rep 32(23):4192–4204 (In Chinese)
16. Mingke DJ, Qin PM, Haibo L (2016) Uniaxial compression constitutive modeling of highly ductile concrete. J Xi'an Univ Architect Technol 48(6):826–831 (In Chinese)
17. Yan L, Liu Z (2014) Uniaxial compressive mechanical properties and constitutive relationship of high toughness PVA-FRCC. J Build Mater 17(4) (In Chinese)
18. Juntao ZZ, Xinling L, Ke W, Weikang L (2021) Modeling of uniaxial tensile principal relationship of cementitious composites for engineering purposes. J Basic Sci Eng (In Chinese)
19. Ghosh D, Abd-Elssamd A, Ma ZJ, Hun D (2021) Development of high-early-strength fiber-reinforced self-compacting concrete. Constr Build Mater 266:121051
20. Yoo D-Y, Oh T, Chun B (2021) Highly ductile ultra-rapid-hardening mortar containing oxidized polyethylene fibers. Constr Build Mater 277:122317
21. Yan W (2016) Experimental study on physical and mechanical properties of high toughness fiber reinforced cementitious composites. Harbin Inst Technol (In Chinese)

Open Access This chapter is licensed under the terms of the Creative Commons Attribution 4.0 International License (http://creativecommons.org/licenses/by/4.0/), which permits use, sharing, adaptation, distribution and reproduction in any medium or format, as long as you give appropriate credit to the original author(s) and the source, provide a link to the Creative Commons license and indicate if changes were made.

The images or other third party material in this chapter are included in the chapter's Creative Commons license, unless indicated otherwise in a credit line to the material. If material is not included in the chapter's Creative Commons license and your intended use is not permitted by statutory regulation or exceeds the permitted use, you will need to obtain permission directly from the copyright holder.

Numerical Simulation Analysis of Mechanical Properties of Semi-rigid Immersed Tube Tunnel Joints

Hai Ji, Yonggang Lv, and Qingfei Huang

Abstract The Hong Kong–Zhuhai–Macau Bridge immersed tube tunnel project is the only deep buried and large siltation immersed tube tunnel in the world. There is no similar case in the world. Its stress deformation characteristics are different from those of conventional shallow buried tunnels. According to the traditional design method of immersed tubes, there will be a series of problems. To address these issues, a semi rigid pipe joint scheme between rigid and flexible pipe joints has been proposed. By comparing the forces acting on the joints with traditional types of pipe joints, the mechanical mechanism of semi rigid immersed pipe structures has been revealed. Compared with flexible immersed pipe structural systems, it improves the load-bearing capacity and water tightness of the segment joints, reduces the internal force of the immersed pipe structure, reduces the risk of concrete cracking, and solves the technical problems of deep buried large sedimentation immersed pipe structures. This can provide reference for similar projects.

Keyword Semi rigid · Immersed tube tunnel · Pipe joint · Stress performance

1 Introduction

The immersed tube tunnel, a tunnel structure that spans rivers and seas, has a history of over 100 years worldwide. The structural forms of its pipe joints are mainly divided into rigid pipe joints and flexible pipe joints. The entire cross-section of the rigid pipe joint structure provides shear and bending bearing capacity, while axial compressive stress is used at the joints to provide bending bearing capacity. Joint friction and shear keys jointly provide shear bearing capacity. Under uneven foundation stiffness and temperature loads, rigid pipe joint structures will generate significant axial tension and bending moments, indicating the presence of significant tensile stress in the structure. In the case of significant differences in siltation load or foundation stiffness,

H. Ji (✉) · Y. Lv · Q. Huang
CCCC Highway Consultants Co., Ltd, Beijing 100010, China
e-mail: 919962750@qq.com

flexible pipe joints release bending moments through the opening of segment joints in the structure, and release bending moments longitudinally through the opening of segment joints, with longitudinal forces controlled at a lower level. At the joint of the segment, only the waterstop and shear key bear the shear load. If the strength of the shear key is insufficient and failure occurs, it may cause serious consequences. Regardless of the type of pipe joint structure, joints are the weakest and most critical link in immersed tube tunnels [1–5], and the study of their mechanical properties is crucial.

There have been some research results on the mechanical properties and failure modes of joints and segments in immersed tube tunnels. Yuan Yong et al. [6] proposed a set of failure test methods and loading schemes to test the compressive shear bearing capacity and failure modes of joints, and conducted a geometric scale 1:10 compression shear test on pipe joints. Cheng Xinjun et al. [7] conducted a 1:4 large-scale model test and found that when subjected to seismic loads, the deformation of the immersed pipe joint is mainly concentrated at the joint and is mainly borne by the shear keys at the joint. Liu Peng et al. [8] conducted stress analysis on the joints of immersed tube tunnels and established a stiffness calculation model for the joints of immersed tube tunnels. Zhang Yong et al. [9] studied and analyzed the material characteristics of GINA rubber waterstops, as well as the stress and deformation mechanism of vertical steel shear keys under shear forces. They obtained the vertical shear stiffness of the joints and the possible failure modes of shear keys under ultimate loads. Wei Gang et al. [10] used a simply supported beam model and an elastic foundation beam simply supported beam composite model to calculate and analyze the longitudinal internal force distribution of the structure during the sinking and docking stage of the immersed tube tunnel section. Yu Haitao et al. [11] established two-dimensional and three-dimensional finite element models for waterstops. By comparing experimental results with finite element calculation results, they analyzed and quantified the influence of support friction, waterstop length, shape, and lateral constraints on experimental results. Su Zongxian et al. [12] studied the longitudinal static analysis calculation model of immersed tube tunnel structures and pointed out that the curvature and curvature change rate of the longitudinal settlement curve of the tunnel affect the longitudinal bending moment and shear force of the structure, respectively. Liu Yuyang et al. [13] studied the mechanical properties and distribution of shear forces at segment joints under ground settlement. In recent years, with the completion and opening of the Hong Kong Zhuhai Macao Bridge, more attention has been paid to the successful application of semi-rigid pipe joints that integrate the characteristics of rigid and semi-rigid pipe joints. However, there is relatively little research on the stress performance of semi-rigid immersed pipe joints.

The semi-rigid immersed tube structure system structure of deep buried immersed tube tunnels is a new immersed tube tunnel scheme. The semi-rigid pipe segment structure of the immersed tube tunnel of the Hong Kong Zhuhai Macao Bridge is shown in Fig. 1. This article takes the technical difficulties encountered in the immersed tube tunnel of the Hong Kong-Zhuhai-Macao Bridge as the background, and focuses on the stress performance of the semi-rigid tube joint through numerical simulation, revealing the stress mechanism of the semi-rigid immersed tube structure. This provides a basis for constructing the semi-rigid immersed tube structure construction system and proposing the calculation theory and design method of the semi-rigid immersed tube structure. To provide strong support for the scientific, rapid, safe, high-quality, and economical construction of the immersed tube tunnel project of the Hong Kong Zhuhai Macao Bridge Island Tunnel, it can effectively guide the promotion and application of semi-rigid immersed tube joint structures in engineering fields such as deep buried immersed tube tunnel semi-rigid immersed tube structural systems in the open sea.

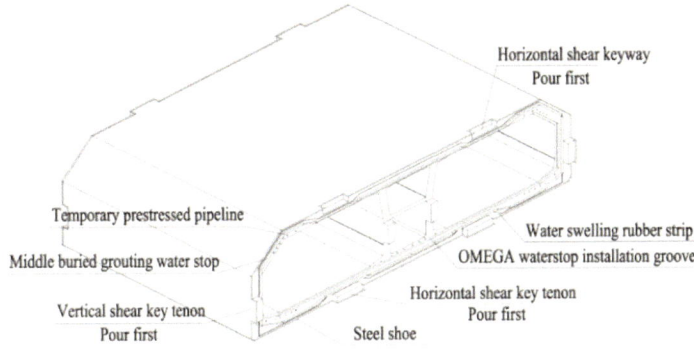

a) Segmental structure

b) Installed immersed tube

Fig. 1 Semi rigid immersed pipe joint

2 Engineering Background

The immersed tube section of the Hong Kong Zhuhai Macao Bridge immersed tube tunnel is 5664 m long and consists of 33 reinforced concrete immersed tubes. The standard immersed tube section is 180 m long, 37.95 m wide, and 11.4 m high. Compared with the already built immersed tube tunnels at home and abroad, the Hong Kong Zhuhai Macao Bridge immersed tube tunnel is the longest, deepest buried, largest single span, and largest scale underwater highway immersed tube tunnel in the world.

The Hong Kong Zhuhai Macau immersed tunnel is located near the island head as a sloping transition section with uneven thickness of silt underneath, and the stress conditions are very complex. At the same time, due to the requirements of channel scale (300,000 ton oil tankers passing through), the middle section of the tunnel adopts a unconventional deep burial scheme under the seabed. Calculated from the original seabed surface to the outer edge of the structural roof, the maximum burial depth is about 23 m. During its service life, it may accumulate thick backfill soil on it, which can cause the tunnel to experience significant overlying loads (5–6 times that of ordinary sunken pipes). Especially during channel dredging, the load on the sunken pipes can undergo drastic changes along the longitudinal direction, resulting in significant internal forces. On the other hand, immersed tube tunnels need to ensure strict waterproofing, and the deformation requirements at the joints are also extremely strict [14]. Due to factors such as uneven settlement of the foundation, siltation, and water pressure, the stress on the pipe joint is different from that of conventional shallow buried tunnels. Both rigid and flexible pipe joints cannot perfectly adapt to this special working condition. Figure 2 shows the structural layout of the immersed tube tunnel of the Hong Kong Zhuhai Macao Bridge.

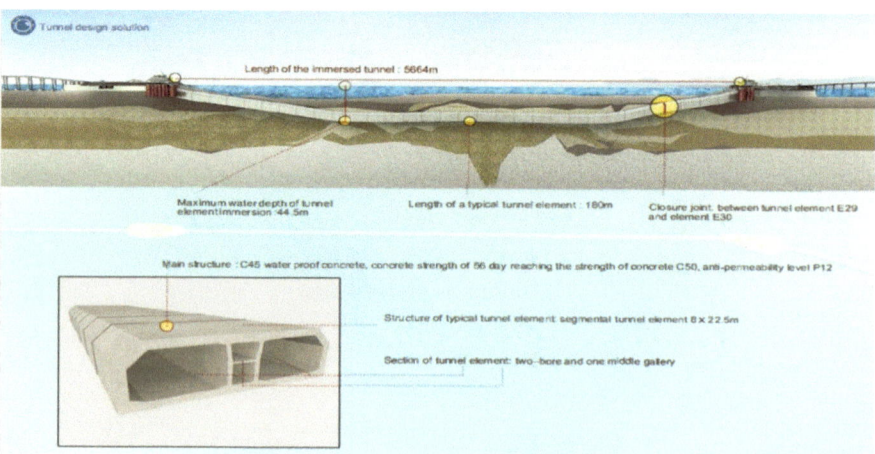

Fig. 2 Structural layout of immersed tube tunnel of Hong Kong-Zhuhai-Macao Bridge

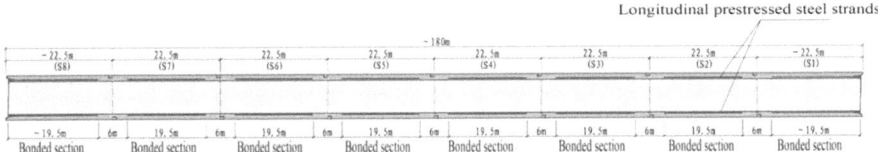

Fig. 3 Layout of prestressed longitudinal section of semi-rigid pipe joint

In response to the characteristics of deep burial and heavy overlying loads in the immersed tunnel project of the Hong Kong Zhuhai Macao Bridge, a semi rigid pipe joint scheme is proposed to obtain the advantages of rigid and flexible pipe joints and reduce the weaknesses of the two systems. This scheme is based on the flexible pipe joint scheme, and the main measure is to transform the temporary prestressed reinforcement of the segment joint into permanent prestressed reinforcement (Fig. 3), and connect 8 sections into one pipe joint with prestressed reinforcement while floating installation. This structural system change retains the main advantages of the original segmental joints, while also controlling the displacement of tunnel segmental joints, improving their safety reserves in normal use. It is necessary to systematically analyze the stress and deformation characteristics of this new type of immersed tube tunnel structure, and reveal its working mechanism and usage range by comparing it with rigid and flexible tube joints, providing reference for similar projects.

3 Force Analysis of Semi-rigid Pipe Joints

The top of the E13 and E14 pipe sections of the Hong Kong Zhuhai Macao Bridge immersed tunnel has a large sedimentation load, and the superimposed channel load is the most unfavorable pipe section under stress. Select these two pipe joints as numerical simulation objects to analyze the mechanical performance of semi-rigid pipe joints.

3.1 Model

Using Plaxis3D to establish a large-scale three-dimensional geological structure model to analyze the forces acting on sunken pipes under different load conditions. Compared with load structure models such as elastic foundation beams and elastic foundation plates, the main advantage of large-scale three-dimensional geological structure models is that they can fully consider the interaction between soil and structure, and can also consider the influence of lateral loading on uneven settlement of the foundation and internal forces of the structure. When the backfill load and siltation load on both sides of the immersed tube tunnel are greater than the load in the immersed tube area and the lateral scale of the immersed tube is larger (38 m), using a general load structure model and the stiffness of the foundation at the center of the immersed tube cannot accurately calculate the internal force of the structure, and the calculation results are often biased towards danger. The establishment of large-scale 3D models can solve this problem through actual backfilling, backfilling, and other processes. Considering the influence of longitudinal length of the tunnel and changes in the force on the immersed tube on the calculation results, the E12-E15 pipe joint is taken along the longitudinal direction of the tunnel for modeling and calculation. The transverse width of the model is taken as 100 m based on symmetry, and the depth direction is taken to −65 m elevation. The stratum is simplified as a horizontal layering situation, and E13-E14 is also horizontally sunk along the longitudinal direction. Divide a total of 267,539 units and extract the calculation results of pipe sections E13-E14.

Spring elements along the longitudinal and vertical directions of the tunnel are used to simulate the contact between immersed pipe joints. The overall model of the joint is shown in Fig. 4.

The mechanical parameter values of the original foundation soil layer and backfill soil layer are shown in Tables 1 and 2.

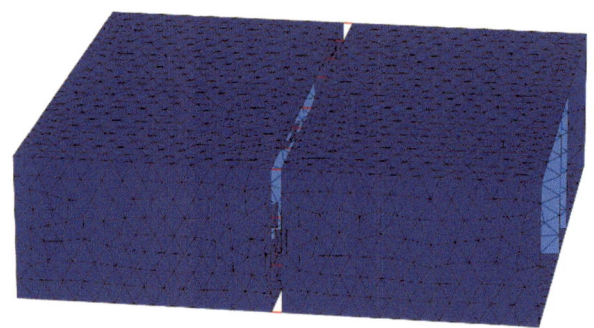

Fig. 4 Joint simulation finite element model

Table 1 Mechanical parameters of soil layer

No.	Solum	γ (kN/m^3)	C (kPa)	Φ (°)	E50ref (MPa)	m
11	Mud and muddy soil	16.2	9	6	2.50	0.76
21	Clay	19.1	23	27	9.30	0.67
31	Clay	18.2	43	20	9.65	0.90
32	Clay interbedded with sand	18.7	36	24	15.25	0.60
33	Fine sand	19.2	1	36.2	29.72	0.60
44	Medium sand	20.3	0	37.8	42.42	0.60

Table 2 Mechanical parameters of main backfill material

Solum	γ (kN.m^{-3})	c (kPa)	φ (°)	E50ref (MPa)	m
Dynamic compaction of crushed stones	20	0	40	100	0.60
Gravel cushion	20	0	40	10	0.60
Backfilling with crushed stones	20	0	40	10	0.60
Backfill soil	15	9	6	2.5	0.76

3.2 Simulation Calculation Steps

In order to analyze the situation of each construction step, numerical simulation is divided into six calculation steps: generating initial ground stress, excavation of foundation trench, construction of foundation cushion layer, tunnel sinking and locking backfill, backfilling of protective surface, backfilling during operation period, and excavation of waterway (Fig. 5).

3.3 Analysis Results

Considering the possible uncertainty of foundation stiffness due to construction deviations and other reasons, a sinusoidal curve with a wavelength of 180 m, a phase of 0 m, and a stiffness difference amplitude of 30% (compared to the benchmark stiffness) is used to consider the variation of foundation stiffness based on the benchmark stiffness. At the same time, the temperature variation condition of ± 10 °C inside and outside the structure is also considered. For semi-rigid pipe joints The finite element calculation results for flexible pipe joints and low pre stress semi rigid cases are shown in Tables 3, 4 and 5.

By comparing the calculation results in Tables 3, 4 and 5, it can be concluded that:

(1) Semi rigid pipe joint

① The initial compression force of 286,025 kN is generated by the prestressed tensioning and hydraulic docking joint. Under unfavorable cooling conditions, the minimum permanent pressure of the segment joint is 149,780 kN. Conservatively considering the concrete friction coefficient of 0.3, it can provide a friction resistance of 44,934 kN, which is greater than the maximum shear force of the segment joint of 28,779 kN. The shear resistance and displacement of the segment joint meet the requirements.

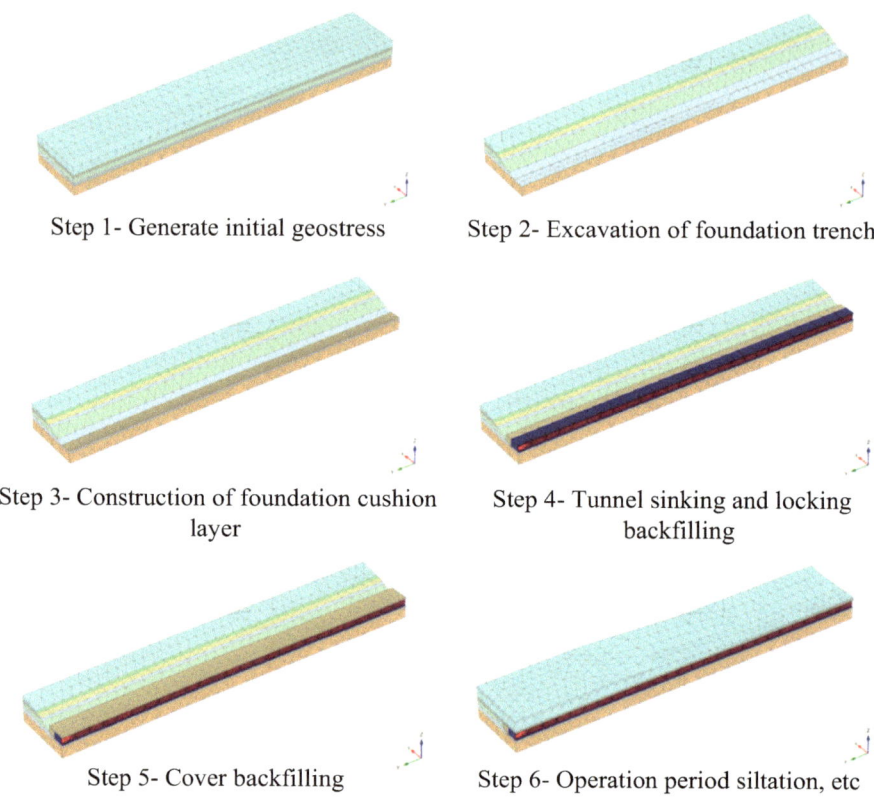

Fig. 5 Simulation calculation steps

② The maximum shear force of the pipe joint is 25,049 kN, which is less than the allowable value of 26,000 kN for the normal use of the steel shear key shear force. The shear resistance of the pipe joint meets the requirements.

(2) Flexible pipe joint

① The hydraulic compression joint generates an initial compression force of 103,558 kN. Under unfavorable cooling conditions, the minimum permanent pressure of the segment joint is 10,089 kN. Conservatively considering the concrete friction coefficient of 0.3, it can provide a friction resistance of 3027 kN, which is less than the maximum shear force of 11,158 kN for the segment joint. The segment joint has slid, and the concrete shear key (allowable shear force in normal use is 16,000 kN) participates in shear resistance.

② The maximum shear force of the pipe joint is 22,140 kN, which is less than the allowable value of 26,000 kN for the normal use of the steel shear key shear force. The shear resistance of the pipe joint meets the requirements.

(3) Low pre stress semi rigid pipe joint

Table 3 Calculation results of semi-rigid pipe joints with moderate prestressing force

Variability of foundation Stiffness		L180-S-30%		
Type of pipe joint		Semi-rigid		
Working condition		Basic	Heating up	Cool down
Joint shear resistance calculation	Section joint pre pressure (kN)	286,025	286,025	286,025
	Maximum tensile increment of segment joint (kN) (tension is positive)	1095	−19,722	136,245
	Minimum permanent pressure at segment joints (kN)	284,930	305,747	149,780
	Friction resistance of segment joints (kN) (Friction coefficient taken as 0.3)	85,479	91,724	44,934
	Maximum shear force of segment joint (kN)	11,228	28,042	28,779
	Section joint shear safety factor	7.6	3.3	1.6
	Maximum shear force of pipe joint (kN)	9980	24,970	25,049
Concrete structure crack verification calculation	Maximum pressure increment of segment joint (kN) (tension is negative)	1381	41,708	−77,554
	Maximum axial pressure (kN)	287,406	327,733	208,471
	Minimum axial pressure (kN)	284,930	305,747	149,780
	Maximum bending moment (kNm)	292,481	720,351	749,881
	Maximum tensile stress of concrete (MPa)	−1.229	−0.552	0.479
Joint opening verification calculation	Maximum opening of segment joints (mm)	0	0	0.1
	Opening increment of pipe joint (mm)	0.2	−6	20
	Initial tightening amount of pipe joint (mm)	154	154	154
	Final tightening amount of pipe joint (mm)	153.8	160	134

① An initial compression force of 194,649 kN is generated by prestressed tensioning and hydraulic docking. Under unfavorable cooling conditions, the minimum permanent pressure of the segment joint is 63,872 kN. Conservatively considering the concrete friction coefficient of 0.3, it can provide a friction resistance of 19,162 kN, which is less than the maximum shear force of the segment joint of 26,764 kN. The segment joint has slid, and the concrete shear key participates in the shear resistance.

Table 4 Calculation results of flexible pipe joints

Variability of foundation stiffness		L180-S-30%		
Type of pipe joint		Flexibility		
Working condition		Basic	Heating up	Cool down
Joint shear resistance calculation	Section joint pre pressure (kN)	103,558	103,558	103,558
	Maximum tensile increment of segment joint (kN) (tension is positive)	611	−22,562	93,469
	Minimum permanent pressure at segment joints (kN)	102,947	126,120	10,089
	Friction resistance of segment joints (kN) (Friction coefficient taken as 0.3)	30,884	37,836	3027
	Maximum shear force of segment joint (kN)	11,159	24,695	11,158
	Section joint shear safety factor	2.8	1.5	0.3
	Maximum shear force of pipe joint (kN)	9978	22,140	10,034
Concrete structure crack verification calculation	Maximum pressure increment of segment joint (kN) (tension is negative)	1382	46,349	−55,904
	Maximum axial pressure (kN)	104,940	149,907	47,654
	Minimum axial pressure (kN)	102,947	126,120	10,089
	Maximum bending moment (kNm)	288,998	559,255	289,561
	Maximum tensile stress of concrete (MPa)	−0.098	0.267	0.483
Joint opening verification calculation	Maximum opening of segment joints (mm)	0	1.6	4.3
	Opening increment of pipe joint (mm)	0.2	−6.3	14.3
	Initial tightening amount of pipe joint (mm)	156.9	156.9	156.9

② The maximum shear force of the pipe joint is 24,802 kN, which is less than the allowable value of 26,000 kN for the normal use of the steel shear key shear force. The shear resistance of the pipe joint meets the requirements.

(4) The difference in opening and compression of pipe joints is relatively small under basic and heating conditions. Under cooling conditions, the local opening of the segment joint of the semi-rigid pipe joint is 0.1 mm, which is significantly smaller than that of the flexible pipe joint, which is 4.3 mm, and both are less than the opening limit of the segment joint of 60 mm (this value is determined by the performance of the Ω water stop). The opening of the segment joint is

Table 5 Low pre stress semi rigid calculation results

Variability of foundation stiffness		L180-S-30%		
Type of pipe joint		Low pre stress semi rigid		
Working condition		Basic	Heating up	Cool down
Joint shear resistance calculation	Section joint pre pressure (kN)	194,649	194,649	194,649
	Maximum tensile increment of segment joint (kN) (tension is positive)	1104	−19,805	130,777
	Minimum permanent pressure at segment joints (kN)	193,545	214,454	63,872
	Friction resistance of segment joints (kN) (Friction coefficient taken as 0.3)	58,064	64,336	19,162
	Maximum shear force of segment joint (kN)	11,105	27,850	26,764
	Section joint shear safety factor	5.2	2.3	0.7
	Maximum shear force of pipe joint (kN)	9981	24,802	23,341
Concrete structure crack verification calculation	Maximum pressure increment of segment joint (kN) (tension is negative)	1381	41,792	−75,740
	Maximum axial pressure (kN)	196,030	236,441	118,909
	Minimum axial pressure (kN)	193,545	214,454	63,872
	Maximum bending moment (kNm)	292,483	705,130	564,275
	Maximum tensile stress of concrete (MPa)	−0.658	−0.010	0.666
Joint opening verification calculation	Maximum opening of segment joints (mm)	0	0.2	0.7
	Opening increment of pipe joint (mm)	0.2	−6	19
	Initial tightening amount of pipe joint (mm)	155.4	155.4	155.4
	Final tightening amount of pipe joint (mm)	155.2	161.4	136.4

far from exceeding the limit. The above verification indicates that the problem of opening quantity is not the main contradiction.

When the stiffness of the foundation changes, the shear force levels of different types of pipe joint segments are shown in Fig. 6.

When the pressure value of the segment joint is small (the segment joint will open), the shear force value increases with the increase of pressure value, and the shear force value without pre-stressed shear is greater than that with pre-stressed shear; However, when the pressure value reaches a certain level (i.e. the segment

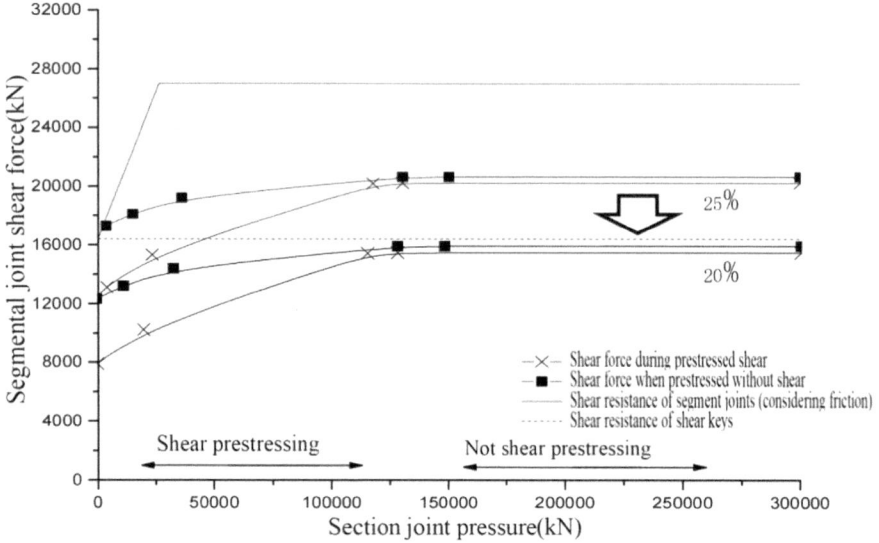

(a) When the coefficient of variation of stiffness decreases by 5%, the change in joint shear force

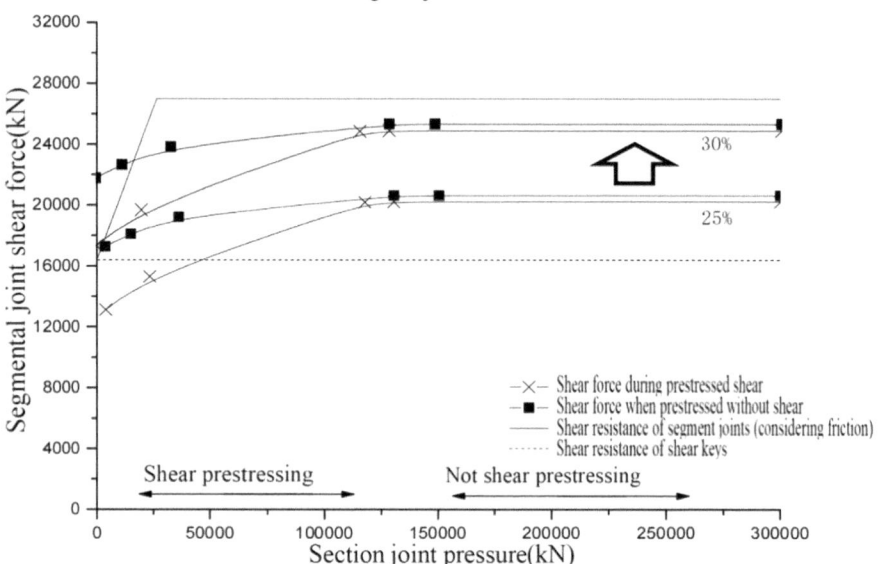

(b) Changes in joint shear force when stiffness coefficient of variation increases by 5%

Fig. 6 Comparison of joint shear forces under different foundation stiffness variation coefficients

joint does not open), the shear force value tends to stabilize, reaching its maximum value. The shear force value of pre-stressed shear is basically equivalent to the shear force value of pre-stressed shear. Considering that the actual internal forces of the structure are all within the demonstrated range in Fig. 6, under the same conditions, the pressure value of the pre-stressed segment joint without shear is greater than that of the pre-stressed shear. The sum of the shear resistance of the segment joint shear key and the frictional resistance is the maximum shear resistance that the structural cross-section can withstand.

4 Outlook and Engineering Application Effects

Due to the presence of significant compression force on the segment joints in the model, it is assumed that the interface is completely rough, and the impact of interface slip was not considered in the analysis of shear stiffness. The model did not consider the influence of the actual longitudinal slope or actual spatial posture of the pipe joint on the force acting on the joint. Subsequent research can further refine the model, improve boundary conditions, and obtain more accurate analysis data.

The research results of this article have been successfully applied in the immersed tube tunnel of the Hong Kong Zhuhai Macao Bridge. In the 10 years from construction to current operation, a total of 219 segment joints of the immersed tube tunnel of the Hong Kong Zhuhai Macao Bridge have not shown any signs of water leakage. Compared with the average leakage rate of about 10% joints in immersed tube tunnels worldwide, the semi-rigid scheme has achieved success and solved the structural problem of deep buried large siltation immersed tube tunnels.

5 Conclusion

(1) With the increase of the pre stress intensity (i.e. stiffness) of the semi-rigid pipe section configuration, the vertical shear resistance of the segment joint is improved.
(2) The calculated value of the horizontal opening of the semi-rigid pipe segment joint is the smallest, and the local opening of 0.1 mm is much smaller than the opening limit of 60 mm. Its waterproof safety is better than that of flexible pipe joints.
(3) For the calculated shear force of pipe joints, semi-rigid pipe joints have the highest value, followed by low pre stress pipe joints, and flexible pipe joints have the lowest value, which is relatively close. Due to the same shear resistance of pipe joints, the shear safety of pipe joints in semi-rigid pipe joints is relatively minimum. However, even under unfavorable variations in foundation stiffness, it does not exceed the limit value of steel shear keys.

(4) In the future, further analysis can be conducted on the force changes of semi-rigid immersed tube structures under earthquake load conditions to improve their stress mechanisms.

References

1. Jun S (2014) Countermeasures for deep and thick soft foundation under serious back-silting condition: case study on artificial island and immersed tunnel of Hong Kong-Zhuhai-Macao Bridge Project [J]. Tunnel Constr 34(09):807–814
2. Haitao Y, Yong Y, Hongzhou L et al (2014) Mechanical modeland analytical solution for stiffness in the joints of an immersed-tube tunnel [J]. Eng Mech 31(6):145–150
3. Anastasopoulos I, Gerolymos N, Drosos V et al (2008) Behaviour of deep immersed tunnel under combined normal fault rupture deformation and subsequent seismic shaking [J]. Bull Earthq Eng 6(2):213–239
4. Journal of China Highway, Editorial Department (2022) Review on China's traffic tunnel engineering research: 2022 [J]. China J Highw Transp 35(04):1–40.https://doi.org/10.19721/j.cnki.1001-7372.2022.04.001
5. Zhigang Z, Hongzhou L (2013) Development and key technologies of immersed highway tunnels [J]. Tunnel Constr 33(05):343–347
6. Yong Y, Haitao Y, Wenhao X et al (2017) Experimental failure analysis on concrete shear keys in Immersion joint subjected to Compression-Shear loading[J]. Eng Mech 34(03):149–154+181
7. Xinjun C, Liping J, Jie C et al (2020) Experimental failure analysis on immersed tunnel joint subjected to shear loading [J]. China J Highw Transp 33(04):99–105. https://doi.org/10.19721/j.cnki.1001-7372.2020.04.010
8. Peng L, Wenqi D, Bo Y (2013) Model for stiffness of joints of immersed tube tunnel [J]. Chinese J Geotech Eng 35(S2):133–139
9. Yong Z, Hailong W, Jun G et al (2016) Vertical compression shear mechanical performance study of immersed tube tunnel Joint [J]. Chinese J Underground Space Eng 12(S1):24–31
10. Gang W, Shijie L, Jianjian X (2018) Study on theoretical calculation model of longitudinal force in immersed tube tunnel [J]. Chinese Journal of Underground Space and Engineering,2018,14(04):912–919+935.
11. Yu Haitao, Xiao Wenhao, Zhao Xu, et al. Compression Performance of Flexible Joints in Immersed Tunnels [J]. China J. Highw. Transp., 2019, 32(05): 115–122+180. https://doi.org/10.19721/j.cnki.1001-7372.2019.05.011.
12. Zongxian S, Shaozhang C, Yue C et al (2018) Discussion on longitudinal static calculation of immersed tunnel [J]. Tunnel Constr (Chinese and English) 38(05):790–796
13. Yuyang L, Yongli X, Hongpeng L et al (2015) Influence of foundation settlement on mechanical performance of shear keys of segment joints in immersed tube tunnels [J]. Chinese J Geotech Eng 37(12):2235–2244
14. Ming L, Wei L, Haiqing Y et al (2018) Memory bearing-problem solutions of immersed tunnel immersion joint differential settlement [J]. China Harbour Eng 38(06):1–8

Open Access This chapter is licensed under the terms of the Creative Commons Attribution 4.0 International License (http://creativecommons.org/licenses/by/4.0/), which permits use, sharing, adaptation, distribution and reproduction in any medium or format, as long as you give appropriate credit to the original author(s) and the source, provide a link to the Creative Commons license and indicate if changes were made.

The images or other third party material in this chapter are included in the chapter's Creative Commons license, unless indicated otherwise in a credit line to the material. If material is not included in the chapter's Creative Commons license and your intended use is not permitted by statutory regulation or exceeds the permitted use, you will need to obtain permission directly from the copyright holder.

Research on Anti-shear Performance of Steel-Mixed Combination Beam Bridge Single-Nail Shear Connector

Haiyuan Yang

Abstract In order to study the shear performance of single spigot shear connectors when applied in practical engineering, the shear performance research is carried out with the engineering background of a waveform steel web steel-mixed girder bridge under construction in a certain area, using the finite element modeling method to establish a local model in the connecting part of the steel box girder and the concrete plate, and combining with the practical engineering to carry out the simulation launching test of the spigot connectors and comparing the results of the load displacement curve with the classical theoretical curve, the results match well and verify the modeling correctness. The results of load displacement curve are compared with the classical theoretical curve, and the results are in good agreement, which verifies the correctness of modeling. On this basis, the influence of steel strength and concrete strength on the shear performance of the peg is investigated, and it is found that the shear capacity increases with the increase of peg strength and concrete strength, and the peg strength has a greater influence on the shear capacity than the concrete strength.

Keywords Steel–concrete composite structure · Bolted connectors · ANSYS · Push-out test

1 Introduction

As a key component of steel-mixed structure, the main function of the shear connector is to combine the steel and concrete structure to form a whole, and its superior performance directly affects the common force and coordinated deformation performance of steel–concrete structural components [1]. The shear performance of shear connectors is generally obtained through the launching test, but it is time-consuming and laborious; and through the numerical analysis method, the launching test can be

H. Yang (✉)
Department of Civil Engineering, Lanzhou Jiaotong University, Lanzhou 730070, Gansu, China
e-mail: 49204829@qq.com

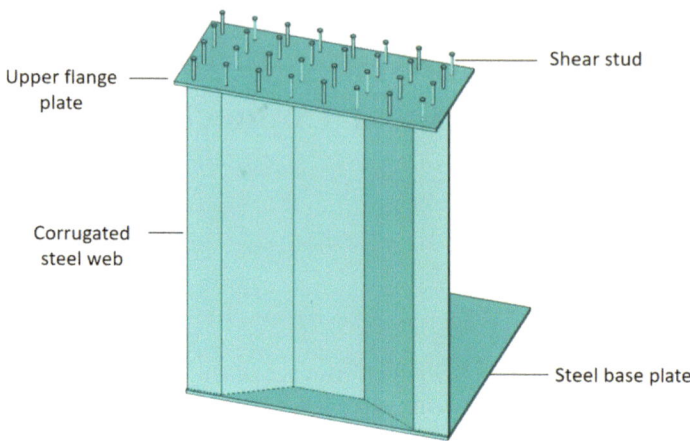

Fig. 1 Schematic diagram of spigot shear connectors

simulated so as to predict the performance of shear parts, and then a small number of tests can be verified, which is a method worthy of popularization.

In the corrugated steel web combined box girder bridge structure, the selection and design of shear connectors is very critical, which combines the two materials of steel and concrete and makes them work together [2]. It has the following characteristics: good force performance, uniform force in all directions and the same shear stiffness, the design of the arrangement does not need to consider the directionality of the force, the pegs are easy to take materials from the ground, and the quality of the welding process requirements are high [3]. The bolted shear connector is formed by welding bolts on the steel flange plate and combining with the corrugated steel web to form a bolted shear connector, the structural form of which is shown in the Fig. 1.

In order to study the performance of bolted connectors in the actual engineering shear performance, a region under construction in the waveform steel web steel-mixed combination of beam bridge steel-mixed combination of parts as the object of study, the use of finite element analysis software ANSYS APDL to establish bolted connectors to launch the specimen of the accurate three-dimensional finite element local model, the study of bolted shear connectors shear slip performance, to study the different concrete strengths, the strength of the bolts, the effect of the shear load bearing capacity of the shear connectors and the slip performance [4].

2 Establishment of Local Finite Element Model of Stud Shear Connection

2.1 Material and Unit Selection

Using finite element software ANSYS APDL programming language to a waveform steel web steel box combination girder bridge as the engineering background, the local model is shown in Figs. 2 and 3, the material parameters, structural dimensions, and reinforcement arrangement is based on the actual project. Since the material properties of the pegs play a dominant role in the shear process of shear nails and steel plates, the material properties of the pegs are uniformly adopted for shear nails and flange plates in modeling, and the strength of the ML15AL in GBT10433-2002 *Cylindrical Head Weld Nails for Arc Stud Welding* is taken. The concrete is taken as C60 and the steel reinforcement is taken as HPB300. the whole is modeled by 3D solid entity unit, and the concrete and steel reinforcement are chosen to be modeled by SOLD65 solid entity unit in one piece, and the real constants are used to define the position of the steel reinforcement as well as the reinforcement rate, and the concrete is used with multi-linear entourage reinforcement model (MKIN) and concrete-specific code of concretes destruction (CONCR). The bolts and steel plates were modeled using SOLID185 solid units, and a multi-linear isotropic hardening model (MISO) was selected [5].

2.2 Material Constitutive Relationship Selection

The model of the concrete is Hongestad formula, as shown in the following formula:

$$\sigma_i = \begin{cases} f_c \left[2\frac{\varepsilon}{\varepsilon_0} - \left(\frac{\varepsilon}{\varepsilon_0}\right)^2 \right], & (\varepsilon \leq \varepsilon_0) \\ f_c \left[1 - 0.15\frac{\varepsilon - \varepsilon_0}{\varepsilon_u - \varepsilon_0} \right], & (\varepsilon_0 \leq \varepsilon \leq \varepsilon_u) \end{cases} \quad (1)$$

In the formula: f_c is peak stress (compressive strength of prism), ε_0 is the strain corresponding to reaching the peak stress, ε_u is the ultimate compressive strain.

The constitutive model of shear nails is selected as the three line model, as shown in the following equation:

$$\sigma_i = \begin{cases} E_s \varepsilon_i, & (\varepsilon_i \leq \varepsilon_y) \\ f_y + 0.01 E_s (\varepsilon_i - \varepsilon_y), & (\varepsilon_y < \varepsilon_i \leq \varepsilon_u) \\ f_u = 1.25 f_y, & (\varepsilon_i > \varepsilon_u) \end{cases} \quad (2)$$

In the formula: σ_i is the equivalent stress of steel. f_y is the yield strength of steel. f_u is the ultimate strength of steel. The yield strength of stud is generally high

Fig. 2 Shear nail and steel flange plate

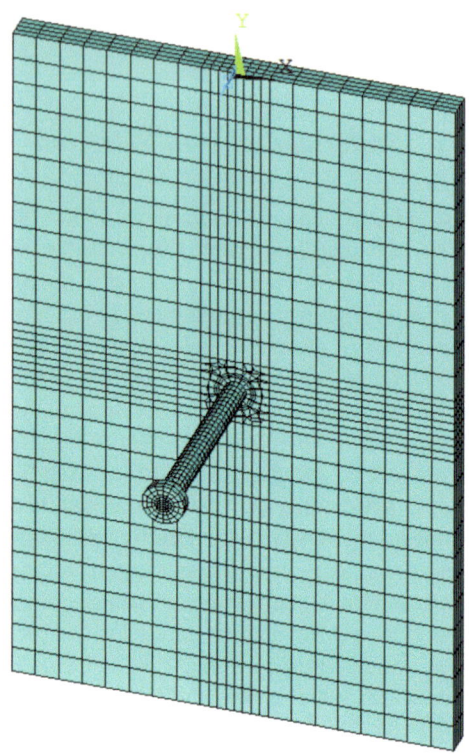

(300,600 MPa), the elastic–plastic strengthening phenomenon is obvious, and the yield strength is relatively high. According to this paper and the existing test results. f_u/f_y is between 1.2 and 1.3. Take the median value of 1.25. $f_u = 1.25 f_y$, E_s is the elastic modulus of steel. Take $E_s = 2.06 \times 10^5$ MPa, ε_i is the equivalent strain of steel. ε_y is the strain of steel when yielding. ε_y is the strain of steel when it reaches its ultimate strength. $\varepsilon_u = 21 \varepsilon_y$.

The material constitutive model of steel bar is rational elastoplastic model, as shown in the following formula:

$$\sigma_i = \begin{cases} E_s \varepsilon_i, (\varepsilon_i \leq \varepsilon_y) \\ f_y, (\varepsilon_i > \varepsilon_y) \end{cases} \quad (3)$$

2.3 Contact Interface Simulation

Because of the relative slip of materials of different properties, the contact surface needs to be defined on the interface to simulate the relative slip on the interface. For

Fig. 3 Overall schematic of the local model

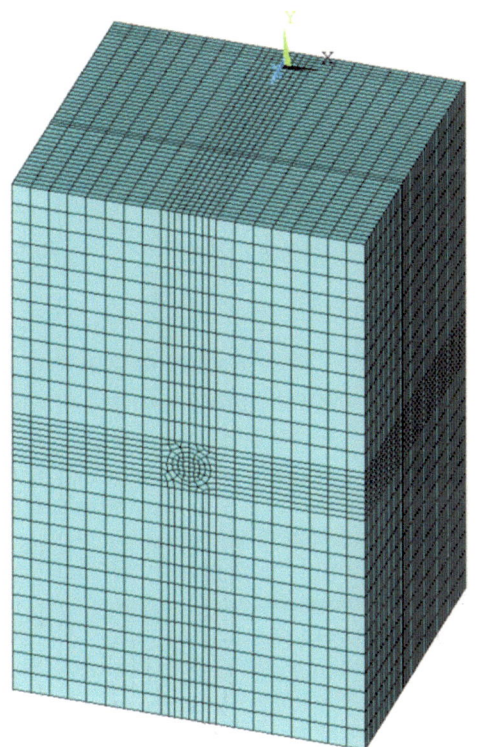

the simulation of the contact interface, according to the characteristics of Solid65 and SOLID185 physical unit, select the Conta173 and Targe170 that matches it. The principle of rigidity defined by ANSYS contact surface, the concrete surface is simulated with Conta173, the shear nails and steel wing edge surfaces are simulated with targe170. In order to simulate the actual situation of the interface interface, the contact mode of the contact interface is defined as a standard contact, the contact surface method is transmitted to the pressure, allowing the contact surface method to be separated by the subject, cutting to the allowed interface to generate relatively smooth movement, the friction coefficient of friction coefficient Take 0.25, to prevent convergence before the material surrender due to excessive invasion during non - linear calculations, and automatically adjust the rigidity of the contact surface on each load step according to the lower surface stress size.

2.4 Boundary Condition Definition and Load Loading

By applying facial constraints to add boundary conditions to the model. For concrete, the surface constraints are applied to the bottom of the vertical bottom surface with the y-axis. The displacement, the surface constraint on the top of the concrete perpendicular to the Z axis limits its displacement in the Z direction. For the steel wing edge, the surface constraints are applied to the sides perpendicular to the X-axis to limit its displacement in the X direction, and the surface constraints that limit the Z direction on the bottom surface of the steel wing plate vertical on the Z-axis. When applying the load, the load F is divided into half in half, and the symmetry is applied to the two ends of the Y-axis. Define the single load load, and the multi-loaded child step is gradually loaded until the structural material is yielded and destroyed.

2.5 Stress Mechanism of Shear Connectors and Results of Finite Element Stress and Deformation Analysis

Scholars at home and abroad show that the maximum pressure on the bottom of the stud shear connector decreases gradually along the height direction, and tensile stress appears at the top. The variation law of compressive stress along the height of the stud is shown in the figure. In the actual composite beam structure, the stud bears biaxial compressive stress, and the stress form is like an elastic foundation beam. The shear strength of the shear connector can be improved by transferring shear force at the steel–concrete composite interface. The elastic foundation beam model is shown in the Fig. 4.

As shown in Fig. 5. The maximum stress on the peg and the flange plate is generated at the root of the peg, and the peg is finally destroyed because the stress at the root of the peg reaches the limit, which is in line with the actual situation of peg destruction, and the stress distribution of the shear peg in the results of the finite element analysis is in line with the stress distribution of the actual peg shown in figure. The deformation of the peg in the shape of the deformation of the peg is also in line with the actual situation, and in conclusion, it has been verified that it is reasonable to carry out the finite element analysis by the above method.

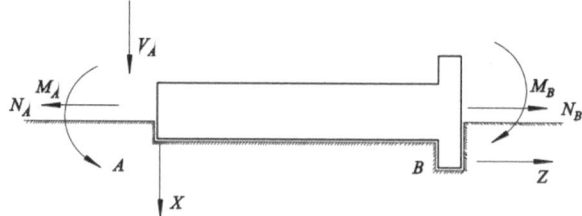

Fig. 4 Schematic diagram of elastic foundation beam model

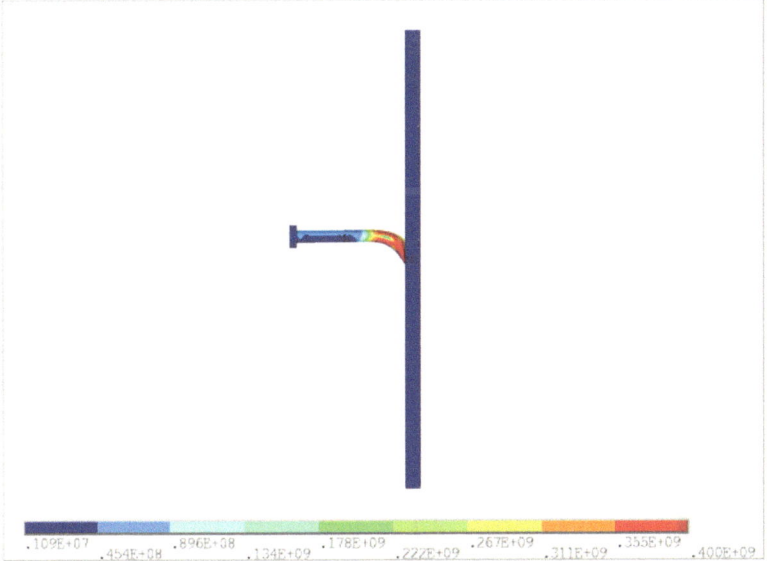

Fig. 5 Peg stress diagram

2.6 Shear Bearing Capacity and Theoretical Values

The formula for calculating the shear bearing capacity of a single stud connector is given in 'Steel Structure Design Standards'.

$$N_u = 0.43A_s\sqrt{E_c f_c} \leq 0.7A_s f_u \tag{4}$$

In the formula: A_s is the cross-sectional area of the stud. f_c is the compressive strength of concrete. f_u is the tensile strength of the stud.

According to 'Steel–concrete combination bridge design specification', the shear bearing capacity of a single bolted connection should be taken as the smaller of the following two equations:

$$\left.\begin{array}{l} N_u = 1.19A_s f_u (E_c/E_s)^{0.2}(f_{cu}/f_u)^{0.1}, \\ N_u = 0.43\eta A_s \sqrt{f_c E_c}. \end{array}\right\} \tag{5}$$

In the formula: η is the reduction factor of group nails. Take 1.0.

According to Eurocode 4, the formula for calculating the shear bearing capacity of bolted connections is:

$$N_u = 0.29\alpha d_s^2 \sqrt{E_c f_{c'}}/\gamma_v \leq 0.8A_s f_u/\gamma_v \tag{6}$$

Table 1 Comparison of shear capacity with code

	Shear capacity	Ratio to the results of this model
Steel structure design criteria	56.3	0.70
Design specification for steel–concrete composite bridges	51.2	0.64
Eurocode4	51.4	0.64
AASHTO2007	68.3	0.86
FEA model calculation results	79.8	1

In the formula: α is the influence coefficient of stud height. $\alpha = 0.2(h_s/d_s + 1) \leq 1.0$. f_c' is the compressive strength of concrete cylinder. γ_v is the partial coefficient of stud resistance, taking 1.25.

The calculation formula for the shear bearing capacity of bolted connections given in the specification AASHTO2007 is:

$$\varphi_{sc}Q_n = \varphi_{sc}0.5A_s\sqrt{E_c f_c'} \leq \varphi_{sc}A_s f_u \qquad (7)$$

In the formula: φ_{sc} is the resistance coefficient, taking 0.85. Q_n is the nominal value of shear capacity.

According to the formula 1–4, the shear bearing capacity of the model is calculated and compared with the simulation results, as shown in Table 1.

2.7 Load–displacement Curve Analysis

Load displacement curves of bolted shear connectors at home and abroad.

The relative slip effect of steel -concrete combination beam interface cannot be ignored, and some scholars are limited to the destruction standard for the breakdown of the nail connector with a sliding amount. Therefore, while studying the bearing capacity of the shear -resistant connector, it is usually accompanied by the relatively slippery research of the interface of the combination beam.

Buttry proposed a fractional load–displacement curve.

$$V = V_{\max} \cdot \frac{3.15s}{1 + 3.15s} \qquad (8)$$

In the formula: $V_{\max}$ is maximum shear force can be carried. S is the displacement value.

Fisher and others put forward an exponential load–displacement curve [6].

$$V = V_{\max} \cdot \left(1 - e^{-0.71s}\right)^{0.4} \qquad (9)$$

N.Gattesco and others modified the exponent and put forward the modified exponent form [7].

$$V = V_{max} \cdot [0.97(1 - e^{-1.34s})^{0.5} + 0.0045s] \tag{10}$$

Finite element deduction results of load–displacement curve.

As shown in Fig. 6. In order to verify the correctness of the results of the finite element model, the load–displacement curves of the model were extracted from ANSYS APDL, and compared with the theoretical curves of the above mentioned scholars, with the horizontal axis as the relative slip value of the model pegs and the concrete, and the vertical axis as the ratio of shear bearing capacity to the maximum shear bearing capacity. The overall trend of the resultant curves of the finite element model is analyzed to be in good agreement with the three theoretical curves, and the best agreement with the curve of Fisher's formula, which verifies the correctness of the finite element model.

Different steel strengths.

As shown in Fig. 7. Taking the shear connector model with a compressive strength of 60 MPa and a diameter of 16 mm for concrete cubes as an example, the yield strength of the bolts is adjusted to 360 MPa, 400 MPa, and 440 MPa, respectively. The load relative slip curves for different yield strengths of the bolts are calculated and plotted as shown in Fig. 8. The failure strength ratio of bolts is 1:1.1:1.2. The corresponding shear bearing capacities are 72.2, 79.8, and 87.7 kN. This indicates that the shear bearing capacity of the bolt increases almost linearly with the increase of the yield strength of the bolt.

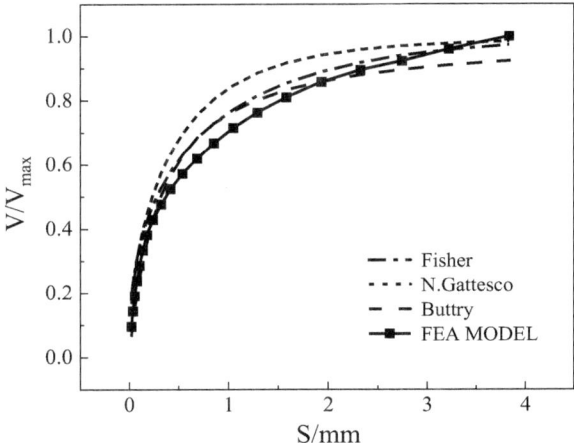

Fig. 6 Finite element deduction results

Fig. 7 Steel strengths

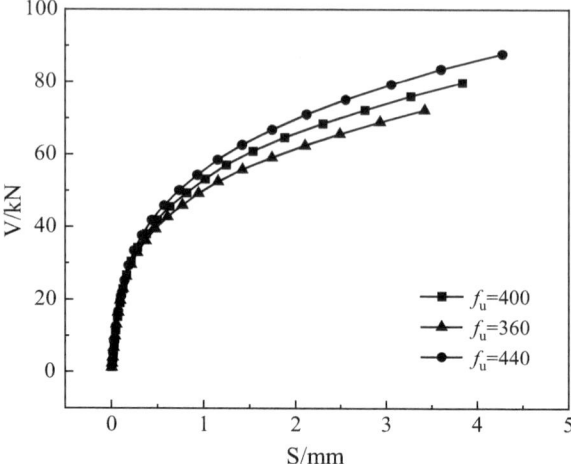

Fig. 8 Concrete strength

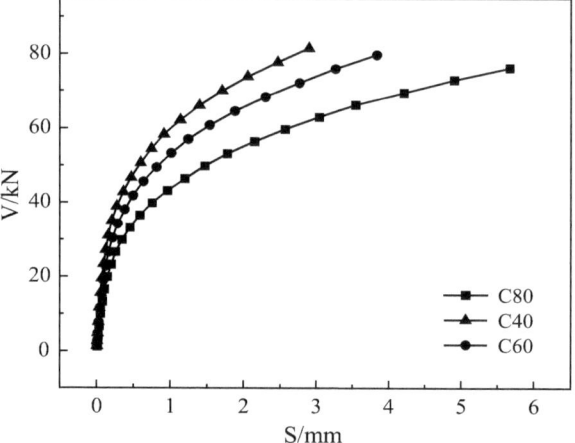

Different concrete strength.

As shown in Fig. 8. Taking the shear connector model with a yield strength of 360 MPa and a diameter of 16 mm as an example, the strength grades of concrete are adjusted to C40, C60 and C80 respectively, and the load-relative slip curves of different concrete strengths are calculated and drawn as shown in Fig. 7. When the strength of concrete increases from 40 to 60 MPa, the corresponding shear bearing capacity increases from 76.3 to 79.8 kN, an increase of about 4.5%. When the strength increases from 60 to 80 MPa, the shear bearing capacity increases to 81.6 kN, which is about 2.3% higher, indicating that the shear bearing capacity of studs increases with the increase of concrete strength, but the growth rate slows down.

3 Conclusion

This paper mainly establishes the three-dimensional finite element model of the push-out specimen of the pegged connectors through the finite element analysis software ANSYS APDL, and studies the influence of the constructional parameters on the anti-shear bearing capacity of the pegged shear connectors, and draws the following conclusions.

(1) The load-relative slip curve calculated by finite element and the predicted value of foreign empirical formulae are in good agreement, therefore, the bolted push-out test can be simulated better by using nonlinear finite element method.
(2) Parametric analysis shows that: the shear bearing capacity of the peg shear connection increases with the increase of concrete strength, but the growth rate slows down; the shear bearing capacity increases with the increase of yield strength of the pegs and increases almost linearly.

References

1. Jianping F, Pingming H, Shulai W et al (2014) Force performance analysis of bolted shear connectors of combined box girders. J Jiangsu Univ 35(04):438–443
2. Jinfeng W, Aiping Z, Wenhao W (2020) Influence of peg height on the shear performance of pegged connectors. J Zhejiang Univ 54(11):2076–2084
3. Xu C, Sugiura K, Wu C et al (2012) Parametrical static analysis on group studs with typical push-out tests. J Constr Steel Res 72:84–96
4. Weichen X, Min D, Hua W et al (2009) Experimental study on shear performance of bolted joints under monotonic loading. J Build Struct 30(1):25–28
5. Fa-xing D, Ming N, Yong-zhi G et al (2014) Experimental study on slip behavior and calculation of shear bearing capacity for stud shear connectors. J Build Struct 35(9):98–106
6. Ollgaard JG, Slutter RG, Fisher JW (1971) Shear strength of stub connectors in lightweight and normal-weight concrete. AISC Eng J 8(2):55–64
7. Gattesco N, Giuriani E (1996) Experimental study on stud shear connectors subjected to cyclic loading. Construct Steel Res 38(1):1–21

Open Access This chapter is licensed under the terms of the Creative Commons Attribution 4.0 International License (http://creativecommons.org/licenses/by/4.0/), which permits use, sharing, adaptation, distribution and reproduction in any medium or format, as long as you give appropriate credit to the original author(s) and the source, provide a link to the Creative Commons license and indicate if changes were made.

The images or other third party material in this chapter are included in the chapter's Creative Commons license, unless indicated otherwise in a credit line to the material. If material is not included in the chapter's Creative Commons license and your intended use is not permitted by statutory regulation or exceeds the permitted use, you will need to obtain permission directly from the copyright holder.

Research on the Fracture Evolution Characteristics and Mesoscopic Fracture Mechanism of Fissured Basalt Based on PFC^{2D}

Jun Chen, Ning Liu, Chaoyi Wang, and Yaohui Gao

Abstract The surrounding rock of large underground engineering projects is often composed of hard rock, and defects such as mineral particles and micro-cracks in the hard rock, significantly affect the mechanical properties of the rock. To investigate the influence of micro-crack distribution on the mechanical behavior, damage evolution, and fracture mechanism of rock, a numerical simulation model of basalt with single crack was established using the particle flow code PFC^{2D}, based on the widely distributed cryptocrystalline basalt in the Baihetan Hydropower Station. In this study, the laws of crack initiation, propagation, and connection in the basalt under loading were deeply studied; the influence of different crack angles on the deformation and strength of the rock mass was discussed; the evolution characteristics and fracture mechanism of the whole process of the basalt under different confining pressures were revealed. The results of this study are significant for further studies into the mechanical properties characterizing fissured rock masses, as well as providing guiding principles for the design of cavern support.

Keywords Microcrack · Hard rock · Particle flow · Crack propagation · Fracture mechanism

1 Introduction

Internal defects in rocks, such as mineral particle boundaries and microcracks, can significantly affect the mechanical properties and failure characteristics of rocks. With the excavation and unloading of underground engineering rock mass, the redistribution of surrounding rock stress causes the continuous propagation of primary

---

J. Chen · N. Liu · Y. Gao
PowerChina Huadong Engineering Co., Ltd, Hangzhou, China
e-mail: chen_j26@hdec.com

C. Wang (✉)
College of Child Development and Education, Zhejiang Normal University, Hangzhou, China
e-mail: cywang@zjnu.cn

cracks inside the rock, and new cracks will continue to initiation, propagation, and connection, ultimately forming macroscopic cracks, leading to the instability and failure of the surrounding rock [1, 2]. Therefore, studying the evolution law of rock damage has important theoretical and practical value for analyzing rock mechanical behavior, revealing rock fracture mechanisms, exploring rock constitutive relationships, and solving the stability of surrounding rocks.

In order to deeply understand and reveal the fracture mechanism of rocks, many scholars have used various experimental methods to study the effects of cracks, voids, and other factors on the strength, and deformation characteristics of rocks. Zhang [3] analyzed the deformation and fracture characteristics, crack propagation, and evolution law of the cryptocrystalline basalt using CT scanning, acoustic emission, and other methods. By using electron microscopy scanning technology, it was revealed that the main microscopic fractures of rocks are intergranular and transgranular fractures of mineral particles. Yang [4] used $RFPA^{3D}$ and micro seismic monitoring systems to study the evolution of stress field and the development, propagation, and connection process of microcracks during rock excavation. Hu [5] analyzed the mechanical and acoustic properties of columnar jointed basalt under uniaxial stress condition, as well as the influence mechanism of microcracks on deformation and failure. Huang [6] studied the influences of medium strain rate on stress–strain, crack tip stress, characteristic stress and damage of rock mass, fracture expansion of crack of sandstone with a single fissure under uniaxial compression based on PFC^{2D}. Research has elucidated that during the compression deformation process of rocks, stress concentration often occurs around internal microcracks, leading to local damage and failure, thereby initiating microcracks, and ultimately culminating in the failure of the rock. Consequently, it is evident that defects, such as microcracks, play an important role in the fracturing of brittle rock.

Because indoor mechanical tests and in-situ testing are limited to observing the macroscopic fracture of rocks from the surface, it is not possible to observe the initiation, propagation, and connection of microcracks inside hard rocks at the microscopic level. The PFC program based on particle flow theory can connect the microscopic mechanical parameters with macroscopic mechanical parameters to simulate the basic mechanical properties of rocks, and can monitor the initiation, propagation, and connection process of microcracks inside rocks in real-time. Reproducing the complex mechanical behavior of rocks at a microscopic level helps to reveal the deformation and fracture mechanism of rocks.

The study chooses the cryptocrystalline basalt of Baihetan Hydropower Station as the research subject based on the outcomes of indoor mechanical tests. This study proceeds to construct a numerical calculation model for basalt containing a single crack, engaging in particle flow simulation analyses of fissured basalt under varying confining pressure levels. It determines the evolution laws of stress, deformation, and energy of fissured basalt under various conditions, studies the initiation, propagation, and connection laws of cracks in fissured rock mass, consequently reveals the influences of crack dip angle on crack propagation and failure mechanism.

2 Numerical Calculation Model for Fissured Basalt

The following four methods can be used to simulate crack production in PFC^{2D} [7]: (1) Use the JSET command to set cracks (Fig. 1a). This method assigns weaker parameters to the particle bonding between the two sides of the crack, but the crack surface generated by this method is rough and uneven, making it difficult to generate sliding along the crack surface; (2) Simulate the fracture surface with a certain thickness of particle units (Fig. 1b). This method assigns weaker parameters to particle bonding within a certain thickness range to simulate cracks, especially suitable for simulating cracks with large opening and containing weak filling materials; (3) Delete a certain thickness of particle units to simulate cracks (Fig. 1c). This method is mainly used to simulate intermittent non closed cracks. After removing particles, the particles on both sides have separated without any interaction; (4) Using a smooth joint model to simulate cracks (Fig. 1d). This method can be used to simulate closed cracks and achieve sliding along the fracture surface, and the theoretical basis of this model conforms to the Mohr–Coulomb strength criterion, making it the most suitable method for simulating closed cracks in PFC^{2D}.

Since the interior cracks of the Baihetan cryptocrystalline basalt are typically firmly sealed, simulating the cracks with a smooth joint model is closer to the actual situation. Therefore, this article uses a smooth joint model to construct single fissured basalt with different dip angles. The angles α are 0°, 30°, 45°, 60°, and 90°, and the confining pressures are 0, 10, 20, 30, 40, 50, and 60 MPa. There are a total of 35 calculation models, and the calculation models for fissured basalt at each dip angle are shown in Fig. 2. The meso-mechanical parameters for PFC^{2D} of basalt are shown in Table 1.

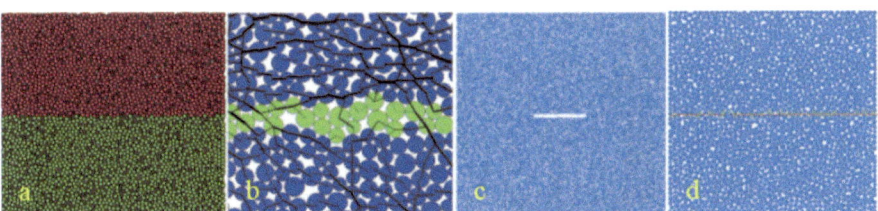

Fig. 1 Crack generation method of PFC^{2D} [7]

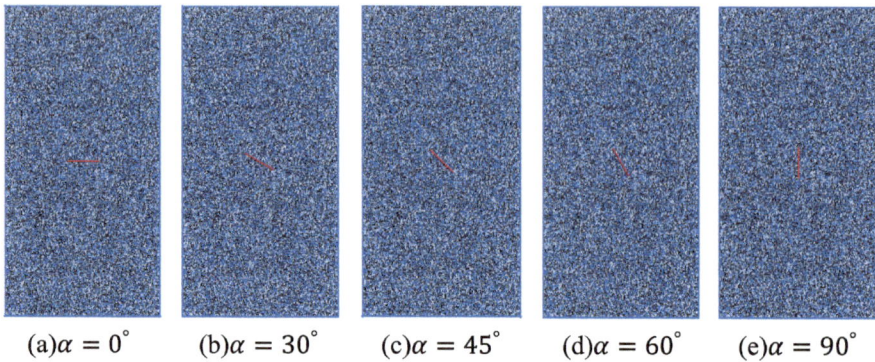

(a) $\alpha = 0°$ (b) $\alpha = 30°$ (c) $\alpha = 45°$ (d) $\alpha = 60°$ (e) $\alpha = 90°$

Fig. 2 Calculation model for fissured basalt with different angles

Table 1 Meso-mechanical parameters for particle flow simulation of cryptocrystalline asalt

Minimum radius of particle/mm	Particle size ratio	Contact modulus/GPa	Particle stiffness ratio	Parallel bonding modulus/GPa	Parallel bonding stiffness ratio	Tensile strength of parallel bonding/MPa	Parallel bonding cohesion/MPa	Internal friction angle of parallel bonding/°
0.2×10^{-3}	1.5	45	0.08	0.9	0.08	17	170	45

3 Numerical Analysis of Macroscopic and Microscopic Mechanical Behavior of Fissured Basalt

3.1 Macromechanical Behavior of Fissured Basalt

Strength characteristics of fissured basalt.

According to the calculation results, Table 2 summarizes the basic mechanical parameters of fissured basalt with different angles under different confining pressures. Figure 3 shows the relationship between the peak strength of fractured basalt and the dip angle of the fractures. It is evident from the figure that within the confining pressure range of 0–20 MPa, the peak strength of basalt basically increases with the increase of crack dip angle, especially in uniaxial compression tests. However, within the confining pressure range of 30–60 MPa, the peak strength of basalt fluctuates with the dip angle of the cracks. For basalts with the same crack dip angle, the increase in confining pressure also increases the peak strength.

Comparing the stress–strain curves of basalt with different crack angles, it can be seen that the pre-peak growth trend of fissured basalt under 10 MPa is basically linear (Fig. 4a), and the difference in elastic modulus is not significant. There are

Table 2 Basic mechanical parameters of fissured basalt with different dip angles

Crack dip angle $\alpha/°$	Confining pressure σ_3/ MPa	Peak strength σ_1/ MPa	Elastic modulus E/ GPa	Residual strength σ_r/ MPa	Cohesion c/ MPa	Internal friction angle $\varphi/°$
0	0	183.84	41.17		53.07	34.75
	10	237.93	43.59	7.15		
	20	288.64	43.99	18.63		
	30	330.90	44.43	21.07		
	40	355.65	44.79	27.27		
	50	386.19	45.09	24.87		
	60	403.45	45.56	77.02		
30	0	187.13	41.87		59.20	30.70
	10	238.85	43.97	6.37		
	20	285.44	44.48	6.53		
	30	315.27	44.92	29.59		
	40	347.87	45.20	26.83		
	50	347.36	45.73	37.03		
	60	382.01	46.02	37.92		
45	0	195.18	42.07		65.67	29.49
	10	270.41	44.55	13.68		
	20	302.56	45.22	19.24		
	30	318.25	45.66	18.15		
	40	345.65	46.19	28.03		
	50	377.02	46.54	56.42		
	60	384.03	47.06	71.61		
60	0	210.27	43.71		74.18	27.14
	10	273.31	45.27	15.54		
	20	330.38	45.77	13.74		
	30	345.66	46.28	36.57		
	40	345.82	46.91	64.39		
	50	357.02	47.43	69.04		
	60	399.24	47.51	77.65		
90	0	247.15	44.72		96.65	18.77
	10	307.92	46.07	1.90		
	20	322.02	46.70	9.34		
	30	332.02	46.70	9.36		
	40	338.13	47.73	34.65		
	50	366.06	48.11	75.02		
	60	384.89	48.65	78.57		

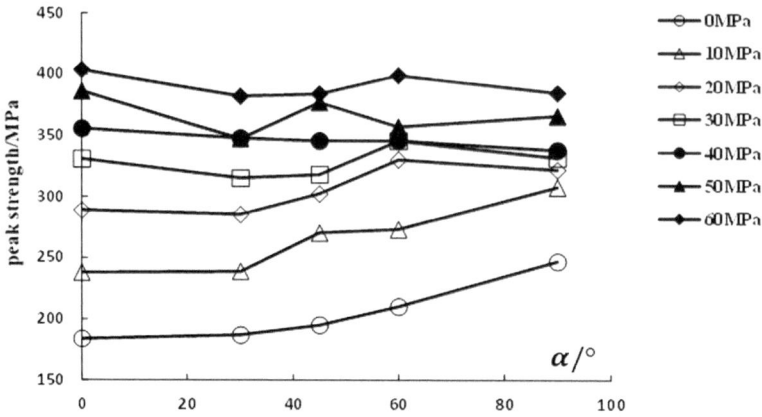

Fig. 3 Effect of crack dip angle on peak strength of fissured basalt

fluctuations near the peak strength when the dip angles are 0° and 30°. There is a small amplitude of stress drop in the stress–strain curve before the peak strength when the dip angles are 45° and 60°, while the stress–strain curve when the dip angles of the cracks are 90° is similar to that of intact basalt. In addition, the specimens with different crack angles exhibit a rapid drop in stress after fracture, exhibiting high brittle failure characteristics. For fissured basalt under 60 MPa, the stress–strain curves of samples with different crack angles fluctuate near the peak value and exhibit varying degrees of yield plateau, leading to rapid drop after failure (Fig. 4b).

Figure 5 shows the relationship curve between the peak strength and confining pressure of fissured basalt. The linear fitting correlation coefficient R of this curve is generally greater than 0.9, indicating that the relationship between the peak strength and confining pressure of fissured basalt can be well characterized using the Mohr Coulomb criterion.

Fracture Morphology and Crack Propagation Mode of Fissured Basalt.

Figure 6 shows the fracture morphology of basalt with different crack angles under uniaxial compression condition. When the crack angle is 90°, the rock appears fragmented and develops multiple vertical macroscopic cracks; In terms of crack type, it is mainly a tension type crack, which is relatively slender. Two macroscopic shear type cracks are developed at the upper and lower ends, and the density of microcracks is relatively high. In addition, the other inclined fissured rock masses are all inclined cracks, with microcracks originating, expanding, and connecting from both ends of the cracks, gradually forming macroscopic fracture surfaces. During the propagation process, multiple small vertical crack surfaces are developed simultaneously.

Figure 7 shows the macroscopic fracture morphology of basalt with different crack angles under low confining pressure (10 MPa). According to the calculation results, the macroscopic fracture surfaces of basalt with different crack angles under confining pressure conditions are all inclined cracks, and the fracture surfaces are connected through microcracks. When the dip angle of the crack is between 45°

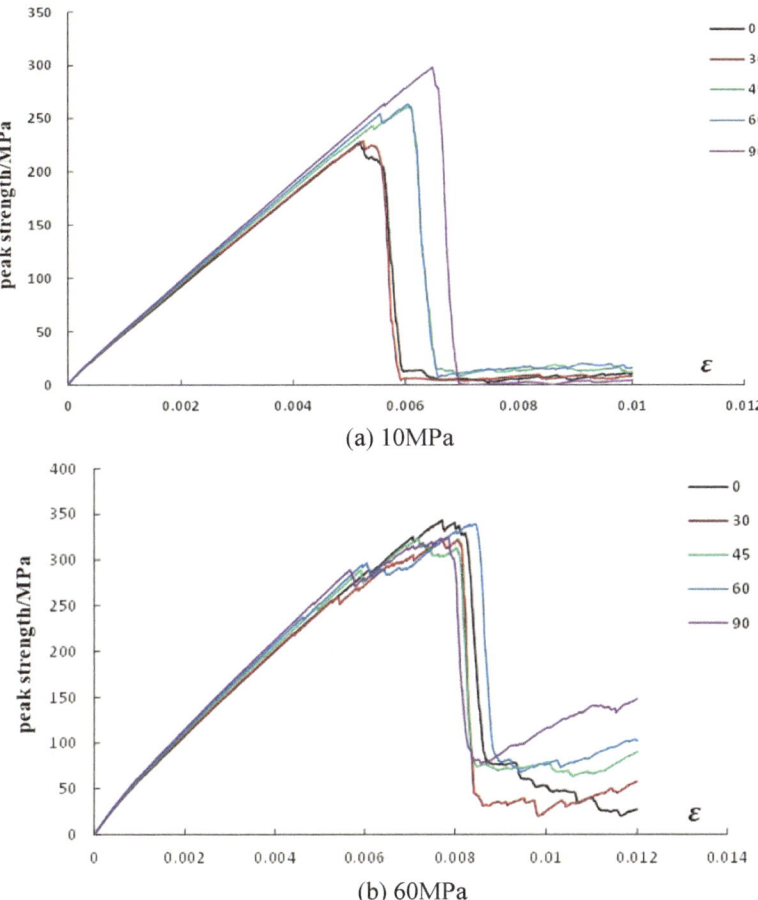

Fig. 4 Stress–strain curves of basalt with different crack angles

and 90°, the particles around the microcracks move with each other, and several secondary cracks develop around the crack. Figure 8 shows the macroscopic fracture morphology of basalt with different crack angles under high confining pressure (60 MPa). At this time, the rock fracture surfaces are all inclined and shear type failure, and the density of microcracks forming the macroscopic fracture surface under high confining pressure is significantly higher than that under low confining pressure. Meanwhile, the mutual dislocation of particles around microcracks is more significant, and the density of microcracks increases with the increase of crack dip angle.

In summary, the fracture mode of fissured basalt under uniaxial compression conditions will change with the change of crack dip angle. When the angle is 90°, it exhibits splitting tensile failure. Under confining pressure conditions, the fractured rock mass mainly exhibits shear failure, and the macroscopic fracture surface is

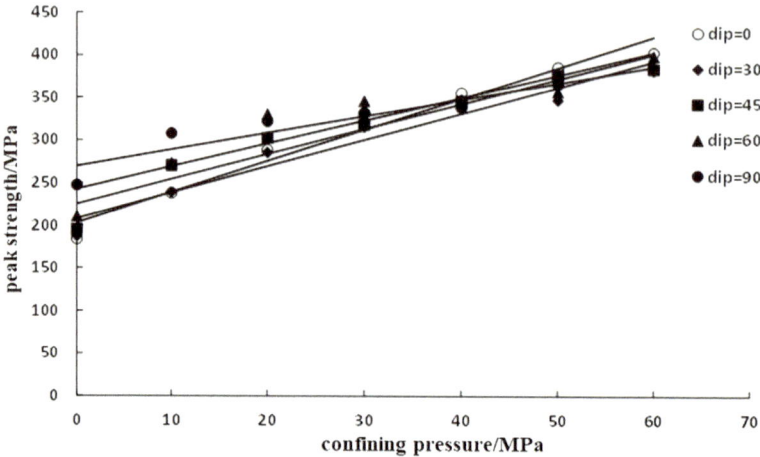

Fig. 5 Relationship between peak strength and confining pressure of fissured basalt

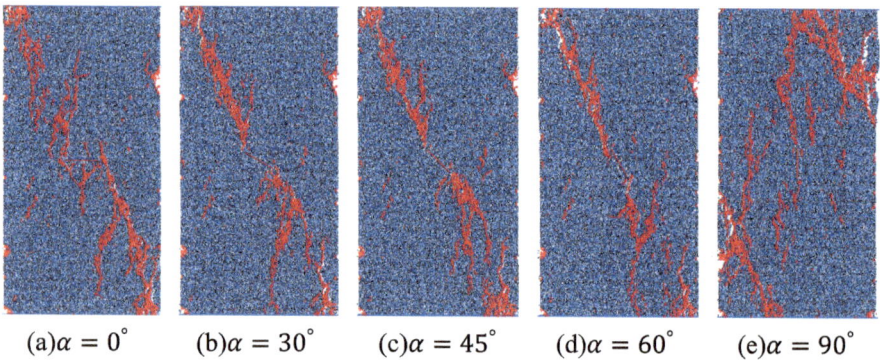

(a) $\alpha = 0°$ (b) $\alpha = 30°$ (c) $\alpha = 45°$ (d) $\alpha = 60°$ (e) $\alpha = 90°$

Fig. 6 Fracture morphology of fissured basalt under uniaxial compression condition

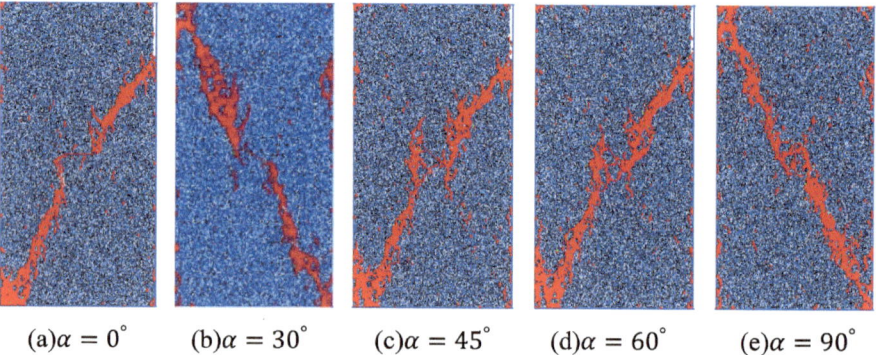

(a) $\alpha = 0°$ (b) $\alpha = 30°$ (c) $\alpha = 45°$ (d) $\alpha = 60°$ (e) $\alpha = 90°$

Fig. 7 Macro fracture mode of fissured basalt (10 MPa)

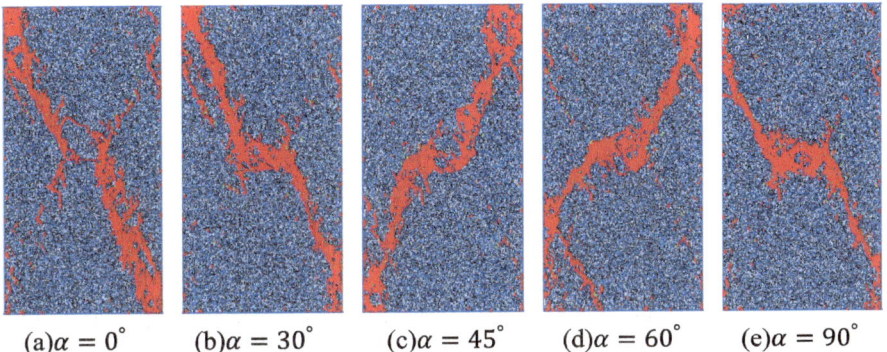

(a) $\alpha = 0°$ (b) $\alpha = 30°$ (c) $\alpha = 45°$ (d) $\alpha = 60°$ (e) $\alpha = 90°$

Fig. 8 Macro fracture mode of fissured basalt (60 MPa)

connected by microcracks as bridges. As the confining pressure increases, the density of microcracks forming the macroscopic fracture surface gradually increases, and as the crack dip angle increases, The dislocation of particles around microcracks is significantly enhanced.

3.2 Microscopic Response Characteristics of Fissured Basalt

Research has shown [8, 9] that rocks mainly develop two types of cracks under load: wing cracks and secondary cracks, both of which crack from the tip of the preset crack. The crack initiation direction of the wing is generally at a certain angle to the preset crack and develops along the main pressure direction; The initiation of secondary cracks is coplanar with the preset cracks (secondary coplanar cracks), or may be parallel to the wing cracks and propagate in the opposite direction (secondary inclined cracks), as shown in Fig. 9.

Secondary inclined cracks can be observed in numerical simulations of fissured basalt under different confining pressures, but secondary coplanar cracks can only be observed in some crack dip angles. Research has shown that the initiation direction of secondary cracks is closely related to material properties, that is, differences in material properties are the main reason for the different initiation directions of secondary cracks. According to the fracture morphology of basalt with different dip angles under different confining pressures, the macroscopic fracture surface generated by basalt fracture is mainly composed of secondary inclined cracks when the crack dip angles are 45° and 60°. When the crack dip angles are 0° and 30°, two types of cracks, wing shaped cracks and secondary inclined cracks, both occur in different confining pressures. However, when the crack dip angle is 90°, the macroscopic fracture surface of basalt is mainly generated by wing shaped cracks.

In order to thorough analyze the evolution law of crack propagation during the loading process of fissured basalt, taking the basalt with a 45° crack dip angle as an

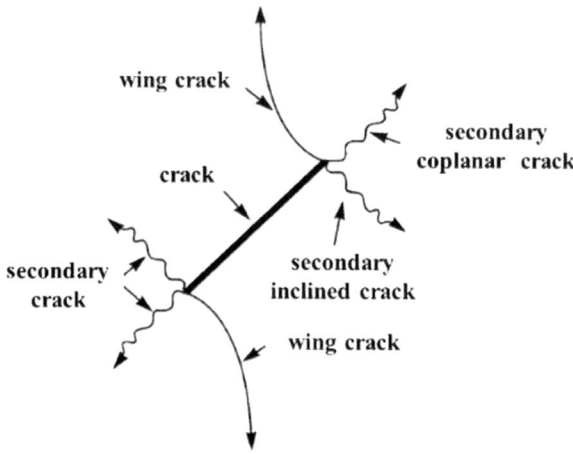

Fig. 9 Crack propagation mode under compression condition

example, combined with the stress–strain curve and crack propagation curve of the entire loading process of the sample, a detailed analysis was conducted on the micro fracture mechanism of basalt under confining pressures of 10 and 60 MPa.

Microscopic evolution process of fissured basalt under 10 MPa confining pressure.

Figure 10 shows the evolution curve of stress and crack with strain in fissured basalt under 10 MPa confining pressure. Based on the process of crack change during sample loading, the fracture morphology of fissured basalt at different loading stages is selected, as shown in Fig. 11. The evolution law of crack propagation in basalt under 10 MPa confining pressure is analyzed.

Point A in the figure ($\sigma = 40$ MPa) is the stage of linear elastic deformation of basalt. The sample is loaded and stores energy, and there are no cracks inside. The cloud map shows that there is tensile stress around the cracks of the sample in this stage, but the stress value is small, and the crack initiation and propagation strength is not reached. Point B ($\sigma = 200$ MPa), when the sample has just reached the initiation strength stage, a small amount of tensile cracks are generated, mainly distributed on both sides of the sample and at both ends of the crack, with a tendency to form secondary inclined cracks. Point C ($\sigma = 240$ MPa), when the specimen reaches the damage strength stage, tensile cracks rapidly increase and shear cracks begin to form. At this time, the stress drops slightly due to crack propagation; The developed cracks are mainly distributed around the microcrack, and two wing shaped cracks and one secondary inclined crack can be clearly observed. D point ($\sigma = 260$ MPa), when the sample reaches the peak strength, the tensile crack pattern rapidly increases, but mainly forms secondary inclined cracks. The two wing shaped cracks are limited in propagation, and the fracture morphology of the sample shows a macroscopic fracture surface. E point ($\sigma = 224$ MPa) is the stress drop stage after specimen failure, with a large number of corresponding tensile and shear cracks developing. The previous

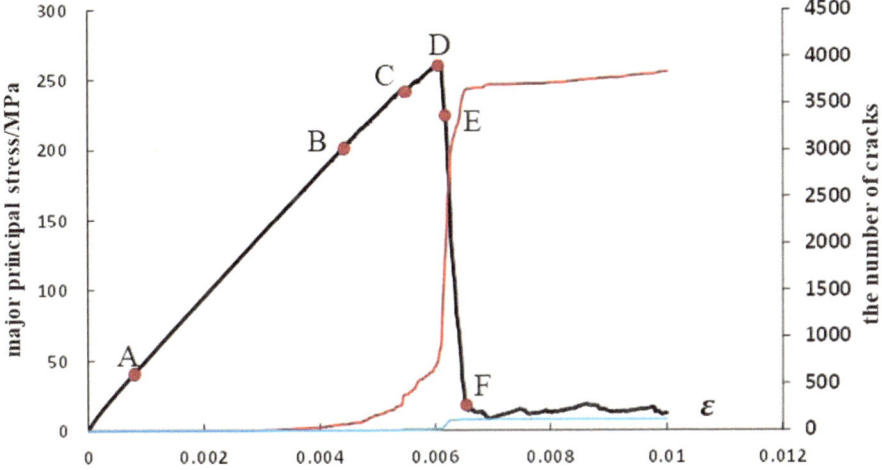

Fig. 10 Stress–strain curve and crack evolution curve of fissured basalt (10 MPa)

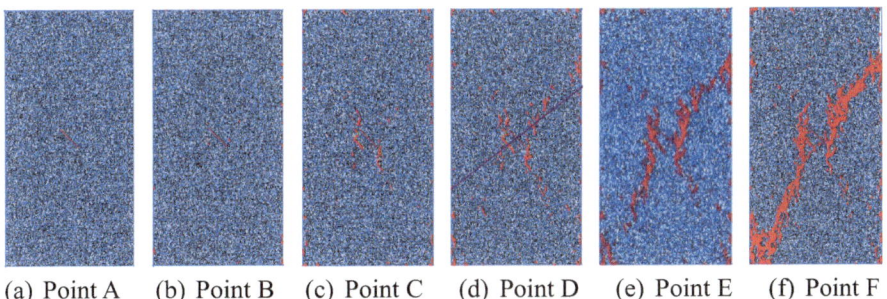

(a) Point A   (b) Point B   (c) Point C   (d) Point D   (e) Point E   (f) Point F

Fig. 11 Fracture morphology of fissured basalt at different loading stages (10 MPa)

inclined crack on the fracture surface further expands, forming a macroscopic shear fracture surface. F point ($\sigma = 14.5$ MPa) is the residual strength stage of the sample, at which time two secondary inclined cracks in the fissured basalt are fully developed, connected by microcracks and forming a macroscopic fracture surface together.

Microscopic evolution process of fissured basalt under 60 MPa confining pressure.

Figure 12 shows the entire loading process curve of 60 MPa fissured basalt. It can be seen that the specimen undergoes multiple stress drops under high confining pressure due to deviatoric stress. Combined with the evolution curve of cracks, six characteristic points corresponding to the fracture morphology of the specimen are selected, as shown in Fig. 13, to analyze the fracture mode of fissured basalt under high confining pressure.

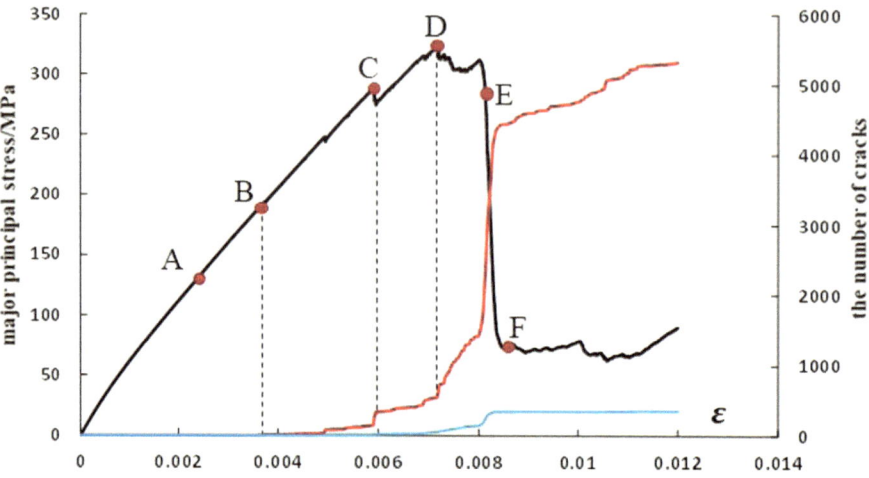

Fig. 12 Stress–strain curve and crack evolution curve of fissured basalt (60 MPa)

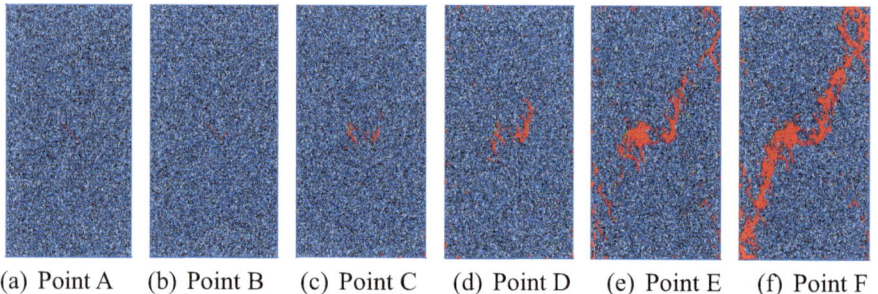

(a) Point A (b) Point B (c) Point C (d) Point D (e) Point E (f) Point F

Fig. 13 Fracture morphology of fissured basalt at different loading stages (60 MPa)

Point A in the figure ($\sigma = 130$ MPa) indicates that the sample is still in the stage of linear elastic loading and there are no cracks generated. Point B ($\sigma = 192$ MPa) is near the initiation strength of the sample, and a small amount of tensile cracks are generated at the lower end of the microcracks, indicating that the crack is a secondary inclined crack. Point C ($\sigma = 288$ MPa) indicates that the sample is in the pre peak damage deformation stage, at which time two secondary inclined cracks develop significantly. At point D of peak intensity ($\sigma = 322$ MPa) and post peak stress drop stage E point ($\sigma = 290$ MPa), secondary inclined cracks further develop and expand, connecting with each other, especially at the beginning of the secondary inclined crack. A large number of particles interact with each other, resulting in a significantly higher crack density than the sample under a confining pressure of 10 MPa. This indicates that under a confining pressure of 60 MPa, not only does the sample fracture form a macroscopic crack surface, but also the shear effect of the crack surface is strong. When the sample reaches residual strength E, two secondary cracks connect and penetrate the surface of the sample, forming a macroscopic shear fracture zone.

4 Discuss

According to the numerical simulation results, two different types of secondary cracks can simultaneously occur in the same sample, and the initiation direction of secondary cracks is closely related to the confining pressure. According to relevant research on fracture mechanics, secondary cracks belong to shear cracks, that is, regardless of the type of secondary crack, its generation is caused by shear stress. However, existing numerical experimental research findings [10] have demonstrated that, secondary cracks can be formed by shear or tensile stress, rather than solely by shear stress. Analysis shows that tensile stress leads to the generation of wing cracks, but when the confining pressure increases, the preset tensile stress at the crack tip decreases. If the confining pressure is large enough, the number of wing cracks will be significantly reduced. The increase in the main stress in the space where the original wing crack is located causes the wing crack to close, and at this time, the secondary crack begins to expand at the point where the tensile stress is high.

Based on the above numerical simulation results, it can be analyzed that the crack propagation mechanism of pre-set single crack rock: under low confining pressure, the specimen is compressed, causing the tensile stress concentration at the pre-set crack tip, leading to wing cracks cracking from the tip and developing towards the main pressure direction; Secondary cracks are also generated due to high tensile stress at the tip, but as loading progresses, secondary cracks are subjected to shear stress; When the stress field near the wing crack and secondary crack is insufficient to propagate the two types of cracks, other cracks appear in the sample to release strain energy in the rock. Under high confining pressure, the tensile stress at the preset crack tip decreases, resulting in a significant reduction in the number of wing cracks. The increase in the main stress in the space where the original wing cracks are located causes the wing cracks to tend to close. At this time, secondary cracks begin to expand at the point where the tensile stress is high, and ultimately develop along the direction of the main pressure under shear stress.

5 Conclusions

This article focuses on cryptocrystalline basalt as the subject of investigation and conducts biaxial compression particle flow simulation tests on fissured basalt at varying confining pressure levels. The study obtained the evolution laws of stress, deformation of specimens under load under different stress levels, analyzed the initiation, propagation, and connection laws of cracks during the loading process of fissured rock masses, and revealed the influence of crack dip angle on crack propagation and fracture mechanism.

(1) In the range of 0–20 MPa, the peak strength of basalt basically increases with the increase of crack dip angle. With the increase of confining pressure, the variation of peak strength of basalt with crack dip angle is not significant. The relationship between the peak strength and confining pressure can be well characterized using the Mohr Coulomb criterion.
(2) Under uniaxial compression conditions, the fracture mode of fissured basalt will change with the change of crack angle. When the crack angle is 90°, it exhibits splitting tensile failure. Under confining pressure conditions, the rock mass mainly exhibits shear failure, and the macroscopic fracture surface is connected through microcracks as bridges.
(3) In the case of the fracture mechanism of single-crack basalt, under low confining pressure, the compression of the sample causes the concentration of tensile stress at the preset crack tip. This leads to the initiation of wing cracks from the tip, propagating towards the principal pressure direction, while secondary cracks are influenced by shear stress. Conversely, under high confining pressure, the tensile stress at the preset crack tip decreases, resulting in a significant reduction in the number of wing cracks. Secondary cracks begin to expand at the point of high tensile stress and develop along the direction of main pressure under shear stress.

Acknowledgements This research is supported by Zhejiang Provincial Natural Science Foundation of China under Grant (No. LQ23E090002). The authors are also grateful to the journal editor and the reviewers for their insightful comments and suggestions for this paper.

References

1. Chen BR, Feng XT, Ming HJ et al (2012) Evolution law and mechanism of rockburst in deep tunnel: time delayed rockburst. Chin J Rock Mech Eng 31(3):561–569
2. Jiang Q, Fan YL, Feng XT et al (2017) Unloading break of hard rock under high geo-stress condition: inner cracking observation for the basalt in the Baihetan's underground powerhouse. Chin J Rock Mech Eng 36(5):1076–1087
3. Zhang CQ, Liu ZJ, Feng XT et al (2019) Experimental study on rupture evolution and failure characteristics of aphanitics basalt. Rock Soil Mech 40(7):2487–2496
4. Yang Y, Xu NW, Li T et al (2018) Stability analysis of left bank rock slope at Baihetan Hydropower station based on RFPA^{3D} code and microseismic monitoring. Rock Soil Mech 39(6):1002–1010
5. Hu W, Wu AQ, Chen SH et al (2017) Mechanical properties of cplumnar jointed basalt rock with hidden fissures under uniaxial loading. Chinese J Rock Mech Eng 6(8):1880–1888
6. Huang D, Cen DF, Huang RQ (2013) Influence of medium strain rate on sandstone with a single pre-crack under uniaxial compression using PFC simulation. Rock Soil Mech 34(2):535–545
7. Chen PY (2018) Research progress on PFC2D simulation of crack propagation characteristics of cracked rock. J Eng Geol 26(2):528–539

8. Fatehi MM, Gholamnejad A, Eghbal M (2011) On the crack propagation mechanism of brittle rocks under various loading conditions. In: Proceedings of the 11th international multidisciplinary scientific geo-conference SGEM2011, pp 561–568
9. Park CK, Bobet A (2009) Crack coalescence in specimens with open and closed flaws: a comparsion. Int J Rock Mech Minin Sci 46(5):819–829
10. Tang Q, Li YA (2015) Particle flow simulation on the influence of confinement on crack propagation in pre-cracked rock. J Tangtze River Sci Res Inst 32(4):81–85

Open Access This chapter is licensed under the terms of the Creative Commons Attribution 4.0 International License (http://creativecommons.org/licenses/by/4.0/), which permits use, sharing, adaptation, distribution and reproduction in any medium or format, as long as you give appropriate credit to the original author(s) and the source, provide a link to the Creative Commons license and indicate if changes were made.

The images or other third party material in this chapter are included in the chapter's Creative Commons license, unless indicated otherwise in a credit line to the material. If material is not included in the chapter's Creative Commons license and your intended use is not permitted by statutory regulation or exceeds the permitted use, you will need to obtain permission directly from the copyright holder.

Proportioning Design and Material Development of High Performance Concrete

Temperature Control Sensitivity Analysis and Research on Thin-walled Hydraulic Tunnel Lining Concrete

Bu Zhang and Zhenhong Wang

Abstract The phenomenon of concrete cracking is a major problem plaguing the engineering community, and one of the major factors causing cracking is temperature loading, while thin-walled concrete, such as hydraulic tunnel lining and other similar projects, makes the impact of this phenomenon more prominent due to its special structure and role. To address this problem based on the concrete temperature field theory, combined with the three-dimensional finite unit method of temperature control methods in engineering analysis and research, it is concluded that surface water conservation and water cooling can effectively reduce the base temperature difference to reduce the temperature stress; early low temperature water flow may cause excessive temperature gradients, resulting in large temperature stress; surface insulation layer can be applied to effectively mitigate the rate of temperature drop and environmental temperature changes on the concrete.

Keywords Lining concrete · Concrete cracks · Finite element method · Temperature stress

1 Preface

Cracking is a common phenomenon in concrete buildings, causing great harm to the use and safety of buildings. It is generally believed that there are two main reasons for causing temperature loading, one is the difference between the internal and external temperature of concrete to produce internal and external temperature, and the other is the difference in base temperature. The difference between internal and external temperature leads to cracks caused by different internal and external deformation, and the difference in temperature of the foundation leads to shrinkage deformation and cracking when heat is dissipated later. In recent years, China's construction of large

---

B. Zhang · Z. Wang (✉)
Institute of Structural Materials, China Institute of Water Resources and Hydropower Research, Beijing 100048, China
e-mail: w3525710468@126.com

concrete buildings slowed down [1], water for living, industrial and ecological water are significantly increased, supporting the deployment of water resources projects are increasing day by day, the ferry, pumping stations, sluice gates, locks, inverted siphon, ground culvert and a variety of cross-building concrete projects have emerged. Such concrete buildings are relatively small in size, but the structure is more complex, resulting in more common cracking phenomenon.

Due to its special role and structure, tunnel concrete is generally made of very high grade concrete and is strongly restrained by the surrounding rock, resulting in high temperature loads. However, the lack of analysis of the specific effects of various temperature control measures on stress development sometimes leads to the phenomenon that the temperature stress cannot be effectively reduced or even counteracted [2–4]. The paper is based on the concrete temperature field theory, combined with the three-dimensional finite element unit method, for the surface water, water cooling and other temperature control techniques for sensitivity analysis, the temperature and stress development of the whole process of the analysis, the effect of different temperature control methods on the concrete temperature and stress changes in various periods, for similar projects for reference.

2 Theory of Computation

2.1 Heat Conduction Equation

In solid heat conduction, the heat flow rate q is proportional to the temperature gradient $\frac{\partial T}{\partial x}$, but the direction of the heat flow is opposite to the direction of the temperature gradient [5].

$$q_x = -\lambda \frac{\partial T}{\partial x}$$

where λ is the thermal conductivity, kJ/(m·h·°C)

From the heat balance know that the heat absorbed by the temperature increase must be equal to the sum of the net heat inflow from outside and the internal heat of hydration.

$$c\rho \frac{\partial T}{\partial \tau} d\tau dxdydz = \left[\lambda \left(\frac{\partial^2 T}{\partial x^2} + \frac{\partial^2 T}{\partial y^2} + \frac{\partial^2 T}{\partial z^2} \right) + Q \right] d\tau dxdydz$$

where c is the specific heat, kJ/(kg-°C); τ is the time, h; ρ is the density, kg/m^3, and Q is the heat kJ/(h-m^3) emitted per unit volume per unit time.

After simplification, we get that

$$\frac{\partial T}{\partial \tau} = a\left(\frac{\partial^2 T}{\partial x^2} + \frac{\partial^2 T}{\partial y^2} + \frac{\partial^2 T}{\partial z^2}\right) + \frac{Q}{c\rho}$$

where a is the temperature conductivity, m²/h.

Due to the effect of heat of hydration, under adiabatic conditions

$$\frac{\partial \theta}{\partial \tau} = \frac{Q}{c\rho} = \frac{Wq}{c\rho}$$

where, θ is the adiabatic temperature rise of concrete, °C; W is the amount of cement, kg/m³; q is the heat of hydration, kJ/(kg-h) released per unit weight of cement in unit time.

The heat conduction equation is obtained by simplifying

$$\frac{\partial T}{\partial \tau} = a\left(\frac{\partial^2 T}{\partial x^2} + \frac{\partial^2 T}{\partial y^2} + \frac{\partial^2 T}{\partial z^2}\right) + \frac{\partial \theta}{\partial \tau}$$

2.2 Water Pipe Cooling

The equivalent heat transfer equation [6–9] considering water pipe cooling is as follows

$$\frac{\partial T}{\partial \tau} = a\left(\frac{\partial^2 T}{\partial x^2} + \frac{\partial^2 T}{\partial y^2} + \frac{\partial^2 T}{\partial z^2}\right) + (T_o - T_w)\frac{\partial \phi}{\partial \tau} + \theta_0 \frac{\partial \psi}{\partial \tau}$$

3 Calculation Model and Scheme

3.1 Calculation Parameters

The ambient temperature in which a dam diversion tunnel is located is shown in Table 1.

The concrete and bedrock parameters are shown in Tables 2 and 3.

Table 1 Ambient temperature

Month	1	2	3	4	5	6	7	8	9	10	11	12	Annual average
Tem	13.2	15.5	18.6	21.9	25.4	27.6	30	29.5	27.1	22.5	18.1	14.3	22

Table 2 Basic parameters

	Volumetric weight (kg/m^3)	Elastic Modulus (GPa)	Poisson's ratio	Thermal diffusivity ($\times 10^{-3}$ m^2/h)	Thermal conductivity 3(kJ/m.h.°C)	Specific heat capacity (kJ/Kg.°C)	Coefficient of linear expansion (10^{-6}/°C)
Bedrock	2580	22	0.18		10.5	0.91	
Concrete	2340		0.203	2.979	7.782	1.093	10.2

Table 3 Concrete adiabatic temperature rise and modulus of elasticity

Strength grade			7 d	28 d	90 d	180 d
C$_{28}$30W10F100	Modulus of elasticity $\times 10^4$ MPa		2.45	2.90	3.42	3.48
	Adiabatic temperature rise °C		41.8	43.9	47.2	

3.2 Calculation Model

The overall model of the calculation and the mesh section are shown in Fig. 1. The length of the tunnel selected for calculation is 18 m, with a total of 35,064 profiled units and 39,884 nodes. The model follows the flow direction in Y direction and the vertical direction in Z direction. The length and width of the foundation are 60 m, All faces of the bedrock are adiabatic boundaries, and the remaining faces are type III boundary conditions, the cave temperature is fitted according to the annual average temperature based on the reduction of the change rate, the stress boundary of bedrock on the left and right sides, bottom and top surface is three-way constraint, the rest of the surface is normal constraint.

The size of the tunnel is much larger in the downstream direction (Y direction) than in the vertical direction (X direction), so the stress is mainly in the downstream direction, i.e. Y direction stress. The middle section of the model is selected as the feature surface, feature point 1 is the internal point (right side of the tunnel) and feature point 2 is the surface point (middle of the tunnel). The exact location is shown in Fig. 2.

Fig. 1 Simulation calculation grid division

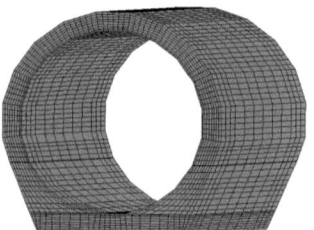

Fig. 2 Typical section and characteristic points

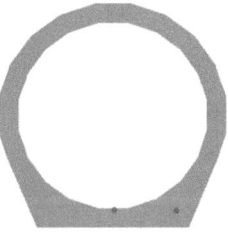

3.3 Calculation of Working Conditions

The dam pouring process generally uses surface flowing water, through the cooling water pipe [10] and surface insulation to reduce the temperature stress, but for the hydraulic thin-walled structure, due to its structural characteristics, the common temperature control measures will be more sensitive to these characteristics to carry out a sensitivity analysis of the parameters, set different working conditions to discuss the degree of influence of the temperature and duration of the flowing water, cooling water temperature, duration and other factors and the law. The specific condition settings are shown in Table 4, conditions 1–3 only change surface flow time, conditions 1, 4 and 5 only change the surface water temperature, conditions 1, 6 and 7 only change the cooling water duration, conditions 1, 8 and 9 only change the cooling water temperature, conditions 1, 10, 11 only change cooling water flow rate, condition 12 based on condition 1, 2 cm of insulation is added to the inner surface of the tunnel, which is achieved by changing the heat dissipation coefficient of the concrete surface.

4 Sensitivity Analysis of Thin-Walled Concrete Structures

4.1 Effect of Surface Flow Time

Figure 3 shows the temperature and stress envelope of the middle section of the tunnel.

Figure 4 shows the temperature stress process line at the characteristic point of internal Fig. 5 shows the temperature stress process line at the characteristic point of surface water flow at different time lengths. From the figure, it can be seen that the highest temperature of concrete appears within 1–2 days after pouring, and the maximum temperature at the characteristic point is not affected when the surface water flow length is 15, 20 and 25 days.

The maximum stresses at feature point 1 are 4.09, 4.04, and 4.01 MPa, respectively. The temperature process line shows that the temperature at feature point 1 rebounded at the end of the flowing water, but the degree of temperature drop enhanced with the increase of the flowing water time, which leads to the increase of tensile stresses at

Table 4 Working conditions

Operating mode	Surface flow		Water pipe cooling				Thermal insulation
	Duration (d)	Water temperature (°C)	Duration (d)	Water temperature (°C)	Flow	Spacing	
1	20	20	10	20	20	1.0 m × 1.0 m	
2	15	20	10	20	20	1.0 m × 1.0 m	
3	25	20	10	20	20	1.0 m × 1.0 m	
4	20	15	10	20	20	1.0 m × 1.0 m	
5	20	25	10	20	20	1.0 m × 1.0 m	
6	20	20	5	20	20	1.0 m × 1.0 m	
7	20	20	15	20	20	1.0 m × 1.0 m	
8	20	20	10	15	20	1.0 m × 1.0 m	
9	20	20	10	25	20	1.0 m × 1.0 m	
10	20	20	10	20	15	1.0 m × 1.0 m	
11	20	20	10	20	25	1.0 m × 1.0 m	
12	20	20	10	20	20	1.0 m × 1.0 m	2 cm

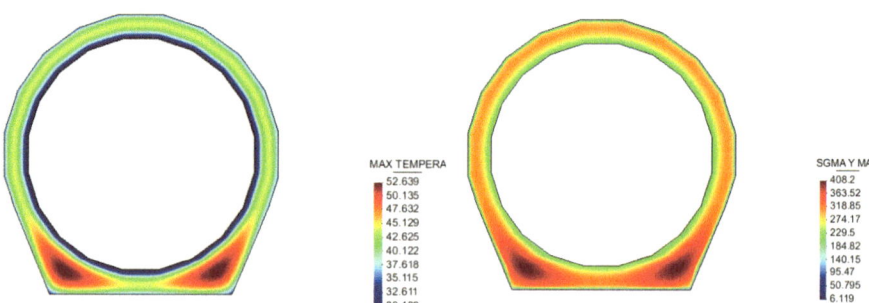

Fig. 3 Temperature stress envelope diagram of tunnel lining

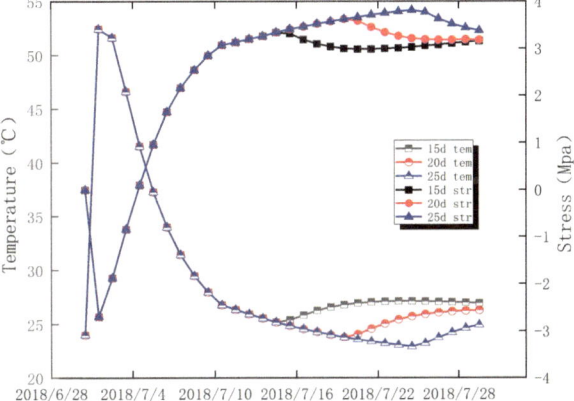

Fig. 4 Temperature stress hydrograph of characteristic point 1 with different flow time

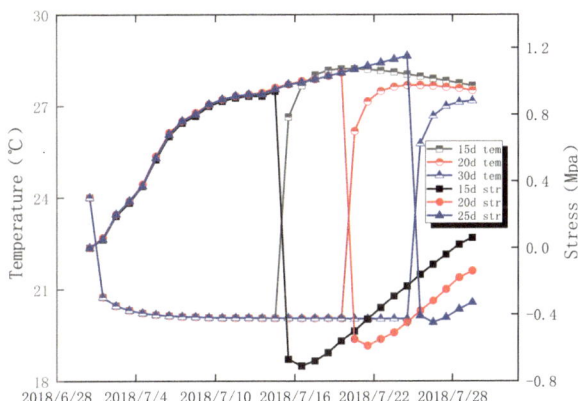

Fig. 5 Temperature stress hydrograph of characteristic point 2 with different flow time

the early flowing water time, but the higher degree of temperature drop is beneficial to the development of stresses at the later stage, which makes the maximum stress is reduced at any time.

The maximum stresses at feature point 1 are 4.09, 4.04, and 4.01 MPa, respectively. The temperature process line shows that the temperature at feature point 1 rebounded at the end of the flowing water, but the degree of temperature drop enhanced with the increase of the flowing water time, which leads to the increase of tensile stresses at the early flowing water time, but the higher degree of temperature drop is beneficial to the development of stresses at the later stage, which makes the maximum stress is reduced at any time.

Overall, the appropriate growth of surface water duration can relieve the late temperature drop pressure of concrete, which can reduce the maximum tensile stress.

4.2 Effect of Surface Cooling Water Temperature

Figures 6 and 7 show the temperature stress process lines of the characteristic points 1 and 2 when the cooling water temperature is 15, 20 and 25 °C.

From the figure data, it can be seen that with the increase of cooling water temperature, the maximum temperature of characteristic point 1 is 51.49, 52.43 and 52.89 °C, and the maximum stress is 3.91, 4.04 and 4.28 MPa, the decrease of running water temperature can reduce the maximum temperature of concrete, so as to reduce the base temperature difference to improve the stress state, but early due to the decrease of surface running water temperature will The early temperature drop rate is large, resulting in the early low temperature water caused by large tensile stress, but after the end of its water cooling, the low temperature conditions due to its small base temperature difference is conducive to reducing the tensile stress.

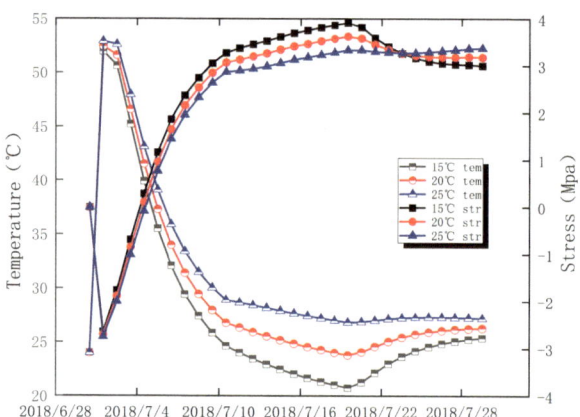

Fig. 6 Characteristic point 1 temperature stress hydrograph of different water temperatures

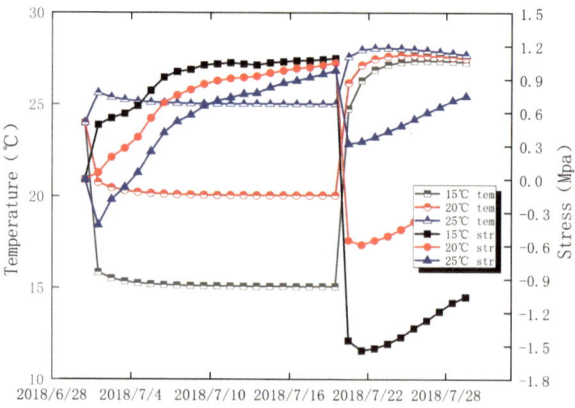

Fig. 7 Characteristic point 2 temperature stress hydrograph of different water temperatures

The temperature of feature point 2 was consistent with the water temperature during the flowing water period and remained consistent with the air temperature after the flowing water, with maximum stresses of 1.47, 2.04 and 2.57 MPa.

The data indicate that lowering the cooling water temperature can reduce its maximum internal temperature and reduce the base temperature difference and stress, but the lowering of the cooling water temperature will cause the early temperature drop rate to become faster, thus making the early stresses larger, but overall the reduced base temperature difference of the concrete is beneficial to the overall stress situation.

4.3 Influence of Internal Water Passage Duration

Figures 8 and 9 show the temperature stress process lines of characteristic point 1 and characteristic point 2 when the length of cooling time through water is 5, 10 and 15 days.

The maximum stresses at feature point 1 are 4.14, 4.04 and 4.01 MPa, respectively. The increase of the internal water passage time prolongs the early temperature drop process and increases the degree of early temperature drop, so the early stresses will appear to increase with the increase of the time, and similarly, the increase of the early temperature drop is beneficial to the later stresses.

The maximum stresses at feature point 2 are 2.02, 2.04 and 2.03 MPa, respectively. The data in the figure show that the internal water passage duration has no effect on the temperature and stress development process at the surface point.

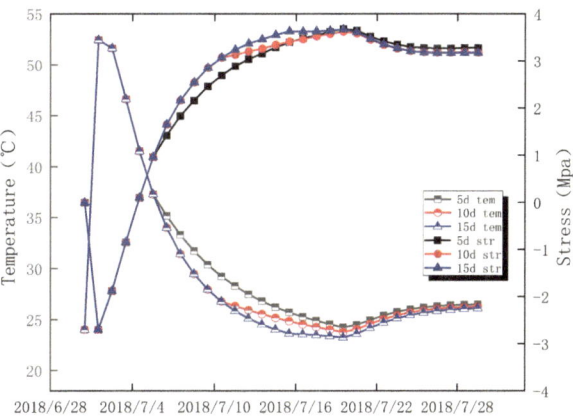

Fig. 8 Characteristic point 1 temperature stress hydrograph of different cooling time

Fig. 9 Characteristic point 2 temperature stress hydrograph of different cooling time

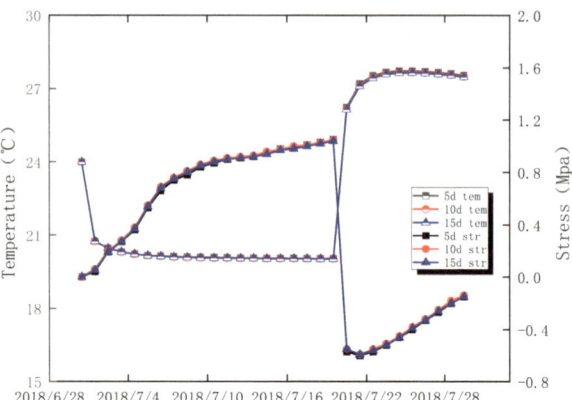

4.4 Internal Water Temperature Influence

Figures 10 and 11 show the temperature and stress process lines for different cooling water pipe water temperature characteristic points 1 and 2.

The temperature of feature point 1 with the cooling water temperature decreases to 52.79, 52.43 and 52.06 °C, and the maximum stresses are 4.15, 4.03 and 3.92 MPa. Again, the lower cooling water temperature in the early stage will produce larger tensile stresses, but the weakening of the base temperature difference will improve the maximum stress situation at the internal point.

Cooling water temperature changes have little effect on the surface point.

Fig. 10 Characteristic point 1 temperature stress hydrograph of different cooling temperature

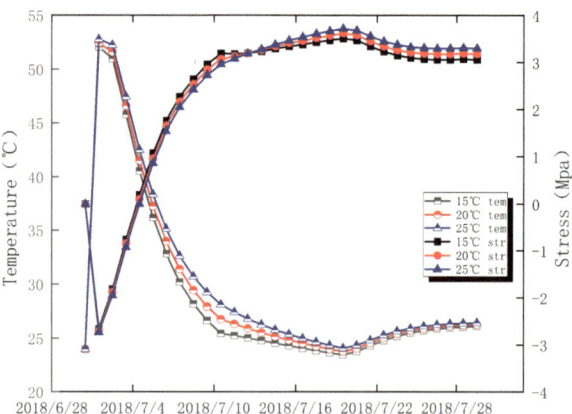

Fig. 11 Characteristic point 2 temperature stress hydrograph of different cooling temperature

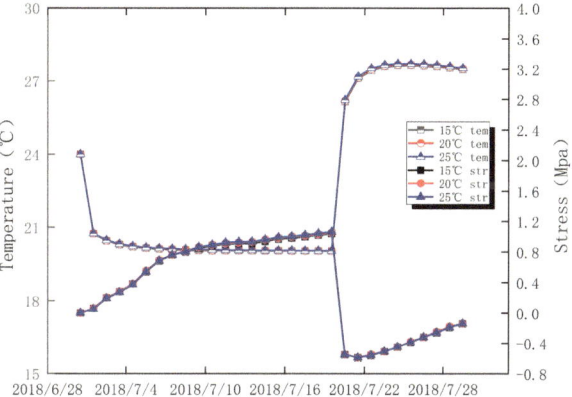

4.5 Cooling Water Flow Analysis

Figure 12 shows the process line of temperature and stress development at feature point 1 under different through-water flow rates, with the increase in cooling water flow rate feature points of the highest temperature of 52.67, 52.43, and 52.25 °C, the maximum stress of 4.12, 4.04, and 3.98 MPa, by the process line can be seen, the increase in through-water flow rate cut the highest temperature, and the temperature drop rate will also becomes faster, so the stress development process is slightly greater in the early stage than in the remaining two cases. The overall data show that the increase of water flow rate has little effect on the internal temperature of concrete, but the improvement of stress is more obvious.

Fig. 12 Characteristic point 1 temperature stress hydrograph of different cooling flow rate

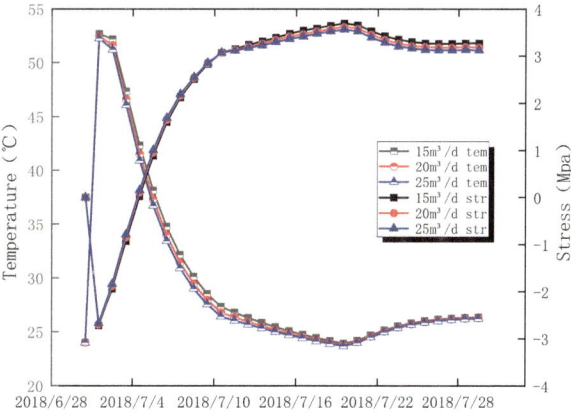

4.6 Surface Insulation Sensitivity Analysis

Figures 13 and 14 show the temperature stress process lines of feature point 1, feature point 2 after adding the insulation layer.

The maximum temperature of feature point 1 does not change after adding the insulation layer, but the maximum stress is reduced from 4.04 to 3.63 MPa. The addition of the insulation layer plays a great role in weakening the excessive temperature drop rate caused by reducing the base temperature difference, thus avoiding the occurrence of excessive early stress or even the situation where the early stress is greater than the allowable stress.

Feature point 2 in the insulation layer cover, the ambient temperature caused by the influence of greatly weakened, the variation rate is significantly reduced, the temperature stress from 2.38 MPa down to 1.83 MPa, and the stress state compared to no insulation layer changes in the undulation reduced.

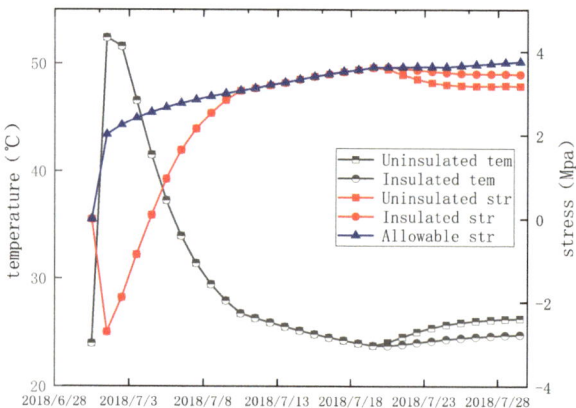

Fig. 13 Temperature stress hydrograph of added insulation layer at feature point 1

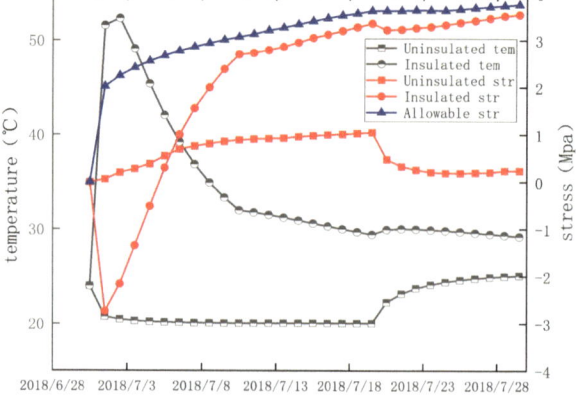

Fig. 14 Temperature stress hydrograph of added insulation layer at feature point 2

4.7 Summary

After calculating the surface flow + water cooling + surface insulation temperature control method can also effectively reduce the temperature stress of thin-walled lining concrete, but because of its special structural characteristics, surface flow and water cooling may cause the early temperature drop rate is too large, the temperature stress exceeds the allowable stress, so the highest temperature after the insulation is a more important temperature control measures.

5 Conclusion

Surface water conservation, through the cooling water pipe, can effectively lower the base temperature difference of concrete, reduce the temperature burden, and is helpful to the thin-walled concrete projects to prevent cracking.

Surface running water conservation, through the cooling water pipe in reducing the base temperature difference at the same time will intensify the early temperature drop rate of concrete, thus early may produce large tensile stress, so in the project should be combined with the actual situation to set a reasonable temperature flow, etc., to prevent the phenomenon of cracking caused by excessive cooling.

After the concrete is poured, adding an insulation layer to the surface can effectively reduce the rate of temperature drop, the temperature gradient generated, as well as the impact of external environmental temperature changes on the surface temperature of concrete, achieving the goal of reducing the temperature load.

Surface flow water temperature, internal water temperature and surface insulation are the three temperature control measures that have the greatest impact on stress.

References

1. Liu J, Bao Z, Liu C et al (2019) Analysis of patterns and causes of changes in water resources and water consumption in China in the past 20 years [J]. J Water Resour Water Transp Eng 176(04):31–41. https://doi.org/10.16198/j.cnki.1009-640x.2019.04.005
2. Wang ZH, Zhang GX, Yu SP et al (2010) Sensitivity analysis of temperature control parameters of hydraulic thin-walled concrete structures[J]. J Wuhan Univ Technol (Transportation Science and Engineering Edition) 34(02):280–284
3. Wang ZH, Zhu YM, Yu SP (2007) Research on temperature control and crack prevention during construction of thin-walled concrete structures [J]. J Xi'an Univ Architect Technol (Natural Science Edition) 39(6):773–778. https://doi.org/10.3969/j.issn.1006-7930.2007.06.007
4. Zhu BF (2003) Equivalent heat transfer equation for water pipe cooling considering the effect of external temperature [J]. J Water Res 3:49–54. https://doi.org/10.3321/j.issn:0559-9350.2003.03.010
5. Zhu BF, Wu LCS, Zhang GX et al (2010) Study on the cooling method of initial water pipe of concrete dam [J]. Hydroelectricity 36(03):31–35

6. Zhu B (1999) Temperature stress and temperature control of mass concrete [M]. China Electric Power Press
7. Bofang Z (2010) On cooling of concrete dams by water pipes [J]. J Water Res 41(5):505–513
8. Zhu B (1991) Equivalent heat transfer equation for concrete considering the cooling effect of water pipes [J]. J Water Resour 28(3)
9. Yeming Z, Zhiqing X, Jinren H et al (2003) Calculation method of cooling temperature field of concrete water pipe [J]. J Changjiang Acad Sci 20(2):19–22. https://doi.org/10.3969/j.issn.1001-5485.2003.02.006
10. Guiyou L, Yueming Z (2006) Study on the arrangement scheme of cooling water pipe for large volume concrete [J]. Red Water River 1. https://doi.org/10.3969/j.issn.1001-408X.2006.01.004

Open Access This chapter is licensed under the terms of the Creative Commons Attribution 4.0 International License (http://creativecommons.org/licenses/by/4.0/), which permits use, sharing, adaptation, distribution and reproduction in any medium or format, as long as you give appropriate credit to the original author(s) and the source, provide a link to the Creative Commons license and indicate if changes were made.

The images or other third party material in this chapter are included in the chapter's Creative Commons license, unless indicated otherwise in a credit line to the material. If material is not included in the chapter's Creative Commons license and your intended use is not permitted by statutory regulation or exceeds the permitted use, you will need to obtain permission directly from the copyright holder.

Breakage Mechanism of Artificial Granular Materials

Longjiang Fan, Enlong Liu, Shijia Tang, and Yanlin Qin

Abstract The artificial cemented granular particles were prepared to simulate the rockfills, and conventional triaxial drainage tests were conducted to explore the mechanical properties of granular materials under different confining pressures. Furthermore, the results were simulated based on the discrete element method. Test results from the laboratory tests indicated the nonlinearity of this materials' strength envelope. The particle flow method could be used to simulate the stress–strain and volumetric deformation properties of artificially prepared granular materials. By the simulation, the number of failure bonds and the variation of the displacement field during the test process were recorded, visually displaying the microscopic displacement and fragmentation of the internal particles.

Keywords Artificial rockfills · Particle breakage · Triaxial test · Discrete element method

1 Introduction

Coarse materials like rockfills are granular materials, which have a large porosity and strong permeability, and have no cohesion. Compared with sand, the particle size of rockfills is larger, and particle breakage can occur under lower stress conditions, which further affects the macroscopic and microscopic mechanical behavior of the materials.

In the classical theory of soil mechanics, it is often assumed that the internal particles of soil are incompressible and unbreakable, and the deformation of soil is mainly caused by the rearrangement and combination of soil particles and the discharge of water and air from soil. Therefore, the strength theory of soil is basically established on the basis of slip and friction between particles. However, because of the

---

L. Fan · E. Liu (✉) · S. Tang · Y. Qin
State Key Lab of Hydraulics and Mountain River Engineering, College of Water Resource and Hydropower, Sichuan University, Chengdu, Sichuan, China
e-mail: liuenlong@scu.edu.cn

particle fragmentation, the coarse materials will have some characteristics different from the classical soil mechanics (such as the strength envelope curve deviates from the straight line and bends downward when the stress is large). The more particle breakage in the soil, the more significant the change of its physical and mechanical properties [1].

The study of particle breakage can be traced back to Terzaghi and Peck's one-dimensional compression tests of sand [2], which showed severe particle breakage under high stress levels. Since then, particle breakage attracted the attention of the other researchers. Subsequently, researchers conducted a series of laboratory experiments and theoretical studies [3–7], exploring the impact of particle breakage on strength and other characteristics from a macroscopic perspective. In recent years, the discrete element method has been frequently used to study the deformation and mechanical properties of coarse-grained soils like rockfills [8, 9], leading to the development of corresponding constitutive equations. However, there is currently little research on the microscopic processes of particle breakage for large-sized particles like rockfills.

This study uses artificial cement-soil spherical particles to simulate easily broken materials, and conducts triaxial tests and numerical simulation under drainage conditions to investigate the impact of particle breakage on the mechanical properties of granular materials and the micromechanics of particle breakage.

2 Laboratory Tests

2.1 Preparation of Samples

The soil material used in the experiment is silty clay from Chengdu, China. The main physical and mechanical properties of the soil are shown in Table 1. The cement used in the experiment was 32.5R composite Portland cement, with a specific gravity of 3.142.

The materials used to prepare cement particles include: cement, silty clay, water, and oil. Using homemade organic glass molds and C-clamps, Spherical particles with diameter of 20 mm were prepared using self-made plexiglass mold and C clamps. The mass ratio of cement, clay, and water was 1:2.5:1.4.

To test the uniformity and integrity of cement products, 20 particles were randomly selected and weighed, and the average density was calculated to be 1.414 g/cm^3. The mass errors of the particles sampled were all within about 1%.

Table 1 The physical–mechanical indexes of silty clay

Specific gravity	Liquid limit	Plastic limit	Plasticity index (%)	Water content (%)
2.684	33	21	12	7.307

Fig. 1 Split tests

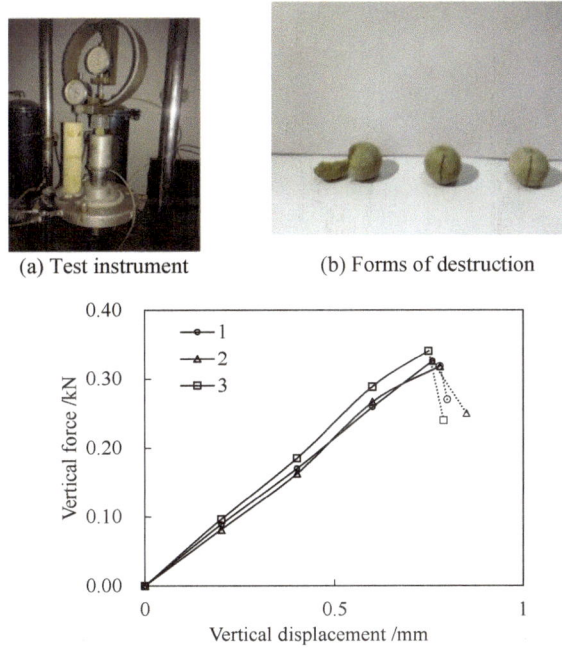

(a) Test instrument (b) Forms of destruction

(c) Result of tests

Meanwhile, a modified triaxial testing instrument (Fig. 1) was used to carry out splitting tests. The test was strain-controlled, with a rate of 0.016 mm/min. The fracturing form and splitting test results are shown in Fig. 1.

2.2 Test Scheme

A series of routine triaxial tests were carried out on the artificially prepared cement ballast material. The samples had a diameter of 10.1 cm and a height of 20 cm, and were composed of 200 cement particles. The confining pressures of 100, 200, 300, and 400 kPa were selected. Considering the high porosity and fast drainage of the samples, the strain rate was set to 0.166 mm/min. To satisfy the test requirements, it was assumed that the strain-hardening samples would fail when the axial strain reached 25%.

2.3 Test Results

Figure 2 shows the deviatoric stress-axial strain and volume strain-axial strain curves of the tests. It can be seen that the all the samples behave strain hardening at different confining pressures, and the soil exhibits plastic flow only under low confining pressure of 100 kPa. The deviatoric stress increases with the increase of confining pressure. When the strain is small (less than 3%), the deviatoric stress-axial strain curve is linear elastic and the larger the confining pressure is, the greater the slope of the curve is, that is, the greater the initial modulus is. When the strain is large, the curve fluctuates, which is due to the fact that the cement particles are almost not broken when the deviatoric stress is small, and particles slip and rotate mostly, while when the stress is larger, the particles break, and the debris of particles fills pores to form a new skeleton, which leads to the stress readjustment, This phenomenon happens throughout the whole process of particle breakage.

The volumetric strain exhibits shear contraction behavior, which is consistent with the study of the initial pore ratio on the volumetric strain of rockfills conducted by Lade [10]. The volumetric strain increases with the increase of axial strain, and the volumetric strain curve with higher confining pressure is located below. The higher the confining pressure, the greater the final volumetric strain of the sample, and the volumetric contraction can even reach 13.67% at a confining pressure of 400 kPa. The confining pressure has great influence on the volumetric strain. At the beginning of the curve, when the axial strain is small, the volumetric strain increases slowly, and then the rate of volumetric deformation increases rapidly and then slows down.

Figure 3 shows the strength envelope obtained from the tests. The artificial cement particles used in the tests are granular materials, and the cohesion force can be assumed to be 0, so the envelope should pass through the origin. We know that

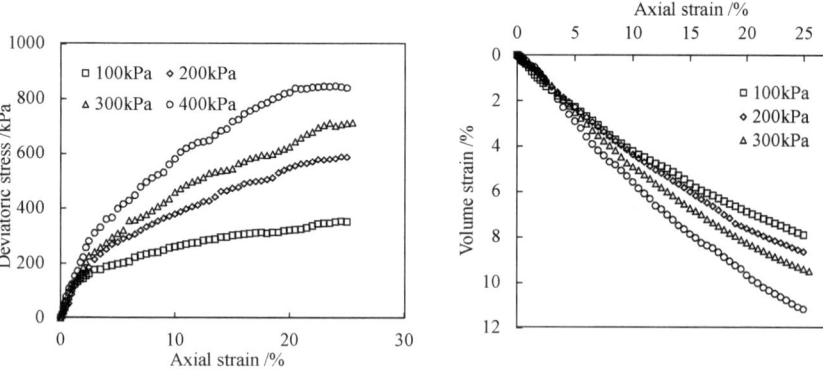

(a) deviatoric stress-axial strain   (b) Volumetric strain-axial strain

Fig. 2 Test results

Fig. 3 Mohr–Coulomb strength envelope

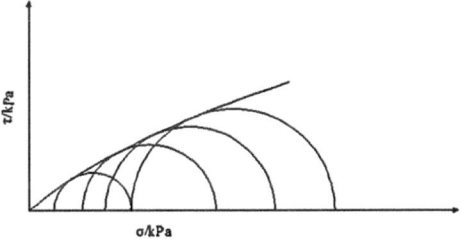

when the particles inside the sample do not break, the envelope should be approximately a straight line. However, the strength envelope obtained from the tests deviates from a straight line and gradually bends downward, indicating that the friction angle decreases continuously. This reflects the strength characteristics where the fragmentation effect gradually dominates with the increase of confining pressure. The interlocking strength between particles is an important component of the strength, which increases linearly with the increase of confining pressure. However, in this study, the breakable cement particles were used, and the cement particles gradually shattered during the testing process. The smaller particles filled the pores, causing the shear-dilation to disappear. Moreover, after the particles breakage, the original interlocking between particles was destroyed, and the interlocking strength gradually weakened, which manifested as a decrease in the friction angle at the macroscopic level.

3 Simulation

3.1 Sample Preparation for Simulation

Based on the Laboratory tests, the two-dimensional Particle Flow Code PFC^{2D} was used to numerically simulate the breaking process of artificial rockfills, to reproduce the breaking process and explore the particle breakage mechanism from a microscopic scale. To observe of the fragmentation of the particle groups, each group is color-coded, as shown in Fig. 4. The model size was 20 cm × 10 cm (length × width), which included 3762 particles, with a particle size range of 1–1.5 mm. There were a total of 50 particle groups, with approximately 75 particles per group.

3.2 Calibration of Model Parameters

In order to make the mechanical properties of the model closer to indoor specimens, a split test of a single particle group was simulated using PFC^{2D}. The split test used

Fig. 4 Model

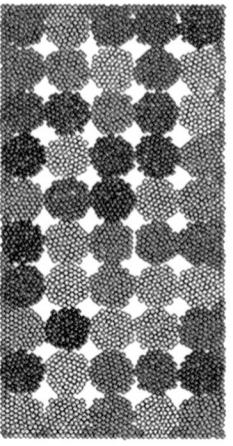

the Brazilian split model in PFC^{2D}, and the particle contacts were described by the parallel bond (PB) model. The result is shown in Fig. 5.

Then we got the parameters of the model (Table 2).

Fig. 5 Sample and results

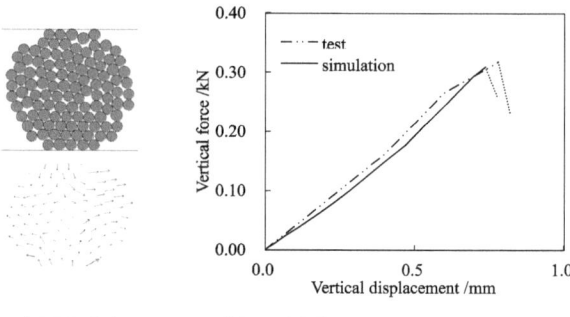

(a) Model (b) Axial force-displacement curve

Table 2 Microscopic parameters of the models

Parameters	Value
Normal stiffness Kn (MPa)	350
Shear stiffness Ks (MPa)	350
Bond normal strength (kPa)	500
Bond shear strength (kPa)	500
Bond radius ratio	0.5
Friction of disk	0.6

Fig. 6 Deviatoric stress-axial strain curves

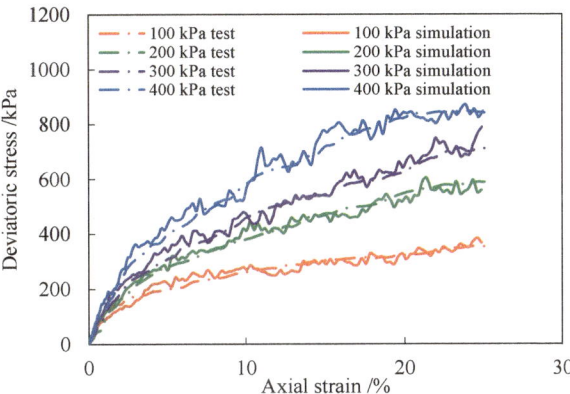

3.3 Results

The deviatoric stress-axial strain relationship obtained from numerical simulation is shown in Fig. 6. We can see that the curves agree well with laboratory tests and both exhibit strain hardening behavior. As the confining pressure in the simulated test increases, the deviatoric stress also increases and the ultimate deviatoric stress increases. The simulated curve exhibits obvious fluctuation, which is because the interaction between particles transitions from the bonding element to the friction element during the simulation due to the destruction of bonds [11]. This is accompanied by stress readjustments, causing fluctuation in the stress curve. The initial section of the simulated curve has little fluctuation because of little bonds destruction, which also confirms this.

Figure 7 shows the volumetric strain-axial strain curves, which exhibits a shear-compression type. The volumetric strain increases with the increase of confining pressure. The simulated volumetric strain values are slightly smaller than the laboratory test results due to the deviation caused by the two-dimensional boundary condition on the model.

During the test, the bonds were continuously broken, and particles were released from the particle groups, achieving the purpose of simulating particle breakage. Here are 5803 bonds in the sample. The bond-breaking curves for fractured bonds are presented for the total bonds, normal bonds, and tangential bonds.

From Fig. 8, it can be seen that the total number of cohesive, normal, and tangential bonds being broken increases as the time step increases and the rate gradually increases. The number of normal bond breakages is greater than the number of tangential bond breakages. In addition, the higher the confining pressure, the more bonds break. This is consistent with the result observed in Laboratory tests. Figure 9 shows the displacement field under the confining pressure of 100 kPa. We note that when the axial strain was very small, the particle displacement in the middle of the sample was smaller than that at the two ends, and a clear cross shearing zone was formed in the middle. As the axial strain increases, the displacements of particles

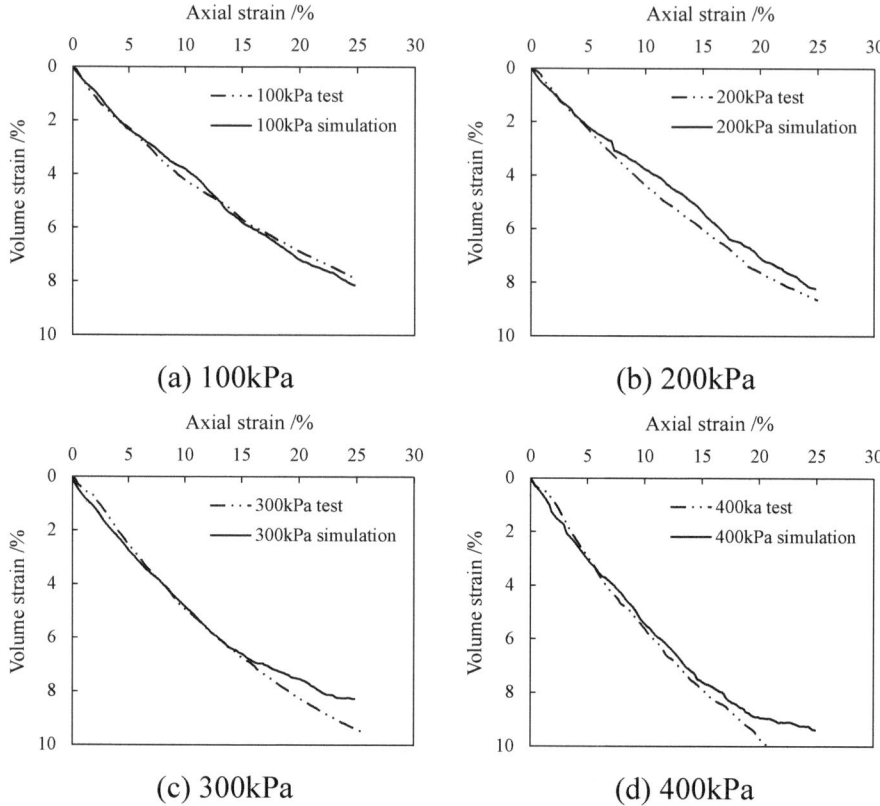

Fig. 7 Volumetric strain-axial strain curves

within the shear band increase and the disorder of displacement direction increases, that is to say, the inconsistency of particle displacements within the shearing zone and the degree of breakage increase significantly. At an axial strain of 9%, the particle groups rotate and break further in the shear band. With further increases in axial strain, the shear band is compressed, and at an axial strain of 15%, the displacements of particles in the middle of sample tend to be consistent, indicating a steady state.

4 Conclusions

(1) The results of laboratory tests show that the deviatoric stress-axial strain curves were strain hardening while the volumetric strain appeared shear contraction during the shearing process. The higher the confining pressure, the bigger the volume-contraction.

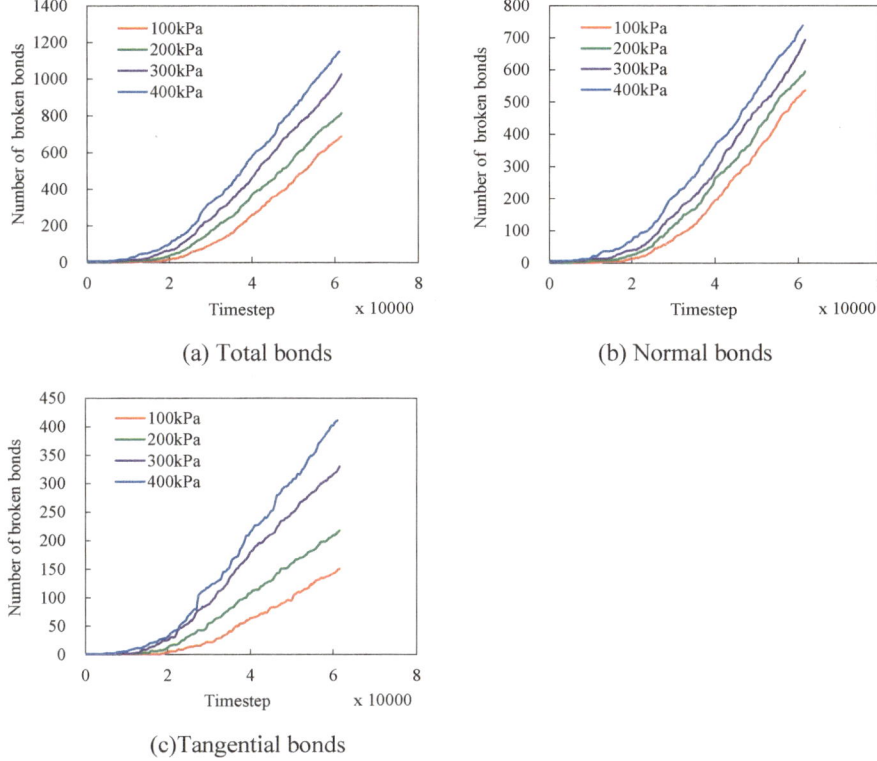

Fig. 8 Number of bonds in simulation

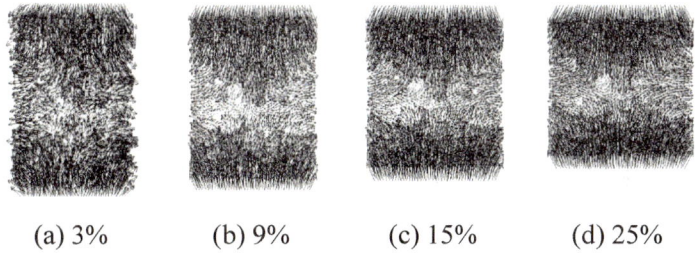

Fig. 9 Displacement field

(2) The strength envelope of laboratory tests is no longer a straight line, which is downwards with the increase of the confining pressure. The strength of the rockfills is considered to be mainly composed of two parts: sliding friction and interlocking friction. With the breakage of particles, the interlocking strength gradually decreases, and ultimately the rockfills exhibit the strength characteristics dominated by sliding friction.

(3) PFC^{2D} can reproduce laboratory triaxial tests well, and the deviatoric stress-axial strain curves agree well with the results of laboratory tests. Though there is a slight deviation in volume strain-axial strain curves, the curves still show the overall trends. The displacement field intuitively display the displacement and fracturing of particles inside the specimen during the triaxial testing process. The results of simulation can provide micro-scale evidence for the study of particle breakage.

References

1. Wei S, Zhu JG, Qian QH et al (2009) Particle breakage of coarse-grained materials in triaxial tests. Chin J Geotech Eng 31(04):533–538
2. Terzaghi K, Peck RB, Mesri G (1996) Soil mechanics in engineering practice, 3rd edn. Wiley, New York, London, Sydney
3. Kjaernsli B, Sande A (1963) Compressibility of some coarse-grained materials. In: Proceedings of the 1st European conference on soil mechanics and foundation engineering. Weisbaden, Germany, pp 245–251
4. Hall E, Gordon B (1964) Triaxial testing with large-scale high pressure equipment. Lab Shear Test Soils ASTM STP 361:315–328
5. Takei M, Kusakabe O, Hayashi T (2001) Time-dependent behavior of crushable materials in one-dimensional compression tests. Soils Found 41(1):97–121
6. Lee KL, Seed HB (1967) Drained strength characteristics of sands. J Soil Mech Found Div 93(6):117–141
7. Cai ZY, Hou HY, Zhang JX et al (2019) Critical state and constitutive model for coral sand considering particle breakage. Chin J Geotech Eng 41(06):989–995
8. Hu SJ, Guo N, Yang ZX et al (2023) Implicit DEM analyses of size and shape effects on crushing strength of rockfill particles. Chin J Geotech Eng 45(02):433–440
9. Shao XQ, Chi SC, Zhang ZL (2021) Numerical simulation of dynamic deformation characteristics of rockfill materials considering particle crushing. Adv Eng Sci 53(04):191–199
10. Lade PV, Bopp PA (2005) Relative density effects on drained sand behavior at high pressures. Soils Found 45(1):1–13
11. Liu EL (2005) Research on breakage mechanism of structural blocks and binary medium model for geomaterials. Tsinghua University, Beijing

Open Access This chapter is licensed under the terms of the Creative Commons Attribution 4.0 International License (http://creativecommons.org/licenses/by/4.0/), which permits use, sharing, adaptation, distribution and reproduction in any medium or format, as long as you give appropriate credit to the original author(s) and the source, provide a link to the Creative Commons license and indicate if changes were made.

The images or other third party material in this chapter are included in the chapter's Creative Commons license, unless indicated otherwise in a credit line to the material. If material is not included in the chapter's Creative Commons license and your intended use is not permitted by statutory regulation or exceeds the permitted use, you will need to obtain permission directly from the copyright holder.

Effect of Mesoscopic Heterogeneity of Concrete on the Macro-mechanical Behavior

Qindong Lin, Chun Feng, Jianfei Yuan, Wenjun Jiao, and Yundan Gan

Abstract Concrete is a typical heterogeneous and multiscale material, and the macro-mechanical response is affected by the meso-scale geometric structure and mechanical parameters. Based on the continuum-discontinuum element method, this study conducts the numerical simulation and quantitively studies the change trend of macro-mechanical response with the different mesoscopic heterogeneity. First, a full-time numerical simulation is conducted, and the elastic modulus and cohesive strength at the meso-scale are assumed to obey the Weibull distribution. Then, the change trend of macro-mechanical parameters is studied. Finally, the change trend of crack evolution characteristic is studied. The results show that the macro-elastic modulus and macro-peak stress gradually increase with the increase of shape parameter k, and the growth rate of macro-mechanical parameters gradually decays. The difference in the shape parameter k causes the value of crack ratio to change, while the change trend of crack ratio-strain curve is similar. As the shape parameter k increases, the final value of crack ratio first decreases and then increases, and the final value of crack ratio when $k = 15$ is the smallest.

Keywords Concrete · Mesoscopic heterogeneity · Macro-mechanical behavior · CDEM · Uniaxial compression

1 Introduction

The multi-scale model of concrete is composed of macro-scale, meso-scale and micro-scale, and the macro-mechanical parameters and damage evolution process are affected by the meso-scale geometric structure and mechanical parameters. At the meso-scale, the concrete is considered as a three-phase material, which is composed

---

Q. Lin · J. Yuan · W. Jiao · Y. Gan (✉)
Xi'an Modern Chemistry Research Institute, Xi'an, Shaanxi 710065, China
e-mail: ganyundan@163.com

C. Feng
Institute of Mechanics, Chinese Academy of Sciences, Beijing 100190, China

of aggregate, mortar and interfacial transition zone (ITZ). Due to the diversity of constituent, complexity of micro-structure and local defect, concrete has a complex nonlinear mechanical behavior.

In the macroscopic homogeneous model established based on the traditional continuum mechanics, the features of meso-scale geometric structure and constituent material are ignored. Therefore, it is difficult to reflect the influence of mesoscopic structural features and constituent material features on the damage evolution process of concrete. With the rapid development of experimental observation technologies, Akbari, Feng, González, Shen and Tian studied the mesoscopic damage evolution process of concrete based on CT, SEM and other technologies [1–5].

Based on the experimental observation results, scholars proposed a variety of mesoscopic mechanical models, such as random aggregate model, MH model and lattice model. With the rapid development of numerical algorithm and computer technology, Maleki [6], Rangari [7], Wu [8], Yang [9], Zeng [10] and Zheng [11] studied the influence of mesoscopic heterogeneity on macroscopic mechanical parameters and damage evolution process based on the numerical simulation. The numerical results indicate that the macroscopic mechanical parameters are affected by the mesoscopic mechanical parameters and geometric characteristic, and the damage evolution characteristic of concrete after introducing the heterogeneity into the meso-scale model is more consistent with the experimental results.

Currently, many scholars have conducted lots of numerical studies on the influence of mesoscopic heterogeneity on macroscope mechanical parameters and damage evolution characteristic of concrete. Since most of the numerical methods belong to the continuum-based method and mesh-free method, they cannot accurately portray the damage evolution process. This study conducts the numerical simulation based on continuum-discontinuum element method, and the change trends of macroscope mechanical parameters and crack evolution characteristic are quantitively studied.

2 Numerical Simulation

2.1 Basic Concept

The continuum-discontinuum element method (CDEM) is a dynamic explicit numerical method, and it is established based on the Lagrangian energy system. The basic model of CDEM includes block and interface, and they represent the continuous feature and discontinuous feature of material, respectively. As a coupled method, CDEM combines the advantages of continuous simulation and discrete simulation, and it can simulate the whole process of material from continuous deformation to crack until movement. Many scholars investigated the static and dynamic mechanical response of brittle material (e.g. rock, concrete), and the test results verify the accuracy of CDEM. [12, 13] Therefore, the mechanical response of concrete under uniaxial compression load is simulated by CDEM, and the change laws are studied.

Fig. 1 Numerical model of concrete specimen

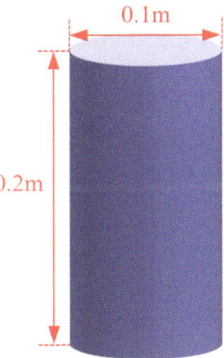

Table 1 Mechanical parameters of concrete

Density (kg/m^3)	Elastic modulus (GPa)	Cohesive strength (MPa)	Tensile strength (MPa)
2400	9	6	4

2.2 Numerical Model

The numerical model of concrete specimen is plotted in Fig. 1. The model is a cylinder, the radius of model is 0.05 m and the vertical height is 0.2 m. The model is meshed by tetrahedral element, the mesh size is set as 5 mm, and there are 46,025 elements in the model. The mechanical parameters of concrete are listed in table 1. To study the effect of heterogeneity of the meso-mechanical parameters on the macro-mechanical behavior, the meso-elastic modulus and meso-cohesive strength are assumed to obey the Weibull distribution. For the meso-elastic modulus, the scale parameter λ_e is 9e9, and the shape parameter k_e ranges from 10 to 30. For the meso-cohesive strength, the scale parameter λ_c is 6e6, and the shape parameter k_c ranges from 10 to 30. To accurately simulate the progressive damage process of concrete in the uniaxial compression experiment, a compression velocity $v_y = 2 \times 10^{-9}$ m/s is applied at the top and bottom boundaries.

2.3 Numerical Results

Macro-mechanical Parameters

The stress–strain curves at the macro scale with different shape parameters k are plotted in Fig. 2, and it is observed that the change trends of stress–strain curves are similar. Before the compressive stress reaches the peak stress, the stress increases linearly with the increase of compressive strain. Once the stress reaches the peak stress, the stress decreases sharply as the strain increases, then the decrease rate of

Fig. 2 Stress–strain curves with different shape parameters k

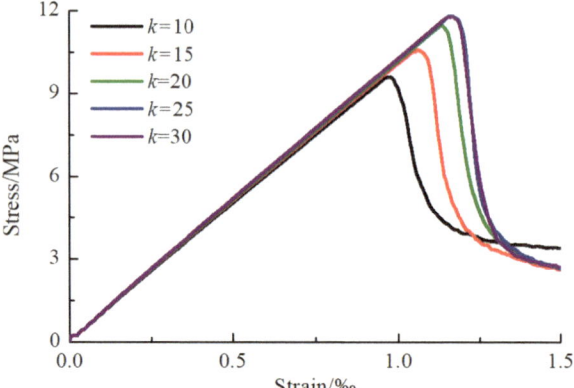

compressive stress decays. Although the change trends of stress–strain curves are similar when the heterogeneity of meso-mechanical parameters changes, some key macro-mechanical parameters (e.g., peak stress, elastic modulus) change.

The change curves of peak stress and elastic modulus when the shape parameter k ranges from 10 to 30 are plotted in Fig. 3. It is observed that with the increase of shape parameter k, the peak stress and elastic modulus gradually increase, and the growth rate gradually decays. As the shape parameter k increases, the heterogeneity of meso-mechanical parameters weakens. The meso-elastic modulus tends to 9e9 Pa, the meso-cohesive strength tends to 6e6 Pa, so the macro-elastic modulus and macro-peak stress gradually increase. In addition, with the increase of shape parameter k, the variation of the heterogeneity of meso-mechanical parameters gradually weakens, so the growth rate of macro-mechanical parameters gradually decays.

Crack Evolution Characteristic

In order to accurately describe the crack evolution characteristic of concrete, a dimensionless index, crack ratio α, is introduced, and it is written as

Fig. 3 Change curve of peak stress and elastic modulus

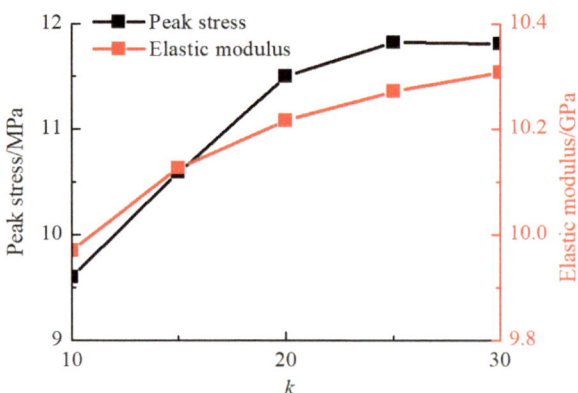

$$\alpha = \frac{S_c}{S_a} \quad (1)$$

where S_c denotes the area of cracked interface, S_a denote the area of all interface.

When the shape parameter k (i.e., k_e and k_c) is equal to 30, the stress–strain curve and crack ratio-strain curve are plotted in Fig. 4. Although the change trends of stress–strain curve and crack ratio-strain curve are different, they have good correspondence. Before the compressive stress reaches the peak stress, with the increase of strain, the stress increases linearly, and the crack ratio α remains zero. When the stress is about to reach the peak stress, the crack ratio α begins to increase, and the increasing amount is small. Once the compressive stress reaches the peak stress, as the strain increases, the compressive stress decreases rapidly, while the crack ratio α increases sharply. Finally, the decrease rate of compressive stress and the growth rate of crack ratio α decays.

When the shape parameter k increases from 10 to 30, the crack ratio-strain curves are plotted in Fig. 5. Although the difference in the shape parameter k causes the value of crack ratio α to change, the change trends of crack ratio-strain curves are similar. The curve has a three-stage characteristic, with the increase of strain, the crack ratio α first remains zero, then increases sharply, and finally increases slowly. With the increase of shape parameter k, the strain value corresponding to the rapid increase of crack ratio α increases gradually, and the strain value when the shape parameter $k = 25$ and 30 are basically the same. In addition, the change in the shape parameter k also affects the final value of crack ratio α. As the shape parameter k increases, the final value of crack ratio α first decreases and then increases. The final value of crack ratio α when $k = 25$ and 30 are basically the same, and the final value of crack ratio α when $k = 15$ is the smallest. According to the crack nephograms and the displacement nephograms (as plotted in Fig. 6), it is concluded that the spatial distribution characteristic of cracked interface is different slightly when the shape parameter k ranges from 10 to 30, and there is a penetrating inclined cracked surface in five cases.

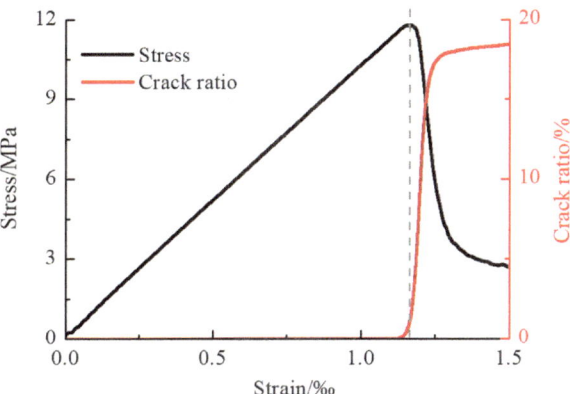

Fig. 4 Change curves of stress and crack ratio when $k = 30$

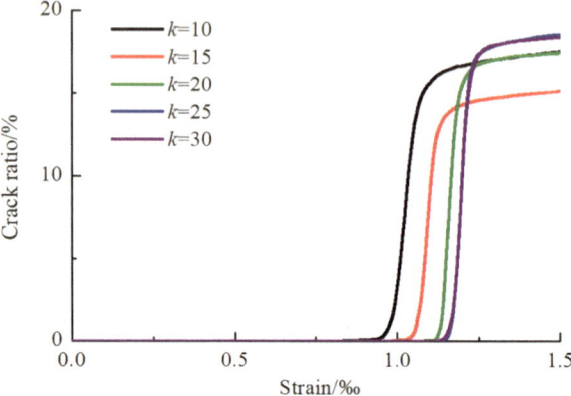

Fig. 5 Change curves of crack ratio with different shape parameters

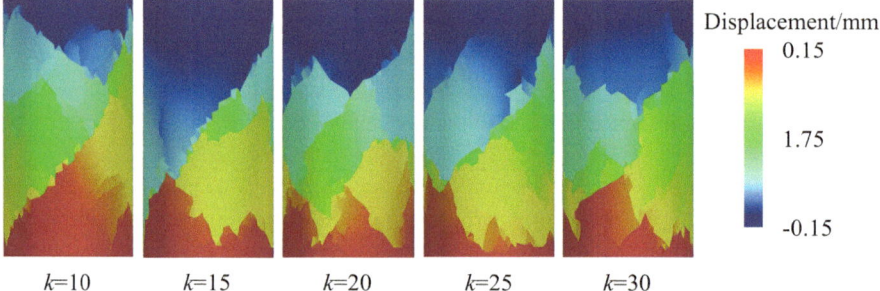

Fig. 6 Displacement nephograms of model with different shape parameters

3 Conclusions

Based on the continuum-discontinuum element method, the change trends of macroscope mechanical parameters and crack evolution characteristic are quantitively studied. First, a full-time numerical simulation is conducted, and the elastic modulus and cohesive strength at the meso-scale are assumed to obey the Weibull distribution. Then, the change trend of macro-mechanical parameters is studied. Finally, the change trend of crack evolution characteristic is studied. The following conclusions can be drawn:

(1) The change trends of stress–strain curves corresponding to different shape parameters k are similar. As the shape parameter k increases, the macro-elastic modulus and macro-peak stress gradually increase, while the growth rate of macro-mechanical parameters gradually decays.

(2) The difference in the shape parameter k causes the value of crack ratio α to change, while the change trends of crack ratio-strain curves are similar. As the shape parameter k increases, the final value of crack ratio α first decreases and

then increases. The final value of α when $k = 25$ and 30 are basically the same, and the final value of α when $k = 15$ is the smallest.

References

1. Akbari M, Tahamtan MHN, Valukolaee SF, Herozi MRZ, Shirvani MA (2022) Investigating fracture characteristics and ductility of lightweight concrete containing crumb rubber by means of WFM and SEM methods. Theoret Appl Fract Mech 117:103148
2. Feng XH, Gong B, Tang CA, Zhao T (2022) Study on the non-linear deformation and failure characteristics of EPS concrete based on CT-scanned structure modelling and cloud computing. Eng Fract Mech 261:108214
3. González DC, Mena Á, Mínguez J, Vicente MA (2021) Influence of air-entraining agent and freeze-thaw action on pore structure in high-strength concrete by using CT-scan technology. Cold Reg Sci Technol 192:103397
4. Shen JR, Xu QJ, Liu MY (2021) Statistical analysis of defects within concrete under elevated temperatures based on SEM image. Constr Build Mater 293:123503
5. Tian W, Cheng X, Liu Q, Yu C, Gao FF, Chi YY (2021) Meso-structure segmentation of concrete CT image based on mask and regional convolution neural network. Mater Des 208:109919
6. Maleki M, Rasoolan I, Khajehdezfuly A, Jivkov AP (2020) On the effect of ITZ thickness in meso-scale models of concrete. Constr Build Mater 258:119639
7. Rangari S, Murali K, Deb A (2018) Effect of meso-structure on strength and size effect in concrete under compression. Eng Fract Mech 195:162–185
8. Wu LW, Huang D (2022) Peridynamic modeling and simulations on concrete dynamic failure and penetration subjected to impact loadings. Eng Fract Mech 259:108135
9. Yang X, Ma S, Dai HZ (2020) Effect of geometric form of concrete meso-structure on its mechanical behavior under axial tension. Constr Build Mater 255:119295
10. Zeng MH, Wu ZM, Wang YJ (2020) A stochastic model considering heterogeneity and crack propagation in concrete. Constr Build Mater 254:119289
11. Zheng ZS, Wei XS, Tian C (2021) Mesoscale models and uniaxial tensile numerical simulations of concrete considering material heterogeneity and spatial correlation. Constr Build Mater 312:125428
12. Feng C, Li SH, Liu XY, Zhang YN (2014) A semi-spring and semi-edge combined contact model in CDEM and its application to analysis of Jiweishan landslide. J Rock Mech Geotech Eng 6:26–35
13. Lin QD, Li SH, Feng C, Wang XQ (2021) Cohesive fracture model of rocks based on multi-scale model and Lennard-Jones potential. Eng Fract Mech 246:107627

Open Access This chapter is licensed under the terms of the Creative Commons Attribution 4.0 International License (http://creativecommons.org/licenses/by/4.0/), which permits use, sharing, adaptation, distribution and reproduction in any medium or format, as long as you give appropriate credit to the original author(s) and the source, provide a link to the Creative Commons license and indicate if changes were made.

The images or other third party material in this chapter are included in the chapter's Creative Commons license, unless indicated otherwise in a credit line to the material. If material is not included in the chapter's Creative Commons license and your intended use is not permitted by statutory regulation or exceeds the permitted use, you will need to obtain permission directly from the copyright holder.

Experimental Study on Splitting Strength of Nano-active Powder Concrete

Hailing Bao, Xiaofeng Ji, and Peibao Xu

Abstract Aiming at the problem of weakening mechanical properties of conventional concrete due to damage, the present paper studies the effect of nano-active powder on the splitting strength of concrete before and after damage. The nano-active powder concrete was prepared by adding 1.5%, 2.0% and 2.5% nano-silica (NS) and nano-zirconia (NZ) respectively. Through the splitting test before and after the damage of the sample, the self-healing performance of the sample was studied with the recovery rate of the splitting strength as the index. The results show that nano-active powders NS and NZ can increase the splitting strength and its recovery rate. Compared with the control group concrete, the addition of nano-active powder effectively improved the recovery rate of concrete splitting strength, and the optimal content of nano-active powders NS and NZ are 1.5% and 2.0%, respectively. This study deepens the understanding of mechanical properties of nano-concrete.

Keywords Nano-active powder · Concrete · Splitting strength · Strength recovery rate · Self-healing

1 Introduction

Concrete is a widely used building material. The raw materials and fossil fuels that concrete consumed aggravate the environmental pollution problem and are also the main source of man-made greenhouse gas emissions [1, 2]. In addition, it is easy to crack under tensile stress, eventually leading to a significant increase in its

---

H. Bao
Anhui Technical College of WaterResources and Hydroelectric Power, Hefei 231603, Anhui, China

X. Ji · P. Xu (✉)
Department of Civil Engineering, Anhui Jianzhu University, Hefei 230601, Anhui, China
e-mail: peibaoxu@qq.com

X. Ji
e-mail: jixiaofeng0411@163.com

© The Author(s) 2025
P. Xiang et al. (eds.), *Frontier Research on High Performance Concrete and Mechanical Properties*, Lecture Notes in Civil Engineering 518,
https://doi.org/10.1007/978-981-97-4090-1_23

poriness, which is the main reasons for the decline of its strength and durability [3, 4]. Therefore, many advanced cementitious systems have been explored and applied to limit the entry of harmful chemicals through pores and cracks, so as to improve the durability of cementitious system [5, 6]. However, most of their working mechanisms rely on external intervention, which is not conducive to cost reduction [7]. To this end, researchers are exploring sustainable economic solutions to reduce costs and environmental problems. A potential approach to actively repair cracks is to develop self-healing concrete technology [8].

There are two main types of self-healing of concrete: natural healing and self/engineering healing. Natural healing means that the concrete itself has the ability to repair, that is, there is a cement paste that has not been fully hydrated at the crack location, which reacts with carbon dioxide and water in the humid environment to generate calcium carbonate and other crystal precipitation to cover the crack location to repair the crack. The self/engineering healing refers to the healing of damaged cracks in concrete by human intervention or non-human intervention. One method of self-healing is to use self-healing admixtures, such as nano-fillers, fibers, etc.

At present, the nano-material has become a kind of special material to improve and manipulate the inherent characteristics of concrete, and the excellent performance in enhancing the mechanical, bending and electromechanical properties of concrete [9]. In addition, the excellent properties of nano-material have promoted their potential applications, from structural strengthening to environmental pollution remediation, the production of self-cleaning materials to the application of self-testing cementitious composites for structural health monitoring [10]. Therefore, scholars at home and abroad began to study the influence of nano-filler on the self-healing of concrete [11, 12].

Beigi [13] found that nano-silica gel can significantly improve the mechanical properties and durability of self-compacting concrete. Salemi [14] found that nano-SiO_2 reduced the permeability and porosity of pavement concrete, enhancing the frost resistance and compressive strength of concrete; Chen et al. [15] found that the impermeability of concrete increased with the increase of nano-SiO_2, and the improvement of permeability of low-strength grade was significantly higher than that of high-strength grade concrete. Noorvand [16] found that the addition of nano-TiO_2 reduced the fluidity and increased the compressive strength of concrete. Hou et al. [17] found that the addition of nano-SiO_2 improved the impermeability and compressive strength of concrete. Eskandari [18] found that the quality of concrete was improved by nanoparticles, and the permeability of chloride ions was significantly reduced by nano-SiO_2.

In order to further study the influence of nano-filler on the self-healing performance of concrete, this paper studies the influence of nano-silica (NS) and nano-zirconia (NZ) on the self-healing performance of concrete through experiments. The nano-active powder concrete was prepared by adding NS and NZ with a mass ratio α of 0, 1.5, 2.0 and 2.5% into concrete, and the splitting strength was tested to explore the effect of the proportion and type of nano-filler on the self-healing properties of concrete.

2 Experiment

2.1 Raw Material and Preparation of Specimen

Raw Material

The raw materials used in this paper include: P.O42.5 ordinary Portland cement with a density of 3050 kg/m3; F class I grade fly ash with a density of 2280 kg/m^3; Medium sand as fine aggregate with fineness modulus of 2.7; Continuous graded natural gravel as coarse aggregate with the diameter of 5–32 mm; Slag powder with the specification of S95; UEA concrete expansion agent; 20nm nano-SiO$_2$ with a density of 2.2–2.6 g/cm^3; 20 nm nano-ZrO$_2$ with a density of 6.0 g/cm^3.

Preparation of Specimen

The mix proportion of nano-active powder concrete in this paper is formulated according to the Specification for Mix Design of Ordinary Concrete (JGJ 55-2011), as shown in Table 1. The size of formed nano-active powder concrete test block is 150 × 150 × 150 mm. Three test blocks were prepared with the content of each nano-filler, and the average measured value of the three specimens was taken as the splitting strength test result. The prepared specimen is shown in Fig. 1, and the specific steps are as follows:

(1) Weigh the raw materials required for each mixing and put them into the container separately; (2) Pour the measured raw materials into the blender, dry mix in the blender for about 5 min, so that the materials are evenly mixed; (3) Pour the mixing water into the mixer and mix for about 5 min; (4) Put the evenly mixed concrete mixture into the prepared mold and put it into the laboratory for curing; (5) Demould after 24 h, and then put the test block into the standard curing box for curing for 28 d.

Table 1 Mix proportion of nano-active powder concrete

Material	Cement	Fly ash	Slag powder	Water	Fine aggregate	Coarse aggregate	Swelling agent	NS/NZ
Kg/m^3	260	26	39	160	680	1200	26	4.9, 6.5, 8.1
Proportion	0.80	0.08	0.12	0.49	2.09	3.69	0.08	1.5%, 2.0%, 2.5%

Fig. 1 Precast concrete specimens

2.2 Experimental Method

This test takes the recovery rate of splitting strength as the index to study the influence of nano-fillers on the self-healing performance of concrete. The splitting strength test is conducted in accordance with the Test Code for Hydraulic Concrete (DL/T 5150-2017) [19]. According to existing research, humid environment is more conducive to concrete repair [20–22]. Therefore, the test blocks in this test are all cured in water with a temperature of (20 ± 2) °C. The specific test steps are as follows:

(1) Take out the test blocks cured for 28 days for splitting strength test, uniformly load the test block at the speed of 1.8 MPa/min ~ 3.6 MPa/min until the test blocks are damaged, and record the failure load. For each group of three test blocks, firstly take a group of test blocks to measure their ultimate splitting strength f_t, the remaining two groups of test blocks are preloaded to $\beta = 60\% f_c$ and $\beta = 80\% f_c$ [23, 24], respectively. (2) After all the test blocks are preloaded, put them into water with a temperature of about 20 °C for curing for 28 days. (3) Take out the test blocks that have reached the curing age to measure their splitting strength. By comparing the splitting strength before and after the damage, analyze the influence of nano-filler on the self-healing performance of concrete.

The data processing formula is as follows:

The splitting strength is defined as

$$f_t = \frac{2F}{\pi A}$$

where f_t is the splitting strength; F is the failure load; A is the pressure bearing area of the test specimen.

The strength recovery rate K_c is

$$K_c = \frac{f_{tR}}{f_{t0}} \times 100\%$$

where f_{tR} is the splitting strength of concrete after preloading and curing in water for 28 days; f_{t0} is the splitting strength of the control group concrete after curing for 28 days.

3 Results and Discussions

Figure 2 shows the relationship between the mass content α of nano-filler and the splitting strength of concrete. It can be seen that the NS and NZ improve the splitting strength of concrete. The optimal mass contents of NS and NZ are 1.5% and 2.0% respectively. For NS and NZ concrete, the splitting strengths are increased by 27.2% and 22.4%, respectively. It can be concluded that the nano-filler can effectively improve the splitting strength of concrete.

Figure 3 shows the influence of nano-filler NS on the splitting strength of concrete with different preloading degrees. It can be seen that the NS improves the splitting strength of concrete, especially when the NS content is 1.5%. The test blocks with preloaded $\beta = 60\% f_c$, $\beta = 80\% f_c$ and ultimate failure were placed in water for curing for 28d, and then the strength test was carried out. It can be concluded that the splitting strength of concrete decreases with the increase of preloading degree; The splitting strength of the concrete is the largest with the NS content of 1.5% and the weakest with the NS content of 2.5%, as shown in Fig. 3. In general, when the local damage occurs under the preload, the best effect to improve the splitting strength is when the content of NS is 1.5%.

Figure 4 shows the influence of nano-filler NZ on the splitting strength of concrete with different preloading degrees. It can be seen that the NZ improves the splitting strength of concrete, especially when the NZ content is 2.0%. The test blocks with preloaded $\beta = 60\% f_c$, $\beta = 80\% f_c$ and ultimate failure were placed in water for curing for 28d, and then the strength test was carried out. It can be concluded that the splitting strength of concrete decreases with the increase of preloading degree; The splitting strength of the concrete is the largest with the NZ content of 2.0% and the weakest with the NZ content of 2.5%, as shown in Fig. 4. In general, when the local

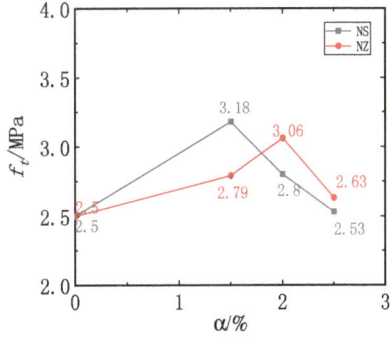

Fig. 2 Relationship between nano-filler content and splitting strength of concrete

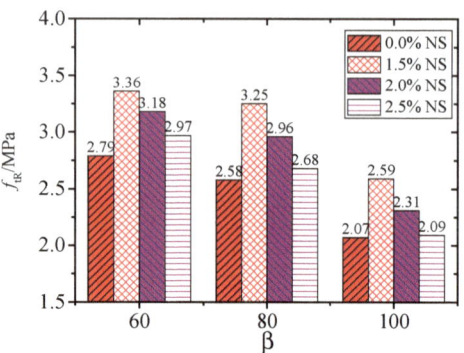

Fig. 3 Effect of NS on splitting strength of concrete with different preloading degrees

damage occurs under the preload, the best effect to improve the splitting strength is when the content of NZ is 2.0%.

Figure 5 shows the effects of different nano-filler NS and NZ and their contents on the recovery rate of concrete splitting strength with different preloading degrees. The test blocks with preloaded $\beta = 60\% f_c$, $\beta = 80\% f_c$ and ultimate failure were placed in water for curing for 28 d, and then the strength test was carried out. It is found that the maximum recovery rate of concrete splitting strength is 111.6%, 103.2% and 82.8% without NS or NZ under different preloading degrees, and the recovery rate of concrete splitting strength is improved after NS and NZ are added. Therefore, the addition of nano-filler has a certain promotion effect on the recovery rate of concrete splitting strength.

Further analysis of Fig. 5 shows that the recovery rate of concrete splitting strength increases first and then decreases with the increase of nano-filler NS and NZ content in general. For $\beta = 60\% f_c$ preloading, the maximum strength recovery rate of concrete is 134.4% with the NS content of 1.5 and 138.4% with the NZ content of 2.0%. For $\beta = 80\% f_c$ preloading, the maximum strength recovery rate of concrete is 130.0% with the NS content of 1.5% and 124.8% with the NZ content of 2.0%. In conclusion, when optimum added contents of NS and NZ are 1.5% and 2.0% respectively, concrete has the maximum recovery rate of splitting strength, and the

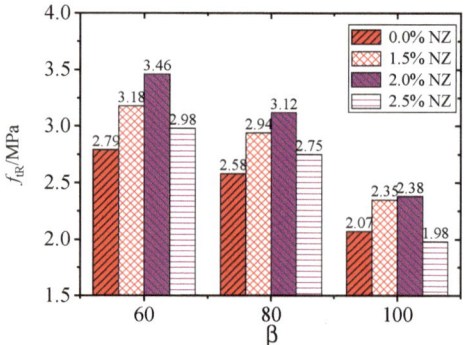

Fig. 4 Effect of NZ on splitting strength of concrete with different preloading degrees

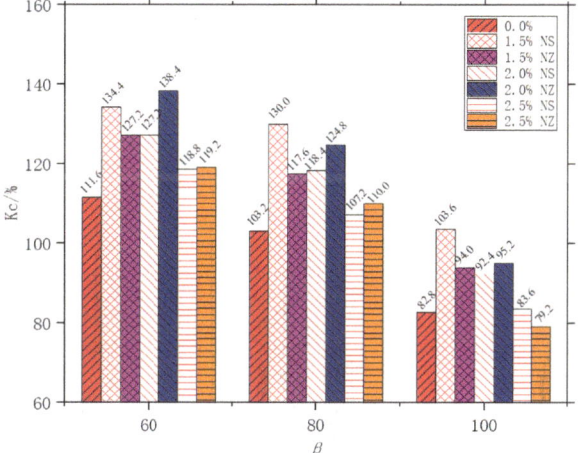

Fig. 5 Effect of nano-filler on recovery rate of concrete splitting strength

effect of nano-filler NZ is slightly better than that of nano-filler NS on improving the recovery rate of concrete splitting strength. In the case of ultimate failure, the recovery rate of splitting strength of concrete with the NZ content of 2.5% is slightly lower than that of the control group. The reason for this phenomenon may be the error caused by concrete cracking. The recovery rate of concrete splitting strength decreases with the increase of preloading degree, which indicates that the recovery ability of concrete splitting strength weakens with the increase of concrete damage degree.

4 Conclusions

The effects of the contents of nano-filler NS and NZ on the splitting strength and recovery rate of concrete with different preloading degrees were studied experimentally. From the analysis of test data, it can be concluded that the addition of nano-filler NS and NZ can improve the splitting strength of concrete. The splitting strength increases first and then decreases with the increase of nano-filler content, in which the optimal contents of NS and NZ are 1.5% and 2.0% respectively. The ultimate splitting strength of the preloaded concrete test blocks increases after adding nano-filler NS and NZ. The splitting strength of the concrete decreases with the increase of the preloading degree. The addition of nano-filler can improve the recovery rate of concrete splitting strength to some extent. In terms of the self-healing effect of splitting strength, the optimal addition contents of NS and NZ are 1.5% and 2.0% respectively. These experimental results can provide data reference for the application of nano-material in self-healing concrete.

References

1. Achal V, Mukherjee A (2015) A review of microbial precipitation for sustainable construction. Constr Build Mater 93:1224–1235
2. Ghosh Kumar S (2008) Self-healing materials: fundamentals, design strategies, and applications, 1–28. https://doi.org/10.1002/9783527625376
3. Schlangen H, Jonkers HM, Qian S et al (2010) Recent advances on self healing of concrete. Civ Eng Geosci
4. Alaloul WS, Musarat MA, Haruna S et al (2021) Mechanical properties of silica fume modified high-volume fly ash rubberized self-compacting concrete. Sustainability 13(10):5571
5. Arbi K et al (2016) A review on the durability of alkali-activated fly ash/slag systems: advances, issues, and perspectives. Ind Eng Chem Res 55(19):5439–5453
6. Tjaronge MW, Musarat MA, Law K et al (2021) Effect of graphene oxide on mechanical properties of rubberized concrete: a review
7. Vaysburd AM, Emmons PH (2000) How to make today's repairs durable for tomorrow—corrosion protection in concrete repair. Constr Build Mater 14(4):189–197
8. Jonkers HM, Thijssen A, Muyzer G et al (2010) Application of bacteria as self-healing agent for the development of sustainable concrete. Ecol Eng 36(2):230–235
9. Sanchez F, Sobolev K (2010) Nanotechnology in concrete–a review. Constr Build Mater 24(11):2060–2071
10. Silvestre J, Silvestre N, De Brito J (2016) Review on concrete nanotechnology. Eur J Environ Civ Eng 20(4):455–485
11. Han B (2018) Long-term machnical, self-healing and impact performances of concrete with nonafillers. Dalian University of Technology (in Chinese)
12. Tang X, Wei X, Liu X (2011) Mechanical properties improvement in concrete of every age by using different contents of Nano-SiO_2. Sci Technol Rev 21:64–69 (in Chinese)
13. Beigi MH, Berenjian J, Omran OL et al (2013) An experimental survey on combined effects of fibers and nanosilica on the mechanical, rheological, and durability properties of self-compacting concrete. Mater Des 50:1019–1029
14. Shirgir B, Alizadeh Goudarzi H, Shirgir V (2016) An experimental study on the abrasion resistance of pervious concrete containing nano SiO_2 in pavement. Q J Transp Eng 8(2):291–302
15. Chen A, Tang X (2013) Influence of nano-SiO_2 on performance of different strength grade concrete. Bull Chin Ceram Soci 32(8) (in Chinese)
16. Noorvand H, Ali A, Demirboga R et al (2013) Incorporation of nano TiO_2 in black rice husk ash mortars. Constr Build Mater 47:1350–1361
17. Hou X, Huang D, Wang W (2013) Recent progress on high performance concrete with nano-SiO_2 particles. Concrete 2013(3) (in Chinese)
18. Eskandari H, Vaghefi M, Kowsari K (2015) Investigation of mechanical and durability properties of concrete influenced by hybrid nano silica and micro zeolite. Procedia Mater Sci 11:594–599
19. DL/T 5150-2017, The test code for hydraulic concrete
20. Huang T, Chen Q, Hu Y et al (2013) Research of XYPEX on the property of concrete crack repair. Concrete 11:88–92 (in Chinese)
21. Roig-Flores M, Moscato S, Serna P et al (2015) Self-healing capability of concrete with crystalline admixtures in different environments. Constr Build Mater 86:1–11
22. Roig-Flores M, Pirritano F, Serna P et al (2016) Effect of crystalline admixtures on the self-healing capability of early-age concrete studied by means of permeability and crack closing tests. Constr Build Mater 114:447–457
23. Yao J (2020) Research on self-healing performance of concrete mixed with nano-modified cementitious capillary crystalline waterproofing. Jiangsu University (in Chinese)
24. Liu S, Zhu D, Guo S (2016) Research on self-healing of concrete cracking sulfate environment. Mater Rep 30(02):108–113 (in Chinese)

Open Access This chapter is licensed under the terms of the Creative Commons Attribution 4.0 International License (http://creativecommons.org/licenses/by/4.0/), which permits use, sharing, adaptation, distribution and reproduction in any medium or format, as long as you give appropriate credit to the original author(s) and the source, provide a link to the Creative Commons license and indicate if changes were made.

The images or other third party material in this chapter are included in the chapter's Creative Commons license, unless indicated otherwise in a credit line to the material. If material is not included in the chapter's Creative Commons license and your intended use is not permitted by statutory regulation or exceeds the permitted use, you will need to obtain permission directly from the copyright holder.

Dynamic Characteristics Test and Microstructure Analysis of Silt Soil Improved by Curing Agent

Qi Lu, Yongzhen Ma, Ganbin Liu, and Fuxin Ni

Abstract With the acceleration of urbanization, the output of construction sludge and waste mud is increasing. Its transportation cost is high and there are potential safety hazards. Therefore, it is necessary to seek effective disposal and utilization methods. Aiming at the silt soil with high water content, the developed YB100 powder curing agent was used to improve the properties of silt soil, and the effects of dynamic properties and water stability of the cured soil were studied by curing agent dosing, surrounding pressure, and dynamic stress ratio. The microstructure of the cured soil was analyzed by SEM scanning electron microscopy to reveal the curing mechanism. The results showed that: the dynamic elastic modulus of the cured soil specimens increased from about 60 MPa to about 150 MPa when the amount of curing agent was increased by 0, 2, 4 and 6%; with the increase of curing agent content, the damping ratio changed less during vibration and the curve was flatter; after 24 h of water immersion, the dynamic elastic modulus of the specimens decreased by about 15–20% and the damping ratio had some growth, and with the dynamic stress ratio. With the growth of dynamic stress ratio, the weakening of dynamic elastic modulus of the specimen is not obvious, and the change of damping ratio weakly decreases less; SEM photos show that the cured soil particles are agglomerated, and the pore space is greatly reduced, because the curing agent produces hydration reaction with water in the silt soil, and the resulting gelling material improves the strength of the soil, and at the same time the complexes generated by the reaction with the soil strengthen the stability of the soil.

Keywords Silt soil · Solidification · Dynamic elastic modulus · Damping ratio · Water stability

---

Q. Lu · G. Liu (✉) · F. Ni
Institution of Geotechnical Engineering, Ningbo University, Ningbo, Zhejiang, China
e-mail: liuganbin@nub.edu.cn

Y. Ma
School of Architecture and Transportation, Ningbo University of Technology, Ningbo, Zhejiang, China

1 Introduction

With the rapid urbanization process, the generation of construction sludge and waste mud from various engineering projects is increasing. The conventional methods of disposing of construction sludge and waste mud, such as dumping or landfilling, incur high transportation costs, pose safety hazards, and exert great pressure on the ecological environment. This situation intensifies the contradiction between production and the capacity for acceptance [1]. Construction sludge, like silty clay found in pits, gives rise to new sources of waste and pollution. Moreover, it hampers the promotion of harmonious development among resources, the environment, and the economy.

Most researchers have increasingly focused on the treatment technology of construction project residues and waste slurry, specifically as roadbed filling materials and composite foundation fillers. They have conducted relevant experimental studies in this area. Li [2] concluded that, after undergoing specific treatment, construction residues exhibit high strength and good stability. They can be directly applied to urban highway road foundation projects using specific construction processes. Qi [3], the basic performance test of construction slag soil demonstrated that construction slag soil can be used as a road base filler by utilizing cement+, cement fly ash+, and lime fly ash+ together with construction slag soil. This study showed that the strength increases with the increase of admixture after modification. However, it should be noted that the permeability of the cured soil remains poor, and its water stability is low, making it susceptible to disintegration. As a result, it is not suitable for long-term deformation and stability in complex environments.

Dynamic properties play a crucial role in the long-term safety performance of roadbed filling materials. Parameters such as dynamic elastic modulus and damping ratio are essential for studying the dynamic properties of soils. Xu [4] conducted extensive vibratory triaxial tests on coarse-grained soils with varying P5 contents to investigate the impact of P5 content on their dynamic properties. Deng [5] examined the dynamic properties of chemically modified soft highway foundation soils and evaluated the corresponding rebound modulus and cumulative plastic strain through cyclic loading tests. Yang [6] introduced rubber powder with different particle sizes (10, 20 and 60 mesh) and contents into red clay, exploring the properties and mechanisms of rubber red clay in road applications. Tan [7] performed dynamic triaxial UU tests on lime-improved roadbed fill using the GDS dynamic triaxial instrument. They analyzed the variations in accumulated axial strain and stress–strain hysteresis characteristics of lime-improved soil by considering lime admixture, surrounding pressure, and dynamic stress amplitude as variables. Qiao [8] investigated the variation patterns of dynamic strength, dynamic elastic modulus, and damping ratio in cured soil with the addition of micronized granulated blast furnace slag powder, using dynamic triaxial tests. However, there is a scarcity of experimental studies on the dynamic properties of soils with added curing agents.

Since the use of silt soil as road base filling material, particularly for high water content silt soil found in regions like Ningbo, has not been extensively implemented.

Table 1 Physical and mechanical indexes of test soil

Specific gravity of soil particles	Liquid limit (%)	Plastic limit (%)	Optimal moisture content (%)	Maximum dry density (g/cm^3)
2.72	49.9	23.0	18.9	1.74

In this study, we employed a developed curing agent to conduct dynamic characteristics and water stability tests on cured soil. Additionally, we examined the influence of the curing agent mixture, surrounding pressure, and dynamic stress ratio on the long-term dynamic characteristics and water stability of the cured road base soil. This research aims to provide valuable insights for the effective curing of high water content silt soil and its application as road fill material.

2 Materials and Methods

2.1 Materials

The test soil is taken from the muddy building slag soil excavated by a foundation pit project in Zhenhai District, Ningbo, and its basic physical properties are shown in Table 1. The initial moisture content of the undisturbed soil sample is 38.0–44.2%, the initial void ratio is 0.98–1.09, and the specific gravity of soil particles is 2.72. According to the Geo-technical Test Specification (SL237-1999), the maximum dry density obtained under standard compaction conditions is 1.74 g/cm^{-3}, and the corresponding water content is the optimal water content of 18.9%. The mineral composition of building slag soil is mainly Illite, with a small amount of kaolinite, chlorite and Aemon mixed layer.

2.2 Testing Apparatus

In this test, a fully automatic dynamic triaxial instrument produced by GDS Company in the UK is used for long-term dynamic characteristic tests, including pressure chamber, confining pressure, axial compression equipment and computer control and analysis system, as shown in Fig. 1a.

In order to study the effect of Solidifying agent on micro-pore structure and mineral composition of soft clay, scanning electron microscope (SEM) is used to analyze it. The Testing Apparatus is shown in Fig. 1b.

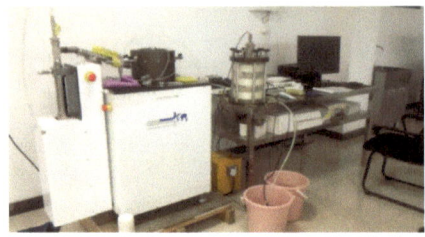

(a) GDS automatic dynamic triaxial instrument (b) scanning electron microscope

Fig. 1 Testing apparatus

2.3 Testing Scheme

The test soil is compacted, turned and screened. Considering the economy of building slag soil treatment, 0, 2, 4 and 6% of solidifying agent content are selected, and the optimal moisture content of improved soil with different amount of solidifying agent content is measured, as shown in Table 2.

The optimal moisture content measured in Table 2 is approximately 18.5%, which is evenly mixed into the soil and sealed for 24 h according to different mixing amounts, and the soil samples under different proportions are controlled at about 18.5% moisture content, and then the compaction is divided into four layers in a triple valve saturator with a high compaction degree of 80 mm and a diameter of 39.1 mm. After the sample is made, it is sealed with plastic film and maintained in the maintenance box for 7 days.

Using the developed YB100 powder Solidifying agent, the solidifying agent is composed of polyacrylamide 0.1–0.5%, AFB additive 1–2%, OFB reinforcer 2–4%, silicate clinker 20–30%, quicklime 10–20%, raw gypsum 1–5%, fly ash 40–60%. According to the previous test results, YB100 powder Solidifying agent has a good curing effect on soft clay. Therefore, the Solidifying agent is used to cure silt soil its dynamic characteristics are studied.

The frequency of traffic load is different on different types of roads. The vibration frequency of this test is 2 Hz, the number of cycles is 10000, the dynamic stress is 10, 20 and 30 kPa, the vibration method is consolidated undrained vibration, and the test confining pressure is 50 and 100 kPa. The test scheme is shown in Table 3.

Table 2 Basic physical properties of solidifying agent improves muddy building slag soil

Solidifying agent content (%)	Maximum dry density (%)	Optimal moisture content (%)
0	1.74	18.9
2	1.75	18.7
4	1.77	18.5
6	1.81	17.8

Table 3 Plan of tests

Solidifying agent content (%)	Consolidation confining pressure (kPa)	Amplitude (kPa)	Dynamic stress ratio	Frequency (Hz)
2	50	10	0.2	2
		20	0.4	2
		30	0.6	2
	100	10	0.1	2
		20	0.2	2
		30	0.3	2
4	50	10	0.2	2
		20	0.4	2
		30	0.6	2
	100	10	0.1	2
		20	0.2	2
		30	0.3	2
6	50	10	0.2	2
		20	0.4	2
		30	0.6	2
	100	10	0.1	2
		20	0.2	2
		30	0.3	2

3 Results and Discussion

3.1 Dynamic Elastic Modulus and Damping Ratio of Slag Soil Without Solidifying Agent

Figures 2 and 3 show the variation of dynamic elastic modulus and damping ratio with vibration times under different confining pressure when the curing age is 7 days and the solidifying agent content is 0%. In Fig. 2, under the condition of the same dynamic stress ratio, the dynamic elastic modulus of the sample increases with the increase of vibration times, but the increase is relatively slow; under the confining pressure of 50 kPa, the increase of dynamic stress ratio in Fig. 2a decreases the dynamic elastic modulus, but Fig. 2b does not have this rule. However, based on the dynamic elastic modulus, the lowest initial dynamic elastic modulus in Fig. 2b is 54.77 MPa, while the highest initial dynamic elastic modulus in Fig. 2a is only 57.55 MPa, and the difference is only 2.78 MPa, indicating that the dynamic elastic modulus is higher when the confining pressure is 100 kPa. At the same time, the greater the confining pressure, the greater the strength of the slag. In Fig. 3, the damping ratio tends to decrease as a whole, and the larger the dynamic stress ratio, the smaller the damping ratio. This is consistent with the phenomenon observed in reference [9].

Compared with Figs. 2 and 3, the dynamic elastic modulus of the sample increases with the increase of confining pressure, and the damping ratio decreases with the increase of confining pressure. Under the same confining pressure, the dynamic elastic modulus decreases with the increase of dynamic stress ratio, and there is no obvious correlation between damping ratio and dynamic stress ratio. On the whole, this is because the constraint caused by the confining pressure on the slag affects the damping ratio and dynamic elastic modulus.

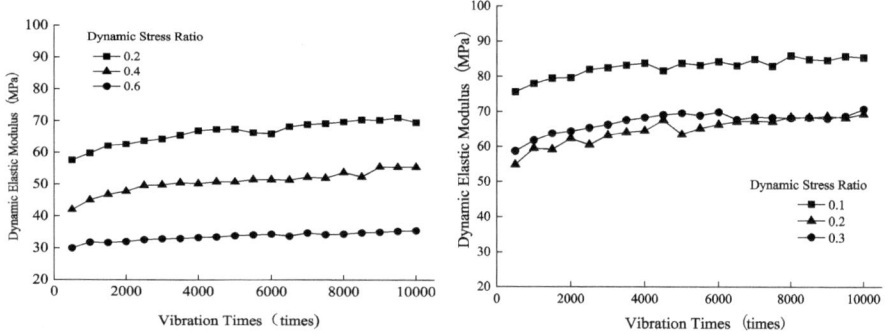

(a) Consolidation confining pressure 50 kPa   (b) Consolidation confining pressure 100 kPa

Fig. 2 Dynamic elastic modulus changes with vibration times (maintaining 7 d, dosage 0%)

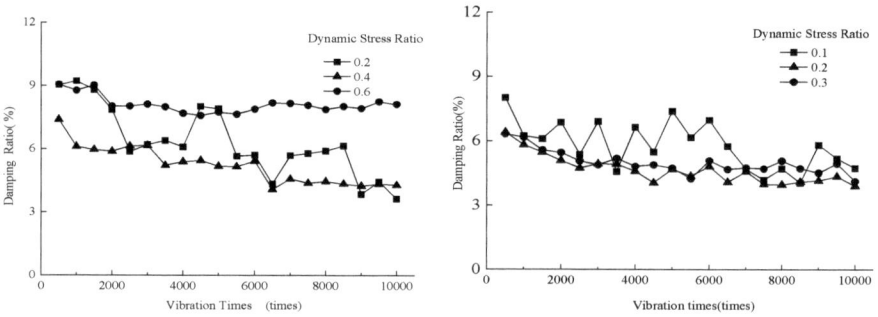

(a) Consolidation confining pressure 50 kPa   (b) Consolidation confining pressure 50 kPa

Fig. 3 Damping ratio changes with vibration times (maintaining 7 d, dosage 0%)

3.2 Dynamic Elastic Modulus and Damping Ratio of Solidified Soil with Different Content

The test results of long-term dynamic characteristics of solidified soil with different solidifying agent content are shown in Figs. 4, 5, 6, and 7. Figures 4 and 5 show the dynamic elastic modulus and damping ratio of solidified soil samples under confining pressure 50 kPa varying with the number of vibrations during the curing period of 7 days. It can be seen from Fig. 4 that after adding Solidifying agent, the dynamic elastic modulus of the sample is obviously higher than that of the sample without Solidifying agent. When the solidifying agent content is 2%, the dynamic elastic modulus of the sample increases by about 60 MPA; when the solidifying agent content is 4%, the dynamic elastic modulus of the sample is close to that of 2% Solidifying agent; when the solidifying agent content is 6%, the dynamic elastic modulus of the sample increases significantly, and the increment is about 90 MPa. As can be seen from Fig. 5, when the solidifying agent content increases, the damping ratio of the sample decreases as a whole, and with the increase of Solidifying agent content, the change of damping ratio in the process of vibration is smaller, and the curve is smoother. This is caused by the strong hydrolysis and hydration reaction between the Solidifying agent and the moisture in the soil. The strength of solidified soil is promoted by the reaction of tricalcium silicate and dicalcium silicate in Portland cement clinker and fly ash. Calcium hydroxide is formed by the reaction of calcium oxide in carbonate and quicklime with water. Calcium hydroxide moves in the soil and continuously absorbs carbon dioxide in the air and moisture in the soil to form calcium carbonate and improve the strength of the soil [10].

Figures 6 and 7 show the dynamic elastic modulus and damping ratio of solidified soil samples with vibration times varying with the curing age of 7 days and confining pressure of 100 kPa, respectively. As can be seen from Fig. 6, when the confining pressure is increased from 50 to 100 kPa, when the solidifying agent content is 2%, the dynamic elastic modulus of the sample is about 40 MPa higher than that without solidifying agent, when the solidifying agent content is 4%, the dynamic elastic modulus of the sample increases by about 60 MPa, and when the solidifying agent content is 6%, the dynamic elastic modulus of the sample increases by about 80 MPa, thus it can be seen that the dynamic elastic modulus of the sample increases with the increase of solidifying agent content. As can be seen from Fig. 7, when the confining pressure increases from 50 to 100 kPa, the damping ratio of the sample decreases accordingly, and with the increase of the solidifying agent content, the damping ratio decreases, and the change of vibration times tends to be smooth, and the stability of the sample is improved to a certain extent. In summary, the solidified slag soil shows better dynamic characteristics when the solidifying agent content is 6%, which indicates that the hydration reaction between 6% Solidifying agent and water is more stable.

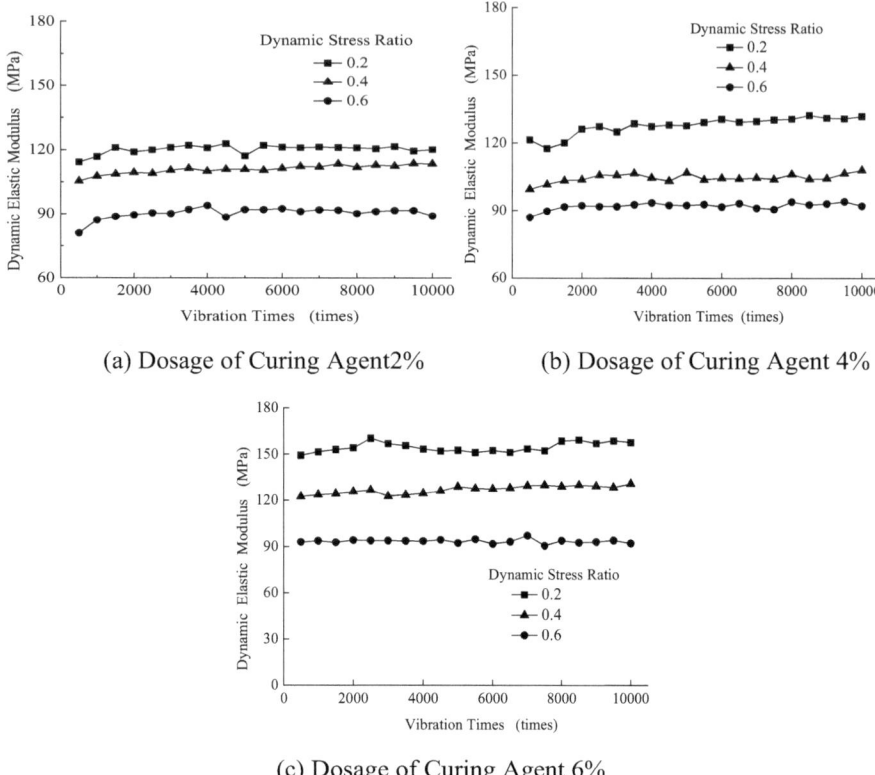

Fig. 4 Dynamic elastic modulus changes with vibration times under stabilized soil of different dosage (maintaining 7 d, confining pressure 50 kpa)

3.3 Water Stability of Solidified Soil

In order to study the water stability of solidified soil, 6% solidified soil samples are selected and immersed in water for 24 h after curing for 7 days. Dynamic experiments are carried out under the confining pressure of 50 kPa. Figures 8 and 9 show that the curing age is 7 days, respectively. The dynamic elastic modulus and damping ratio of solidified soil samples under confining pressure 50 kPa change with the number of vibrations before and after immersion. After immersion, the dynamic elastic modulus is reduced by about 15–20%, and the damping ratio increases to a certain extent. It can be seen from Fig. 8 that the dynamic elastic modulus weakens to a certain extent when the dynamic stress ratio is 0.2, but the weakening phenomenon is not obvious when the dynamic stress ratio is 0.4 and 0.6. Therefore, when the solidifying agent content is 6%, the water stability of solidified soil is better. Figure 9 shows that there is no obvious change in the damping ratio of the sample before and after immersion in water, but the stability of the damping ratio decreases slightly with the increase

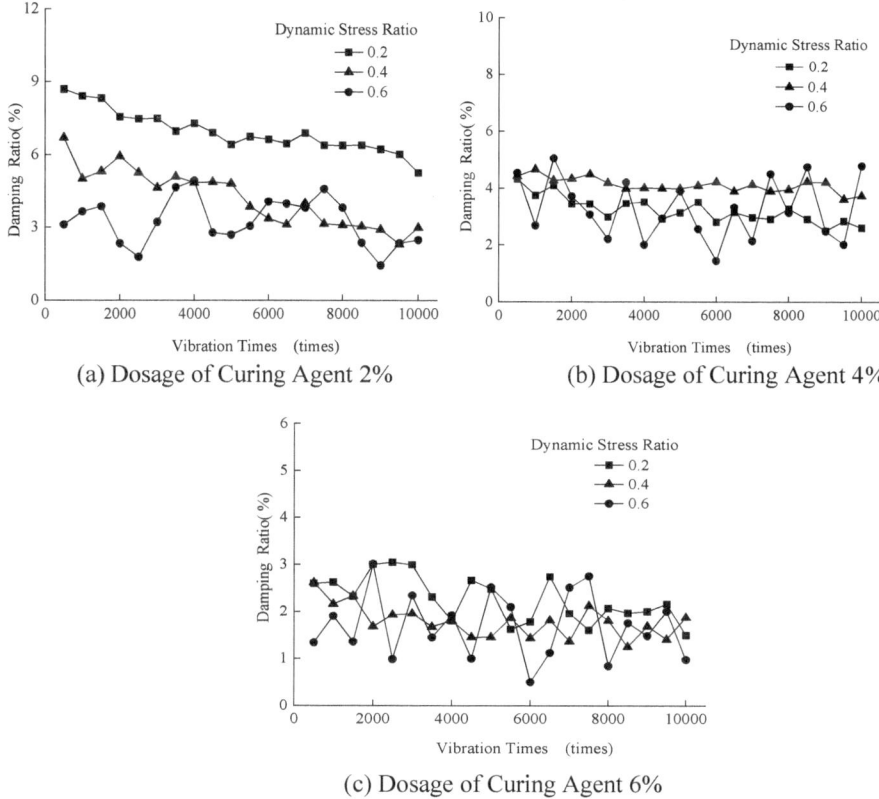

Fig. 5 Damping ratio changes with vibration times under stabilized soil of different dosage (maintaining 7 d, confining pressure 50 kpa)

of dynamic stress ratio. When the dynamic stress ratio is 0.6, the damping ratio fluctuates to a certain extent with the increase of vibration times. This is because polymer polyacrylamide and OFB have good water resistance and bonding ability, which improve the water stability of solidified soil.

3.4 Microstructure Analysis of Solidified Soil

Scanning Electron Microscope (SEM) shows the microstructure and morphology of a specimen and works based on the physical effects that occur when X-rays irradiate the surface of the specimen. YB100 powder Solidifying agent is mixed with muddy building slag soil with high moisture content to destroy the structure of the original soil and increase the soil colloid. It can be polymerized by physical, mechanical and chemical methods of stirring and compaction. Sodium ion and sulfate

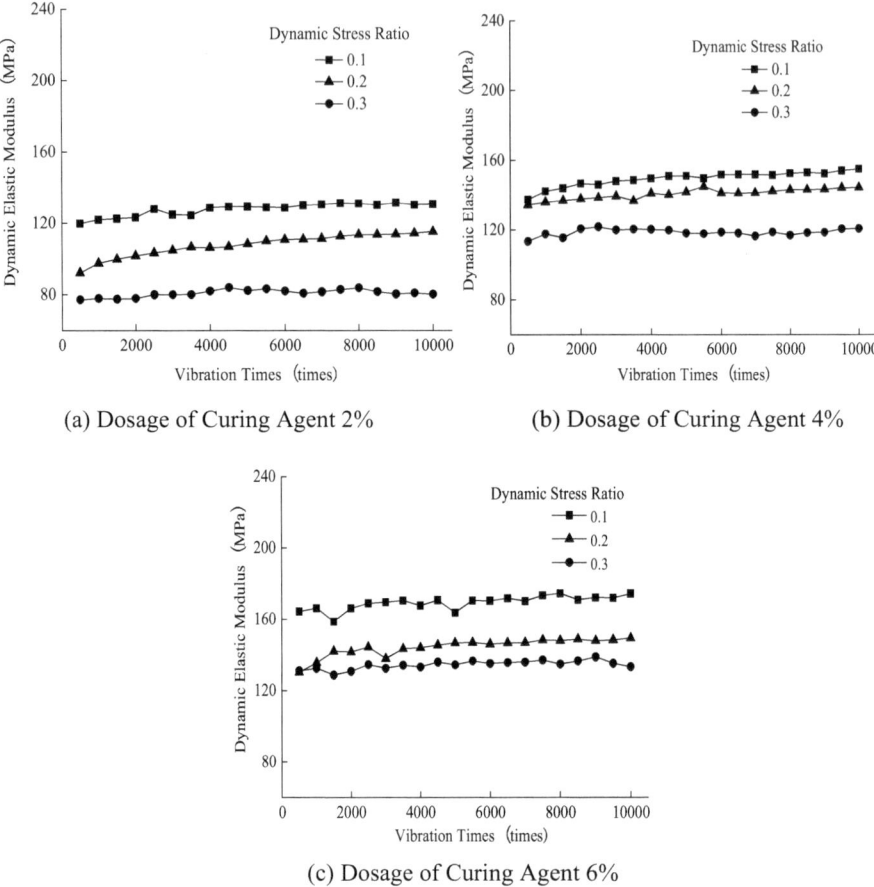

Fig. 6 Dynamic elastic modulus changes with vibration times under stabilized soil of different dosage (maintaining 7 d, confining pressure 100 kpa)

corrosion are treated by adding raw gypsum, blast furnace powder and AFB III additive, setting it with silicate clinker and lime to improve compressive strength, and high molecular polyacrylamide and OFB reinforcer are used to solve water stability and expansion cracks. SEM technique is used to study the micro-pore structure and mineral composition evolution of solidified soil. The sample without solidifying agent and samples with 2, 4, 6% solidifying agent content are selected to carry out SEM scanning after 7 days of curing, and the results are shown in Figure 10.

According to the comparative analysis, Fig. 10a is a sample without solidifying agent, the aggregates are obvious and the pores between the aggregates are very large. Figure 10b, c, d show the internal structure of the sample mixed with Solidifying agent, the macropores between aggregates are obviously reduced, the small pores are increased, and the agglomerated soil particles are aggregated. Quicklime

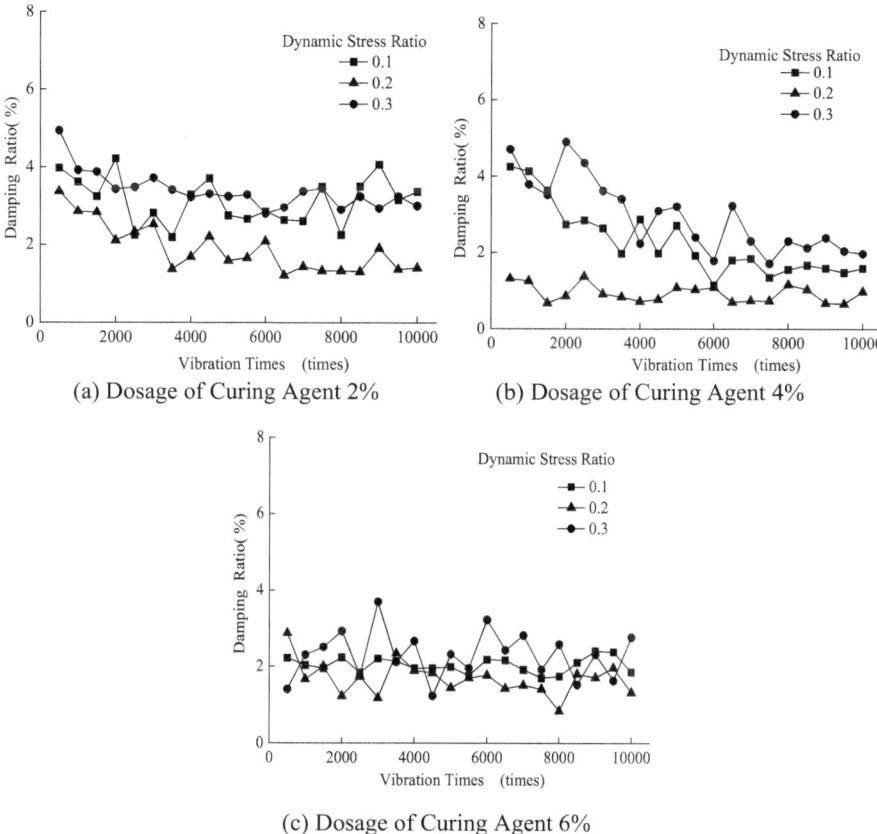

Fig. 7 Damping ratio changes with vibration times under stabilized soil of different dosage (maintaining 7 d, confining pressure 100 kpa)

absorbs water after hydration to produce calcium hydroxide. Calcium hydroxide itself has a strong adsorption, so a large number of soil particles form large soil masses, which gradually seal the pores in the soil, connect the soil, and form a stacked structure, as shown in Fig. 10b–d. After the soft clay is mixed with Solidifying agent, network cementitious substances such as hydrated calcium silicate and other network cementing substances can be produced in the sample, which can enhance the cementation between clay particles and enhance the strength of the soil; the reaction of dicalcium silicate, tricalcium silicate, tricalcium iron aluminate and water also strengthens the whole strength of the soil; calcium hydroxide reacts with carbon dioxide in the air to form calcium carbonate, while calcium carbonate has high strength and water stability [10].

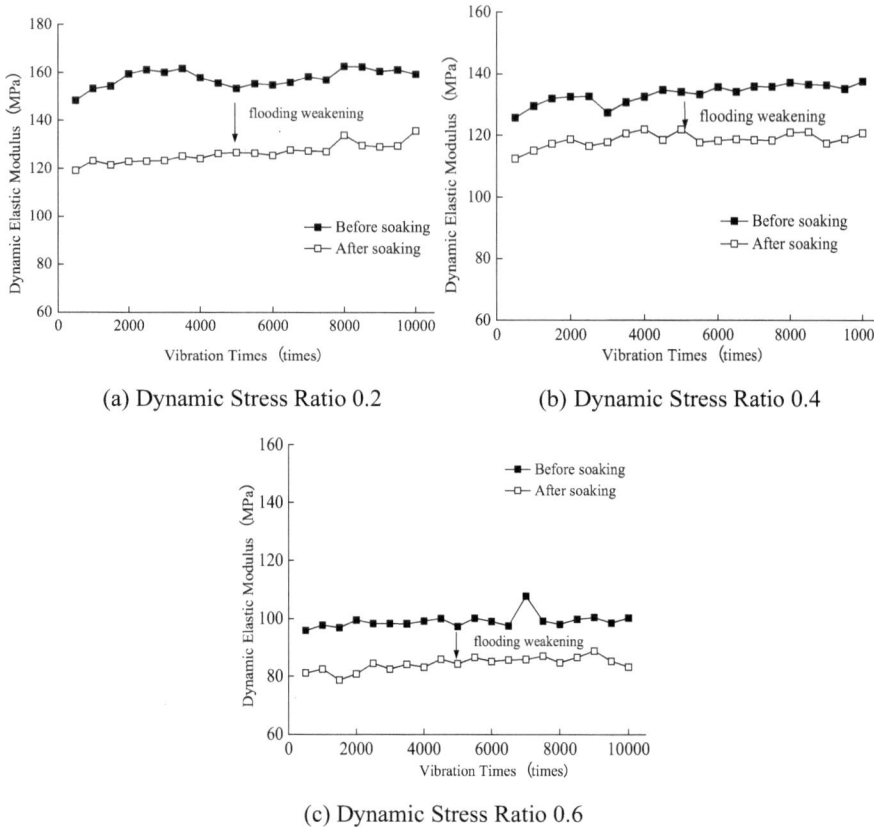

Fig. 8 Dynamic elastic modulus changes with vibration times of sample before and after soaking under different dynamic stress ratio (maintaining 7 d, confining pressure 50 kpa)

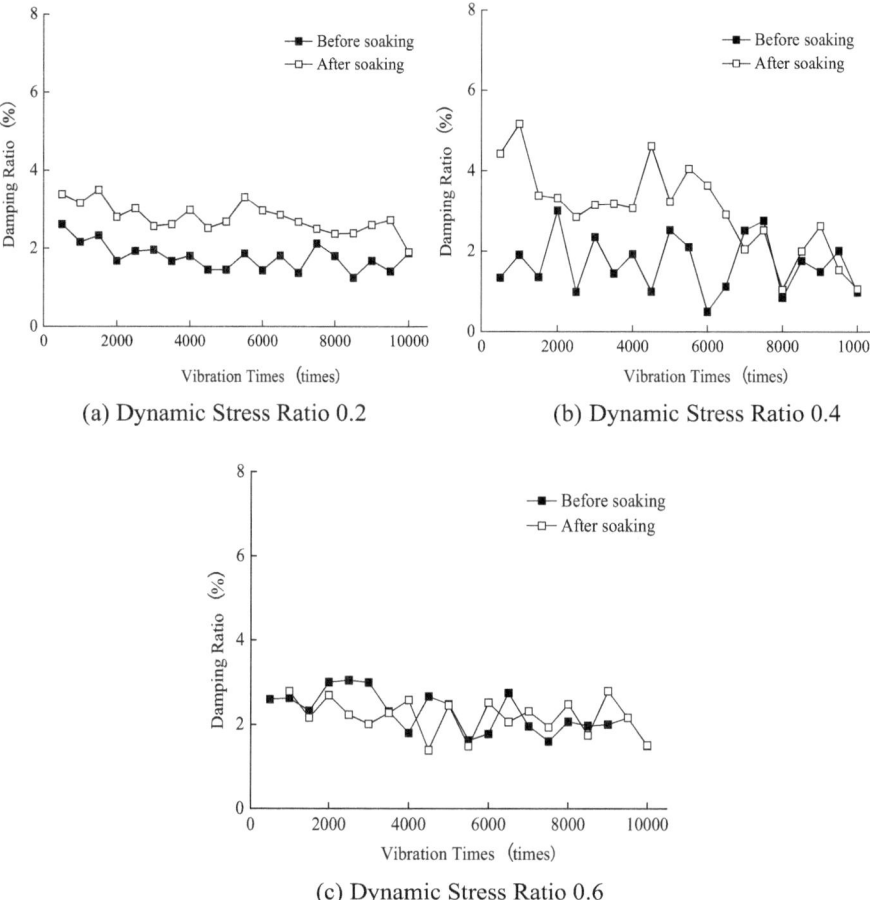

Fig. 9 Damping ratio changes with vibration times of sample before and after soaking under different dynamic stress ratio (maintaining 7 d, confining pressure 50 kpa)

Fig. 10 Microscopic structure under stabilized soil of different dosage

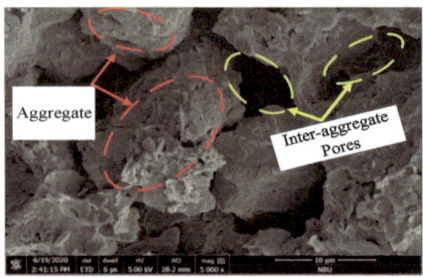

(a) Dosage of Curing Agent 0%

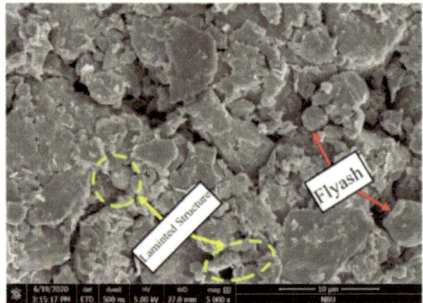

(b) Dosage of Curing Agent 2%

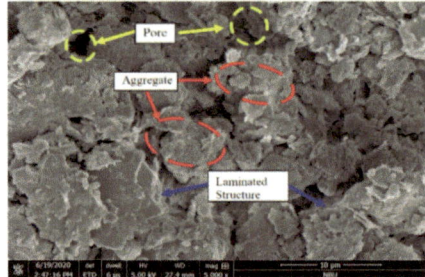

(c) Dosage of Curing Agent 4%

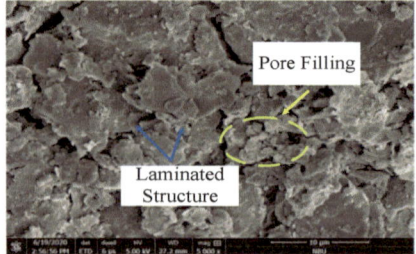

(d) Dosage of Curing Agent 6%

4 Conclusion

(1) Without Solidifying agent, the dynamic elastic modulus increases and the damping ratio decreases with the increase of confining pressure; under the same confining pressure, the dynamic elastic modulus decreases with the increase of dynamic stress ratio, and there is no obvious correlation between damping ratio and dynamic stress ratio. If the solidifying agent content increases at 0, 2, 4, 6%, the dynamic elastic modulus of building slag soil solidified soil increases from about 60 MPa to about 150 MPa, and the curve of damping ratio is smoother and the change is smaller.

(2) After the solidified soil sample is immersed in water for 24 h, the dynamic elastic modulus decreased by about 15–20%, and the damping ratio increased slightly; with the increase of dynamic stress ratio, the weakening phenomenon of dynamic elastic modulus and damping ratio is not obvious, this is because polymer polyacrylamide and OFB have good water resistance and adhesion ability, which improves the water stability of solidified soil.

(3) According to the SEM photos, the solidified soil particles gather with each other and the macropores decrease, which improves the strength and stability of the soil. The reason is that the Solidifying agent reacts with the water in the soil to form a cementitious substance and a complex, which enhances the adhesion, ion exchange and granulation between the aggregates.

Acknowledgements National Natural Science Foundation of China (51778303).

References

1. Lu K (1999) Status quo and comprehensive utilization of ref-use produced from construction and removal of buildings in China. Constr Technol 28(5):44–45
2. Li YY, Li Z (2013) Study on the application of construct-ion waste to the municipal road. Highway 007:235–240
3. Qi SZ, Fu CM, Qu ZW (2015) Experimental study on modification of construction waste as road filling material. J China Foreign Highway 35(1):262–267
4. Xu P, Jiang GL, Ren SJ et al (2019) Experimental study of dynamic response of subgrade with red mudstone and improved red mudstone. Rock Soil Mech 40(02):678–683+692
5. Deng ZQ, Zeng C (2018) Study on dynamic characteristics of subgrade chemically improved soil under controlled CBR condition. J China Foreign Highway 38(04):253–258
6. Yang DZ, Chen KS (2021) Study on road performance of rubber red clay. J China Foreign Highway 41(04):319–326
7. Tan YQ, Peng C, Tian ZK (2020) Experimental study of dynamic characteristics of lime modified roadbed filler. J Univ South China (Sci Technol) 34(01):81–87
8. Qiao JS, Wang XY, Wang GH et al (2021) Dynamic characteristics and microscopic mechanism of muddy clay solidified by ground granulated blast-furnace slag. Bull Chin Ceram Soc 40(07):2306–2312
9. Liu DP, Yang XH, Liu HG et al (2017) Experimental study on influence factors of gravel soil dynamic elastic modulus and damping ratio. J Rail Sci Eng 14(02):264–270

10. Geng YJ (2009) Study on action mechanism and road performance of improved red sandstone by EN-1 soil agent. South Jiangtong University

Open Access This chapter is licensed under the terms of the Creative Commons Attribution 4.0 International License (http://creativecommons.org/licenses/by/4.0/), which permits use, sharing, adaptation, distribution and reproduction in any medium or format, as long as you give appropriate credit to the original author(s) and the source, provide a link to the Creative Commons license and indicate if changes were made.

The images or other third party material in this chapter are included in the chapter's Creative Commons license, unless indicated otherwise in a credit line to the material. If material is not included in the chapter's Creative Commons license and your intended use is not permitted by statutory regulation or exceeds the permitted use, you will need to obtain permission directly from the copyright holder.

Numerical Analysis of Soil-Rock Mixture Subgrade Based on High Density Resistivity Surveys

Jiangong Chen and Diming Lou

Abstract A new method to obtain the distribution of rock stone in RSM subgrade by high density resistivity surveys is proposed. Using image processing technology, the true resistivity image could be transformed to a vector format which can imported into finite element software. The finite models of RSM subgrade are established based on the vector image. The RSM subgrade bearing capacity and differential settlement under foundation's load are analyzed. Numerical analysis result shows that the bearing capacity of RSM subgrade is bigger than the soil subgrade. The differential settlement is effected mainly by subgrade stiffness under two corners of the foundation and the location of the interface of rock and soil.

Keywords Soil-rock mixture · High density resistivity surveys · Numerical analysis · Subgrade

1 Introduction

Landslide deposits and collapse deposits are mainly composed of hard stones mixed with relatively soft soil, and the stones are generally hard, irregular in shape and large in particle size. Medley et al. [1] named it BIMrocks (block in matrix rocks), and You [2, 3] and other scholars called it SRM (soil-rock mixture).Soil-rock mixture is a kind of complex and discontinuous multiphase medium material, which is composed of high strength rocks of a certain size, filling components of relatively low strength soil and corresponding pores. Soil-rock mixture (SRM) is widely distributed in the three Gorges reservoir area, Qinghai-Tibet Plateau and other areas (Fig. 1).

J. Chen (✉) · D. Lou
College of Civil Engineering, Chongqing University, Chongqing 400045, China
e-mail: cjg77928@126.com

J. Chen
Key Laboratory of New Technology for Construction of Cities in Mountain Area, Ministry ofEducation, Chongqing University, Chongqing 400045, China

© The Author(s) 2025
P. Xiang et al. (eds.), *Frontier Research on High Performance Concrete and Mechanical Properties*, Lecture Notes in Civil Engineering 518,
https://doi.org/10.1007/978-981-97-4090-1_25

Fig. 1 A large stone in situ

The physical and mechanical property of this subgrade is very unpredictable so that the ground improvement is very hard and expensive. The key problem is how to obtain the inner structure of SRM. The common method is to use image processing and recognition technology. Through the image of SRM section, the distribution of rock stones in SRM can be located. With computer processing, some finite element model or discrete element model can be established to simulate the load procedure or the failure process to obtain the physical and mechanical properties of materials. In recent years, Lebourg [4] obtained the morphological characteristics of glacial moraines in Aspe valley based on image processing technology and established the relations between morphological characteristics and mechanical-physical properties. Also Yue [5] use the image processing technology to measure the scale, shape and distribution of coarse aggregate in asphalt Concrete. Xu [6, 7] applied image processing into soil rock mixture (SRM). The digital image of SRM section had been processed into binary image and the outline of rock stones had been determined. The binary image had been converted into FEM sketch file and some analysis of SRM had been done. Also, random generation is a popular way to get the inner distribution of rock in SRM. Based on SRM section image obtained in the three gorges Yangtze river, Zhang [8] use Maximum Likelihood Estimating method to match the distribution of rock in the image and build the random generation models of SRM. Besides, CT image is another processing technology applied in SRM. Yuan [9] monitored the regularity of SRM inner change under uniaxial loading with computer CT method.

The electrical resistivity survey method is one of the most important methods among geophysics surveys. The study of it begins during the nineteenth century and gains a great development in the twentieth century. In 1987, Shima and Sakayama [10] firstly use the resistivity method to acquire the electrical properties underground and present the way of inversion. With inversion data, an image can be established to find the failure surface in structure and formation interface or abnormal region in terrane.

In this research, we intended to present a way to use high density resistivity surveys in SRM ground area to identify the rock stone and then establish the SRM subgrade model in FEM with some appropriate processing.

2 High Density Resistivity Surveys in SRM Ground Area

2.1 The Principle of High Density Resistivity Surveys

The fundamental physical law used in resistivity surveys is Ohm's Law that governs the flow of current in the ground. The equation for Ohm's Law in vector form for current flow in a continuous medium is given by

$$J = \sigma E \tag{1}$$

where σ is the conductivity of the medium, J is the current density and E is the electric field intensity. In geophysical surveys the medium resistivity ρ, which is equals to the reciprocal of the conductivity ($\rho = 1/\sigma$), is more commonly used. The resistivity measurements in homogeneity, infinite and isotropy medium are made by injecting current into the ground through the two current electrodes, and measuring the voltage between two measuring electrodes. From the current I and voltage ΔU values, a resistivity ρ value is calculated.

$$\rho = K \frac{\Delta U}{I} \tag{2}$$

where K is the coefficient of device depends on the arrangement of four electrodes. The calculated resistivity value ρ is not the true resistivity of subsurface, but is an apparent value that is the resistivity of a homogeneous ground that will give the same resistance value for the same electrode arrangement. There are many way to measure the apparent resistivity value. One of them is high resistivity survey. High resistivity survey is one of most accurate and efficient way of resistivity surveys. Through high density electrode disposal and automatic control of electrode type, multiple devices can be achieved. In this paper, we chose Winner arrangement as measurement mode. As a widely applied device in high resistivity surveys, the electrode arrangement of Winner is shown in Fig. 2: A, B and M, N are respectively current electrodes and measuring electrode. Spaces between electrodes are equal and arrangement of electrodes is in the order of AMNB. When measuring, electrodes space increased by the law of isolation coefficient. This device is profile surveys way. Measured data section shape tends to be trapezoid. The accuracy of resistivity surveys depend on moisture content of ground and the contact condition of electrode to soil.

To determine the true subsurface resistivity from the apparent resistivity values can be obtained by the cell-based inversion method. The model parameters are the resistivity values of the model cells, while the data is the measured apparent resistivity values. The formulation is given by

$$(J^T + \lambda F_R)\Delta q_k = J^T R_d g - \lambda F_R q_k \tag{3}$$

Fig. 2 Wenner device and instrument of survey

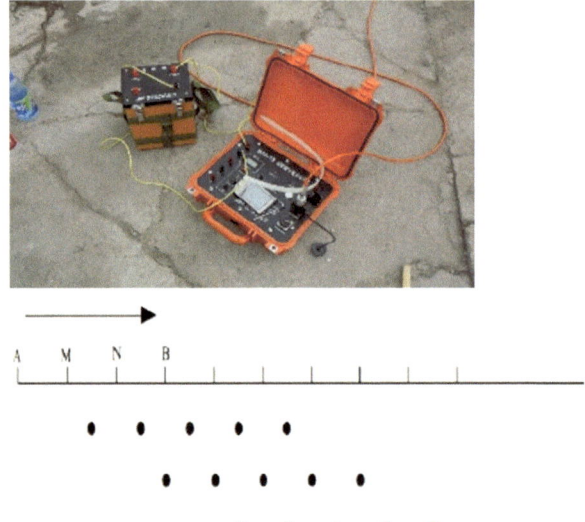

where q is the model parameter, and J is the Jacobian matrix (of size m by n) of partial derivatives. The factor λ is known as the Marquardt or damping factor, and where R_d is weighting matric. This method is also known as an l_1-norm or robust or blocky inversion method which can be more consistent with the known geology in the situations like heterogeneity of measurement surface and tends to produce models that are piecewise constant.

2.2 Fieldwork Data and Processing

Fieldwork data of high resistivity surveys is acquired near the river Zhu Xi in Wan Zhou district of Chong Qing, China. The ground was backfilled with silty clay and limestone in about 2 decades. The size of many limestones is bigger than 0.5 m and some of rocks can reach 4 m or even more. The instrument of electrical resistivity survey is WDA-1B, produced by Chongqing Geological Instrument plant. The parameters of fieldwork are shown in Table.1. The survey line movement is the rolling measurement, an efficient way to measure a long section in a lack of cable. Resistivity data shows that there are large resistivity variations near the surface. So we use the robust inversion model which cell width is half the unit electrode spacing in order to get better results [11]. Thus a true resistivity image can be obtained.

Before the FEM model is established, true resistivity image should be transfer to vector format of boundary image which can be input into AutoCAD. AutoCAD can export ACIS file which can be imported into ABAQUS as a sketch file. So there are some problem should be solved: (1) boundary location; (2) boundary smooth; and (3) vectorization transformation. Compare with digital image processing, the

Table 1 The parameters of fieldwork

Array	Polar distance	Total electrodes	Survey line movements
Wenner array	1 m	40	2

process of true resistivity image is easier. The boundaries between high resistivity body and low resistivity body can be identified apparently from the true resistivity image as shown in Fig. 3. The resistivity range of limestone is 100–1000 Ωm due to the porosity of the rock and of silty clay is 1–100 Ωm due to the water content. So we can confirm that the high resistivity bodies are the limestone and the lower bodies are soil. Then the true resistivity image can be turned into a binary image with clear outline of rock stones based on image processing technology. The binary image can be expressed to a boundary image with MATLAB code using boundary smooth method. The boundary image can be transformed to vectorised figure. Figure 3 shows the processing procedure of a true resistivity image.

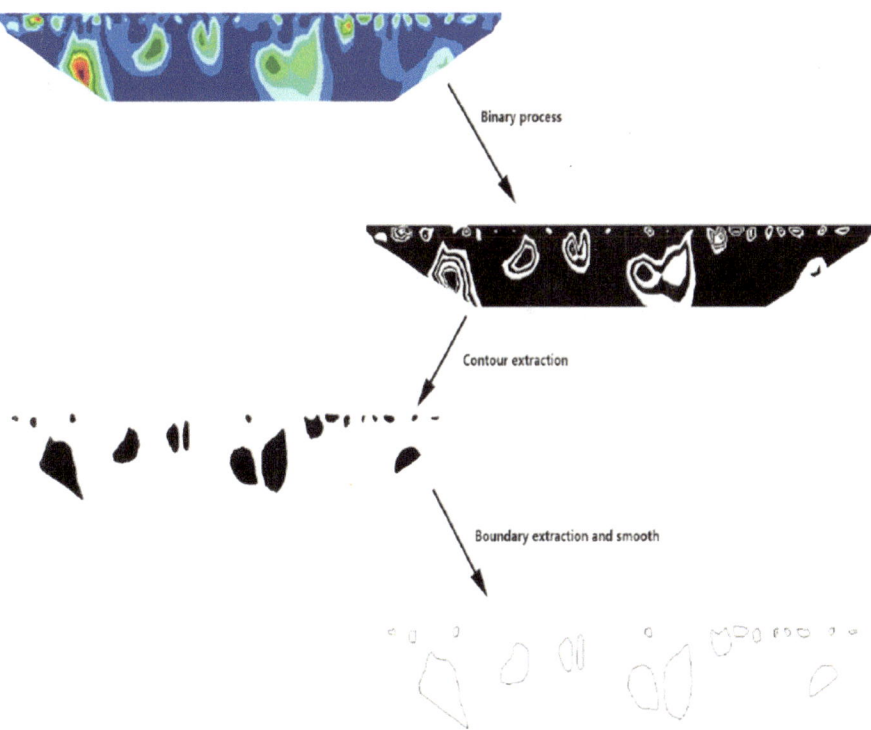

Fig. 3 Image processing

3 Finite Element Model Build and Bearing Capacity Analysis

3.1 FEM Modeling

In order to analyse the bearing capacity of the RSM subgrade, some finite element models has been built. Our research us ABAQUS software to conduct finite element analysis. The parameters of the rock and soil in this area are shown in Table 2.

Two FEM models of soil subgrade and RSM subgrade model are established. Mohr–coulomb constitutive model with linear elastic is the constitutive relation applied in these models. Mohr–coulomb constitutive model in ABAQUS is given below.

$$F = R_{mc}q - p\tan\varphi - c = 0 \qquad (4)$$

where R_{mc} is:

$$R_{mc} = \frac{1}{\sqrt{3}\cos\varphi}\sin(\theta + \frac{\pi}{3}) + \frac{1}{3}\cos\left(\theta + \frac{\pi}{3}\right)\tan\varphi$$

where ϕ is frictional angle and c is cohesion. θ is deviatoric polar angle. p is equivalent pressure stress and q is the Mises equivalent stress. Both models have the same soil parameters listed in table 2. The foundation is set as rigid body above the ground and the contact condition between foundation and ground, soil and rock is frictional contact. The frictional coefficient is 0.5. As to simulate the nonlinear and elastic–plastic of soil and rock, the mesh element type of both models are set as CPE4R. A pressure had been loaded on top of the foundation. The size of subsoil is 52.7 × 15.8 m and of foundation is 2 × 0.5 m. Loading method is uniform distributed load on whole top surface of foundation until the calculation is non-convergence. The model had been shown in Fig. 4.

Table 2 Parameter of the rock and soil

	Density (10^3 kg/m^3)	Elasticity modulus (Kpa)	Poisson ratio	Cohesive (Kpa)	Internal friction angle	Dilation angle
Limestone	2.42	10.09e6	0.3	12e3	36	18
Silty clay	1.80	7e4	0.26	27	14	7

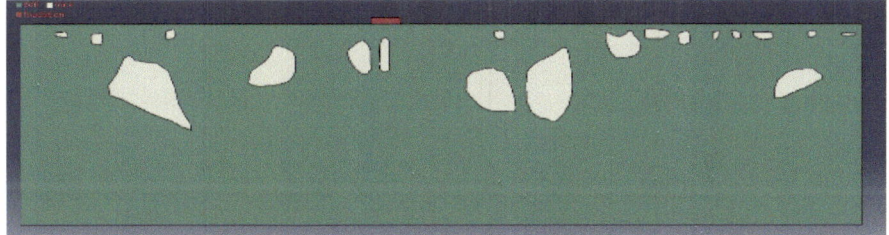

(a) RSM subgrade model

(b) RSM subgrade meshed model

Fig. 4 Subsoil model of boulder ground

3.2 Result Analysis

Figure 5 are the plasticity deformation nephogram of RSM subgrade model as the subgrade began to failure. The plastic zone is divided into three part:

(1) Wedge-shaped zone beneath the foundation.

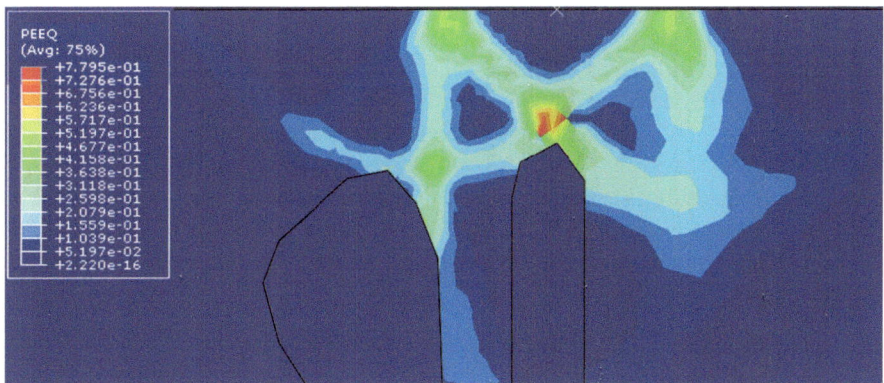

Fig. 5 Bearing capacity of models

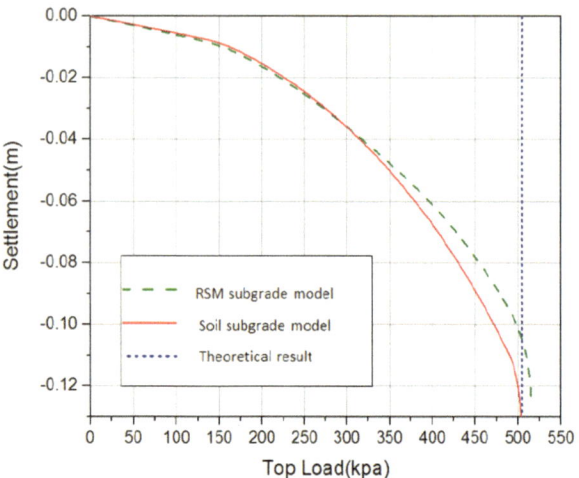

Fig. 6 The plasticity deformation nephogram of RSM subgrade model

(2) Radial shear zone. The right ground radial shear zone has more plastic deformation area than left zone. And there is another plastic zone located besides left radial shear zone.
(3) Asymmetry passive Rankine zone. Two parts of Rankine zone is with big difference and the left part has more plastic deformation development than the right part.

The plastic zone of the RSM subgrade ground has some similarity with soil ground bearing capacity model based on traditional foundation bearing capacity theory. Soil-rock contact is the bottle neck of this model so that the additional plastic zone is on top of the rock surface and cause slip between rock and soil. But the rocks of high stiffness will enhance the bearing capacity of ground. Figure 6 shows the bearing capacity of soil and RSM subgrade.

The soil subgrade model tend out to have an approximate capacity compare to the result of theory. In the beginning of the load procedure, the settlement of RSM subgrade model is little faster than the soil subgrade model. There is a cross point of two capacity curves near the 300kpa load. After then the settlement increase slowly and the RSM subgrade can bear greater loads than that of the soil subgrade with the same settlement. During this period, compression settlement the RSM subgrade is completed and the subgrade begin to act as an overall structure. That's the reason why the RSM subgrade can bear more loading. Because of the rock, differential settlement will arise when the load increase. Figure 7 shows the differential settlement of the RSM subgrade model.

In the SRM subgrade, the interface of rock and soil is the most destructible zone when the rock stone is in the bearing region of subgrade. To analyse the effect of rock to differential settlement, we moved the location of foundation. Another six models with different location of foundation have been calculated. The settlement

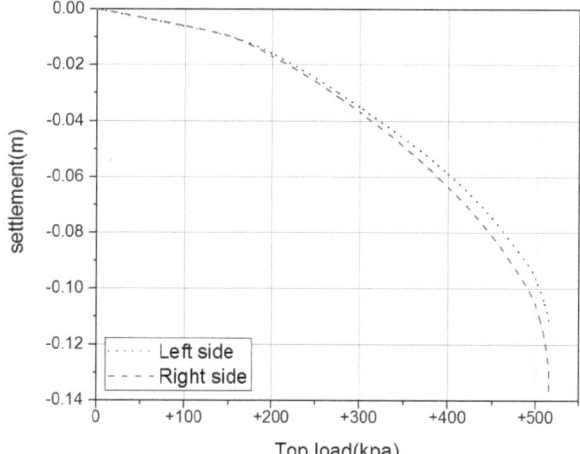

Fig. 7 Differential settlement of BG model

under each vertical side of foundation had been taken into analysis. The analysis results are shown in Fig. 8 (the foundation's location 1 is shown in Fig. 5).

Differential settlement rate (DS):

$$DS = RD - LD \tag{5}$$

where **RD** is settlement under right foot of foundation and **LD** is settlement under left foot. With diagram shown in Fig. 8, it could be seen that the highest value of plastic zone is always appeared around the surface of rock beneath the foundation or near the corners of the foundation. The slip line of the foundation is deformed with the change of relative location of the foundation and rock. If the slip line is one side partly overlap to the rock top surface, the development of slip line will grow faster than the other side. It is believed that the difference of slip line growth will significantly affect the differential settlement. Figure 9 shows the differential settlement of each seven location.

From curves and the plasticity deformation nephogram of location 1 and 4, when the subgrade's stiffness under two corners of foundation is approximate equal, the side which slip line is more growth, the settlement of it will be larger.

From curves and the plasticity deformation nephogram of location 1 and 4, when the subgrade's stiffness under two corners of foundation is approximate equal, the side which slip line is more growth, the settlement of it will be larger.

When the subgrade's stiffness at the two corners of foundation are not approximate equal, just like the case in location 2, the slip line of left foot is more growth than the right foot for the subgrade's stiffness at right side is much larger than that at the left side. Curve shows that the settlement of right foot is much bigger than the left foot. Another case is location 7. It is similar to the case in location 2 but the stiffness difference is smaller. The differential settlement turned out to be approximated same.

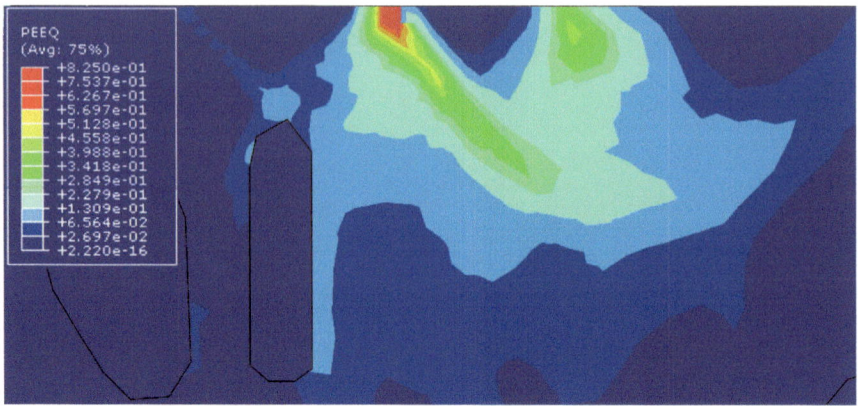

The plasticity deformation nephogram of foundation's location 2

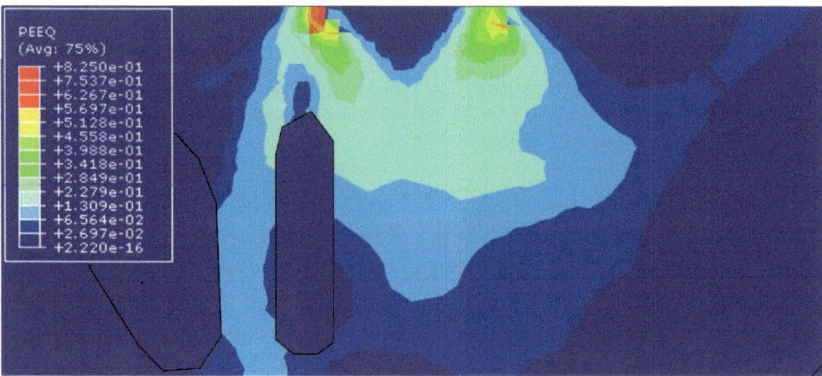

The plasticity deformation nephogram of foundation's location 3

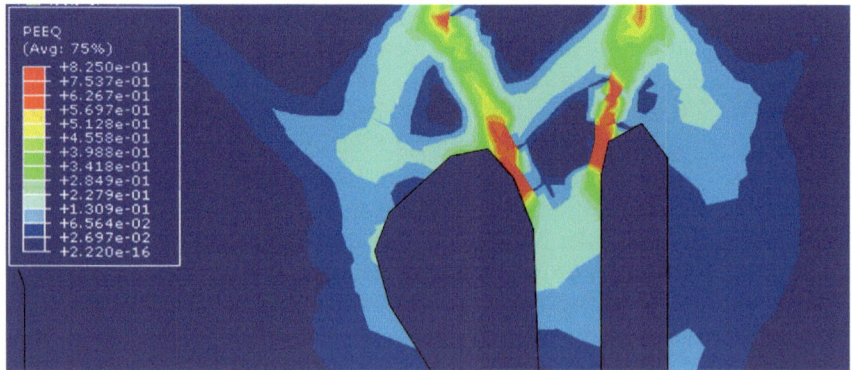

The plasticity deformation nephogram of foundation's location 4

Fig. 8 The plasticity deformation nephogram of different locations of foundation

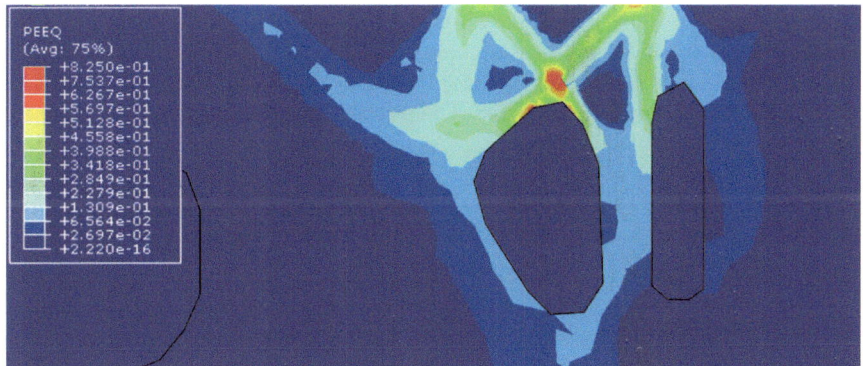

The plasticity deformation nephogram of foundation's location 5

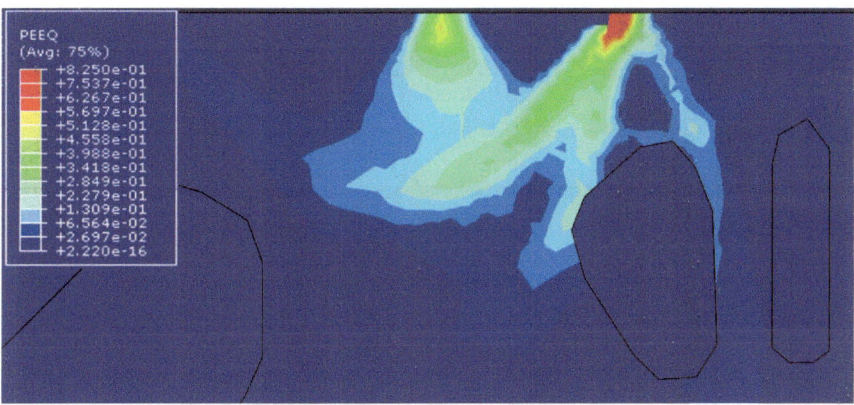

The plasticity deformation nephogram of foundation's location 6

Fig. 8 (continued)

Fig. 9 Differential settlement growth of different foundation location models

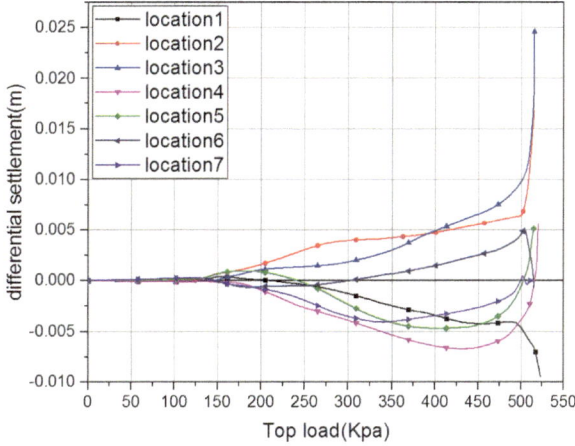

During the increasing of settlement, the difference of stiffness affect the differential settlement the most just like the case in location 4, 5, 7. From Fig. 9 it can be seen that the left foot settlement of these three models are all bigger than the right foot Because of the left foot stiffness of these models all larger than the right foot stiffness.

4 Conclusions

A new method to obtain large size rock stone in subgrade based on high density resistivity method is proposed. The two-dimensional image containing the distribution of large rock stones can be obtained easily by carrying on technical processing to the true resistivity image. Then the image can be transferred to vector format which can be imported into FEM program. The FEM models are established to calculate the capacity and differential settlement under foundation's load. The plastic zone of SRM subgrade is analysed. The numerical analysis result shows that the bearing capacity of RSM subgrade is bigger than the soil subgrade. The differential settlement is effected mainly by subgrade stiffness under two corners of the foundation and the location of the interface of rock and soil.

References

1. Medley E, Goodman RE (1994) Estimating the block volumetric proportions of melanges and similar block-in-matrix rocks (bimrocks). In: 1st North American rock mechanics symposium, pp 851–858
2. You X-H (2001) Stochastic structural model of the earth-rock aggregate and its application. Beijing Jiao Tong University, Beijing
3. You X-H, Tang J-S (2002) Research on horizontal push-shear in-situ test of soil and rock-mixture. Chin J Rock Mech Eng 21(10):1537–1540
4. Lebourga T, Rissb J, Pirardc E (2004) Influence of morphological characteristics of heterogeneous moraine formations on their mechanical behavior using image and statistical analysis. Eng Geol 73:37–50
5. Yue ZQ, Morin I (1996) Digital image processing for aggregate orientation in asphalt concrete mixtures. Can J Civ Eng 23:480–489
6. Xu W, Hu R, Yue Z-Q (2006) Numerical simulation on stability of right bank slope of Longpan in Tiger-Leaping gorge area. Geotech Eng 28:996–2004
7. Xu W, Hu R, Yue Z (2007) Mesostructural character and numerical simulation of mechanical properties of soil-rock mixtures. Rock Mech Eng 26:300–311
8. Zhang S, Tang M (2015) Research of Soil-rock mixture random model based on Meso-structural statistical rules. Yangtze River 46:48–52+79
9. Yuan W, Li X, He J (2013) Structure effect study of deformation and failure of rock and soil aggregate with CT technique. Rock Mech Eng 32:3134–3140
10. Shima H, Sakayama T (1987) Resistivity tomography: an approach to 2-D resistivity inverse problems. 57th SEG, Expanded Abstract, pp 204–207
11. Yue ZQ, Chen S, Tham LG (2003) Finite element modelling of geomaterials using digital image processing. Comp Geotech 30:375–397

Open Access This chapter is licensed under the terms of the Creative Commons Attribution 4.0 International License (http://creativecommons.org/licenses/by/4.0/), which permits use, sharing, adaptation, distribution and reproduction in any medium or format, as long as you give appropriate credit to the original author(s) and the source, provide a link to the Creative Commons license and indicate if changes were made.

The images or other third party material in this chapter are included in the chapter's Creative Commons license, unless indicated otherwise in a credit line to the material. If material is not included in the chapter's Creative Commons license and your intended use is not permitted by statutory regulation or exceeds the permitted use, you will need to obtain permission directly from the copyright holder.

Experimental Study of the Fiber Improvement Test on the Unsaturated Soil

Lina Guo, Yun Chen, and Minmin Luo

Abstract At present, the engineering properties of unsaturated soil have been studied intensively in the soil mechanics community. The shear strength of unsaturated soil is particularly important for the engineering properties of soil. In this study, GDS unsaturated back pressure shear apparatus (UBPS) is employed to control the matrix suction to 50 kPa to conduct the direct shear test of consolidation and drainage on the unsaturated soil and the unsaturated fiber soil. Through the analysis of the test data, the fiber material is found to be able to improve the shear strength of unsaturated soil when controlling the matrix suction, but the improvement effect is not as good as in the case without considering the matrix suction. It can be concluded that the matrix suction is an important factor that must be taken into account in soil improvement engineering since most soils are unsaturated under natural condition.

Keywords Fiber improvement · Unsaturated soil · Matrix suction · Direct shear test

1 Introduction

The common practice of soil property chemical improvement is adding gravel powder, fly ash, lime and chloride salt in the soil [1], which leads to problems such as high cost, potential environmental pollution. Physical soil improvement methods, however, can sometimes overcome the problems mentioned above of the chemical methods [2, 3]. In North America, Europe, Japan and other developed countries, engineers have begun using the physical methods to enhance fiber reinforced plastic.

L. Guo (✉) · Y. Chen · M. Luo
The Architectural Design and Research Institute of Zhejiang University, Hangzhou 310028, China
e-mail: 549244857@qq.com

Center for Balance Architecture, Zhejiang University, Hangzhou 310027, China

They tried to solve the durability problem caused by steel corrosion. The fiber reinforced plastic technology has been widely used in foreign mining, tunnel construction, highway slope shoring since the end of 1990s [4]. In China, the research and application of using fiber materials to improve the properties of soil has just started and is developing rapidly [5].

Fiber as a new type of reinforcement materials has many advantages. Firstly, it has high strength; secondly, it can be mixed uniformly with soil and make the sample strength isotropically; thirdly, not like grilles and geotextiles which may form some potential weak surfaces in the soil because of the reinforcement direction and spacing [6]. In addition, fiber material also has good insulation, low cost, low thermal conductivity, less pollution, etc. Consequently, the fiber application is increasingly used in the geotechnical engineering recently. There are also plenty of conventional tests of soil improvement using fiber material [7–9]. All of these studies did not consider the effect of the matrix suction which is ubiquitous in soil. To the best of our knowledge, the experimental study under controlled the matrix suction of unsaturated fiber soil is rarely seen, where the unsaturated fiber soil referred to the unsaturated soil with added fiber. Most soils are in the unsaturated state under natural condition, so it is particularly significant to study the shear strength of unsaturated fiber soil. For this purpose, this study took the advantages of the fiber materials and employed the advanced GDS unsaturated back pressure shear apparatus (UBPS) to conduct laboratory tests to study the fiber improvement effect on unsaturated soil under controlled the matrix suction to 50 kPa. The reason for choosing the matrix suction of 50 kPa is as follows. The soil water characteristic curve (SWCC) indicated that the moisture content of soil in the matrix suction of 50 kPa (22.02%) is most close to the natural moisture content (20.79%) in our samples.

2 Basic Cognition of the Matrix Suction

The matrix suction is one of the most important state parameters of the unsaturated soil. Therefore, it is necessary to introduce the basic concept of the matrix suction before the study.

It is generally acknowledged that the surface tension of the interface between water and air (just like Fig. 1) generates the matrix suction among unsaturated soil. The surface tension makes the interface between pore water and pore air has the property of elastic film in unsaturated soil. It will make the interface be curving and emerge tension inside the shrink film to maintain balance. Thus, the pore air pressure can be greater than the pore water pressure to create the matrix suction in unsaturated soil. Just as the formula below:

$$S = u_a - u_w \tag{1}$$

where u_a and u_w are pore air pressure and pore water pressure, S is on behalf of the matrix suction.

Fig. 1 Unsaturated soil units linked together with the gas phase

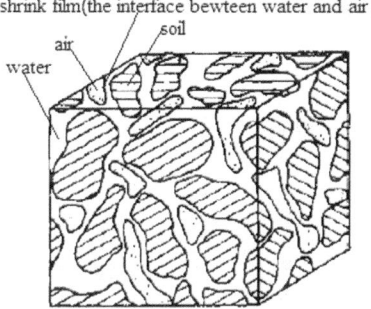

Table 1 Physical properties of the soil

Natural moisture content (%)	Severe (kN/m³)	Dry density (g/cm³)	Void ratio	Saturation (%)	Liquid limit (%)	Plastic limit (%)	Compression modulus (MPa)
20.79	21.14	1.55	0.680	87.7	31.46	19.23	4.96

The matrix suction is through the surface tension of the water–air interface (shrink film) to impact the shear strength of the soil.

3 Sample of Test

The soil of the test takes from landslide I in Majiagou, located in the left bank of the Yangtze River and apart from the estuary of Yangtze River 2.1 km. The landslide soil mainly consists of silty clay which mixed with fragment stones. The composition of the fragment stone is fine sandstone and mudstone. Silty clay mainly consists of accumulations. The size of the fragment stone is uneven, which are into angular or subangular. The material structure of the landslide is loose and has strong permeability. The corresponding indexes of soil are shown in Table 1.

4 Experimental Scheme

4.1 Test Objective

The objective of the test is to research the changes of the vertical displacement and the shear strength in the consolidation process and shearing process between the unsaturated soil and unsaturated fiber soil under the matrix suction of 50 kPa. Thereby, you can comparative analysis the improvement effect of the fiber under the controlled matrix suction. It is a reference to the further research on the properties of the unsaturated soil.

4.2 Preparation of Test Specimens

The GDS unsaturated back pressure shear apparatus is used in the tests and the method of consolidation drainage shear is adopted. The specimens are cube with the size of 75 × 75 × 30 mm. The specimens are in accordance with the initial moisture content of 20.79% and the maximum dry density of 1.55 g/cm^3. First, you should crush the specimens retrieved from the field and screened by the 2 mm sieve, then put them inside an oven at the temperature of 105 °C to dry the soil. Then, you can weigh the weight of the soil and the water which is in accordance with the initial moisture content of 20.79% and maintain the homogeneous mixture between soil and water for 1 or 2 days. After them wetting uniformly, you can prepare the specimens. Last, exhaust and saturate the prepared specimens to make sure they are saturated and you will control the matrix suction of the samples in the GDS UBPS.

In the test, the content of the fiber is 0.2%, the length is 30 mm. Because the specimens are disturbed soil, you can use the stratified hammering method. The fiber should be uniformly distributed in the soil when preparing the specimens.

4.3 Test Method

Before the test, you should saturate the clay plate using the water pressure of 20 kPa.

The unsaturated soil direct shear test (consolidation and drainage shear and controlled the matrix suction) includes three stages. First, the stage of matrix suction equalization; second, the consolidation stage under the same suction (the two stages can be combined into one stage); third, the shear stage under the same matrix suction. After installing the specimen, you can change the original matrix suction through controlling the pore air pressure and the pore water pressure. The pore air pressure and the pore water pressure must be uniformly distributed in the whole specimen to ensure reaching the balance of the matrix suction, this is the stage of matrix suction equalization. Before the first stage of the test, you could imposed the vertical pressure of 25 kPa to the specimen to ensure the specimen well contacts with the clay plate.

The identification of the matrix suction equalization is very important, which directly affects the reliability of the subsequent test results [10]. The standard for judgment of the matrix suction equalization in the test is that the drainage volume (or the absorbent volume) is smaller than 0.02% of the sample volume in 24 h (the standard of my test is that the drainage volume of the specimen is not more than 33.75 mm in 24 h). The required time depends on the D-value between the target suction and the initial suction in the stage of matrix suction equalization. The specimen carries out consolidation in the vertical pressure and the time of the consolidation is generally not less than 24 h. The requirement of the consolidation is that the vertical deformation is less than 0.005 mm/h. At the same time, the pore air pressure and the pore water pressure must be controlled in a fixed value. Finally, you can start the shear stage under the condition of the constant matrix suction. The shear rate is 0.003 mm/min and the maximum shear displacement is 12 mm in my test.

5 Experimental Results and Analysis

5.1 Experimental Results

The tests are divided into two groups. One group is unsaturated soils which controlled the matrix suction to 50 kPa and the vertical loads are respectively for 50 kPa, 100 kPa, 200 kPa and 400 kPa. The other is unsaturated fiber soils which controlled the matrix suction to 50 kPa and the vertical loads are respectively for 50 kPa, 100 kPa, 200 kPa and 400 kPa. The content of the fiber is 0.2% and the test conditions were shown in Table 2.

Curves of Consolidation Deformation

From Fig. 2 we can see that with the vertical load increasing the consolidation of the samples are also increasing and with the increasing of time they are no longer increasing. It shows the consolidation of the samples has already met the test requirement.

Vertical Displacement During the Shearing Process

Figure 3 is the relationship curves of vertical displacement and shear displacement in unsaturated soils under different vertical loads in the shearing process. It shows

Table 2 Log sheet of the direct shear tests

Specimen Type	No	Matrix suction (kPa)	Vertical load (kPa)	Vertical stress (kN)	Time (d)	
					Stage of matrix suction equalization	Shear stage
Group one	EZ2-1	50	50	0.281	4	3
	EZ2-2		100	0.563	4	3
	EZ2-3		200	1.125	4	3
	EZ2-4		400	2.25	4	3
Group two	FZ2-1	50	50	0.281	4	3
	FZ2-2		100	0.563	4	3
	FZ2-3		200	1.125	4	3
	FZ2-4		400	2.25	4	3

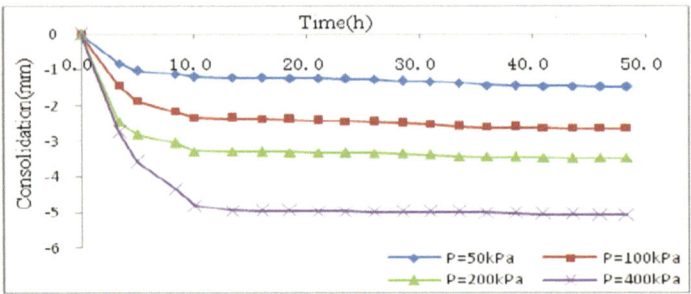

Fig. 2 The relationship curves of consolidation and time in unsaturated soils

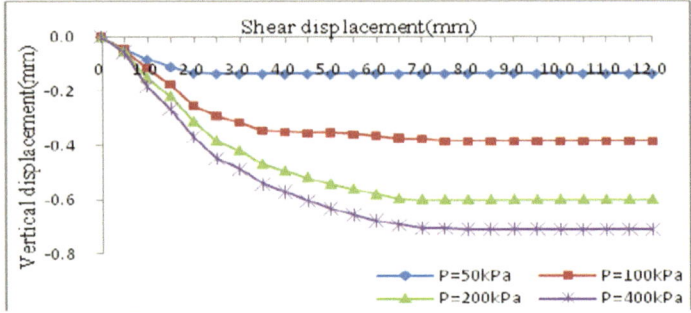

Fig. 3 The relationship curves of vertical deformation and shear displacement in unsaturated soils

all the four samples appear shear shrink phenomenon in the matrix suction of 50 kPa and the value of the shear shrink increases with the increasing of the vertical load. The shear shrink values of the two samples under the vertical load of 200 and 400 kPa are about 2–4 times than the other two samples which are under lower vertical load.

In the shearing process, the soil skeleton of the shear surface is destroyed, some soil particles flip or broke into smaller particles and filling the gap of the soil skeleton. So the samples appear shear shrink phenomenon under the combined effect of the vertical stress and the matrix suction. As the shear displacement increasing, the soil particles of the shear surface do not flip and destroy, the shear shrink value tends to the steady state. The extent of the damage on the shear surface is different under different vertical loads. The greater the vertical loads are, the greater the extent of damage on the shear surface of the samples. Therefore, the extent of flipping and filling of the soil particles on the failure surface is increasing with the increase of the vertical load.

Relationship Curves of Shear Stress–Strain

The shear stress increases rapidly when starting shear. When the shear displacement reaches to a certain extent, the shear strength tends to stability, the relationship curves of shear stress–strain show shear harden type. When the shear displacement reached

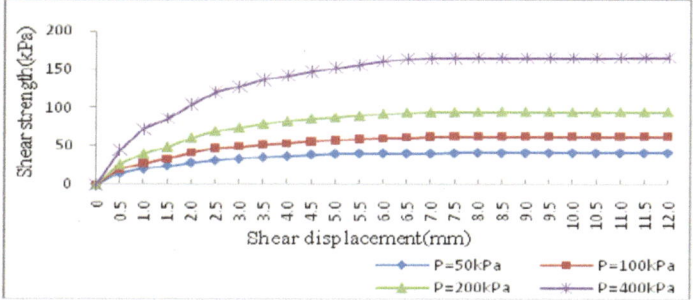

Fig. 4 The relationship curves of shear stress–strain in unsaturated soils

a certain level, the samples do not appear to the shear brittle failure and maintaining a certain strength tend to stability, just as Fig. 4 shown. The soil density increases with the increasing of the vertical stress, thus the bite friction between the soil particles gets greater and the shear stress also increases.

5.2 Fiber Improvement and Comparative Analysis

Relationship Curves of Shear Stress–Strain on Unsaturated Fiber Soil

Figure 5 is the relationship curves of the shear stress–strain on unsaturated fiber soils under different vertical loads and the matrix suction of 50 kPa in the shearing process. From Fig. 6 we can see the shear stresses increase rapidly when starting shear, and then, as the shear displacement increasing the shear forces remain basically unchanged. It illustrates that the shear force has reached the maximum shear strength.

Contrast and analysis Figs. 4 and 5 we can obtain that most relationship curves of the shear stress–strain are the shear harden type in the shear process, the shear

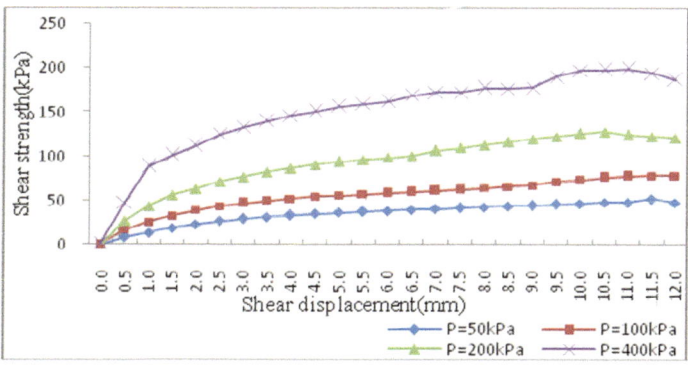

Fig. 5 The relationship curves of shear stress–strain in unsaturated fiber soils

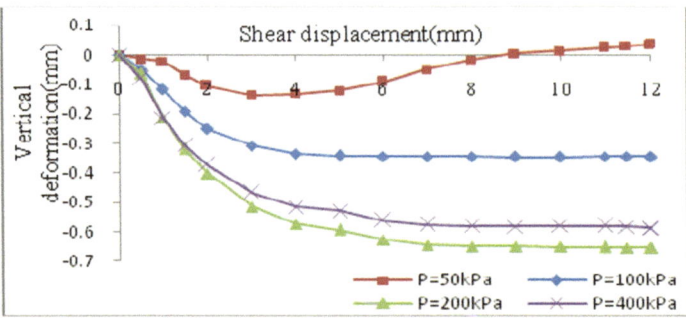

Fig. 6 The relationship curves of vertical deformation and shear displacement in unsaturated fiber soils

stresses with the increasing of the shear displacement are not changed. But the sample which is under the vertical load of 400 kPa displays shear softening type. There is a clear peak shear force and certain fluctuations in this sample. The reasons for this phenomenon may be the fiber's uniformity. The fibers maybe not so uniform when mixed with the soil which leads the sample occur brittle fracture in the shear stage.

Contrast Figs. 4 and 5 we can also know that the shear stress of unsaturated fiber soil is significantly higher than the shear stress of unsaturated soil under different vertical loads. As Table 3 shows, the shear strength of samples in each vertical load increases by 20–30% after fixing fiber with the unsaturated soil. When the ups and downs fibers are pulled out or pulled on the shear surface, the samples should overcome the friction resistance among soil particles and overcome the occlusal frictional resistance between the soil particles and the surface of the fiber. When the fiber samples occur the tension cracks or deformation under the vertical loads, the fibers which fill in the cracks or pores evenly woven together with the soil particles and then a space structure (Three-dimensional) would be formed [11]. The occlusal frictional resistance and the space constraint force between the soil and the fiber make the sample can assume a greater tensile stress (the occlusal frictional resistance between fibers and soil particles is much higher than it among the soil particles). Thereby the fiber can improve the strength and the toughness of the soil, and increase the ability of the soil to resist deformation. Thus it shows that the fiber material can delay the deformation and the destruction of unsaturated soil when controlling the matrix suction.

Relationship Curves of the Vertical Deformation and the Shear Displacement on Unsaturated Fiber Soil

Contrast Figs. 3 and 6 we can know the vertical deformation of the samples is increasing with the increase of the vertical load. Most samples occur shear shrink phenomenon under the matrix suction of 50 kPa. Only the sample which is in the vertical load of 50 kPa, occurs shear shrink phenomenon at first and then occurs the characteristic of dilatancy. The reasons why there is such a phenomenon may be that the fibers within the sample appear folding in the later part of the shear process and

Table 3 The correlation table of shear strength between unsaturated soils and unsaturated fiber soils in different vertical loads

Vertical load (kPa)	Shear force of unsaturated soil (kPa)	Shear force of unsaturated fiber soil (kPa)	Percentage of improved strength (%)
50	41.4111	52.1006	25.81
100	61.9873	78.0459	25.91
200	94.5231	122.4216	29.51
400	164.7687	198.2064	20.29

Fig. 7 The uplift soil pulled by fiber

they pull the soil particles uplift when they are stretched by large shear force just as Fig. 7.

Contrast of the Shear Strength Between Unsaturated Fiber Soil and Unsaturated Soil

Figure 8 is the curves of shear strength in unsaturated soil and unsaturated fiber soil under different vertical loads with making the linear fit to them. From Fig. 8, we can see the cohesion of unsaturated soil is 25.25 kPa and the friction angle is 19.19°; while the cohesion of unsaturated fiber soil is 35.32 kPa and the friction angle is 22.39°. Contrast them we can know the shear strength has been improved by the fiber. The cohesion improves 29.88% and the friction angle increases 16.68%. In some experiments, all of their cohesions improved no less than 50% when fixed fibers with the soils but no controlling the matrix suction in their tests [12]. So, if you do not consider the matrix suction in the reinforcement engineering with fiber material, you will overestimate the improvement effect of it. That's why some reinforcement buildings cannot bear the expected intensity and would go to damage. Therefore the matrix suction must be recognized in reinforcement engineering.

In Fig. 8, R^2 is an index of the fitting degree of trend line. Its value can reflect the fitting degree between the estimated value of the trend line and the corresponding actual data. The higher the fitting degree is, the higher the reliability of the trend line. Where the R^2 are 0.996 and 0.999 in Fig. 8, they mean the estimated value of the

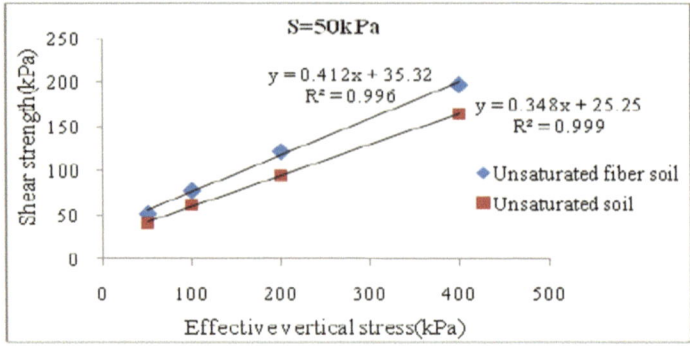

Fig. 8 The curves of shear strength in unsaturated soil and unsaturated fiber soil

trend line is almost exactly the same with the corresponding experimental data and the fitting degree is extremely high.

The index of the shear strength of the soil is the cohesion and the friction angle. After fiber is mixed with the unsaturated soil, the cohesion increases significantly and the friction angle is also improved but not so obvious. So the inference can be made that the cohesion is the main factor to affect the shear strength of the unsaturated soil. The author considers this may be highly related to the mechanical properties of the fiber material itself. As we all know, the elastic modulus of the fiber is greater than the elastic modulus of the soil, making the occlusal frictional resistance between the fibers and the soil particles is higher than the frictional resistance among the soil particles. Thus the fiber significantly enhances the shear strength of the unsaturated soil. Most soils in engineering practice are unsaturated, so it is particularly meaningful to research on the shear strength of unsaturated soil and the fiber improvement effect on unsaturated soil under controlling the matrix suction. The results of this study have particularly important application value in engineering practice.

6 Conclusion

This study researches the direct shear tests on the unsaturated soil and the unsaturated fiber soil using the GDS unsaturated back pressure shear apparatus (UBPS) under controlled the matrix suction to 50 kPa. The following conclusions can be drawn:

(1) After fiber is mixed with unsaturated soil, the shear surface on the sample overcomes the frictional resistance among the soil particles, and overcomes the occlusal frictional resistance and the frictional resistance between the fibers and the soil particles. Thereby the fiber can enhance the strength and toughness of the soil and increase the capability of resisting deformation of the soil.

(2) The fiber material can significantly improve the shear strength of unsaturated soil, especially in the cohesion. At the same time, fiber material has a certain delayed effect on the deformation and failure of unsaturated soil, which reaches a certain improvement effect.
(3) Most soils in engineering practice are unsaturated, so it is particularly meaningful that research on the shear strength of unsaturated soil and the fiber improvement effect. The matrix suction in soil is often neglected in conventional practice of the reinforcement engineering, so the improvement effect of the fiber is always overrated, which could cause serious engineering problem. This study emphasizes the importance of matrix suction in engineering practice.

The test requires a longer time, so we just studied the samples of unsaturated soil and unsaturated fiber soil which are under the matrix suction of 50 kPa at present. Further study on the effect of the fiber on unsaturated soil under higher matrix suction is underway. Thus it can provide a reliable basis for reasonable selection of the content of fiber on unsaturated soil under high matrix suction in later reinforcement engineering.

Acknowledgements The authors gratefully acknowledge the funds from the Construction Research Project of Zhejiang Province (Nos: 2021K103).

References

1. Kong L-W, Zhou B-C, Bai H et al (2010) Experimental study of deformation and strength characteristics of Jingmen unsaturated expansive soil. J Geotech Eng 31(10):3037–3043
2. Gao D-Y, Zhu H-T, Xie J-J (2004) Applications of fiber reinforced plastic (FRP) bolts. Chin J Rock Mech Eng 23(13):2205–2210
3. Lei S-Y, Ding W-T (2005) Experimental investigation on restraining the swell of expansive soil with fiber-reinforcement. Chin J Geotech Engi 27(4):482–485 (in Chinese)
4. Hu B, Wang X-G, Lian B-Q (2010) Applicability exploration of engineering properties improving of expansive soil with fiber materials. Chin J Geotech Eng 32(Suppl. 2):615–618
5. Wang C, Zhang Z-C, Zhou Z et al (2006) Application study of FRTP materials in the civil engineering products. Fiber Compos 6(2):35–40
6. Tang Z-S, Shi B, Gao W et al (2007). Study on effects of sand content on strength of polypropylene fiber reinforced clay soil. Chin J Rock Mech Eng 26(Suppl. 1):2969–2974
7. Temel Y, Omer S (2003) A study on shear strength of sands rein-forced with randomly distributed discrete fibers. Geotext Geomembr 21(2)
8. Prabakar J, Sridhar RS (2002) Effect of random inclusion of sisal fiber on strength behaviour of soil. Constr Build Mater 16(2)
9. Zhang X-P, Shi B (2008) Experimental study on reinforced fiber expansive soil. J Yangtze River Sci Res Inst 25(4):61–63
10. Zhan L-T, Ng Charles WW (2007) Effect of suction on shear strength and dilatancy of an unsaturated expansive clay. Chin J Geotech Eng 29(1):83–88
11. Zhao Y-Y, Zhao Y-R, Li C et al (2009) Study on triaxial experiment of fibrous soil. J Water Resour Archit Eng 7(1):127–128
12. LI G-X, Chen L, Zheng J-Q et al (1995) Experimental study on fiber-reinforced cohesive soil. J Hydraul Eng 6:31–36

Open Access This chapter is licensed under the terms of the Creative Commons Attribution 4.0 International License (http://creativecommons.org/licenses/by/4.0/), which permits use, sharing, adaptation, distribution and reproduction in any medium or format, as long as you give appropriate credit to the original author(s) and the source, provide a link to the Creative Commons license and indicate if changes were made.

The images or other third party material in this chapter are included in the chapter's Creative Commons license, unless indicated otherwise in a credit line to the material. If material is not included in the chapter's Creative Commons license and your intended use is not permitted by statutory regulation or exceeds the permitted use, you will need to obtain permission directly from the copyright holder.

Experimental Study on Physical and Mechanical Characteristics and Microstructure of Sandstone After High Temperature-Water Cooling Treatment

Jinbin Lu, Lifeng Zheng, Feng Chen, Liang Yang, and Qiang Zhang

Abstract In geothermal energy development, high-temperature rock mass will go through the process of water cooling. It is of great significance to study the physical and mechanical characteristics and microstructure of high-temperature rock after water cooling for the long-term stability analysis of underground engineering. Based on this, the surface characteristics, mass, and volume variation of sandstone cooled by water at high temperatures (100, 200, 400, 600, 800, and 1000 °C) were investigated. Using the Rock Top multifield coupling tester, a series of axial compressions and longitudinal wave velocity tests of the sandstone after a high temperature-water cooling treatment are performed. The microstructure characteristic obtained by X-ray diffraction and scanning electron microscope was studied, and the effect of high temperature-water cooling behavior on the mechanical properties of sandstone was investigated. The results show that: (1) The mass loss rate, volume expansion rate and peak strain of sandstone increase with increasing temperature, while peak strength decreases gradually. When the temperature exceeds 400 °C, the physical and mechanical parameters of sandstone change markedly. (2) When the temperature is less than 400 °C, corresponds to the compressive and line-elastic phases, the stable crack propagation phase, the rapid crack propagation phase and the destructive phase during the

---

J. Lu · L. Yang
Yunnan Dianzhong Water Diversion Engineering Co. Ltd, Kunming 65000, China
e-mail: 86784334@qq.com

L. Yang
e-mail: 285268747@qq.com

L. Zheng · Q. Zhang (✉)
China Institute of Water Resources and Hydropower Research, Beijing 100048, China
e-mail: zhangq@iwhr.com

L. Zheng
e-mail: zhenglf@iwhr.com

F. Chen
Jiangxi Province Survey and Design Research Institute Co. Ltd, Nanchang 330224, China
e-mail: 631295350@qq.com

failure process of sandstone, the wave velocity of sandstone are steadily increasing, oscillating, and sharply decreasing, respectively. While the temperature is below 1000 °C, the wave velocity of sandstone is oscillation increases, slows down and drops sharply, respectively. (3) When the temperature is below 400 °C, the mineral content of sandstone varies less. While the temperature exceeds 400 °C, there is an overall increasing trend in the sodium feldspar content of sandstone. (4) The increase in temperature promotes the development of pore fractures within the sandstone, especially at higher temperature states where microcracks expand along the intergranular to form microcrack networks, leading to an increase in the scale and number of defects such as fractures within the sandstone.

Keyword Sandstone · High temperature-water cooling · Physical mechanical properties · Microstructures

1 Introduction

As a clean energy source, dry thermal geothermal is playing a more prominent role in the national energy restructuring with the implementation of China's "double carbon target" strategy [1, 2]. In the process of geothermal energy development, deep rock mass will experience a series of physico-chemical changes due to the intense temperature drop during contact with normal temperature drilling fluid. The temperature shock caused by water cooling will aggravate the damage of rock mass and easily induce engineering disaster, which will seriously affect the safety and long-term stability of the project [3]. Therefore, it is of great significance, for the development of deep geothermal energy, to study the intrinsic relationship between the physical and mechanical properties of rocks after high temperature- water cooling treatment, as their microscopic damage mechanisms.

At present, a great deal of research has been carried out on the physical mechanics of high-temperature rocks, including the effects on rock strength [4–6], deformation properties [4, 7], wave velocity [8, 9], thermal conductivity [8, 10, 11], porosity [12, 13], etc. For example, Jia et al. [7] and Zhu et al. [4] carried out axial compression test of granite after high-temperature-water cooling treatment, and found that the peak strength and elastic modulus of granite decreased with the increase of temperature. Cui et al. [9] tested the longitudinal and transverse wave velocities of sandstone after high-temperature-water cooling treatment, and found that the longitudinal and transverse wave velocities of granite decrease with the increase of temperature. Wu et al. [8] and Zhao et al. [11] tested the thermal conductivity of granite after high-temperature-water cooling treatment, and the results showed that the thermal conductivity decreased nonlinear as the temperature increased. S. Chaki et al. [12] tested the porosity, permeability and longitudinal wave velocities of sandstone after high-temperature-water cooling treatment, and the results show the porosity of sandstone increased with the increase of temperature, while permeability and longitudinal wave velocities decreased. However, the above work mainly focuses on the physical and mechanical properties of rocks after high temperature -water cooling treatment, but it is not enough to fully reveal the evolution of thermal damage in rocks after high temperature water cooling treatment. Meanwhile, Zhang et al. [14] considered that it is necessary to study this thermal damage characteristic from the microscopic structure.

Methods used to determine the microstructure of rock include X-ray diffraction spectroscopy (XRD) [3, 15], scanning electron microscopy (SEM) [16, 17], magnetic resonance imaging analyzer (NMR) [18], and so on. Zhang et al. [14] conducted XRD of compacted sandstone after 1000 °C, and found that the quartz content increases with the increase of temperature, while the sodium content decreases, which is the cause of the increase of compressive strength. Huang et al. [17] and Huang et al. [3] investigated the distribution of thermal cracks on the granite surface after water cooling treatment by electron microscopy, the results show that the number and size of microcracks in granite increase gradually with the increase of temperature. However, these scholars have only studied the micro-damage mechanism of the rock, and they lack the connection between the macrophysics and microstructure, so it is necessary to study the relationship between the microdamage mechanism and the micromechanical properties of the sandstone after high temperature—water cooling treatment.

In view of this, the axial compression and longitudinal wave velocities of sandstone after high temperature -water cooling treatment were tested. And the physical properties such as mass, volume and color of sandstone subjected to heat damage were measured. Finally, the XRD and SEM techniques were used to obtain the microstructure of sandstone and then reveal the mechanism of influence of microstructure on its physical and mechanical properties. The results can provide theoretical and experimental basis for the development of deep geothermal energy.

2 Test Principle and Method

2.1 Preparation of Tested Rock Samples

The tested sandstone is a standard cylindrical shape of ϕ 50mm × H100 mm (see Fig. 1), with an irregularity of less than 0.02 mm at both ends of the sandstone, which was taken from Sichuan Province. According to X-Ray Diffraction test results, the main composition of sandstone is chlorite, sodium feldspar, potash feldspar, and contains some quartz, dolomite and calcite, as shown in Fig. 2. Prior to the test, the sandstone was dried in a 105 °C temperature oven for 24 h, and then the acoustic testing on sample was conducted on ELB-UTD400 non-metallic ultrasonic detector to remove samples with abnormal wave speeds.

Fig. 1 Standard sandstone samples

Fig. 2 X-ray diffraction of sandstone

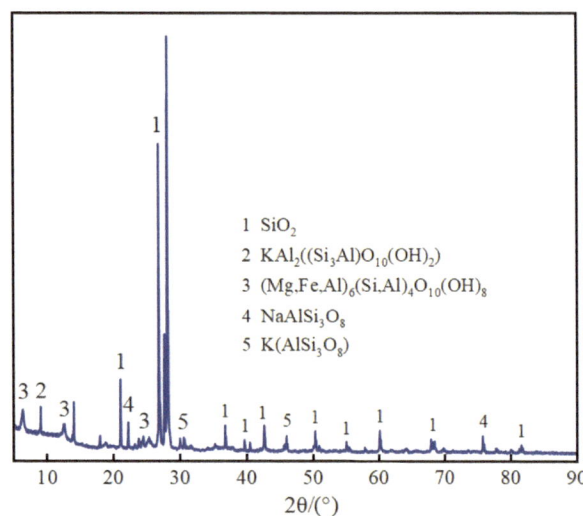

1 SiO_2
2 $KAl_2((Si_3Al)O_{10}(OH)_2)$
3 $(Mg,Fe,Al)_6(Si,Al)_4O_{10}(OH)_8$
4 $NaAlSi_3O_8$
5 $K(AlSi_3O_8)$

2.2 High Temperature Process

Heating equipment is a high-temperature oven with a maximum temperature of 1200 °C. The sandstone is first placed in high-temperature oven, where it is heated to 4h after increasing from room temperature to the target temperature (100, 200, 400, 600, 800 and 1000 °C, respectively) at a rate of 10 °C/min. The sandstone is then quickly cooled in distilled water, the diagram of the high temperature-water cooling treatment as shown in Fig. 3. Finally, once the sandstone is completely cooled, it is placed in a dryer (105 °C) for 24h drying to avoid the presence of water inside the sample that could affect the experiment.

Fig. 3 Standard sandstone samples

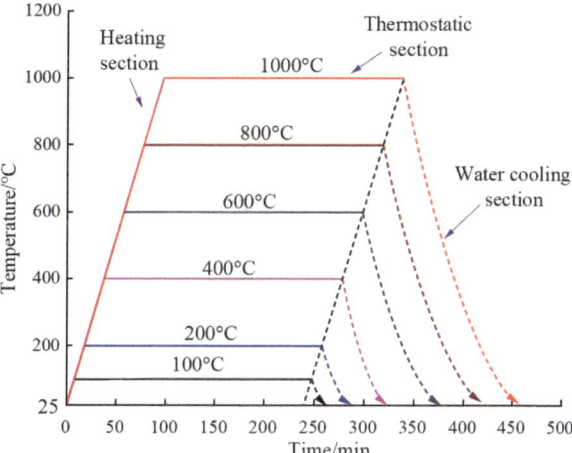

2.3 Instrument Equipment

The test loading device adopts Rock Top multi-field coupled triaxial tester, which is composed of axial pressure system, confining pressure system and seepage system, the pressure is provided by a high-precision brushless servo electronically controlled high-pressure pump, which can apply a maximum axial stress of 500 MPa, a maximum confining pressure of 60 MPa, a control accuracy of ± 0.01 MPa, and two LVDT displacement sensors (measuring range of 12 mm, accuracy of 0.001 mm), a radial deformation sensor (accuracy of 0.001 mm, Resolution 0.0001 mm), etc., can measure the axial and circumferential deformation of the sample in real time.

2.4 Test Options and Steps

The axial compression and longitudinal wave velocity test of sandstone after high temperature-water cooling treatment are carried out as follows:

(1) Apply coupler HVG, encapsulate sandstone, and test for clarity of waveform.
(2) Load the stress differential ($\sigma 1 - \sigma 3$) gradually at a displacement of 0.02 mm/min, when the stress differential increases by 4 MPA, convert the test loading mode into a stress pressure loading mode. After the stress differential curve stabilizes and flattens, conduct a wave velocity test and record the waveform data, and repeat this until the rock failure. The test loading schematic was shown in Fig. 4.

Fig. 4 Test loading schematic

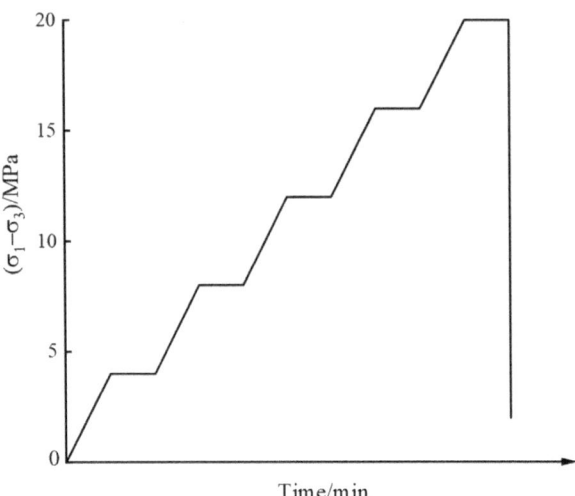

2.5 Wave Speed Test Principle

$$VP = \frac{L(1 - \varepsilon_1)}{trp - top} \times 10^{-3} \qquad (1)$$

where VP is longitudinal wave speed (m/s), L is specimen height (m), ε_1 is axial strain (%), trp is travel time for longitudinal wave testing, and top is the delay for longitudinal wave transducers and instrumentation systems (μs).

3 Physical Properties of Sandstone After High Temperature—Water Cooling Treatment

3.1 Morphological Changes in Sandstone Surfaces

The morphology of sandstone surface changes after high temperature water cooling is shown in Fig. 5. It can be found that when the temperature is between 100 and 200 °C, there are no obvious cracks on the sandstone surface, and the surface morphology does not change much. When the temperature increases from 400 to 1000 °C, the surface cracks of sandstone gradually change from microcracks to larger cracks, and the color changes from taupe to yellowish brown. This is mainly due to an increase in the number of thermally induced microcracks generated within the sandstone after high temperature-water cooling treatment. In addition, the temperature gradient formed by cooling in water is large, and the heat shock force is large in a short time.

Fig. 5 Morphological characteristics of sandstone after high temperature water cooling

Heat stress destroys the original cementation between particles, which results in more microfractures in the rock samples. Iron oxidation of iron-containing materials such as green mudstone is the main cause of color change in this phase [19].

3.2 Mass, Initial Longitudinal Wave Velocities

In this paper, the mass loss rate Rm and volume expansion rate RV are used to analyze the mass and volume change characteristics of sandstone after high temperature-water cooling treatment, and the results are shown in Fig. 6, which are calculated as follows:

$$RN = \frac{\Delta N}{N} \times 100\% \qquad (2)$$

where ΔN is the amount of change of the physical parameter, and N is the initial value of the physical parameter.

As can be seen from Fig. 6, both the mass loss rate Rm and volume expansion rate RV of sandstone increase exponentially with the increase of temperature, and the R^2 is greater than 0.97. The changes in the mass loss rate Rm and volume expansion rate RV of sandstone with temperature can be discussed in two stages:

Stage 1 is from 100 to 400 °C, Mass loss at this stage is caused by the escape of free and structural water as water vapour from within the sandstone, resulting in relatively little change, while the thermal expansion of the sandstone mineral particles results in a slight increase in volume.

Stage 2 is from 400 to 1000 °C, new microfractures develop inside the sandstone after high temperature-water cooling treatment. The external volume expands which causing the sandstone surface to break off debris and a rapid increase in mass loss and volume expansion, indicating that more severe damage to sandstone after water cooling at temperatures above 400 °C.

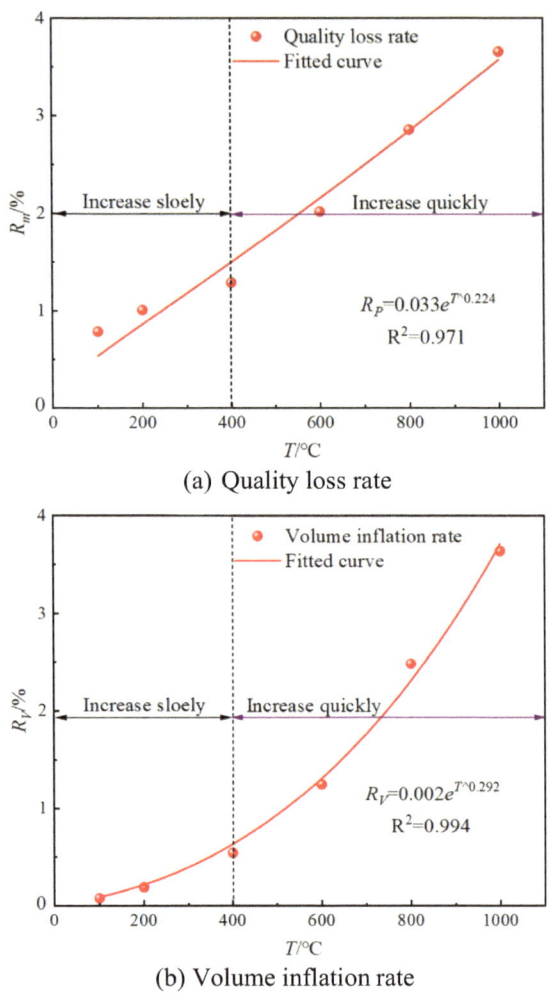

Fig. 6 Changes of sandstone physical parameters after water cooling at high temperature

4 Mechanical Properties of Sandstone After High Temperature—Water Cooling Treatment

4.1 Stress–strain Curves

Figure 7 shows the stress–strain curves of sandstone after high temperature-water cooling treatment. As can be seen from Fig. 7, high temperature has a significant effect on the overall stress–strain curve of sandstone. The whole curve can be divided into compaction phase, linear elasticity phase, crack stable propagation phase, crack unstable propagation phase and post-peak failure phase. The stress–strain curve of sandstone dropped rapidly in the post-peak failure stage after the treatment of water

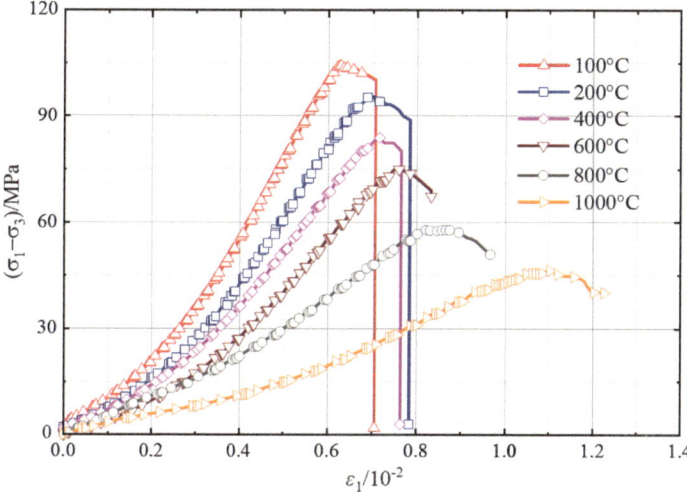

Fig. 7 Stress–strain curves of sandstone after water cooling at high temperature

cooling under 600 °C, which showed obvious brittleness. The yield phase of the curve is obvious after water cooling above 600 °C, which means that the brittle-ductility transition of sandstone occurs due to the increase of microcracks in sandstone and the expansion and connection of original microcracks when water cooling above 600 °C.

4.2 Strength Characteristics

Figure 8 shows Relationship between peak strength and temperature of sandstone after high temperature-water cooling treatment. It can be seen from Fig. 8 that the peak strength of sandstone decreases with the increase of temperature, and the effect of strength weakening is obvious. Compared to 100 °C, the temperature increased from 200 to 1000 °C, with strength decreasing by 8.02, 19.53, 27.69, 44.28 and 55.26%, respectively. This is mainly due to the high temperature gradient formed when the sandstone is exposed to water, the heat stress damages the original cementation between particles, which leads to more microfractures, less overall compactness and more internal damage, which reduces the strength of sandstone.

4.3 Peak Strain Characteristics

The variation curve of the peak strain of the sandstone with temperature after high temperature-water cooling treatment is given in Fig. 9. It can be found from the

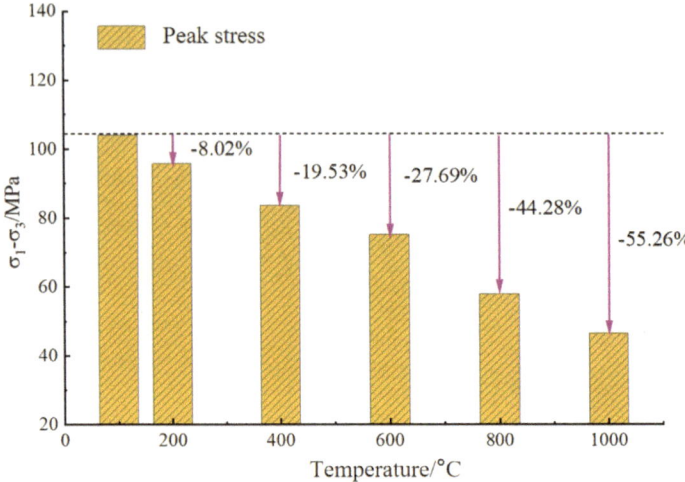

Fig. 8 Relationship between peak strength and temperature of sandstone after high temperature-water cooling treatment

figure that the peak strain of sandstone increases rapidly in a non-linear form with increasing temperature, it means that there is a significant strain strengthening effect. Compared to the 100 °C, the increase in temperature from 200 to 1000 °C, the increase in the peak strain by 9.88, 13.63, 19.59, 29.39 and 74.89% respectively. This is mainly due to the gradual development and expansion of pores and microcracks inside sandstone when the temperature rapidly reduced from the target temperature to water temperature, which resulting in an increase in the ductile characteristics of the sandstone, corresponding to an increase in peak strain.

4.4 Characteristics of Longitudinal Wave Velocities

In the loading process of sandstone after high temperature-water cooling treatment, longitudinal wave monitoring is used simultaneously. The results at temperatures of 100, 400 and 1000 °C were selected for analysis, as shown in Fig. 10. It can be found from Fig. 10 that with the increase of temperature, the longitudinal wave velocity shows an overall decreasing trend. This is mainly due to the increase in the number of microcracks within the sandstone caused by the sudden drop in temperature, which increases the scattering effect of the acoustic waves between the cracks, resulting in a decrease in the overall wave speed of the sandstone.

In addition, the longitudinal wave velocity of sandstone is closely related to its stress state during loading. During the compaction and linear elasticity phase, the wave velocity showed a steady increase when the temperature was less than 400 °C, while it is an oscillatory increase when the temperature was 1000 °C. The main reason

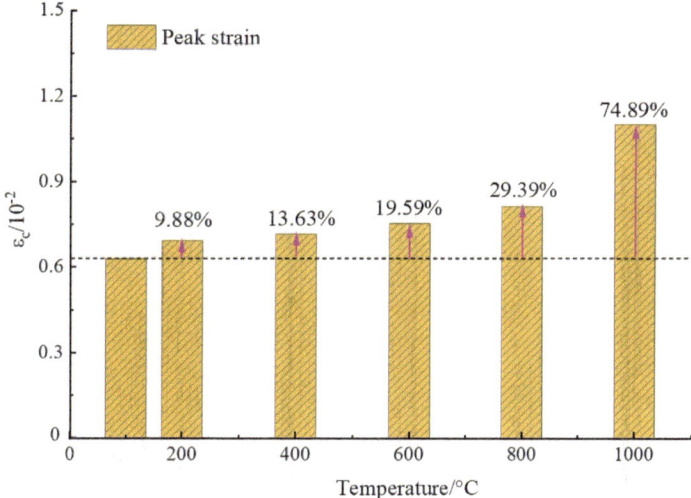

Fig. 9 Relationship between peak strain and temperature of sandstone after high temperature-water cooling treatment

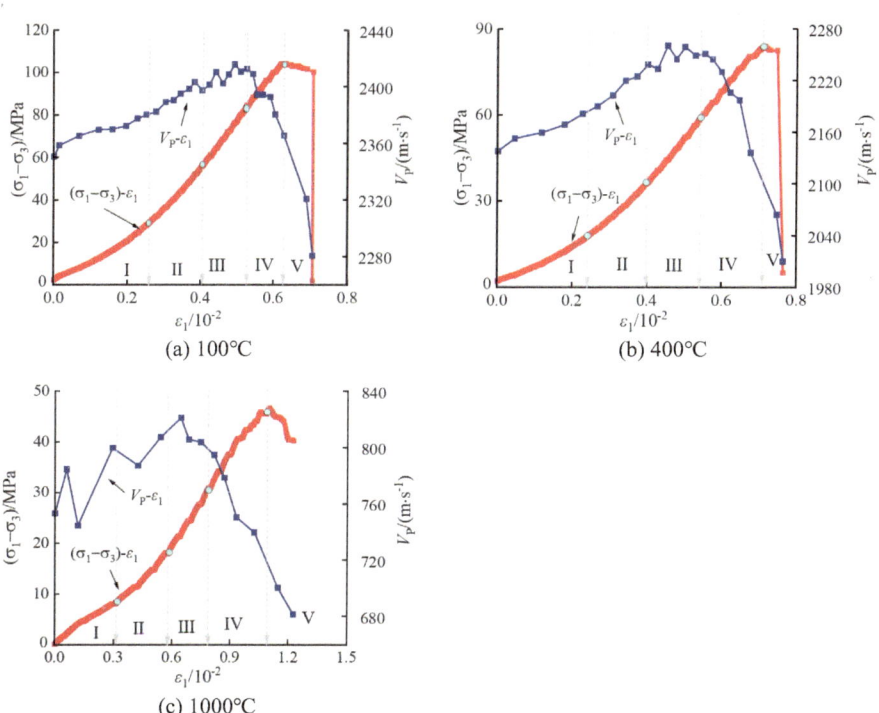

Fig. 10 Stress-strain-longitudinal wave curves of sandstone after different high temperature-water cooling treatment

is that when the temperature is less than 400 °C, the sealing effect of microcracks in sandstone increases, the density increases, and the wave velocity increases. When high temperature is 1000 °C, Heat stress caused by high temperature-water cooling treatment, which increase the damage inside rock, with more pronounced cracks appearing on the surface, causing the wave velocity to oscillate. During the stable crack propagation, the wave velocity starts to oscillate obviously when the temperature is less than 400 °C, while it begins to slow down when the high temperature is 1000 °C. The results show that the microcracks in the rock are more extensive when the temperature is higher, as the damage of the sample is more serious and more cracks are produced. During the unstable crack propagation and post-peak failure, the wave velocity decays rapidly, which shows that the formation of macrocracks causes the wave velocity to decline sharply. It can be seen that the change of wave velocity can reflect the damage of the sandstone.

5 Microstructure Characteristics of Sandstone After High Temperature-Water Cooling Treatment

5.1 Mineral Composition Analysis of Sandstone

According to the X-ray diffraction results, the variation of mineral composition of sandstone samples after high temperature-water cooling treatment is analyzed, as shown in Fig. 11. In general, when the temperature is below 400 °C, the mineral content of sandstone varies less. When the temperature exceeds 400 °C, the sodium feldspar content of sandstone tends to increase as a whole. It expands easily under high temperature and has a large expansion pressure. This causes the volume of mineral particles to expand, which causes them to loosen. This is one of the main reasons for the deterioration of the strength and softening of sandstone.

5.2 Microstructure Analysis of Sandstone

SEM tests were conducted to extrapolate the micromorphological changes of sandstone after high temperature-water cooling treatment, as shown in Fig. 12. As can be seen from Fig. 12a, when the temperature is 100 °C, there are more primary microcracks and pores in the sandstone. Some of the particles have smoother surfaces, that is, the cross-section of the particles remains relatively complete and shows cracks along the crystal surface, but the other part of the particles has a clear cut surface and shows a penetrating crack, at which point the maximum crack width is only 2 μm. When the temperature is 200–400 °C (Fig. 12b–c), the primary crack widens gradually due to thermal stress caused by different particle expansion differences, mainly by penetrating cracks, while at 400 °C, the particle implodes and forms a

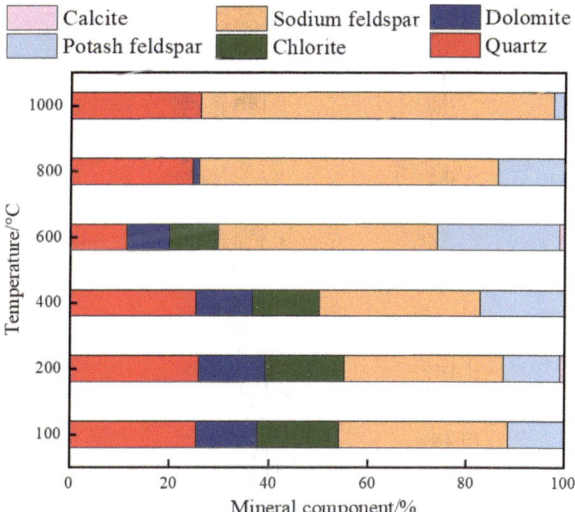

Fig. 11 Changes in mineral composition of sandstone after high temperature-water cooling treatment

distinct layered crack. When the temperature is 600 to 800 °C (Fig. 12d–e), the number of pores and cracks increases significantly, some microcracks developed and penetration occurred. When the temperature is 800 °C, there appears extruded blocks in the sandstone section, with a heat-induced crack spreading along the edge of the particle to form a crack network, at which point the maximum crack width is 20 μm. When the temperature is 1000 °C (Fig. 11f), the crack expanded further into a network of large-scale fractures, at which point the maximum crack width is larger than 41 μm. Thus, the ultimate bearing capacity (peak strength) of sandstone can be seriously weakened by increasing the number and scale of internal fractures with increasing temperature, but it can effectively provide sufficient space for sandstone deformation, that is, increase the ultimate deformation capacity of sandstone (peak strain) effectively.

6 Discussion

The test results show that high temperature-water cooling treatment has an important effect on the internal damage of sandstone. It is due to the cooling rate is high, the surface temperature of sandstone drops sharply when it is quickly cooled in distilled water, while the internal temperature of the specimen drops relatively late. So that the temperature difference between the inside and outside of sandstone makes its surface produce tensile stress, while the internal produce compressive stress, when these two stresses exceed the structural stress of sandstone, it will cause microcracks to develop, expand and penetrate. In addition, the internal fracture of sandstone is further developed under high temperature, and the softening effect when water entering

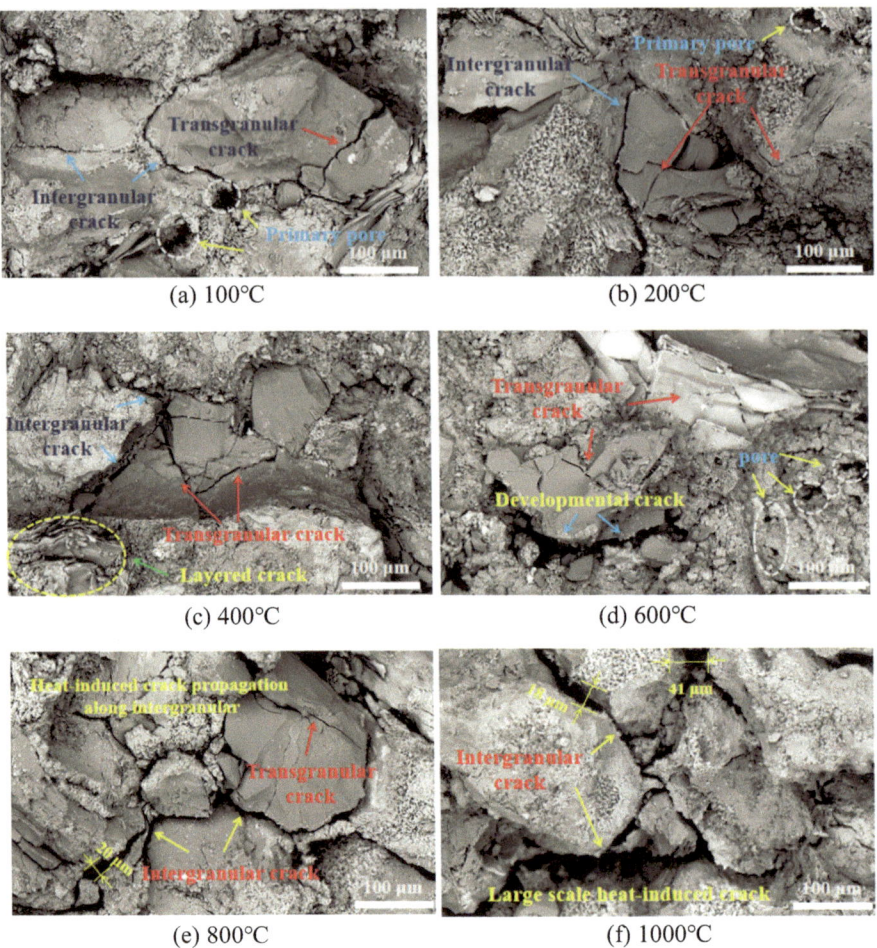

Fig. 12 Microfeatures of sandstone section after high temperature-water cooling treatment

the fracture which results in more obvious mechanical deterioration of sandstone. Therefore, high temperature rapid cooling induced the developing of crack inside rock has been applied in engineering practice. For example, during the development of enhanced geothermal systems, the cold impact on high-temperature rocks can produce a large number of cracks in the high-temperature reservoir, thus increasing the porosity and permeability of the reservoir and realizing the efficient exploitation of dry-heated rock geothermal resources.

7 Conclusion

(1) With the increase of temperature, the surface color of sandstone changed from grayish-white to tawny, the length of surface crack increased significantly, the mass loss rate, volume expansion rate and peak strain of sandstone increased, while the peak strength decreased gradually.
(2) At temperatures below 400 °C, the variation of longitudinal wave velocity during failure process of sandstone after high temperature-water cooling treatment can be divided into a steady increase, oscillation and a sharp decrease, while there is an oscillation increase, a slow decrease and a sharp decrease at temperatures below 1000 °C, respectively.
(3) When the temperature is below 400 °C, the mineral content of sandstone varies less. When the temperature exceeds 400 °C, the sodium feldspar content of sandstone tends to increase as a whole. It expands easily under high temperature and has a large expansion pressure. This causes the volume of mineral particles to expand, which causes them to loosen.

Acknowledgements This research was financially supported by the Major Science and Technology Projects of Yunnan Province (No. 202102AF080001), the National Natural Science Foundation of China (Nos. U1965204) and the IWHR Research and Development Support Program (Nos. GE110145B0022021).

References

1. Xie HP, Ren SH, Xie YC et al (2021) Development opportunities of the coal industry towards the goal of carbon neutrality. J China Coal Soc 46(7):2197–2211
2. Pan ZH, Luo J, Cheng ZY et al (2020) Evaluation of geology conditions for the development of geothermal energy in China. Earth Sci Front 27(1):134–151
3. Huang YH, Tao R, Chen X et al (2023) Study on fracture behavior and thermal cracking evolution law of granite specimens after high temperature treatment. Chin J Geotech Eng 45(04):739–747
4. Zhu ZN, Tian H, Dong NN et al (2018) Experimental study of physico-mechanical properties of heat-treated granite by water cooling. Rock Soil Mech 39(S2):169–176
5. Xi BP, Wu YC, Zhao YS et al (2020) Experimental investigations of compressive strength and thermal damage capacity characterization of granite under different cooling modes. Chin J Rock Mech Eng 39(2):286–300
6. Deng LC, Li XZ, Wu Y et al (2021) Study on mechanical damage characteristics of granite with different cooling methods. J China Coal Soc 46(S1):187–199
7. Jia P, Yang QY, Liu DQ et al (2021) Physical and mechanical properties and related microscopic characteristics of high-temperature granite after water-cooling. Rock Soil Mech 42(6):1568–1578
8. Wu XH, Cai MF, Ren FH et al. (20220) P-wave evolution and thermal conductivity characteristics in granite under different thermal treatment. China J Rock Mech Eng 41(3):457–467
9. Cui HB, Tang JP, Jiang XT et al (2019) Experimental study on the mechanical and acoustic characteristics of high-temperature granite after natural and water cooling. Chin J Solid Mech 40(6):571–582

10. Cho W, Kwon S (2010) Estimation of the thermal properties for partially saturated granite. Eng Geol 115(1):132–138
11. Zhao Z (2016) Thermal influence on mechanical properties of granite: a microcracking perspective. Rock Mech Rock Eng 49(3):747–762
12. Chaki S, Takarli M, Agbodjan WP et al (2008) Influence of thermal damage on physical properties of a granite rock: porosity, permeability and ultrasonic wave evolutions. Constr Build Mater 22(7):1456–1461
13. Janio DCLJ, Paraguass A (2004) Linear thermal expansion of granitic rocks: influence of apparent porosity, grain size and quartz content. B Eng Geol Environ 63(3)
14. Zhang Y, Li G, Wang XY et al (2021) Microfabric characteristics of tight sandstone of Xujiahe formation in western Sichuan after high temperature and the effect on mechanical properties. Chin J Rock Mech Eng 40(11):2249–2259
15. Su CD, Wei SJ, Qin BD et al (2017) Experimental study on the influence mechanism of high temperature on mechanical properties of fine sandstone. Rock Soil Mech 38(3):623–630
16. Ping Q, Su HP, Ma DD et al (2021) Experimental study on physical and dynamic mechanical properties of limestone after different high temperature treatments. Rock Soil Mech. 42(4):932–942
17. Huang Y, Yang S, Bu Y et al (2020) Effect of thermal shock on the strength and fracture behavior of pre-flawed granite specimens under uniaxial compression. Theor Appl Fract Mec 106:102474
18. Gu QX, Huang Z, Zhong W et al (2023) Study on the variations of pore structure and physical and mechanical properties of granite after high temperature cycling. Chin J Rock Mech Eng 1024(3):1–16
19. Hu SH, Zhang G, Zhang M et al. (2016) Deformation characteristics tests and damage mechanics analysis of Beishan granite after thermal treatment. Rock Soil Mech 37(12):3427–3436+3454

Open Access This chapter is licensed under the terms of the Creative Commons Attribution 4.0 International License (http://creativecommons.org/licenses/by/4.0/), which permits use, sharing, adaptation, distribution and reproduction in any medium or format, as long as you give appropriate credit to the original author(s) and the source, provide a link to the Creative Commons license and indicate if changes were made.

The images or other third party material in this chapter are included in the chapter's Creative Commons license, unless indicated otherwise in a credit line to the material. If material is not included in the chapter's Creative Commons license and your intended use is not permitted by statutory regulation or exceeds the permitted use, you will need to obtain permission directly from the copyright holder.

Investigating Effect of Bentonite Support Fluid on Soil-Pile Interface Behavior by Ultra-Weak Fiber Bragg Gratings

Zhihong Li, Yu Gu, Xiaonan Jia, Xuehui Hu, Xuqun Zhang, and Zhaofeng Li

Abstract Engineering practice indicates that the use of bentonite support fluid in borehole may significantly reduce the pile bearing capacity. To estimate the reduction quantitively, the static load test in field was performed on two kinds of piles with and without bentonite fluid, i.e., the slurry displacement (SD) pile and continuous flight auger (CFA) pile. Test site was selected in Huangpu District of Guangdong, and two CFA piles and one SD pile were installed. Sensing technology of ultra-weak fiber Bragg grating was applied to measure the soil-pile interface behavior, with a series of sensors attached to the longitudinal steel bar. Evolution of shaft resistance with the static load was thereby captured. Test results show that the bentonite support fluid reduces the shaft resistance by 52 and 61% in granitic residual soil and completely decomposed granite soil, respectively. Field excavation suggests that the reduction is attributed to the existence of residual bentonite layer between the pile concrete and soil, instead of the filter cake formed by bentonite filtration.

Keyword Continuous flight auger pile · Slurry displacement pile · Ultra-weak Fiber Bragg gratings · Weathered granitic soil

---

Z. Li
Bureau of Municipal Housing and Urban-Rural Development of Huangpu District of Guangzhou (Bureau of Construction and Transportation of Guangzhou Development District), Guangzhou, China

Y. Gu · X. Jia · X. Hu
Guangzhou Huangpu No. 2 Light Rail Construction and Investment Co. Ltd, Guangzhou, China

X. Zhang
Guangzhou Metro Design and Research Institute Co., Ltd, Guangzhou, China

Z. Li (✉)
School of Civil and Environmental Engineering, Harbin Institute of Technology (Shenzhen), Shenzhen, China
e-mail: lizhaofeng@hit.edu.cn

1 Introduction

Slurry displacement (SD) pile, as a kind of cast-in-drilled-hole pile, is most commonly used in foundation engineering, where the bentonite/polymer support fluid is introduced into the borehole simultaneously with the boring process. In this process, a thin layer of bentonite filter cake may be formed by filtration, especially in sandy soil, which acts as a membrane to stabilize the borehole [1].

However, according to the engineering practice, the use of bentonite support fluid may significantly undermine the soil-pile interface resistance. According to the foundation design code of Guangdong Province (DBJ 15-31-2016), the reduction may be severest in weathered granitic soil. For instance, the code suggests that the ultimate shaft resistance of the SD pile in completely decomposed granite (CDG) soil may be the same of that in cohesive soil, i.e., in a low range of 28 ~ 48 kPa. Yet, such an estimation is roughly based on the engineering experience, without a direct comparison to other kinds of cast-in-drilled-hole pile without using the bentonite support fluid, such as the continuous flight auger (CFA) pile [2].

A few studies have been conducted to understanding the effect of bentonite fluid on the soil-pile interface behavior. Lam et al. [3] modified direct shear test to imitate the bentonite filtration process in sand, and they revealed that the interface shear strength reduces linearly with the square root of filtration time. As a continuing study, Lam et al. [4] conducted field trial test and observed that the polymer fluid yields a much stiffer load-settlement response than the bentonite fluid does. O'Dwyer et al. [5] further investigate the influence of mix proportion between sand and bentonite on the interface shear strength, and reported a transition from sand-governed strength to bentonite-governed strength. Instead of filtration, Chen et al. [6, 7] directly added a layer of bentonite slurry between pile concrete and soil in the direct shear test, and found that the interface strength was dramatically reduced with the increase of the slurry thickness. Still, most of these studies are limited to the laboratory test on sandy soil, and cannot effectively provide a guidance to the engineering practice. Moreover, the works may not provide an explanation on the strength reduction in the other non-sandy soils with little bentonite filtration, such as the CDG soil mentioned in the Guangdong design code (DBJ 15-31-2016).

With the above regards, this study takes the initiative to quantify the effect of bentonite support fluid on the soil-pile interface strength by field test. To achieve this goal, two kinds of cast-in-drilled-hole piles were installed, i.e., the SD pile and CFA pile, and their performances under the static load test were compared. In addition, to provide a direct measurement on the interface behavior in each soil layer, the sensing technique of fiber Bragg grating (FBG) was used. In the following texts, the descriptions of test site and test configuration are delineated first. Then, the test results are given to elucidate the effect of bentonite fluid quantitively. These works and test data here are significant to update the design guideline for cast-in-drilled-hole pile.

2 Detials of Field Test

2.1 Site of Field Test

As the foundation design code of Guangdong Province (DBJ 15–31-2016) has been aware of the influence of bentonite fluid, the field test site was chosen in Huangpu District, Guangzhou. The specific location of the site is near the Kaichuan Avenue, with the latitude of ~ 23°8′7″ N and the longitude of ~ 113°32′35″ E, as shown in Fig. 1. Soil stratum of the site is shown in Table 1. Thickness of each soil layer and the associated physical properties (e.g., moisture content and SPT N value) are also summarized in Table 1. As can be seen, on the top of the ground are the plain fill and silty sand, where the filtration of bentonite can occur to some extent. Beneath are two kinds of weathered granitic soils, i.e., the granite residual soil (GRS) and the complete decomposed granite (CDG) soil. Such a stratum is representative and can be similarly observed in other places of Guangdong Province. Besides, it allows to examine the aforementioned estimation on the effect of bentonite fluid in the design code of DBJ 15-31-2016.

Fig. 1 Site of field test and pile layout

Table 1 Soil strata and properties of the site

Soil type	Layer thickness	Bulk density	Moisture content	Liquid limit	Plastic limit	SPT N value
	(m)	(g/cm³)	(%)	(%)	(%)	(−)
Plain fill	3.67	1.95	24.31	39.20	23.10	6.8
Silty sand	3.03	1.88	31.50	-	-	10.5
GRS	3.55	1.91	25.04	36.21	23.68	20.4
CDG	> 7.90	1.97	21.05	32.90	21.58	36.7

Table 2 Testing program

Pile label	Length (m)	Diameter (mm)	UBC (kN)
SD-1	13	600	1560
CFA-1	13	600	3300
CFA-2	17	800	6020

2.2 Test Program

Three different piles were installed in the site, and the pile layout are again shown in Fig. 1. All the piles were arranged not too far from each other, and therefore, the penetrated soil profiles were similar among them. Details of the three piles are summarized in Table 2. Specifically, SD-1 is a slurry displacement pile with the usage of bentonite support fluid. Its pile length and diameter are 13 m and 600 mm, respectively. CFA-1 is a continuous flight auger pile without the usage of bentonite fluid, and it has the same pile dimensions to SD-1. Considering that SD pile is widely used in Guangdong but the CFA pile is rarely used, CFA-2 with a length of 17 m and a diameter of 800 mm was also installed, thereby providing a more convincing comparison between the two kinds of piles.

Static load test was performed on the piles to measure the entire interface response between the soil and pile, as illustrated in Fig. 2. The test was conducted following the code for testing of building foundation in Guangdong Province (DBJ/T 15-60-2019). The load was applied to the pile head in an incremental way, and the resultant settlement of pile head was recorded under each load. The test was terminated and the load was retrieved incrementally, when the settlement of 60 mm was reached or all weights had been used. The ultimate bearing capacity (UBC) along the axial direction, which characterizes the entire pile-soil interface response, was determined as the load corresponding to the settlement of 40 mm. Note that there are other kinds of criteria for the test termination and determination of UBC in the code, which are however not applicable and not considered in the test here.

2.3 Deployment of UwFBG

The sensing technique of ultra-weak fiber Bragg grating (uwFBG), which has a very low reflection rate as -20 dB, was used to measure the interface response in each soil layer during the static load test. Such a technique enables a quasi-distributed strain measurement, with an assurance of the high signal-to-noise ratio. It is reliable and cost-effective to the object with a great number of gauging points, such as the piles in this study [8]. As shown in Fig. 2, the uwFBG sensors were attached firmly on a longitudinal bar of the steel cage (see the red line). The space resolution of the sensors was 1.0 m, which is capable to capture the response in each soil layer (see

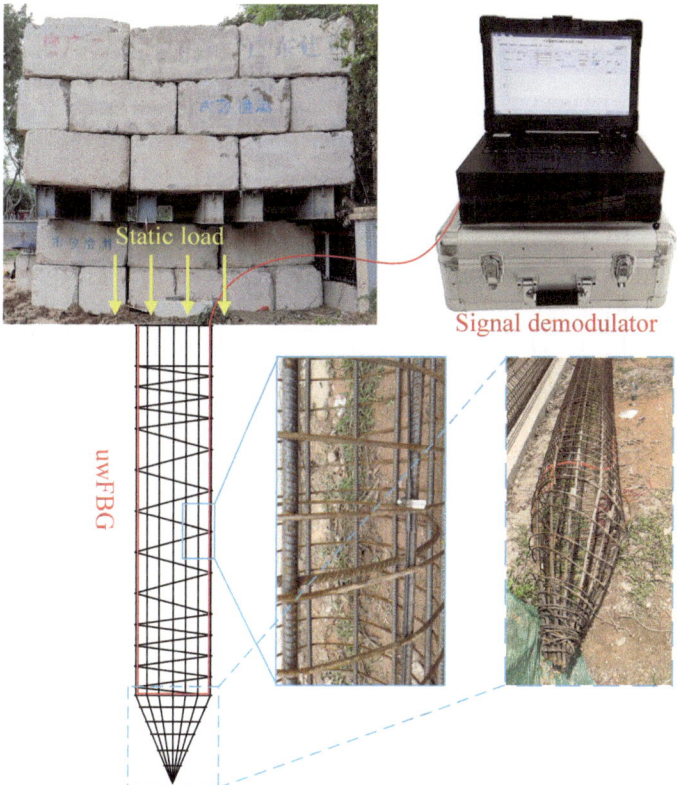

Fig. 2 Static load test with uwFBG sensing technique

the layer thickness in Table 1). Near the base of the pile, the fiber was attached to the ferrule and then turned to another longitudinal bar. With such a configuration, the interface response can be better estimated by averaging the responses of the sensors on the two bars, eliminating the bias from the pile bending.

Signal demodulator was connected to the fiber head and measured the axial strain at the location of the sensor via the change of the wavelength of the reflected light as in Eq. (1)

$$\varepsilon_{i,t} = \mu \cdot (\lambda_{i,t} - \lambda_{i,0}) \tag{1}$$

where $\varepsilon_{i,t}$ is the axial strain of the ith sensor at time t; μ is the calibrated strain coefficient, which is 831 με/nm here; and $\lambda i,t$ and $\lambda i,0$ are the measured wavelength of the ith sensor at time t and before loading, respectively. Then, according to the elasticity theory and the static equilibrium of cylindrical section [9], the shaft resistance between soil and pile can be derived by Eq. (2) as

$$q_{ij,t} = \frac{ED}{4l_{ij}}(\varepsilon_{i,t} - \varepsilon_{j,t}) \quad (2)$$

where $q_{ij,t}$ is the resistance between the ith and jth sensors at time t; E is the elastic modulus of the pile which is 31.5 GPa here, D is the pile diameter; l_{ij} is the axial spacing between the sensors, which is 1.0 m here. Note that following Eq. (2), the distribution of shaft resistance along the pile axis can be obtained, and the base resistance can be further estimated by deducting the shaft resistance from the total bearing capacity.

2.4 Test Result Discussion

In this section, the results from the static load test, i.e., the load-settlement responses, are first discussed to demonstrate the effect of bentonite support fluid on the entire pile-soil interface response. Then, the results from the measurement by the uwFBG sensing technique, i.e., the distributions of shaft resistance and the value of base resistance in CDG soil, are discussed to elucidate the effect in the individual soil layer.

2.5 Load-Settlement Response

The load-settlement curves for the three piles are given in Fig. 3. For the ease of comparison, the section of unloading is not shown here. Note that after unloading, the residual settlements of all pile are greater than 90%, demonstrating that all piles are frictional, instead of being end-bearing. As can be seen in Fig. 3, each pile follows the same development pattern; that is, the settlement gradually increases together with the load.

Still, difference can be observed among the piles. At the initial stage with a small load, probably within the elastic range, the response of CFA-1 is much stiffer than that of SD-1; that is, the usage of bentonite support fluid can undermine the soil-pile stiffness. As expected, CFA-2 yields an initial response stiffer than CFA does, since the pile length and diameter are both larger in CFA-2. All piles reach a settlement of 40 mm. Therefore, the UBCs can be determined accordingly (see the denotations of red stars). The data are also summarized in Table 2. Specifically, the UBC of SD-1 is 1560 kN, while that of CFA-1 is 3300 kN. These data indicate that, in the engineering site in Guangdong Province, the use of bentonite support fluid may almost halve the bearing capacity of the pile. Note that which soil layer dominates such a reduction is revealed later using the uwFBG sensing technology. The value of CFA-1 is 6020 kN, i.e., a double of that of CFA-1; that is, the pile bearing capacity is proportion to the pile diameter and length. In fact, it is suggested that the interface response in each soil layer is moderately affected by the pile dimensions here.

Fig. 3 Load-settlement curves of three test piles

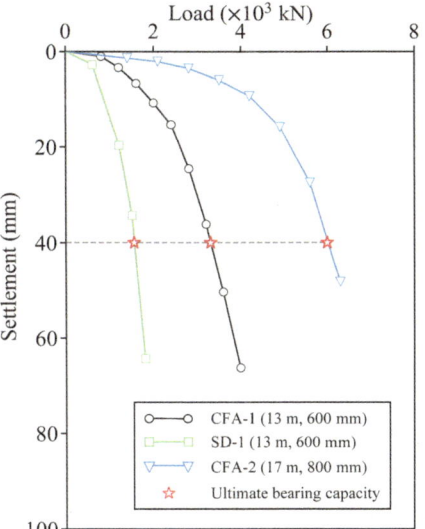

2.6 Shaft Resistance and Base Resistance

The interface behavior in each soil layer, as mentioned, was captured using the sensing technique of the ultra-weak fiber Bragg grating. Using Eq. (2), the distribution of shaft resistance can be measured for the slurry displacement pile SD-1, as shown in Fig. 4. Generally, in each soil layer, as the load increases, the shaft resistance increases correspondingly; that is, the strain-hardening behavior occurs among the soil layers in the test field. The dash lines indicate the upper bound and lower bound of each layer. It can be seen that the peak of the distribution is generally located in the granitic residual soil (GRS) layer.

The UBC of 1560 kN is between the load of 1500 and 1800 kN. Therefore, the distribution of ultimate shaft resistance (see the red line) is obtained by linear interpolation of the distribution under these two loads (see the blue lines). Thereby, a mean value of the ultimate shaft resistance can be measured over each layer. As summarized in Table 3, the ultimate resistance in plain fill and silty sand, i.e., the top of the ground, are 12 kPa and 27 kPa, respectively. The ultimate resistance is 62 kPa in the GRS, but it is lower in the CDG soil as 59 kPa. The reason for a lower resistance in a stronger soil layer here may be that the pile base prohibits the deformation along the pile shaft to some extent, and then the shaft resistance around therein cannot be fully developed. Such a phenomenon may be severer, when the base resistance takes a greater proportion in the total bearing capacity. Here, by deducting the shaft resistance from the UBC, the ultimate base resistance is given as 1670 kPa and contributes to the UBC by 30.4%. Note that these measured values in weathered granitic soils are over the range of 28–48 kPa in the Guangdong design

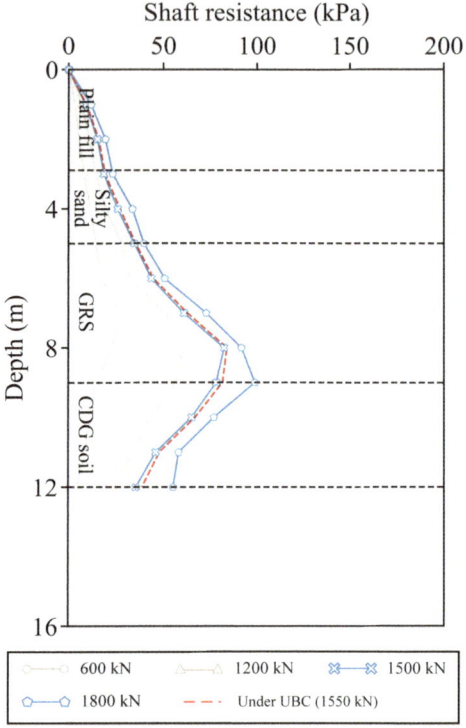

Fig. 4 Evolution of shaft resistance in Pile SD-1

code (DBJ 15–31-2016). Therefore, the experience-based values in the code can be updated using the measurement-based values here.

Using the uwFBG sensing technique, the evolution of shaft resistance is also obtained for the continuous flight auger pile CFA-1, as displayed in Fig. 5. Obviously, the evolution pattern of shaft resistance is unaffected by the bentonite support fluid; that is, the shaft resistance is continuously increased together with the load in each soil layer and no strain softening occurs. Moreover, the peak location is maintained in the CRS layer. Nevertheless, comparing Fig. 5 with Fig. 4, the effect of bentonite fluid can be clearly seen. By interpolation, the distribution of ultimate shaft resistance

Table 3 Comparison on shaft and base resistances

Soil layer	SD pile (kPa)	CFA pile (kPa)	Difference (kPa)	Difference (%)
Plain fill	12	30	18	60
Silty sand	27	64	37	58
GRS	62	128	66	52
CDG[a]	59 (1670)	154 (3080)	65 (1410)	62 (46%)

[a]Data in bracket belongs to the base resistance

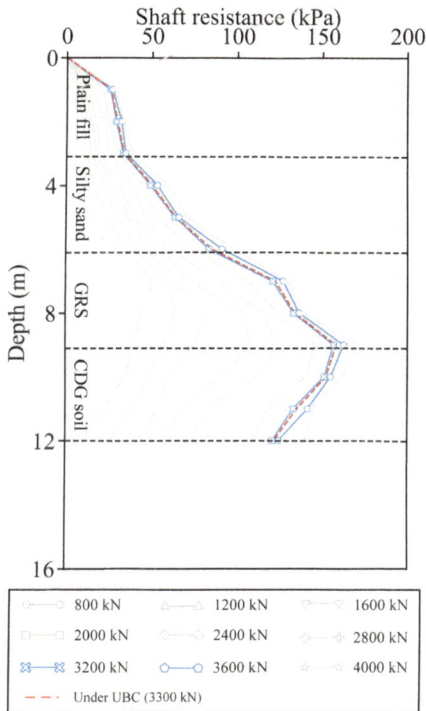

Fig. 5 Evolution of shaft resistance in Pile CFA-1

is given for CFA-1 as the red line. Then, it is observed that the ultimate resistance is much higher in CFA-1 than SD-1 in every soil layer. For instance, in the CDG soil, the ultimate shaft resistance is over 100 kPa for CFA-1, while that for SD-1 is only 59 kPa. In the CRS, the ultimate resistance is between 80 and 150 kPa, while that for SD-1 is 62 kPa. In essence, the usage of bentonite fluid can significantly reduce the shaft resistance in different soil, no matter it is sandy or not.

Similar observations on the effect of bentonite support fluid are attained in CFA-2, as shown in Fig. 6. First, the general evolution pattern is identical to those of CFA-1 and SD-1, with little effect from the bentonite. Second, the ultimate shaft resistance (see the red line in Fig. 6) is much higher than that of SD-1 (see the red line in Fig. 4). Again, for instance, the ultimate resistance in CDG soil is over 100 kPa in CFA-2. Corresponding to the aforementioned indication in Fig. 3, the distribution of CFA-2 is similar to that of CFA-1 in some degree; that is, the interface response in each soil layer is moderately affected by the pile dimensions here, compared with the bentonite fluid. Therefore, to provide a more reliable estimate, the ultimate resistance of the CFA pile is measured by averaging between the values of CFA-1 and CFA-2, as summarized in Table 3.

As shown in Table 3, the ultimate shaft resistances in plain fill and silty soil are 30 kPa and 64 kPa, respectively. The differences of the SD pile in reference to the CFA pile in these two layers are 60% and 58%, respectively. For the two weathered

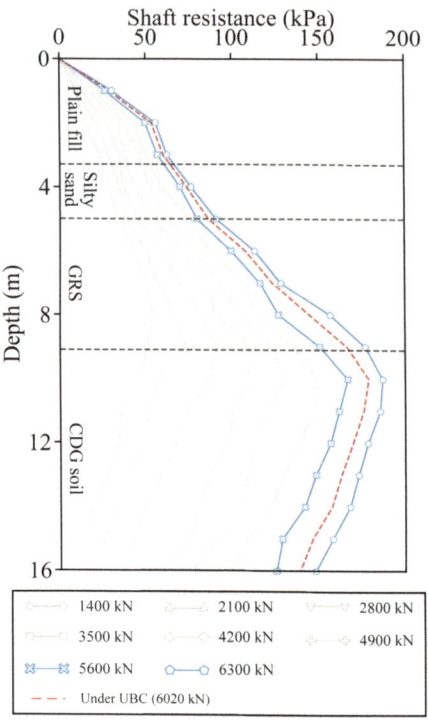

Fig. 6 Evolution of shaft resistance in Pile CFA-2

granitic soils, i.e., the CSR and CDG soils, the shaft resistances are estimated as 128 kPa and 154 kPa, respectively. The differences are 52 and 62% in these two layers, respectively, similar to those values in the upper two layers. Yet, they indeed contribute mostly to the difference on the total bearing capacity since the absolute difference is much higher therein. By deducting the shaft resistance from the total bearing capacity, the ultimate base resistance account for 29.7 and 22.4% of the UBC in CFA-1 and CFA-2, respectively. The mean value is measured as 3080 kPa, with a difference of 46% to that of the SD pile, i.e., less severely affected by the bentonite fluid. Nevertheless, it demonstrates that bentonite cannot be fully removed by the pumping of concrete in the borehole and may be remained at the pile base, softening the base resistance response.

2.7 Mechanism of Bentonite Fluid Effect

As mentioned, a thin layer of filter cake may be formed by bentonite filtration in sand, and Lam et al. [5] have revealed that the interface strength reduces linearly with the square root of filtration time, since the bentonite-over-sand ratio is increased accordingly. That is why in the plain fill and silty sand layer here, the shaft resistance

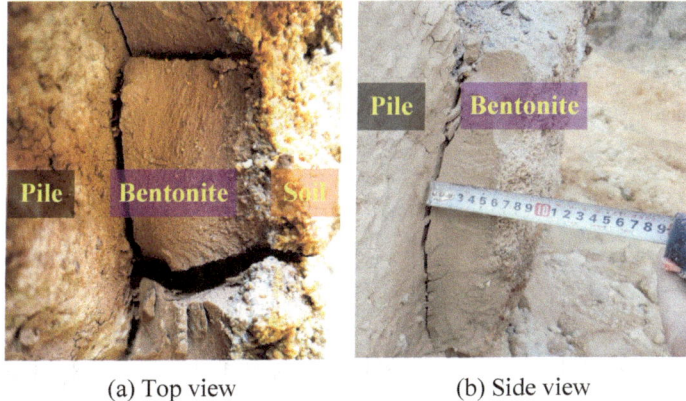

(a) Top view (b) Side view

Fig. 7 Disclosure of soil-pile interface in the field

is lowered by the usage of bentonite support fluid. However, the filtration cannot considerably occur in the weather granitic soil using the pore-throat analysis proposed by Li et al. [10]. To elucidate the mechanism of bentonite effect in such a kind of soil, the field excavation is carried out, as shown in Fig. 7. Beyond expectation, a viscous layer of residual bentonite with a thickness of 80 mm was found therein, instead of a thin layer of filter cake. Possibly, it may be formed under the effect of double-layer attraction and gravity, and cannot be fully squeezed out from borehole by the pumping of concrete. Then, the soil-pile interface is changed to soil-bentonite-pile interface. The surface texture indicates that this layer was sheared under loading, contributing to the settlement largely. That is why less load is required to reach the settlement of 40 mm for SD pile, and the ultimate bearing capacity and resistance in each soil layer is reduced accordingly.

3 Conclusions

The influence of bentonite support fluid on the soil-pile interface behaviour has been investigated by field test in Huangpu District, Guangzhou. One slurry displacement pile and two continuous flight auger piles were installed. Static load test was performed and the sensing technique of ultra-weak fiber Bragg grating was applied to measure the soil-pile interface response. The salient findings from this study are as follows:

- With the same pile dimensions, the axial bearing capacity of SD pile is halved from that of CFA pile in the typical site of Guangdong Province by the usage of bentonite support fluid.

- Applying the uwFBG sensing technique, the ultimate shaft resistance was found to be reduced by over 50% in each soil layer, while the ultimate base resistance in the complete decomposed granite soil was reduced less severely by 46%.
- From the field excavation, a viscous layer of residual bentonite was found between the pile concrete and soil, typical in the weathered granitic soil. The reduction on the bearing capacity is attributed to the existence of such a layer, instead of the filter cake formed by the bentonite filtration.

Acknowledgements This research was supported by National Natural Science Foundation of China (Grant No. 52209126 and 52239008), Guangdong Basic and Applied Basic Research Foundation (Grant No. 2023A1515012860), Shenzhen Science and Technology Program (Grant No. GXWD20220818152909001, GXWD20231130125225001, and GXWD20231129105817002).

References

1. Qin S, Xu T, Zhou WH, Benzuijen A (2023) Infiltration behaviour and microstructure of filter cake from sand-modified bentonite slurry. Transp Geotech 40:100963
2. Farrell ER, Lawler ML (2008) CFA pile behaviour in very stiff lodgement till. Proc Inst Civ Eng 161:49–57
3. Lam C, Jefferis SA, Martin CM (2014) Effects of polymer and bentonite support fluids on concrete–sand interface shear strength. Géotechnique 64(1):28–39
4. Lam C, Jefferis SA, Suckling PT, Troughton VM (2015) Effects of polymer and bentonite support fluids on the performance of bored piles. Sois Found 55(6):1487–1500
5. O'Dwyer KG, McCabe BA, Sheil BB (2022) Critical state shear strength at concrete–sand–bentonite slurry interfaces: mix proportions and rate effects. Geotech Lett 12(4):281–287
6. Chen C, Leng WM, Qi Y, Jin ZH, Nie RS, Qiu J (2018) Experimental study of mechanical properties of concrete pile-slurry-sand interface. Rock Soil Mech 39(7):2461–2472
7. Chen C, Leng WM, Yang Q, Dong JL, Xu F, Ruan B (2022) Effect of a filter cake on shear behavior of sand-concrete pile interface. J Cent South Univ 29(6):2019–2032
8. Li Z, Zhang Z, Tai P, Shen P, Li J (2024) Investigation of morphological effects on crushing characteristics of calcareous sand particle by 3D image analysis with spherical harmonics. Powder Technol 433:119204
9. Liu X, Li Z, Tai P, Chen R, Fu W (2021) In-situ experimental investigation on stress distribution of grout body of tension-type ground anchor. Chinese J Undergr Space Eng 17(S1):63–70
10. Li Z, Wang YH, Chow JK, Su Z, Li X (2018) 3D pore network extraction in granular media by unifying the Delaunay tessellation and maximal ball methods. J Petrol Sci Eng 167:692–701

Open Access This chapter is licensed under the terms of the Creative Commons Attribution 4.0 International License (http://creativecommons.org/licenses/by/4.0/), which permits use, sharing, adaptation, distribution and reproduction in any medium or format, as long as you give appropriate credit to the original author(s) and the source, provide a link to the Creative Commons license and indicate if changes were made.

The images or other third party material in this chapter are included in the chapter's Creative Commons license, unless indicated otherwise in a credit line to the material. If material is not included in the chapter's Creative Commons license and your intended use is not permitted by statutory regulation or exceeds the permitted use, you will need to obtain permission directly from the copyright holder.

Study on Improvement Measures of Hydraulic Engineered Cementitious Composites Layer Bonding Performance

Yupu Wang, Jiazheng Li, and Yan Shi

Abstract The layer bonding performance is very vital for the utilization of hydraulic engineered cementitious composites (HECC). The layer bonding performance involves the interlayer performance of HECC and HECC, and the layer performance of HECC and normal mortar (NM). The bonding strength of the untreated layer is very low, only half of that of the whole specimen without layer. Based on the fact, this article proposes three different improvement measures to improve the bonding performance of the layers: pickling surface, using interface agents, and using high-speed stirring devices to stir layer. The layer bonding performance is tested by the layer flexural strength. The test results show that: muriatic acid and acetic acid pickling can improve the interlayer bonding performance by 10% to 30%, while citric acid will reduce it by more than 90%; different interface agents have different improvement effects, among which The epoxy resin mortar interface agent has the most obvious improvement effect, reaching 44.9%; high-speed stirring of the layer eliminates the layer in a certain sense, improving the layer bonding performance by 349%, and has the most obvious effect among all methods.

Keywords Hydraulic engineered cementitious composites · Normal mortar · Layer · Flexural strength · Acid · Interfacial agents · Stir

1 Introduction

Engineered Cementitious Composites (ECC) is a new type of engineered cementitious composites optimized and designed based on microphysical and mechanical principles, and was proposed by Li and Leung [1]. Polyvinyl alcohol (PVA) is

---

Y. Wang · J. Li (✉) · Y. Shi
Institute of Materials and Structure, Changjiang River Scientific Research Institute, Wuhan 430010, Hubei, China
e-mail: 1659959395@qq.com

Research Center of Water Engineering Safety and Disaster Prevention of Ministry of Water Resources, Wuhan 430010, Hubei, China

with high strength, high modulus of elasticity, non-toxic and other characteristics of organic fibers, can be well dispersed in the cement matrix and relatively inexpensive, to enhance the toughness of the control cracks, to improve the deformation capacity, and significantly improve the tensile and compressive strength and flexural strength, toughness, ductility of the concrete material, PVA-ECC is widely used in construction projects.

The Changjiang River Scientific Research Institute came up with the concept of Hydraulic Engineered Cementitious Composites (HECC) applicable to hydraulic construction [2]. Li [2] incorporated the features of HECC and put forward the idea of its utilization in water conservancy facility. In some specific concrete dams, the former concrete was substituted with the arrangement of the HECC. Above mentioned various applications in various hydraulic buildings, are involved in the layer of HECC bonding problems, HECC layer bonding performance is good or bad directly determines the application effect of HECC [3], so this paper explores the layer of performance improvement measures, aimed at providing guidance for water conservancy engineering practice.

Concrete layer often weak bonding performance, layer bonding effect makes the concrete layer to produce a layer of water film, hindering the contact between the layers, leading to the emergence of a weak surface to form a transition zone [4–5]. Xiong et al. [6], Jiang [7] pickled the surface with hydrochloric acid, so that the surface roughness was changed, exposing active SiO_2, which led to increased matrix activity, thinning of the weak transition layer. Interfacial agents can improve the porosity and microcracks of the interface layer, and their reaction products can grow in the pores and microcracks, forming a dense structure and increasing chemical bonding strength [8]. The use of mortar interfacial agent [9], epoxy resin-based interfacial agent [10], and expanded cement interfacial agent [11] will improve the layer bonding performance to different degrees. An [12] compared the influence of non-interface agent, cement paste, silica fume paste and expanded cement mortar on the interface bond strength through the UHPC-NC splitting test, and found that they can improve the interface bond strength. Ganesh et al. [15] studied the effect of interfacial agents on the bonding strength of UHPC and NC, the bonding strength of specimens using epoxy resin agent was increased by 12.2 to 60%. Rashid et al. [13] selected four commonly used interfacial agents, namely cement mortar, epoxy resin, styrene butadiene lotion (SBR) and CFRP impregnated epoxy resin. The bond performance under tensile stress is significantly improved compared with agent. Wu et al. [14] also obtained the same experimental phenomenon.

At present, HECC is still in the preliminary application stage, and there is almost no research on the HECC layer issue. This paper explores the layer bonding performance of HECC and HECC, as well as the bonding performance between HECC and normal mortar (NM), and improves the layer bonding performance through surface pickling with muriatic acid and acetic acid, the use of several types of interfacial agents such as epoxy resin interfacial agent and nano-silicon dioxide interfacial agent, as well as high-speed stirring.

Table 1 Chemical compositions of cement and fly ash (wt.%)

Oxide	SiO_2	Al_2O_3	Fe_2O_3	CaO	MgO	K_2O	Na_2O	SO_3	LOI	Na_2O_{eq}
Cement	21.41	4.95	3.81	59.36	0.94	0.75	0.13	3.11	2.59	0.62
Fly ash	48.33	17.58	8.63	8.73	2.89	1.41	0.75	1.86	3.44	1.68

Table 2 Quality indicators of fiber

Type	Diameter (μm)	Length (mm)	Density (g/cm^3)	Breaking strength (MPa)	Elastic modulus (GPa)	Fracture elongation (%)
PVA	37	12	1.3	1800	34	6.6

Table 3 Artificial sand particle distribution (%)

Particle size range	1.25 ~ 0.63	0.63 ~ 0.32	0.32 ~ 0.16	< 0.16
Distribution	25.5	22.1	20.6	19.2

2 Test Materials and Test Method

2.1 Test Raw Materials and Properties

The materials utilized consist of PO42.5 grade cement, F grade fly ash, artificial sand, superplasticiser (SP), viscosity modified admixture (VMA) and polyvinyl alcohol (PVA) fibers. The chemical composition of the cement and fly ash is presented in Table 1 and the quality specifications of the fibers supplied by the manufacturer are given in Table 2. The particle distribution of the artificial sand is shown in Table 3.

2.2 Materials Mixing Ratio

HECC and NM Mixing Ratio

The mix ratio of HECC and NM is shown in Table 4.

Interfacial bonding agents mix ratio

Six kinds of interfacial agents were used for the experimental selection, with a purpose to facilitate the experimental operation, the appropriate adjustment of the water reducing agent dosage and unit water consumption, among which, the cement

Table 4 The mixing ratio of HECC and NM (kg/m^3)

Type	Cement	Fly ash	Artificial sand	Water	SP	VMA	PVA	W/b
HECC	485	485	733	320	7.76	0.48	26	0.33
NM	409	409	1014	270	6.55	\	\	0.33

Table 5 The mixing ratio of bonding agents (kg/m^3)

Type	Cement	MgO	SiO$_2$	Fly ash	Silica fume	Artificial sand	Water	SP	w/b	Emulsion
Mor	1067	\	\	\	\	761	320	6.55	0.3	\
Exp	1040	62.4	\	\	\	805	312	9.36	0.3	\
Nano	1037	\	51.8	\	\	810	311	8.29	0.3	\
Fasf	894	\	\	99	99	744	298	8.94	0.3	\
Er	1135	\	\	126	126	667	227	7.57	0.18	252

paste interfacial agent (Pas) (not listed in the table), the cement mortar interfacial agent (Mor), the expansion mortar interfacial agent (Exp), the nano-silicon dioxide interfacial agent (Nano), the fly ash and silica fume double mixing interfacial agent (Fasf), the epoxy resin interfacial agent (Er), 20% epoxy resin emulsion added (the epoxy resin emulsion made by mixing epoxy resin and water 1:1), and the specific mixing ratio is shown in Table 5.

2.3 Test Methods

HECC mixing process

The mixing process lasts 6–8 min, starting with a dry mix of cement, fly ash and artificial sand at low speed for 1 min, followed by a full mix of superplasticiser and viscosity modified admixture with water at low speed for 2 min, and eventually the addition of PVA fibres for a further 3–5 min at high speed.

Preparation of specimens and testing method of layer performance

Unlike casting the whole specimen in one go, the laminated specimen needs to be moulded in two stages, for illustration purposes the former casting layer is so-called substrate layer and the latter is so-called overlay layer. To make the specimen, a steel mould of half the length of the specimen is inserted before pouring the base layer. Then, the steel mould was removed according to the experimental requirements and the remaining half of the layer was cast. All specimens were tested according to the method for determining flexural strength in GB/T 17,671-2021.

Method of pickling the surface

Three different acids were selected: 95% concentration of acetic acid (Ace), 5% concentration of muriatic acid (Mur), 95% concentration of citric acid (Cit). The specific pickling method is as follows: pour 4 g of acid onto the surface, let it sit for 3 min, wash with water for 5 min, dry the surface, and the remaining half of the overlay layer of HECC was cast.

Method of using interfacial bonding agent

Dip the prepared interface bonding agent with a brush and apply it evenly on the bonding surface of the substrate. The thickness is about 2 mm, and then a covering layer is poured for bonding.

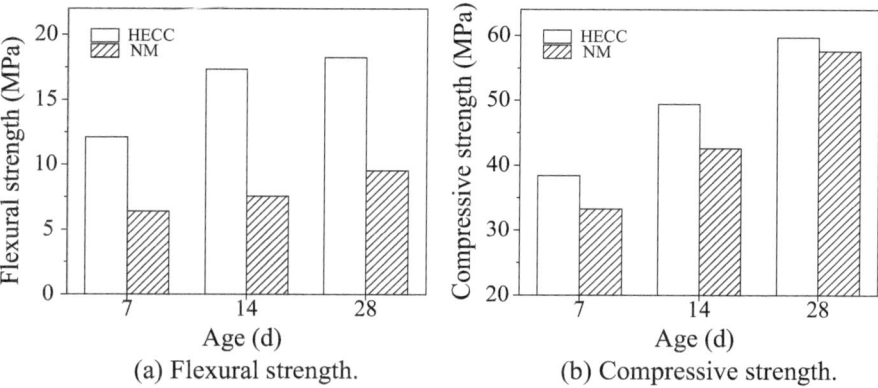

Fig. 1 HECC and NM basic mechanical properties

Method of high-speed stirring

Choose a 60W laboratory electric mixer with maximum speed 3000 r/min, and the mixing blades are selected to be three-bladed paddles with a diameter of 30 mm. The blades of the mixer is placed at the layer, the blades sink to the depth of half of the specimen mold, turn on the mixer, mixing at the maximum speed for 10 s, and then take out the blades, and re-layering the layer of the specimen.

3 Results and Discussion

3.1 Basic Flexural Strength and Compressive Strength of Materials

The flexural strength and compressive strength of HECC and NM are shown in Fig. 1. The mechanical properties of the six different interfacial agents are shown in Fig. 2.

3.2 Layer Bonding Strength of Untreated Surfaces

The specimens without later treatment were divided into two pouring interval times, 60 min (substrate not yet hardened) and 28 days (substrate already hardened), and the flexural strengths of the layers are shown in Fig. 3.

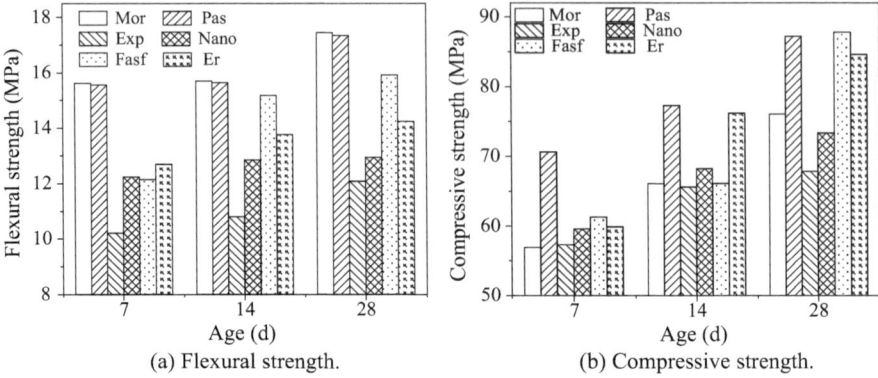

Fig. 2 The flexural strength and compressive strength of bonding agents

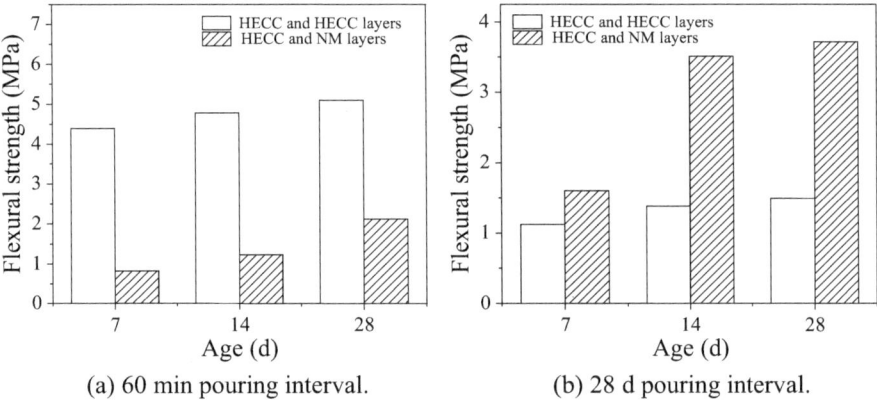

Fig. 3 Layer bonding strength of untreated surfaces

3.3 Research on Measures to Enhance the Bonding Properties of the Surface

Improvement of layer bonding properties by pickling the surface

HECC and NM were selected to maintain the substrate at the age of 28 d. Figure 4 demonstrate HECC and HECC layers flexural strength and the HECC and NM layers flexural strength, respectively, and show that acetic and hydrochloric acids showed a 10% to 30% increase in the layer flexural strength, whereas the surface acid-washed by citric acid showed a severe decrease in the layer flexural strength, with a maximum reduction of 91%. In the figure, N-acid represents the flexural strength of the layer without acid washing.

Acid washing of the surface by muriatic acid will corrode the surface to a certain extent, and the constant shedding of fine grit particles is visible to the naked eye

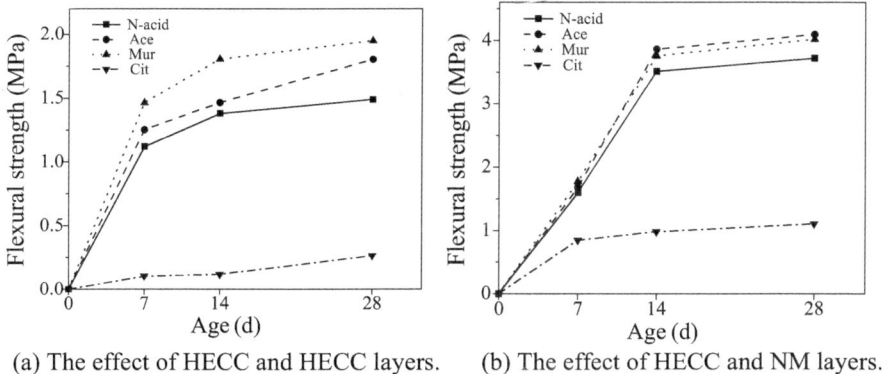

(a) The effect of HECC and HECC layers. (b) The effect of HECC and NM layers.

Fig. 4 Effects of different acid pickling on the properties of layers

during acid washing, and many holes are added to the interface, and these holes and micropores significantly increase the specific surface area of the substrate [6]. Acid washing with vinegar reacts to produce calcium acetate, which on the one hand produces tiny crystals or precipitates on the surface of the matrix, thus increasing the surface roughness, and calcium acetate facilitates the refinement of the pore structure, with an increase in the innocuous porosity and microstructural densities [16]. And citric acid, as a more acidic organic acid, can complex with calcium ions in cement to form soluble calcium citrate complexes, reducing the concentration of calcium ions and decreasing the layer bond strength.

Improvement of layer bonding properties by using interfacial agents

Figure 5a and b show the improvement effect of interfacial agent on the substrate of HECC and NM maintained at the age of 28 d (i.e., 28 d pouring interval), and it can be found that compared with the layer without interfacial agent, all types of interfacial agents have a certain degree of improvement on the flexural strength of the layer, among which, the improvement effect of Er interface agents on the HECC and HECC layers is the most obvious, with an increase of 27.5–44.9%, followed by Exp and Nano interface agents, which also have an improvement of about 20%. For the HECC and NM layers, several interface agents improvements are also around 30%.

The newly generated C-S-H gel as well as the small needle-like calcium alumina radiate outward into the capillary pores of the substrate enhance the chemical force and mechanical occlusion of the interface. After incorporating fly ash and silica fume, the secondary hydration reaction consumes calcium hydroxide, which further generates C-S-H and fills the pores, making the structure of the interfacial transition layer dense [17]. Silica fume can reduce the accumulation of moisture at the interface, and reduce the enrichment of calcium hydroxide crystals at the interface. MgO expansion effect produced will offset the drying shrinkage in the hydration process of the cement, thus making the structure of the transition layer denser [18]. Nano-silica can have a secondary hydration reaction, forming more calcium silicate

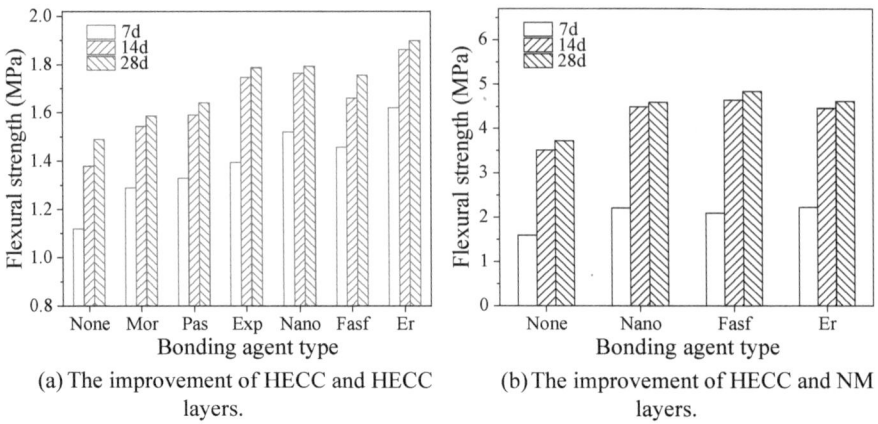

Fig. 5 The improvement of flexural strength of layers by different bonding agents

hydrides, and the high dispersibility of nano-silica can sufficiently wet and encapsulate the cement particles, fill the pores and increase the homogeneity [19]. The epoxy resin will form a uniform, dense and non-porous three-dimensional network structure inside the mortar and in the layer, which will substantially enhance the bond strength of the layer [20].

Improvement of layer bonding properties by high-speed stirring

Select the specimen with a pouring interval of 60 min for the experiment. Figure 6a demonstrates that high-speed mixing of the improvement effect on the layer of the HECC and the HECC, it can be found that the improvement is 168%. Figure 6b demonstrates the improvement effect of high-speed mixing on the layer of HECC and NM, and the strength is improved by 349% compared to that of the unmixed specimen. In the figure, N-layer (HECC) represents HECC flexural strength of no layers, N-layer (NM) represents NM flexural strength of no layers, and N-stir represents layers without stirring. Through the high-speed mixing of the layer, the water film formed due to the time between pouring is eliminated in a certain sense [21], and when the layer is stirred, the layer can be broken, and the fibers are able to penetrate through the layer to better play the role of bridging.

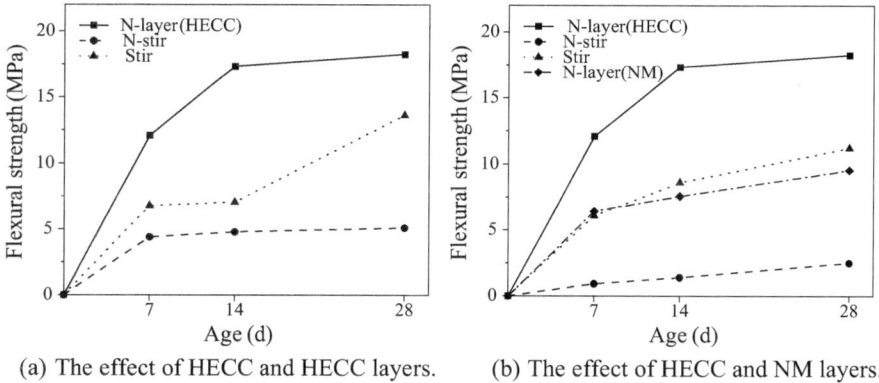

(a) The effect of HECC and HECC layers. (b) The effect of HECC and NM layers.

Fig. 6 The improvement of flexural strength of layer by high-speed stirring

4 Conclusion

In this paper, the improvement measures of the layer bonding properties of HECC were investigated and evaluated by the layer flexural strength, and the conclusions drawn are as follows: Muriatic acid pickling and acetic acid pickling can enhance the layer flexural strength by 10–30%, while the surface strength of citric acid pickling shows a serious decrease. The improvement effect of different interfacial agents is different, among which nano-silica mortar interfacial agent and epoxy resin mortar interfacial agent have the most obvious improvement effect, with a maximum improvement of 200%. High-speed mixing of the layers can significantly improve the flexural strength of the layers, with a maximum improvement of 168% and 349% for the two types of layers, respectively.

Acknowledgements This research was funded by the Hubei Provincial Natural Science Foundation (grant No. 2022CFD026), National Natural Science Foundation of China (grant No. 52179122 and U2040222) and the Basic Scientific Research Funds Programs in National Non-profit Scientific Research Institutes (grant No. CKSF2023304/CL).

References

1. Li VC, Leung C (1992) Steady state and multiple cracking of short random fiber composites. J Eng Mech 188(11):2246–2264
2. Li J (2023) Material properties of hydraulic engineered cementitious composites (HECC) and its application in hydraulic structures. J Yangtze River Sci Res Inst 40(02):1–6+26
3. Wang Y, Li J, Shi Y (2023) Study on influencing factors of hydraulic engineered cementitious composites layer bonding performance. Mater 16(20):6693
4. Baloch WL, Siad H, Lachemi M, Sahmaran M (2021) A review on the durability of concrete-to-concrete bond in recent rehabilitated structures. J Build Eng 44:103315

5. Li G, Xie H, Xiong G (2001) Transition zone studies of new-to-old concrete with different binders. Cem Concr Compos 23(4):381–387
6. Xiong GJ, Jiang H, Chen LQ, Xie HC (2002) Improvement of interfacial structure between old and repair concrete. J Chin Ceram Soc 30(2):4
7. Jiang H (2001) Improved method and mechanism of micro fine structure of new-to-old concrete repaired interface layer. Shantou University, Shantou
8. Zhang Y, Zhu P, Liao Z, Wang L (2020) Interfacial bond properties between normal strength concrete substrate and ultra-high performance concrete as a repair material. Constr Build Mater 235:117431
9. Manzur T, Yazdani N, Emon MAB (2016) Potential of carbon nanotube reinforced cement composites as concrete repair material. J Nanomater 2016:2–2
10. Santos DS, Santos PMD, Dias-da-Costa D (2012) Effect of surface preparation and bonding agent on the concrete-to-concrete interface strength. Constr Build Mater 37:102–110
11. Sun C, Zhang J, Jia L, Liu Y (2016) Experimental study on the influence of different interface agents on the splitting tensile bond behaviour of the new and old concrete interface. Constr Tech 45(S2):504–508
12. An N (2018) Effect of interfacial properties on bonding behaviors of UHPC-NC. Beijing Jiaotong University, Beijing
13. Rashid K, Ahmad M, Ueda T, Deng J, Aslam K, Nazir I, Sarwar MA (2020) Experimental investigation of the bond strength between new to old concrete using different adhesive layers. Constr Build Mater 249:118798
14. Wu X, Zhang X (2018) Investigation of short-term interfacial bond behavior between existing concrete and precast ultra-high performance concrete layer. J Build Struct 39(10):156
15. Ganesh P, Murthy AR (2020) Simulation of surface preparations to predict the bond behaviour between normal strength concrete and ultra-high performance concrete. Constr Build Mater 250:118871
16. Lyu B, Guo L, Wu J, Fei X, Bian R (2022) The impacts of calcium acetate on reaction process, mechanical strength and microstructure of ordinary portland cement paste and Alkali-activated cementitious paste. Constr Build Mater 359:129492
17. Xie H, Li G, Xiong G (2003) The mechanism formed the bonding force between new and old concrete. Bull Chin Ceram Soc 03:7–10+18
18. He W, Peng B (2004) Experimental study on the new-to-old concrete interfacial bond. Concr 5:3
19. Abhilash PP, Nayak DK, Sangoju B, Kumar R, Kumar V (2021) Effect of nano-silica in concrete; a review. Constr Build Mater 278:122347
20. Newlands M, Khosravi N, Jones R, Chernin L (2018) Mechanical performance of statically loaded flat face epoxy bonded concrete joints. Mater Struct 51(2):1–14
21. Lee HS, Jang HO, Cho KH (2016) Evaluation of bonding shear performance of ultra-high-performance concrete with increase in delay in formation of cold joints. Mater 9(5):362

Open Access This chapter is licensed under the terms of the Creative Commons Attribution 4.0 International License (http://creativecommons.org/licenses/by/4.0/), which permits use, sharing, adaptation, distribution and reproduction in any medium or format, as long as you give appropriate credit to the original author(s) and the source, provide a link to the Creative Commons license and indicate if changes were made.

The images or other third party material in this chapter are included in the chapter's Creative Commons license, unless indicated otherwise in a credit line to the material. If material is not included in the chapter's Creative Commons license and your intended use is not permitted by statutory regulation or exceeds the permitted use, you will need to obtain permission directly from the copyright holder.

Microscopic Investigation of Granular Materials in Filter Layer Based on LBM-DEM Method

Qirui Ma, Xing Peng, and Congpeng Zhang

Abstract Soil erosion is one of the most serious problems that threaten the safety of embankments and earth-rock dams. When water flows through the base soil, fine particles subjected to hydrodynamic action will be washed away, which will lead to internal erosion of the embankment project. The inverted filter structure is one of the effective methods to eliminate or reduce the risk of infiltration and erosion, with a history of nearly a century of engineering practice. This article is based on the three-dimensional LBM-DEM fluid solid coupling simulation method. By setting different particle size ratios and hydraulic gradients in the filter medium, the particle contact erosion and subsequent migration and transportation of fine particles in the filter layer structure were comprehensively reproduced, verifying the effectiveness of traditional empirical formulas. This study demonstrates the effectiveness and feasibility of the three-dimensional LBM-DEM coupling method in practical engineering applications from a microscopic view, laying the foundation for the next large-scale simulation of embankment filter layers and the revelation of microscopic mechanisms.

Keywords Granular materials · Filter layer · DEM · LBM · Microscopic

1 Introduction

In hydraulic geotechnical engineering, the inverted filter layer is usually composed of well graded sand or gravel [1]. Due to its importance and complexity, the design of inverted filters has been an active research topic since the early 1990s. The empirical criteria widely adopted in engineering practice come from a series of physical experiments, and the experimental methods are still the most important method in the design of inverted filters [2]. However, less attention was paid to the micro effective mechanisms of such structures, especially since empirical standards have been proven to

---

Q. Ma (✉) · X. Peng · C. Zhang
Yangtze River Survey, Planning, Design and Research Co., Ltd, Wuhan 430010, Hubei Province, China
e-mail: maqirui1101@foxmail.com

be overly conservative and, in some cases, may even contradict each other. Because all outdoor and laboratory experiments are based on the exploration of macroscopic phenomena, and soil erosion is a problem that occurs at the particle level (micro scale). Some scholars have used particle level fluid solid coupling methods such as CFD-DEM and LBM-DEM to study the performance of inverted filters [3, 4]. Nevertheless, the microscopic mechanism of particle migration and transportation in the inverted filter layer has not been well explained, as the entire process can be divided into two main parts: the occurrence of contact erosion between fine and coarse particles, and the subsequent development of particle erosion, migration and transportation in the inverted filter layer. This has not been taken seriously in previous research.

Unlike external erosion, internal erosion often occurs inside damaged embankments or buildings, making it more concealed and difficult to observe visually. For embankments, this slow and difficult to observe damage to the structural load-bearing capacity is crucial. It may remain stable under conventional loads, but it is highly likely to cause serious consequences when the water level rises rapidly or when very special loads are encountered. Therefore, revealing the principle of the inverted filter structure from a microscopic perspective can help us better understand the failure mechanism of embankments, and more effectively take preventive measures and optimize design.

Therefore, the purpose of this article is to conduct research on the effectiveness of the inverted filter structure at the micro level based on the three-dimensional LBM-DEM fluid–solid coupling simulation method.

2 Simulation Process

2.1 Method

The three-dimensional LBM-DEM coupling method based on two open-source software, LIGGGHTS responsible for the solid part in DEM [5], Palabos solve the fluid region in LBM [6]. As for the fluid/grain coupling part, the immersed moving boundary is selected to solve the interaction between solid and fluid [7]. The detail of the coupling method will be briefly introduced respectively.

In LBM, the governing formula can be written as,

$$f_i(\mathbf{x} + \mathbf{c}_i \delta_t, t + \delta_t) - f_i(\mathbf{x}, t) = -\frac{1}{\tau}\left[f_i(\mathbf{x}, t) - f_i^{eq}(\mathbf{x}, t)\right] \quad (1)$$

where the f_i is denotes the density distribution function whose coordinate is x directing in the i-th path at time t; $\mathbf{c}_i$ represents 19 discrete velocity vectors, because the D3Q19 LBM model is used in this study; and τ is a relaxation coefficient, which controls the stability of LBM simulation. The equilibrium distribution functions $f_i^{eq}(\mathbf{x}, t)$ in the right side of Eq. (1) is defined as,

$$f_i^{eq} = w_i \rho_{\text{fluid}} \left[1 + \frac{\mathbf{c}_i \cdot \mathbf{u}_{\text{fluid}}}{c_s^2} + \frac{(\mathbf{c}_i \cdot \mathbf{u}_{\text{fluid}})^2}{2c_s^4} - \frac{|\mathbf{u}_{\text{fluid}}|^2}{2c_s^2} \right] (i = 0, \ldots, 18) \quad (2)$$

where $w_i (i = 0, \ldots, 18)$ are the weighting factors, $w_0 = 1/3, w_{1-6} = 1/18, w_{7-18} = 1/36$; ρ_{fluid} and $\mathbf{u}_{\text{fluid}}$ are the density and velocity of fluid, respectively. The sound speed $c_s = 1/3$ in the lattice units in the 3D LBM model.

According to the transfer method, the fluid velocity, fluid density, and fluid pressure in the macroscopic scale can be calculated through LBM quantities as follows:

$$\rho_{\text{fluid}} = \sum_{i=0}^{18} f_i \quad (3)$$

$$\mathbf{u}_{\text{fluid}} = \frac{1}{\rho_{\text{fluid}}} \sum_{i=0}^{18} \mathbf{c}_i f_i \quad (4)$$

$$p_{\text{fluid}} = c_s^2 \rho_{\text{fluid}} \quad (5)$$

In DEM, the spherical particles' displacement is controlled by the Newton's second law of motion:

$$m_i \mathbf{a} = \mathbf{F}_c + \mathbf{F}_{\text{fluid}} \quad (6)$$

$$I_i \dot{\omega} = \mathbf{T}_c + \mathbf{T}_{\text{fluid}} \quad (7)$$

where I_i is the moment of inertia of particle, m_i is the mass of the particle; $\mathbf{a}$ is the motion acceleration; ω is the angular velocity; $\mathbf{F}_c$ are the contact forces between particles and $\mathbf{T}_c$ is corresponding torques provided by particle collision; $\mathbf{F}_{\text{fluid}}$ and $\mathbf{T}_{\text{fluid}}$ are the hydrodynamic force and the corresponding torque provided by the fluid part, respectively. The normal force $\mathbf{F}_n$ and tangential force $\mathbf{F}_t$ between two particles are calculated by the Hertz-Mindlin contact model.

Fluid–solid interaction calculations are achieved by adding a collision term Ω_i^s to Eq. (1), the govern equation becomes

$$f_i(\mathbf{x} + \mathbf{c}_i \delta_t, t + \delta_t) - f_i(\mathbf{x}, t) = -\frac{1}{\tau} [f_i(\mathbf{x}, t) - f_i^{eq}(\mathbf{x}, t)](1 - B) + B\Omega_i^s] \quad (8)$$

$$\Omega_i^s = f_{-i}(\mathbf{x}, t) - f_{-i}^{eq}(\rho_{\text{fluid}}, \mathbf{u}_{\text{fluid}}) + f_i^{eq}(\rho_{\text{fluid}}, \mathbf{u}_s) - f_i(\mathbf{x}, t) \quad (9)$$

where $B = \frac{\varepsilon(\tau-0.5)}{(1-\varepsilon)+(\tau-0.5)}$, which is a weighting function of relaxation coefficient τ and solid ratio ε of the LBM cell. $\varepsilon = V_{\text{solid}}/V_{\text{cell}}$, and thus, $\varepsilon = O(1)$ yields $B = O(1)$. The subscript $-i$ in the Eq. (9) represents the opposite direction of i.

Taking the effect of grain rotation into consideration, the velocity of the solid part $\mathbf{u}_s$ is calculated as,

$$\mathbf{u}_s = \mathbf{u}_p + \boldsymbol{\omega}_p \times \left[(\mathbf{x} + 0.5\mathbf{c}_i\delta_t) - \mathbf{x}_p\right] \tag{10}$$

where $\mathbf{u}_p$ and $\boldsymbol{\omega}_p$ are translational and angular velocities of solid grain, and $\mathbf{x}_p$ is the position of solid grain.

Lastly, the hydrodynamic force $\mathbf{F}_{\text{fluid}}$ and the corresponding hydraulic torque $\mathbf{T}_{\text{fluid}}$ applied by fluid motion can be calculated as:

$$\mathbf{F}_{\text{fluid}} = \sum_{j=1}^{n} B_j \sum_{i=0}^{18} \Omega_i^s \mathbf{c}_i \tag{11}$$

$$\mathbf{T}_{\text{fluid}} = \sum_{j=1}^{n} \left[B_j(\mathbf{x}_j - \mathbf{x}_p) \times \sum_{i=0}^{18} \Omega_i^s \mathbf{c}_i \right] \tag{12}$$

where n is the total number of lattice cells covered by the solid part; B_j refer to the weight coverage function in j-th lattice cell among n lattice cells.

The above three-dimensional LBM-DEM fluid structure coupling method has been validated through experiments such as particle descent [9], and multiple studies have been conducted based on this method, which can refer to the author's previous research [10, 11].

2.2 Simulation

In engineering practice, the particle size distribution of the inverted filter layer should be appropriately controlled to avoid potential risks of infiltration and erosion. Generally speaking, the grading should meet two conditions [4]:

(1) The pore size in the particle material of the inverted filter layer should be small enough, and under the action of the fluid, the larger particles in the protected layer particle material will not be washed away and will remain in the protected layer;
(2) The inverted filter material should have a high hydraulic conductivity to prevent the accumulation of large permeability and hydrostatic pressure in the inverted filter layer.

In previous studies, people often focused on the first condition, which is the classic design standard for inverted filters. Terzaghi and Peck [8] used experience to

Fig. 1 Simulated particle size distribution diagram of inverted filter layer

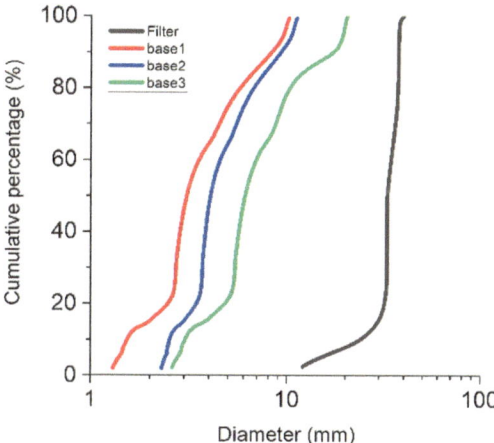

represent the representative ratio of typical particle size of inverted filter particles to typical particle size of soil particles in the protected layer of the foundation,

$$D_{15}/d_{85} < 4 \tag{13}$$

Among them, d_{85} is the particle diameter corresponding to the 85% mass fraction of fine particles within the protected fine particle category; And D_{15} is the particle diameter corresponding to particles with a mass fraction of 15% within the particle category in the inverted filter layer.

In fact, after preliminary research and analysis, all the gradation of the inverted filter structure meets condition 2. Therefore, in this article, condition 1 is still considered as a controllable condition quantity. The particle size distribution of this simulation is shown in Fig. 1.

The grading of the inverted filter layer is shown in black, taking into account three types of protective particle grading, corresponding to base1, base2, and base3 in the figure, with a representative particle size ratio of $R_1 = 4.34$; $R_2 = 4.01$; $R_3 = 3.13$, only base3 meets the grading design criteria for the inverted filter layer. The area size of protected particles is 0.10 m × 0.10 m × 0.02 m, where the x and y directions are equal, and the z direction is 0.02 m. To ensure comparability, the porosity of the protected layer particles with different particle size distributions is set to around 0.5. The space size of the entire simulation is 0.10 m × 0.10 m × 0.25 m, leaving ample space for the flow of particles in the vertical direction. To avoid boundary effects, periodic boundary conditions are set in both the x and y directions of the sample. The fluid pressure difference in the z-direction is achieved by applying an equivalent hydraulic gradient, considering the operating conditions of $i = 1.0, 2.0, 4.0, 5.0$, and 10.0. In this study, the number of particles in the protective layer of the substrate was 5000, and the number of particles in the filter layer was 150. However, periodic boundary conditions were used in the simulation, allowing for the exploration of the micro mechanism of the effectiveness of the inverted filter layer even for particle

Table 1 Parameters of the LBM-DEM simulations

	Parameters	Units	Values
DEM	Particle density, ρ	kg/m^3	2600
	Particle number, N_{total}		5000/150 (Protected particles/filter layer)
	Contact model		Hertz-Mindlin
	Young's module, Y	GPa	25
	Poisson ratio, v		0.3
	Maximum diameter, d_{max}	mm	30
	Minimum diameter, d_{min}	mm	3
	Particle size ratio		10
	Sliding friction, μ		0.5
	Rolling friction, μ_r		0.1
	DEM timestep, Δt_{DEM}	s	8.33×10^{-7}
LBM	Fluid density, ρ_{fluid}	kg/m^3	1000
	Kinematic viscosity, v_{fluid}	m^2/s	1.01×10^{-6}
	LBM timestep, Δt_{LBM}	s	8.33×10^{-6}

aggregates that are not particularly large in scale. The specific parameters used in this simulation are shown in Table 1.

3 Results and Discussions

Firstly, we provide a process diagram of contact erosion and migration transportation of protected particle gradations base1 and base3 in the inverted filter layer. The diagram shows the particle samples of two different protected substrates at $t = 1.0$ s and 5.0 s under $i = 2.0$ working conditions. As shown in Fig. 2a, b, when the gradation of the base material is base3 ($R_3 = 3.13$) corresponding to the gradation, the particles in the protected layer can be effectively blocked in the inverted filter layer until the simulated final $t = 5.0$ s. At this time, the internal system has basically stabilized, but only a few particles can pass through the inverted filter layer and undergo erosion damage, which means that the inverted filter layer has played a good protective role.

When the protected particle grading is base1 ($R_1 = 4.34$), as shown in Fig. 2c, d, particles have already broken through the inverted filter structure and eroded since $t = 1.0$ s. By the time the $t = 5.0$ s system had reached a stable state, a large number of fine particles had already eroded and accumulated at the top of the sample. The above results indicate that when the representative particle size ratio meets Eq. (13), the inverted filter layer can effectively protect the bottom particles washed by the fluid; On the contrary, when the representative particle size ratio does not meet Eq. (13), the ability of the inverted filter structure to prevent erosion of the base soil

Fig. 2 Evolution curve of erosion rate of bottom materials with different grades

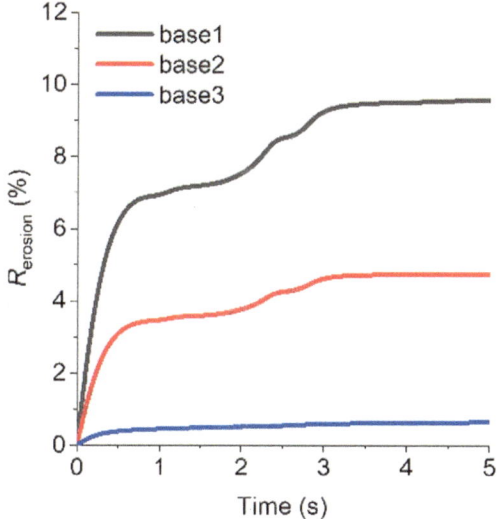

particles decreases. Therefore, the results of this study once again demonstrate the effectiveness of empirical formulas at the micro scale.

We consider the process of particles being washed out of the protective layer and reaching the top of the sample as erosion, and the erosion rate can be calculated by the following equation,

$$R_{erosion} = \frac{Mass_{eroded}}{Mass_{total}} \quad (14)$$

As shown in Fig. 3, the erosion rate distribution of three representative particle size ratios under the condition of $i = 2.0$ is shown. The erosion rate distribution in the figure corresponds to the condition shown in Fig. 2.

From the Fig. 3, it can be seen that the erosion rate shows significant different trends under different representative particle size ratios. When the base particle grading is base1, i.e. $R_1 = 4.34$, the erosion rate significantly increases and is higher than the other two working conditions; When the base particle grading is base3, i.e. $R_3 = 3.13$, the erosion rate is maintained below 1%, achieving good filtration effect. The results correspond to the schematic diagram of the erosion process of particles in the inverted filter layer in Fig. 2.

To consider the impact of different hydraulic gradients on erosion rate, Fig. 4 shows the variation curve of erosion rate under different hydraulic gradients when the base is base2 ($R_2 = 4.01$).

Because the representative size ratio $R_2 = 4.01$ is in the critical state specified by empirical criteria, we investigated the influence of different hydraulic gradients on the samples at this time, which can better highlight the role of hydraulic conditions. As shown in Fig. 4, with the increase of hydraulic gradient, the erosion rate also

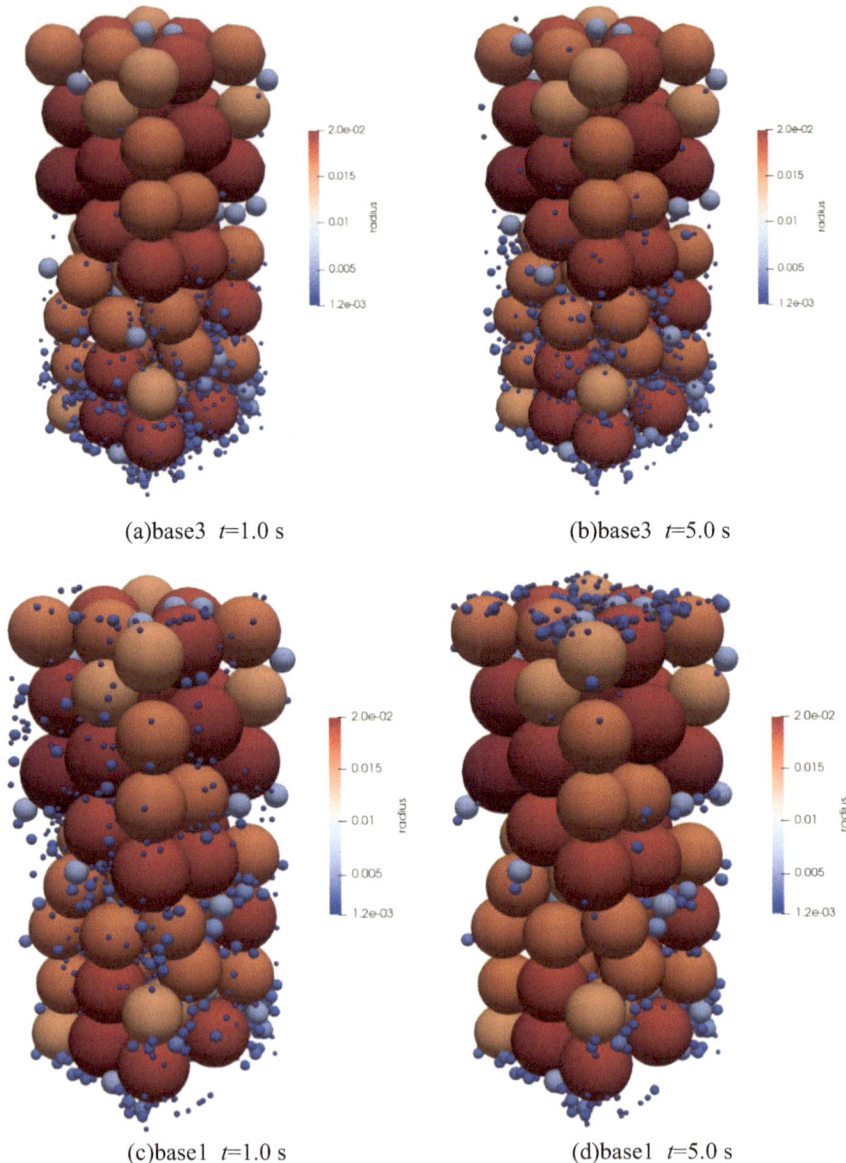

Fig. 3 Comparison of particle conditions in different grades of protected layers

Fig. 4 Evolution curve of erosion rate under different hydraulic gradients

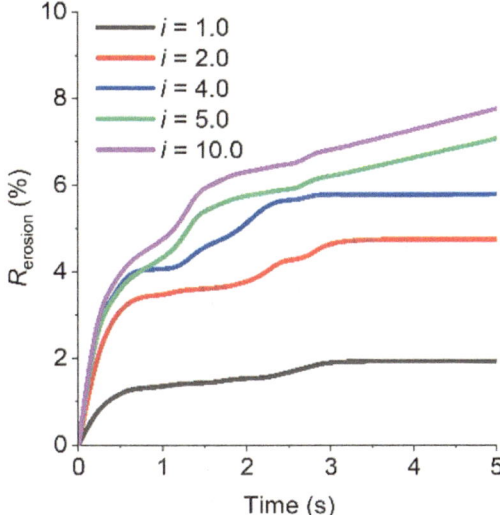

shows an increasing trend. When $i = 1.0$, the erosion rate is not high at this time, and the inverted filter layer can still play a certain protective role on particles. This is also consistent with Terzaghi's critical theory, as the critical water head for soil damage caused by piping flow has not been reached at this time. But as the hydraulic gradient increases, the erosion rate instantly increases significantly. When $i = 5.0$ and $i = 10.0$, the difference in erosion rates is not very significant, indicating that the promotion of hydraulic gradient on particle erosion is also limited. An increase in hydraulic gradient within a certain range can lead to a significant increase in erosion rate, and some research also reached a similar conclusion in the simulation of inverted filters [3].

4 Conclusions and Outlook

This paper mainly uses the LBM-DEM method to study the transport and migration of fine particles in the inverted filter structure, verifies the rationality and feasibility of the LBM-DEM fluid solid coupling method, and reveals the effective mechanism and failure mechanism of the inverted filter layer from a microscopic perspective. Erosion simulation was conducted on three different particle size ratios of the bottom material of the protected layer, and traditional empirical formulas were validated. The main conclusions can be summarized as follows:

(1) When the particle size ratio between the protected layer and the particle filter layer meets the empirical formula $D_{15}/d_{85} < 4$, the protective effect of the filter layer structure on particles is very obvious. However, when the particle size ratio does not meet the empirical formula, the protected particles will pass through

the filter layer and form erosion damage. The results of numerical simulation once again verify the effectiveness of the empirical formula, and also prove that the LBM-DEM method is feasible for the micro exploration of inverted filter erosion.

(2) As the representative particle size ratio R between the protected layer and the particle filter layer increases, the erosion rate of soil particles gradually increases. In addition, under higher hydraulic loads, more particles can penetrate the inverted filter structure, resulting in higher erosion rates. However, the influence of hydraulic gradient on erosion rate is also limited by the distribution of particle structure, which has certain limitations.

We have to admit that the work of this article is only the beginning of our subsequent research work, but it is also an important part of our research plan. In the near future, we will further utilize LBM-DEM coupling methods to simulate the entire process of erosion and its subsequent effects at a microscale, reveal its microscale effective mechanism. Multiscale methods will be used to establish the relationship between micro mechanisms and macro phenomena in soil filters and other structures, and related work is already underway. A similar work has been published on the study of soil internal erosion based on the multi-scale H-model method [12]. After establishing the connection between macro and micro scales, the effective mechanism of the inverted filter layer and the damage mechanism of the embankment soil will be better revealed, which will greatly benefit the construction and operation of embankment engineering and provide new ideas and inspiration for engineering design, flood control and emergency response.

Acknowledgements This work was financially supported by the National Key R&D Program of China (Grant No. 2021YFC3000102).

References

1. Craig RF (2004) Craig's soil mechanics. CRC Press
2. Zou YH, Chen Q, Chen XQ, Cui P (2013) Discrete numerical modeling of particle transport in granular filters. Comput Geotech 47:48–56
3. Huang QF, Zhan ML, Sheng JC, Luo YL, Su BY (2014) Investigation of fluid flow-induced particle migration in granular filters using a DEM-CFD method. J Hydrodyn Ser B 26(3):406–415
4. Wang M, Feng YT, Pande GN, Chan AHC, Zuo WX (2017) Numerical modelling of fluid-induced soil erosion in granular filters using a coupled bonded particle lattice Boltzmann method. Comput Geotech 82:134–143
5. Kloss C, Goniva C, Hager A, Amberger S, Pirker S (2012) Models, algorithms and validation for opensource DEM and CFD–DEM. Progress in Computat Fluid Dynam Int J 12(2–3):140–152
6. Latt J, Malaspinas O, Kontaxakis D, Parmigiani A, Lagrava D, Brogi F, Chopard B (2021) Palabos: parallel lattice Boltzmann solver. Comput Math Appl 81:334–350
7. Noble DR, Torczynski JR (1998) A lattice-Boltzmann method for partially saturated computational cells. Int J Mod Phys C 9(08):1189–1201

8. Terzaghi K, Peck RB (1948) Soil mechanics. In: Engineering practice. Wiley and Sons, Inc., New York
9. Seil P, Pirker S (2017) Lbdemcoupling: open-source power for fluid-particle systems. In: Proceedings of the 7th international conference on discrete element methods. Springer, Singapore, pp 679–686
10. Zhou W, Ma Q, Ma G, Cao X, Cheng Y (2020) Microscopic investigation of internal erosion in binary mixtures via the coupled LBM-DEM method. Powder Technol 376:31–41
11. Ma Q, Wautier A, Zhou W (2021) Microscopic mechanism of particle detachment in granular materials subjected to suffusion in anisotropic stress states. Acta Geotech 16:2575–2591
12. Ma Q, Wautier A, Nicot F (2022) Mesoscale investigation of fine grain contribution to contact stress in granular materials. J Eng Mech 148(3):04022005

Open Access This chapter is licensed under the terms of the Creative Commons Attribution 4.0 International License (http://creativecommons.org/licenses/by/4.0/), which permits use, sharing, adaptation, distribution and reproduction in any medium or format, as long as you give appropriate credit to the original author(s) and the source, provide a link to the Creative Commons license and indicate if changes were made.

The images or other third party material in this chapter are included in the chapter's Creative Commons license, unless indicated otherwise in a credit line to the material. If material is not included in the chapter's Creative Commons license and your intended use is not permitted by statutory regulation or exceeds the permitted use, you will need to obtain permission directly from the copyright holder.

Experimental Study on Strength of Luminous Concrete with Double Admixture of Fly Ash and Slag Powder

Meng Li, Guangxiu Fang, Haonan Wu, Chunming Wang, Huaiyu Li, and Zhoutong Li

Abstract Luminescent concrete is based on ordinary concrete, in which zinc sulfide luminescent material is added to make ordinary concrete with luminescent function of concrete, and its mechanical properties are greatly affected by the dosage of luminescent powder and mineral admixture. In order to study the mechanical properties and optical properties of luminescent concrete, luminescent concrete composite adding different dosages of fly ash and slag powder compressive test and flexural test, obtained different fly ash and slag powder dosage of luminescent concrete compressive strength and flexural strength with the curing time of the change curve and based on this proposed luminescent concrete compressive strength and flexural strength of the correction coefficient, for the subsequent light-emitting concrete. The research and engineering application of luminous concrete provides theoretical basis.

Keyword Luminous concrete · Zinc sulfide luminous substance · Compressive test · Flexural test

1 Introduction

In recent years, with the boom of urbanization around the world driving the rapid growth of the construction industry, it has brought great challenges to the low-carbon and environmentally friendly development of the construction industry [1]. In the carbon emission of the construction industry, the carbon pollution generated by the building materials themselves is as high as 38%, and the annual carbon emission of concrete, as one of the most common building materials, accounts for about 8% of the total global carbon emission. Therefore, the development of new energy-saving cementitious materials is an important way to reduce building energy consumption and operational carbon emissions in order to achieve the goal of energy saving and emission reduction, and is one of the hot spots in the current research of building

---

M. Li · G. Fang (✉) · H. Wu · C. Wang · H. Li · Z. Li
Civil Engineering, College of Engineering, Yanbian University, Yanji, Jilin, China
e-mail: gxfang@ybu.edu.cn

materials [2]. Luminescent concrete is a new type of building material formed by combining long afterglow materials and concrete, which possesses the characteristics of energy saving, environmental protection, decoration, and illumination [3], and promoting its further application will inevitably play a role in promoting the low-carbon and environmental protection development of the building energy-saving industry.

At present, the research on luminous concrete is mainly on the luminous concrete production process and luminous material mixing on the luminous concrete compression, folding and other mechanical properties of the research [4–8], in the mineral admixture mixing on the luminous concrete compression, folding performance of the lack of systematic research. In the actual engineering application of luminous concrete, it is often necessary to add fly ash, slag powder and other mineral admixtures in luminous concrete to reduce the project cost and improve the environmental benefits of the project. Therefore, it is very necessary to study the effect of fly ash and slag powder on the compressive and flexural properties of luminous concrete, and to find out the optimal mixing amount in luminous concrete.

To this end, this paper, through the compound mixing of fly ash and slag powder, take the control test method, carried out the compressive test and flexural test research of luminous concrete and ordinary concrete with different dosages of fly ash and slag powder (the dosages of fly ash and slag powder are 0%, 0%, 15%, 25%, 20%, 25%, 25%, 25%, 20%, 20%, 20%, 20%, 30%) [9, 10], and established the compressive and flexural test research of the respective The change curves of compressive and flexural strength with curing time under each dosage were established, the optimum dosage of fly ash and slag powder in luminous concrete was obtained, and the correction coefficients of the strength formula of the current specification of luminous concrete were proposed through relevant theoretical analysis.

According to the optimal mixing amount of fly ash and slag powder to replace the raw materials for the production of luminescent concrete, can reduce the amount of cement, reduce the cost, which is also important for reducing carbon emissions in the construction industry. However, whether the addition of fly ash and slag powder will have a greater impact on the luminous effect of luminous concrete requires further experimental research, which makes its application in the luminous concrete project subject to certain limitations.

2 Test Raw Materials and Programs

2.1 Experimental Setup

Cement: Huaxue brand P-O 42.5 grade white cement provided by Northern Cement Company.

Fly ash: first-class scientific research fly ash.

Slag powder: S95 grade slag powder provided by Dignity Industrial Products Specialized Shop.

Luminous Powder: Zinc Sulfide Luminous Powder provided by Yaohua Pigment Dye Chemical.

Fine Aggregate: Yanji river sand with fineness modulus of 2.8, which is medium sand.

Coarse aggregate: continuous grading gravel, particle size 5–20 mm.

Water reducing agent: standard polycarboxylic acid water reducing agent (PCA) mother liquor provided by Weike Building Material Sales Center, with a water reducing rate of 45%.

Water: Yanji city tap water.

2.2 Specimen Program

Firstly, the water-cement ratio and the total amount of gel material were fixed; secondly, double mixing of fly ash and slag powder to replace cement by equal mass; finally, compressive and flexural tests were carried out on luminous concrete specimens. The effects on 3d, 14d, 28d compressive strength and 28d flexural strength of luminescent concrete were investigated respectively: ① when fly ash admixture (20%) was constant and slag powder admixture was different (20%, 25%, 30%) [10]; ② when slag powder admixture (25%) was constant and fly ash admixture was different (15%, 20%, 25%) [9]; the effects on 3d, 14d, 28d compressive strength and 28d flexural strength of luminescent concrete were investigated respectively. The mixes used for each group of matrix of the test are shown in Table 1.

In Table 1, the water-cement ratio of the concrete was 0.46, the amount of cementitious material was 456.52 kg/m^3, and the water reducing agent dosage was 0.1%. group A is the base group without mineral admixture, B, C and D are the control group with 25% constant dosage of slag powder; and C, E and F are the control group with 20% constant dosage of fly ash.

The specific instructions are as follows:

B, C, D: luminous concrete with 25% slag powder admixture and 15%, 20% and 25% fly ash admixture respectively.

C, E, F: Luminous concrete with 20% fly ash admixture and 20%, 25% and 30% slag powder admixture respectively.

2.3 Specimen Fabrication

Luminescent concrete specimens were produced using a layered casting method. The thickness of luminescent concrete luminescent layer was taken as 2 cm for fabrication [11], in order to reduce the color difference between luminescent concrete matrix and luminescent layer, the fabrication material of luminescent layer was exactly the

Table 1 Illuminated concrete matrix ratios

Serial number	Water-to-cement ratio	Water reducing agent /%	Coal ash /%	Slag powder /%	Concrete mixing ratio/(kg/m^3)						
					Clinker	Coal ash	Slag powder	Water	Granule	Stone
A	0.46	0.1	0	0	456.52	0	0	210	658.72	1074.76
B	0.46	0.1	15	25	273.91	68.48	114.13	210	658.72	1074.76
C	0.46	0.1	20	25	251.09	91.30	114.13	210	658.72	1074.76
D	0.46	0.1	25	25	228.26	114.13	114.13	210	658.72	1074.76
E	0.46	0.1	20	20	273.92	91.30	91.30	210	658.72	1074.76
F	0.46	0.1	20	30	228.26	91.30	136.96	210	658.72	1074.76

same as that of the matrix except for the luminescent powder which was mixed with 14% [11] of the gel material to replace some of the coarse aggregates [12]. In the test, six groups of specimens were made, each containing nine 150 × 150 × 150 mm and three 150 × 150 × 600 mm specimens, of which one group of specimens with dimensions of 150 × 150 × 150 mm was used to determine the compressive strength of 3d, 14d and 28d in units of three, respectively, and the dimensions were 150 × 150 × 600 mm specimens for the determination of 28d flexural strength. The specimen production and strength determination is based on GB/T 50,081–2019 "Standard Test Methods for Physical and Mechanical Properties of Concrete". The specific method of layered casting is as follows: firstly, the well-mixed base concrete is loaded into the standard test mold of 150 × 150 × 150 mm, with the filling thickness of 130 mm, and the test mold with concrete is vibrated with the vibrating table for 1 min and then left to stand for 2 h, and after the concrete is gradually beginning to solidify, then the concrete of the light-emitting layer is loaded into the test mold and then put into the curing room for curing after vibrating for 20 s.

3 Results and Discussion

The effect of double mixing of fly ash and slag powder on the compressive strength and flexural strength of luminous concrete is shown in Tables 2 and 3, respectively. From Tables 2 and 3, it can be found that the 3d, 14d and 28d compressive strengths as well as the 28d flexural strengths of almost all luminescent concretes double-mixed with fly ash and slag powder increased to different degrees compared with those of the baseline group.

Table 2 Compressive strength of luminous concrete

Serial number	Coal ash /%	Slag powder /%	Compressive strength/MPa		
			3d	14d	28
A	0	0	22.68	30.95	33.57
B	15	25	21.78	29.91	35.93
C	20	25	25.68	34.08	36.28
D	25	25	23.70	32.26	35.60
E	20	20	25.76	33.31	36.25
F	20	30	25.51	32.88	35.89

Table 3 Flexural strength of luminous concrete

Serial number	Coal ash /%	Slag powder /%	Flexural strength/MPa 28d
A	0	0	3.52
B	15	25	4.06
C	20	25	4.61
D	25	25	3.93
E	20	20	4.57
F	20	30	4.19

3.1 Luminous Concrete Compression Test

The effect of double mixing of fly ash slag powder on the strength of luminous concrete is shown in Fig. 1;When the dosage of slag powder (25%) is unchanged, the effects of different fly ash dosages on the 3d, 14d and 28d compressive strength of luminous concrete are shown in Fig. 2a; when the dosage of fly ash (20%) is unchanged, the effects of different slag powder dosages on the 3d, 14d and 28d compressive strength of luminous concrete are shown in Fig. 2b.

As can be seen from Fig. 1: except for group B (15% and 25% of fly ash and slag powder mixing, respectively), the compressive strengths of 3, 14 and 28d of double-mixed fly ash and slag powder luminous concrete are higher than those of the reference concrete, and the degree of increase is 3d, 14d and 28d in descending order. In addition, comparing the compressive strengths of luminous concrete with fly ash and slag powder mixing as variables, respectively, it can be seen that luminous concrete has a better improvement effect on strength than fly ash. It can be seen that slag powder has better improvement effect on concrete strength than fly ash. From Fig. 2a, it can be seen that: when the slag powder mixing amount (25%) is unchanged, with the increase of fly ash mixing amount, the compressive strength of luminescent concrete 3, 14d show the trend of increasing and then decreasing, and

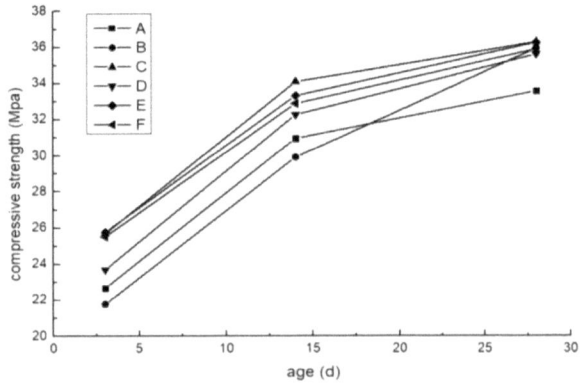

Fig. 1 Effect of double mixing of fly ash and slag powder on the strength of luminous concrete

Fig. 2 Effect of variation of individual mineral admixture on compressive strength

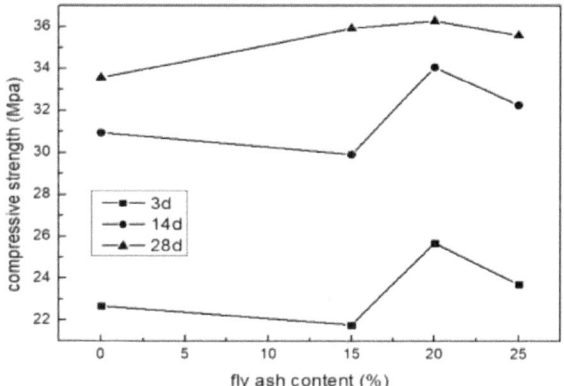

(a) Different Fly Ash Incorporation Amounts

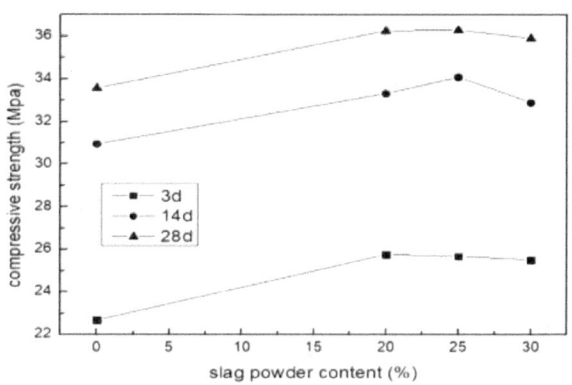

(b) Different slag powder mixing amount

it reaches the maximum value when the fly ash mixing amount is 20%, and is higher than the benchmark group when the fly ash mixing amount is 20% and 25%. From Fig. 2b, it can be seen that: when the fly ash dosage (20%) is unchanged, with the increase of slag powder dosage, the compressive strength of luminescent concrete 3d basically remains unchanged, while the compressive strength of 14d shows a tendency to increase and then decrease, and reaches the maximum value when the dosage is 25%, and its compressive strength in 3d and 14d is higher than that of the benchmark group.

This is because: in the early stage, mainly clinker hydration generated by Ca(OH)2 as the alkaline exciter of slag micropowder, and with the interaction of the active components in the slag micropowder, to generate hydrated calcium silicate, hydrated calcium thioaluminate or hydrated calcium sulfur ferrate, while most of the particles of the slag micropowder in the early stage of hardening like the core to participate in the process of structure formation, Calcium alumina that is, in the slag micropowder around the surface of the growth around the surface. In the late stage, mainly the Ca(OH)2 precipitated by the hydration of cement clinker, through the liquid phase diffusion to the surface of the spherical vitreous body of fly ash, chemical adsorption and erosion, and generate hydrated calcium silicate and hydrated calcium aluminate. Most of the hydration products began to appear in gel form, and gradually transformed into fibrous crystals with increasing number and crossing each other to form interlocking structures with the growth of age [13]. And in the fly ash admixture of 20%, the admixture of slag powder of 25% can not only make the hydration products produced by the early cement clinker minerals can be exactly allotted to the surface of the slag microfine powder, it can increase the contact opportunity of the hydration products and the original particles, so as to obtain better early strength, and it can make the cement clinker precipitation of Ca(OH)2 and the surface of the spherical vitreous body of the fly ash to react fully, let the better formation of interlocking structure inside the concrete, to get the best late strength.

3.2 Luminous Concrete Flexural Test

The effect of different fly ash admixture on 28d flexural strength of luminescent concrete when slag powder admixture (25%) is constant is shown in Fig. 3a; the effect of different slag powder admixture on 28d flexural strength of luminescent concrete when fly ash admixture (20%) is constant is shown in Fig. 3b.

From Fig. 3, it can be seen that the 28d flexural strength of luminous concrete with double mixing of fly ash and slag powder are higher than that of the base group concrete. From Fig. 3a, it can be seen that: when the slag powder mixing amount (25%) is unchanged, with the increase of fly ash mixing amount, the 28d flexural strength of luminous concrete is gradually reduced, the magnitude of the reduction is gradually increased, and its strength is lower than that of the benchmark group. From Fig. 3b, it can be seen that: when fly ash mixing (20%) is unchanged, with the increase of slag powder mixing, the 28d flexural strength of light-emitting concrete is gradually reduced, the magnitude of the reduction gradually becomes larger but smaller than that of the fly ash change group, and its strength is also lower than that of the benchmark group.

It can be seen that the flexural strength of luminescent concrete in the early and late period has the same change rule as its compressive strength in the corresponding period. This shows that in the case of the same compressive strength, the mixing of a certain proportion of slag powder and fly ash (fly ash: 20%, slag powder: 25%) can complement each other. Slag powder can well make up for the problem of low strength of luminous concrete in early stage caused by the dosage of 20% of fly ash, and at the same time, it can effectively improve the strength of luminous concrete in late stage when the dosage of 20% of fly ash is used.

Comparing the flexural strength of luminous concrete with fly ash mixing as the variable and slag powder mixing as the variable, it can be seen that the effect of fly ash on the flexural strength of luminous concrete is greater than that of slag powder,

Fig. 3 Impact on flexural strength when varying the amount of single-mixed fly ash and slag powder

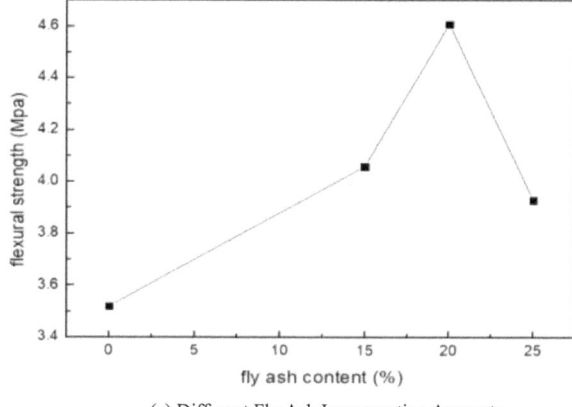

(a) Different Fly Ash Incorporation Amounts

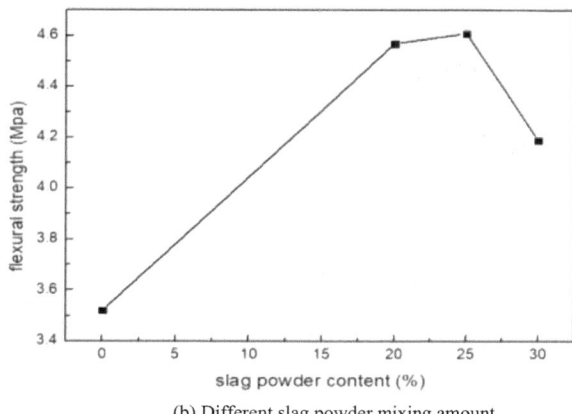

(b) Different slag powder mixing amount

and after the fly ash and slag powder reach their optimal mixing amounts (fly ash: 20%, slag powder: 25%), the effect of increasing or decreasing the mixing amount of fly ash by 5% on the reduction of the flexural strength of luminous concrete is three times as much as that of increasing or decreasing the mixing amount of slag powder by 5% on the reduction of the flexural strength of luminous concrete.

This is because: fly ash is a powdery substance with a finer particle size than slag powder. Therefore, fly ash has a larger surface area than slag powder, which can react more fully with the hydration in cement and produce more cementitious substances.

3.3 Optical Performance Testing

Test apparatus: light source for the Jiangxi U.S. Guiya Lighting Co., Ltd. produced removable lamps, model MTD4.5-M/K-07, rated voltage of 5 V, rated power of 4.5 W (16 × 0.2 W / LED module), homemade dimensions for the 200 × 200 × 450 mm black wooden box, photometric instruments for the Hangzhou Delixi Group Limited production of split-type Ltd., model DLY-1801C, the light-emitting concrete specimen is a standard specimen of 150 × 150 × 150mm with 20 and 25% of fly ash and slag powder respectively, which has the best strength performance in the mechanical test. The specific optical properties of luminous concrete were tested as follows:

Luminescent concrete to the darkness, the illumination will be consumed (with a split-type illuminance meter measured luminous layer illuminance and the substrate illuminance is the same), and then the test block into the black box, and irradiated with a light source for 1 h [11], remove the light source, with a split-type illuminance meter to measure its luminous time length of 0 h, 2 h, 4 h, 6 h when the illuminance. Test will be the light source irradiation position and split-type illuminance meter measurement position is set to the same point, placed in the light-emitting concrete surface layer of 300 mm, at this time can be more accurately measured illuminance of light-emitting concrete [14]. The change curve of the illuminance of the test block with the luminous time is shown in Fig. 4.

As can be seen from Fig. 4: the initial maximum illuminance of the luminous concrete can reach 105 lx and its luminous 6 h after the illuminance is still retained 0.6 lx, higher than 0.003 lx can be observed by the human eye, i.e., the luminous time of the luminous concrete is longer than 6 h, which indicates that the luminous concrete mixed with fly ash and slag powder can achieve the luminous effect of the unadulterated mineral dopant [9], and it can be used for practical production use.

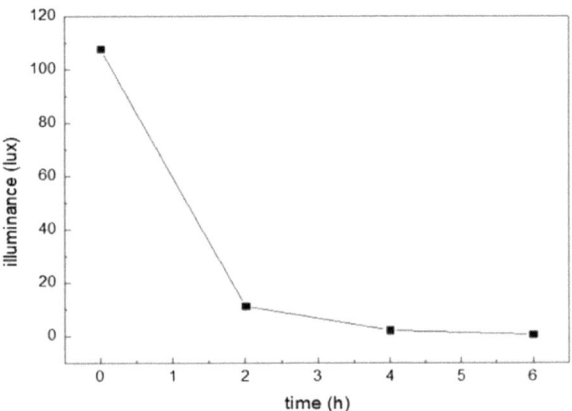

Fig. 4 Effect of luminescence time on illuminance

4 Propose a Correction Factor for the Strength Formula of the Current Relevant Code

According to the luminescent concrete compressive test data can be derived from the benchmark group and the best strength group (fly ash: 20%, slag powder: 25%) compressive strength with age change curve, that is, Fig. 1 in the A, C two curves.

The luminous concrete strength correction factor β is introduced, so that the standard value of compressive strength of luminous concrete cube is the product of β and the standard value of compressive strength of benchmark concrete cube.

In Fig. 1, the ratio of the compressive strength value of the benchmark group to the compressive strength value of the best strength group is the correction factor for its corresponding age. Therefore, the relationship between the correction coefficient of luminous concrete strength and the age of concrete curing can be drawn as in Fig. 5.

As can be seen from Fig. 5, the final correction factor β = 1.08 for luminous concrete is substituted into Eq. when the age of concrete curing reaches 28d:

$$f'_{cu,k} = \beta f_{cu,k} \tag{1}$$

Gotta:

$$f'_{cu,k} = 1.08 f_{cu,k} \tag{2}$$

The current code compressive strength formula is:

$$f_{cu,0} = f_{cu,k} - t\sigma \tag{3}$$

Substituting (2) into (3) yields the adjusted luminous concrete strength formula:

$$f_{cu,0} = f_{cu,k} - t\sigma = f'_{cu,k}/1.08 - t\sigma \tag{4}$$

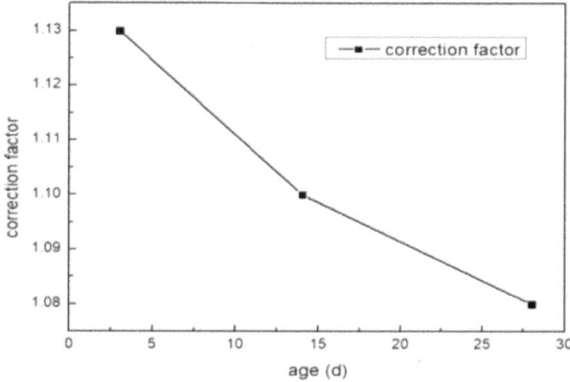

Fig. 5 Correction factor for compressive strength of luminous concrete versus age

5 Conclusion and Analysis

(1) The optimum mixing amount of double-mixed fly ash and slag powder for light-emitting concrete is as follows: fly ash: 20%, slag powder: 25%.
(2) At the same dosage of slag powder mixing (25%), 3d, 14d and 28d compressive strengths of luminescent concrete reached their maximum values when the dosage of fly ash was 20%. The 3d compressive strength of luminescent concrete did not change with the change of slag powder dosage when the dosage of fly ash (20%) was kept constant, while the 14d and 28d strengths reached the maximum value when the dosage of slag powder was 25%.
(3) The flexural strength of luminous concrete was maximum at 20% constant fly ash dosage and 25% slag powder dosage; the flexural strength of luminous concrete was maximum at 25% constant slag powder dosage and 20% fly ash dosage.
(4) In this paper, on the basis of the strength formula of the current specification of concrete, by analyzing the compressive test values of luminous concrete, the strength modification coefficient of luminous concrete is proposed, and the formula for calculating the strength of luminous concrete preparation is proposed as:

$$f_{cu,0} = f_{cu,k} - t\sigma = f'_{cu,k}/1.08 - t\sigma \tag{5}$$

References

1. Xianlong YANG, Fengyang ZHU, Xun XU et al (2022) Study on the luminescence performance of alkali magnesium sulfate self-luminous concrete. J Wuhan Univer Technol 44(12):14–21
2. Yingli GAO, Liangchen QU, Bei HE et al (2019) Study on the properties and mechanism of action of superhydrophobic-self-luminous cementitious composites. Silicate Bulletin 38(1):70–76
3. Pengfei H, Dawei Y (2015) Research progress and prospect of luminescent concrete. Jiangxi Building Mater 17:2–3
4. Andrea M, Hossein E, Fatemeh J et al. (2023) Reliability estimation of the compressive concrete strength based on non-destructive tests. J Sustain 15(19)
5. Kazemi R, Naser MZ (2023) Towards sustainable use of foundry by-products: Evaluating the compressive strength of green concrete containing waste foundry sand using hybrid biogeography-based optimization with artificial neural networks. J Build Eng 76
6. Varma B, Prasad EV, Singha S et al (2023) Study on predicting compressive strength of concrete using supervised machine learning techniques. J Asian J Civil Eng 24(7):2549–2560
7. Kashem A, Das P (2023) Compressive strength prediction of high-strength concrete using hybrid machine learning approaches by incorporating SHAP analysis. J Asian J Civil Eng 24(8):3243–3263
8. Albina N, Askhat B (2022) Research on machine learning algorithm for predicting concrete compressive strength. J Progress in Civil Eng 4(10)
9. Xiaobo L, Yu S, Qiming G et al. (2021) Experimental study on the effect of fly ash admixture on mechanical properties of concretp. Concrete 08:88–90+5

10. Hongtu S, Qiang Z, Zhiyuan C et al. (2023) Research on the effect of fly ash and slag powder on concrete properties. Concrete World 07:53–56
11. Qian W (2012) Research on the preparation and performance of luminescent concrete. Shenyang University of Architecture, Liaoning
12. Jialong YANG (2018) Preparation of long afterglow luminescent powder and its application in luminescent concrete. Qingdao University of Science and Technology, Shandong
13. Qianyun W, Qinyong MA, Ying W (2021) Tensile strength test and fine structure of basalt fiber-slag powder-fly ash concrete under freeze-thaw cycle. J Composite Mater 38(03)
14. Peijun C, Xiaodong Z, Chunao W et al. (2020) Application and exploration of the inverse square law of illuminance in LED light sources. Laboratory Res Explorat 39(05):63–66

Open Access This chapter is licensed under the terms of the Creative Commons Attribution 4.0 International License (http://creativecommons.org/licenses/by/4.0/), which permits use, sharing, adaptation, distribution and reproduction in any medium or format, as long as you give appropriate credit to the original author(s) and the source, provide a link to the Creative Commons license and indicate if changes were made.

The images or other third party material in this chapter are included in the chapter's Creative Commons license, unless indicated otherwise in a credit line to the material. If material is not included in the chapter's Creative Commons license and your intended use is not permitted by statutory regulation or exceeds the permitted use, you will need to obtain permission directly from the copyright holder.

Research on Design Method and Optimization of New Epoxy Resin Concrete Mix Ratio

Haoran Xu, Guangxiu Fang, and Baiyang Xue

Abstract In order to obtain epoxy concrete that can adapt to different strengths and different environmental conditions, this paper innovatively constructed a new type of accurate epoxy concrete mix calculation logic and calculation method through literature investigation, analysis, and test, which can make the process of calculating epoxy concrete mix ratio more concise and precise. In this paper, a new type of epoxy concrete mix with a strength class of C30 is designed, which can provide a reference for engineers and technicians to design accurate epoxy concrete mix better. It also provides some reference value for improving the mix ratio design method and engineering application of epoxy resin concrete.

Keywords Epoxy concrete · Proportioning design · Precision · Calculation method construction

1 Introduction

The research of new epoxy concrete started earlier in foreign countries. It was first used as a commercial application in the United States in the 1950s and began to be applied to the reinforcement and repair of structural members in the mid-1970s [1]. At present, in the engineering application of epoxy concrete in China, Yanming [2] believes that epoxy concrete can be applied in the repair of cement pavement; Qingfen [3] explained the application of epoxy resin concrete in building structures through examples; Xinyin et al. [4] applied epoxy resin concrete to the connection at the fabricated truss joints to strengthen the mechanical properties of the fabricated truss joints. Domestic epoxy concrete is gradually transforming from laboratory to engineering.

H. Xu · G. Fang (✉) · B. Xue
Department of Civil Engineering and Hydraulic Engineering, Yanbian University, Yanji, Jilin 133002, China
e-mail: gxfang@ybu.edu.cn

Through 2023, the design of epoxy resin mix ratios relies heavily on empirical formulas. Although this design method can provide a certain degree of accuracy, it often fails to provide a stable mix ratio due to the differences in the epoxy resin models selected. When preparing epoxy resin concrete in practice, if the materials used differ from those in the empirical formula, continuous experiments are needed to find the optimal mixture ratio that can avoid paste bleeding while maintaining the fluidity of the epoxy resin concrete mortar. This trial-and-error process not only prolongs the construction schedule but also generates additional costs, resulting in the waste of manpower, material resources, and financial resources.

Therefore, this paper constructs a novel method for designing epoxy resin concrete mix ratios to improve the accuracy and stability of the mix ratios.

In this paper, a novel method of epoxy resin concrete proportion design is constructed. This method is based on the nature of epoxy resin and the strength requirements of concrete, and through a series of calculations, a new type of epoxy resin concrete proportion with a strength class of C30 is designed. This ratio design method not only broadens the de-sign ideas of epoxy resin concrete ratio but also provides a possibility for the large-scale application of epoxy resin concrete.

2 Construction of Computational Logic Theory and Computational Method

2.1 Construction of Computing Logic

In the study of epoxy concrete mix, it is usually necessary to consider the following aspects:

(1) Strength requirements: According to the specific engineering needs, determine the required compressive strength, tensile strength, and other performance indicators;
(2) Adhesion of epoxy resin and aggregate: to ensure a good bond between epoxy resin and aggregate, you can consider using an appropriate amount of powder filler to increase the interface bonding force;
(3) Fluidity: Adjust the fluidity of concrete to adapt to the specific construction method, which can be adjusted by controlling the ratio of epoxy resin and aggregate and the use of powder fillers;
(4) Construction performance: To ensure that t-he concrete has good plasticity and self-levelling during the construction process, you can consider adding an appropriate amount of polymer water reducer or thickener.

In the specific mix ratio research process, it is necessary to conduct laboratory tests and engineering field tests, test the performance of different combinations of concrete by adjusting the ratio, and select the optimal mix scheme. In addition, factors such as environmental factors and cost control need to be taken into account [5].

Considering the above four preconditions, the calculation logic of constructing the mix ratio of epoxy resin concrete is carried out.

(1) The density of the mixture of epoxy resin and admixture (including curing agent and diluent) should be measured.
(2) The packing density and apparent density of aggregates are determined by experimental method.
(3) The bulk density and apparent density of aggregate are used to calculate its porosity. On this basis, the weight can be calculated when the pores in the aggregate are filled with an epoxy resin mixture.
(4) The weight of the obtained epoxy resin mixture can only fill the pores of the aggregate, and cannot completely wrap the aggregate particles. In order to ensure the construction performance of epoxy concrete, it must be made to have good fluidity. Therefore, it is necessary to calculate the weight of the epoxy resin mixture when the aggregate particles are wrapped according to the specific surface area of the aggregate and the coating thickness of the epoxy resin mixture on the surface of the aggregate particles and to correct the weight of the initial obtained epoxy resin mixture.
(5) Sum the weight of the corrected epoxy resin mixture with the weight of the epoxy resin mixture when filling the pores to obtain the theoretical weight of the epoxy resin mixture.

2.2 Construction of Calculation Methods

Based on the established calculation logic, the calculation method of the mix ratio of new epoxy resin concrete can be clearly constructed.

The first step is to determine the number of epoxy resin, curing agent, and diluent dosage form, the type of coarse aggregate and fine aggregate, the density of mixed epoxy resin and admixtures (curing agent and diluent), the bulk density and apparent density of coarse aggregate and fine aggregate are measured through experiments. After the above data are measured, a new design method for the mix ratio of epoxy resin concrete is constructed.

The second step is to calculate the porosity per unit volume of aggregate by using the accumulated density and apparent density of aggregate obtained through the experiment:

$$n = \frac{\rho_1 - \rho_2}{\rho_1} \quad (1)$$

In formula (1), n is the porosity of aggregate per unit volume, ρ_1 is the packing density of fine aggregate (unit: kg/m^3), and ρ_2 is the apparent density of fine aggregate (unit: kg/m^3).

The weight of epoxy resin filling aggregate pores is calculated by using the porosity of aggregate:

$$m_{e1} = n \times \rho_e \qquad (2)$$

In formula (2), m_{e1} is the weight of epoxy resin filling the pore of aggregate per unit volume (unit: kg), ρ_e is the density of epoxy resin and admixture (unit: kg/m^3).

After determining the weight of the epoxy resin mixture required to fill the aggregate pores, the weight used to wrap the aggregate with the epoxy resin is further calculated and the weight of the epoxy resin mixture is corrected accordingly.

The coating weight correction formula of aggregate is calculated according to the thickness of the theoretical epoxy resin coating, the specific surface area, and the weight of sand per unit volume. The formula is as follows:

$$m_e = S_0 \times m \times d_0 \times \rho_e \qquad (3)$$

In formula (3), m_e is the weight of aggregate particles coated by epoxy resin (unit: kg), S_0 is the specific surface area of aggregate (unit: m^2/kg), m is the weight of aggregate under unit volume (unit: kg), and d_0 represents the coating thickness of epoxy resin mixture on the surface of sand particles (unit: m). ρ_e is the density of epoxy resin and admixture (unit: kg/m^3).

The weight of the epoxy resin mixture can be modified by formula (3), and the theoretical weight of the epoxy resin mixture can be obtained by summing it with the weight of the epoxy resin mixture filled with pores.

Through the calculation of the process, we can get the theoretical weight of the epoxy resin mixture and aggregate weight ratio. Next, the theoretical epoxy weight and the aggregate weight need to be mixed in a calculated ratio to prepare the epoxy concrete. Next, the fluidity of the concrete needs to be tested. If the fluidity reaches the normative threshold, then we can proceed to the next step, which is to correct the volume coefficient to obtain the final mix ratio of epoxy concrete.

If the fluidity of the concrete does not reach the normative threshold, the amount of the epoxy mixture should be increased. The amount of each increase should be 5% of the pulp to set ratio. Then, the configuration and fluidity of the epoxy concrete were tested again. This process needs to be repeated until the fluidity of the concrete meets the normative threshold.

3 Application of Computational Logic Theory and Computational Methods

3.1 Materials Used and Their Parameters

In the mix ratio design, this paper uses E44 epoxy resin, whose parameters are shown in Table 1.

Table 1 Material parameters of E44 epoxy resin Materials

Materials name	Production manufacturers	Amount of epoxy (g/mol)	Density (g/cm^3)
Phoenix brand E44 epoxy resin	Jiangsu Nantong Star synthetic Material Co., LTD	210–230	1.5–2.0

Table 2 Parameters of admixture materials

Material name	Manufacturer	Density
T31 Curing agent	Wuxi Pinhua Chemical Co., LTD	1.2 g/cm^3
Ethanol diluent	Suzhou Haibai Chemical Co., LTD	0.75 g/cm^3

Table 3 Aggregate parameters

Materials	Production manufacturers	Factory pile density (g/cm^3)	Apparent density (g/cm^3)
Ceramisite	Ceramisite Yichang bright ceramisite Products Co., Ltd	753.1	1356.2
Experimental sand	Unknown	1.653	2.4067

The parameters of the curing agent and diluent of epoxy resin concrete used were measured under the conditions of standard humidity, standard air pressure, and experimental temperature of 20 °C. See Table 2.

Coarse aggregate is selected with 900 grade light sintered ceramics, fine aggregate is selected with laboratory sand. Aggregate parameters are measured through experiments, as shown in Table 3.

The weight ratio of E44 epoxy resin, T31 curing agent, and ethanol used in the preparation of epoxy resin mixture [6] is 1:0.25:0.25. Through experiments, the density of the epoxy resin mixture obtained by using the preparation ratio of the above epoxy resin mixture is 1250 kg/m^3

3.2 Determination of Theoretical Cement-Sand Ratio

Sand packing density $= 1653$ kg/m^3, sand apparent density $= 2406.7$ kg/m^3, formula (1) can be used to calculate sand porosity per unit volume:

$$n_1 = \frac{\rho_1 - \rho_2}{\rho_1} = 0.456$$

If the porosity of sand needs to be completely filled, the volume of epoxy resin required per cubic meter of epoxy resin mortar is $v_{e1} = 0.456$, and the measured density of epoxy resin mixture is $\rho_e = 1250$ kg/m^3. Formula (2) can be used to calculate the weight of the epoxy resin mixture required to fill the pores of a unit volume of sand

$$m_{e1} = n_1 \times \rho_e = 569.4 \text{ kg}$$

Then there is a theoretical mortar ratio:

$$m_{e1} : m_{s1} = 569.4 : 1653 = 0.344$$

At the current ratio of cement to sand, the amount of epoxy concrete can only fill most of the gaps between the sand particles, but it cannot make the sand particles fully wrapped to the required thickness, resulting in insufficient fluidity of the epoxy mortar. Therefore, it is necessary to adapt the cement mortar to determine the best cemented-sand ratio.

In the design method of the epoxy concrete mix ratio mentioned in this paper, it is written that in order to ensure the fluidity of epoxy concrete mortar, the coating thickness of epoxy resin per sand grain should reach 10 microns. Therefore, in order to ensure the fluidity of concrete, the existing theoretical cement-sand ratio is modified.

First of all, the specific surface area of fine aggregate sand is measured as follows: $S_0 = 2.826$ m^2/kg. In order to obtain the weight of the epoxy resin coating layer on the surface of the sand, the relationship between formula (3), volume, and density is used to correct the weight of the epoxy resin mixture. Then the volume of the epoxy resin mixture is modified:

$$V_{e1} = S_0 \times m_{s_1} \times d_0 = 0.04671 \text{ m}^3$$

The quality of modified epoxy resin is:

$$m_{e1}{'} = V_{e1} \times \rho_e = 58.39 \text{ kg}$$

Then there is the corrected cement-sand ratio:

$$(m_{e1} + m_{e1}{'}) : m_{s1} = 597.79 : 1653 = 0.361$$

The coarse aggregate is made of a new type of light aggregate ceramite, the bulk density of which is 753.1 kg/m^3, and the apparent density of the ceramite is 1356.2 kg/m^3

Formula (1) can be used to calculate the porosity of ceramics:

$$n_2 = \frac{\rho_1 - \rho_2}{\rho_1} = 0.44$$

It is calculated that the porosity of ceramides per cubic meter $= 0.44$, and the porosity and density of epoxy resin mixture $= 1250$ kg/m^3. Through these two parameters, we can calculate that the quality of epoxy resin required to fill the void of each cubic meter ceramide is as follows:

$$m_{e2} = n_2 \times \rho_e = 550\,\text{kg}$$

Then the ratio of ceramic particle to epoxy resin mixture is:

$$m_{e2} : m_{t1} = 550 : 753.1 = 0.73$$

3.3 Mix Ratio Correction

After completing the above steps, the weight ratio of the sand to the epoxy resin mixture and the weight ratio of the clay to the epoxy resin mixture should be corrected to the ratio between the volume of the epoxy resin mixture, the volume of the sand and the volume of the light grade 900 sintered ceramics. The method of correction is as follows:

First of all, we need to calculate the ratio of the weight of the fine aggregate to the packing density, which can be obtained by the fine aggregate volume, the calculation process is as follows:

$$V_{s0} = \frac{m_{s1}}{\rho_{2s}} = 1\,\text{m}^3$$

In the above formula, V_{s0} is the volume of fine aggregate (unit: m^3), m_{s1} is the mass of epoxy resin required to fill the void of each cubic meter of ceramic particle (unit: kg). ρ_{2s} is the packing density of sand (unit: kg/m^3).

Secondly, the volume of epoxy resin mixture used when the cement-sand ratio of fine aggregate is satisfied is calculated as follows:

$$V_{e0} = \frac{(m_{e1} + m_{e1}\prime)}{\rho_e} = 0.46\,\text{m}^3$$

In the above formula, V_{e0} is the volume of the epoxy resin mixture (unit: m^3) $(m_{e1} + m_{e1}\prime)$ is mass of the epoxy resin required to fill the void of each cubic meter of sand and the weight of the epoxy resin under the condition that the thickness of the fine aggregate coating layer is satisfied (unit: kg). ρ_e is the density of the epoxy resin mixture (unit: kg/m^3).

According to the above calculation process, the ratio between the volume of fine aggregate and the theoretical volume of epoxy resin mixture can be obtained according to the above steps, and the relationship is shown in the following equation:

$$V_{e0} : V_{s0} = 0.46 : 1$$

Make the theoretical volume of the epoxy resin mixture is reduced to 1, the following formula can be obtained:

$$V_e : V_s = 1 : 2.18$$

Similarly, he ratio between the volume of epoxy resin mixture and the volume of 900 grade light sintered ceramics can be obtained, and the relationship is as follows:

$$V_e : V_t = 1 : 2.73$$

The ratio of the volume of fine aggregate to the theoretical volume of the epoxy resin mixture and the ratio of the volume of the epoxy resin mixture to the volume of the 900 grade light sintered ceramic can be obtained to the new epoxy concrete with strength grade C30:

$$V_e : V_s : V_t = 1 : 2.18 : 2.73$$

In the above formula, V_e is the volume of the epoxy resin mixture, V_s is the volume of the sand, and V_t is the volume of the ceramic particle.

3.4 Experimental Verification of Mix Ratio

The epoxy resin mixture was prepared with E44 epoxy resin, T31 curing agent, and ethanol. The light aggregate was prepared with 900-grade light-sintered ceramic particles for coarse aggregate, and the fine aggregate was prepared with laboratory sand according to the above mix ratio. The epoxy resin concrete was fully vibrated, the aggregate was fully wrapped and there was no pulp.

In order to make a 150 × 150 × 150 mm standard epoxy concrete compressive test block for the standard compressive test, epoxy concrete does not need to be cured in the curing room, but it is necessary to pay attention to avoid external interference before the epoxy concrete sets. After the final solidification of epoxy resin concrete for 12 h, the conditions of the concrete compressive test were obtained.

The standard test block of epoxy concrete is placed in the center of the lower pressure plate of the pressure testing machine, so that the pressure surface is perpendicular to the top surface of the specimen, and then the pressure is applied evenly at a speed of 0.3–0.5 MPa until the specimen is damaged.

Finally, the experimental results show that the compressive strength of the epoxy concrete compressive specimen is 42.05Mp.

The experimental result is obviously higher than that of C30-grade concrete. In line with the expected design strength of this design method, it can better reflect the feasibility of the mix ratio designed by this method and provide the possibility for large-scale application of epoxy resin concrete.

4 Conclusion and Prospect

4.1 Conclusion

(1) By considering four preconditions, such as strength requirements, adhesion, fluidity, and construction performance of epoxy resin and aggregate, this paper constructs a calculation underlying logic of a new type of epoxy concrete mix design.
(2) Through literature investigation, analysis, and testing, an innovative accurate calculation method for epoxy concrete mix design is proposed.
(3) E44 epoxy resin, T31 curing agent, ethanol diluent, sand as fine aggregate, grade 900 light sintered ceramics as coarse aggregate, and the design method proposed in this paper, accurately designed a strength grade of C30 epoxy concrete mix ratio:

$$V_e : V_s : V_t = 1 : 2.18 : 2.73$$

It provides reference for engineering application.

4.2 Outlook

In this study, we designed a precise mix of epoxy concrete using only E44 type epoxy, T31 type curing agent, and ethanol diluent, with sand as fine aggregate and 900 class light sintered ceramisite as coarse aggregate.

Although this research has made some achievements, there are still some promising directions, which will play a role in further improving the design method of epoxy resin concrete mix. For example, the effects of different epoxy resin types, curing agent types, and aggregate types on the performance of epoxy resin concrete can be further studied in future research, so as to obtain more epoxy resin concrete mix ratios suitable for different scenarios. Based on the above prospects, the following three suggestions are given:

(1) Other types of epoxy resins, such as E51, E64, etc., can be used to carry out mix design research with the help of the calculation logic and innovative methods proposed in this paper;
(2) The curing agent can be used:T40, D4 and other models;
(3) New aggregates can be used for aggregates, such as slag, steel slag, expanded perlite, etc.

However, it should be noted that the computational logic and methods constructed in this study are not applicable to all cases, but are limited to epoxy concrete design and engineering applications under similar test conditions and environments as those in this paper. Therefore, in future research, we need to continuously improve, upgrade and refine these calculation logics and methods so that they can be better applied to a wider range of engineering application scenarios.

References

1. Beeldens A, Monteny J, Vincke E, De Belie N, Van Gemert D, Taerwe L, Verstraete W (2001) Resistance to biogenic sulphuric acid corrosion of polymer-modified mortars. Cement and Concrete Composites, Elsevier Ltd. 23(1):47–56. https://doi.org/10.1016/S0958-9465(00)00039-1
2. Gong YM, Yan WT (2022) Research on the application of epoxy resin mortar in the repair of cement concrete pavement. Northern Construct 7(05):53–56
3. Cai QF (2019) Application of epoxy resin concrete in construction. Sichuan Build Mater 45(09):25–26+48
4. Xie XY, Li HW, Jin YJ (2022) Study on influence factors of K-shaped joints of new epoxy concrete prefabricated truss. Concrete 02:148–154
5. Mostafizur R, Akhtarul M (2022) Application of epoxy resins in building materials: progress and prospects. Polym Bulletin 3:79
6. Jin YJ, Fan YL, Xie XY. (2022) Effect of different Admixtures on the strength of epoxy resin concrete. In: 7th Symposium on engineering oriented foundation technology. Beijing, pp 5. https://doi.org/10.26914/c.cnkihy.2022.086358

Open Access This chapter is licensed under the terms of the Creative Commons Attribution 4.0 International License (http://creativecommons.org/licenses/by/4.0/), which permits use, sharing, adaptation, distribution and reproduction in any medium or format, as long as you give appropriate credit to the original author(s) and the source, provide a link to the Creative Commons license and indicate if changes were made.

The images or other third party material in this chapter are included in the chapter's Creative Commons license, unless indicated otherwise in a credit line to the material. If material is not included in the chapter's Creative Commons license and your intended use is not permitted by statutory regulation or exceeds the permitted use, you will need to obtain permission directly from the copyright holder.

Research on Service and Crack Control of Concrete in Ultra-High Altitude Environment

Zaifeng Yao, Lei Liu, Shuanye Han, Yaning Wang, and Xiang Lv

Abstract This paper investigates the cracking prevention and control of concrete engineering in ultra-high altitude areas. Combined with the hydration-temperature-humidity-constraint coupling model, the cracking risk assessment of bridge piers under extreme environment was carried out. The effect of deformation compensation crack control in concrete cracking risk control was revealed. Finally, the concrete cracking risk was assessed after long-term temperature changes. The results show that wind speed, air temperature, light, humidity and freeze–thaw greatly affect the cracking control of concrete. The cracking risk in the surface layer of the bridge piers is maximum around 2–4 days. The maximum cracking risk coefficient is between 0.6 and 0.95. And the risk of core cracking increases progressively after 14 days. When the HME-V® crack-resistant product is added, the unit expansion deformation of concrete increases during the temperature rise phase. Furthermore, the unit volume shrinkage is decreased during the temperature drop phase. A significant deformation compensation is produced. The risk of early and long-term cracking in the core and surface layers of the concrete structure is significantly reduced. The effect of crack control is remarkable. In summary, pre-cracking control in the surface layer of bridge piers is crucial. The risk of long-term cracking in the core is significantly higher than that in the early stage, and long-term cracking control should also be emphasized.

Keywords Ultra high altitude · Concrete · Service environment · Cracking risk · Deformation compensation · Cracking control

---

Z. Yao (✉) · S. Han · Y. Wang · X. Lv
China Construction Second Engineering Bureau Co., Ltd., Beijing, China
e-mail: 498778959@qq.com

L. Liu
China Construction Second Bureau Civil Engineering Group Co., Ltd., Beijing, China

1 Introduction

The Tibetan Plateau is the world's highest average altitude plateau, above 3500 m. A large number of construction projects are located in ultra-high altitude areas (3500–5500 m), far exceeding the neighboring areas at the same latitude. Due to the unique geographic structure and location, the Tibetan Plateau has extreme environments such as large temperature difference, low air pressure, dryness, strong ultraviolet rays, high wind speed, sudden hail and heavy snowfall. The service safety of concrete engineering faces serious challenges. At present, there is insufficient experience in the bearing capacity and durability of concrete in ultra-high altitude environments [1–3]. Ensuring the safe service of concrete has become a major challenge for engineering construction.

Concrete cracking prevention and control has been a key component of concrete research. The research shows that under the low air pressure of 60 kPa on the plateau, the compressive strength and splitting tensile strength decrease by 1.6–14.8% and 1.5–10.8%, respectively, compared with the normal pressure. Low air pressure leads to a decrease in bubble spacing and an increase in bubble surface tension. Bubbles will constantly appear to overflow, dissolve and fuse [3], and the properties of concrete are reduced. Nassif and Petrou et al. [4] found that when concrete is cured at −5 to 20 °C. The stiffness was reduced by about 20% and microcracks appeared in the concrete structure. Wang et al. [5] determined that large temperature difference can cause crack damage on concrete surfaces. Emborg and Bernander [6] explained the cracking problem of high performance concrete due to temperature and shrinkage. Obvious cracks were produced within the interfacial transition zone. The content of calcium hydroxide increased. And the strength of concrete was reduced under the effect of temperature fatigue. Chen et al. [7] revealed that dry and low temperature environments are the main factors inducing the concrete porosity structure and mechanical properties deterioration. Therefore, the environment has a significant impact on concrete performance. The importance of studying concrete crack control in extreme ultra-high altitude environments cannot be overstated.

A large number of existing studies on crack control in concrete have mostly been carried out in conventional environments such as low altitude and atmospheric pressure. The few studies are aimed at initially revealing the effects of air pressure on strength, loss of collapse, low temperature on mechanical properties and hydration characteristics of concrete [7]. However, the extreme environment unique to the plateau has not been taken into account. This leads to the fact that crack control studies do not fully reflect the real environment of concrete. Due to the limitations of the conditions, there are still fewer studies on concrete cracking resistance in plateau environments.

2 Engineering Situation

The project includes mass concrete construction of bridges, culverts and retaining walls. The altitude is 3600–3750 m and the air pressure is 64.8 kPa. The environment is characterized by significant low air pressure, strong ultraviolet rays, large temperature differences, dry cold, strong winds, hail and snowfall. The average wind speed in the region is about 1.5–3.8 m/s, with the maximum wind speed exceeding 20 m/s. The wind speed is lower from July to October and higher from November to December. As shown in Fig. 1, the project is located in a canyon area between mountains.

Daily temperatures in the region are highly variable, with temperature differences exceeding 20 °C. The temperatures in January to April and October to December are low, with the lowest temperature below -15 °C. According to the definition of the annual number of freeze–thaw times for concrete in the Code for Durability Design of Railway Concrete Structures [8], the average number of freeze–thaw times for three years in the region from 2020 to 2022 is 88. Extremely low temperatures and high frequency of freeze–thaw cycles accelerate concrete cracking damage, which further reduces concrete strength and durability. The extremely complex temperature environment poses a great threat to concrete curing and cracking. And cracks are often induced by the ambient temperature. The average light intensity of the project site is 25,000 lx, and the maximum light intensity reaches 202,308 lx. Strong light easily causes water volatilization in the process of concrete demolding and curing. This causes cracks and fissures in the concrete. At the same time, strong light often induces a temperature difference between the sunlit side and the back side of the piers. According to on-site monitoring, the temperature difference can reach more than 10 °C. Long-term temperature unevenness is bound to have an adverse effect on the concrete.

The average humidity range of this project is 20–60% RH. The setting process of cement, concrete structures and plastering mortar requires sufficient moisture to promote hydration reactions. The material is difficult to solidify in dry climates. This leads to concrete disease problems such as reduced strength. At the same time, water loss from cement and concrete is further exacerbated by direct sunlight. These environmental factors are what affect the durability of concrete in the long term. It is

Fig. 1 Environment of the project

necessary to consider these factors to comprehensively assess the effects of cement concrete cracking.

3 Cracking Analysis

3.1 Methodology for Cracking Risk Assessment

Whether there is a risk of cracking and the extent of cracking in the extreme environment at high altitude is not known. These factors will directly affect the construction process strategy of concrete. A shrinkage model based on "hydration-temperature-humidity-constraint" multi-field coupling mechanism was constructed to solve this technical problem [9]. The model took into account material, structural, construction, and environmental factors. The cracking risk of concrete structures (bridge piers for example) was quantified. The risk factor for concrete cracking is calculated as shown in the following equation:

$$\eta = \sigma(t)/f(t) \tag{1}$$

where t is the time, s; σ(t) is the maximum tensile stress of concrete at time t, MPa, and f(t) is the tensile strength of concrete at time t, MPa. The early elastic modulus and tensile strength of concrete are necessary parameters for the calculation of shrinkage stresses and cracking risk. And the calculation formulas are as follows:

$$E(\alpha) = E^\infty \left(\frac{\alpha - \alpha_0}{\alpha^\infty - \alpha_0} \right)^p \tag{2}$$

$$f_t(\alpha) = f_t^\infty \left(\frac{\alpha - \alpha_0}{\alpha^\infty - \alpha_0} \right)^q \tag{3}$$

where E^∞ is the final modulus of elasticity, GPa; f_t^∞ is the final tensile strength, MPa; α_0 and α^∞ denote the initial and final values of the degree of hydration α; and p, q are exponential constants taking the values of 0.5 and 1.

3.2 Numerical Modeling and Construction Process

The numerical model for half of the piers and bearing platforms was established. The bearing platform and the range of 1 m above the bearing platform were C45 concrete, and the part above 1 m was C35 concrete. The concrete unit for the numerical model was selected as SOLID 65. The model is shown in Fig. 2.

Fig. 2 The established numerical model

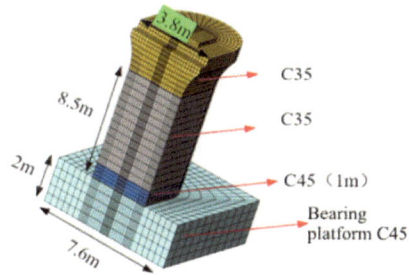

Table 1 Parameters of the model

Parameter	C35	C45
Final tensile strength f_t^∞ (MPa)	2.0	2.3
Final modulus of elasticity E^∞ (GPa)	32	36
Adiabatic temperature rise (°C)	49	53
28 days of self-contraction (με)	150	200

According to the actual conditions, the concrete molding temperatures were selected as 16 and 24 °C. Then, the cracking risk under two different molding temperatures was analyzed. Considering the actual pouring process, the pier body was constructed after 7 days of building the bearing platform. The C45 and C35 concrete for the pier body was poured together and the pier cap was poured 7 days later. The concrete was demolded after 7 days and a water energy film was posted. Subsequently, covering measures were taken to take thermal insulation measures. The surface heat dissipation coefficient was taken as 12 kJ/(m² h K), and the steel mold heat dissipation coefficient was taken as 70 kJ/(m² h K). The parameters of concrete mechanics and deformation are shown in Table 1.

4 Analysis of the Cracking Risk

4.1 C45 Concrete for Piers

Figure 3 shows the cracking risk of the pier body with C45 concrete. The difference between surface and core cracking risk is significant. When the casting temperature is 16 °C, the surface cracking risk of the pier body reaches a maximum value around 0.8 after 4 days. At a casting temperature of 24 °C, the maximum cracking risk occurs after 3 days, about 0.6. It is significantly lower than the cracking risk at a casting temperature of 16 °C. The main reason is that the high molding temperature reduces the temperature gradient of the surface concrete. Overall, the surface cracking risk shows a trend of increasing and then decreasing.

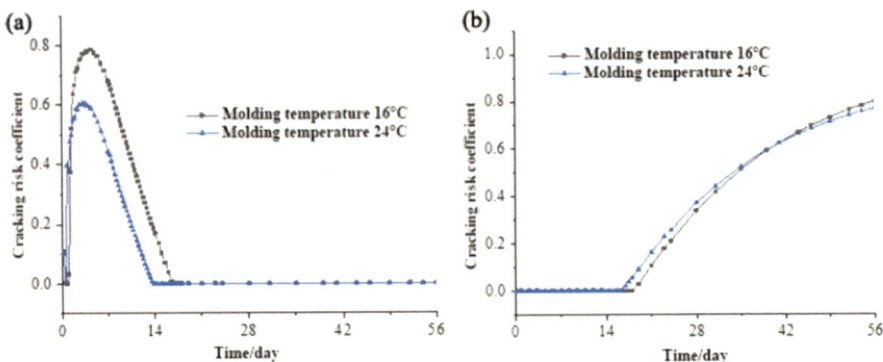

Fig. 3 Cracking risk of pier body with C45 concrete, **a** Surface, **b** Core

After about 14 days, the surface temperatures are equal to ambient and no longer posed a risk of cracking. Because of the low temperature gradient in the core, the temperature is largely maintained at a high level and there is no cracking risk until 20 days. After 20 days, the cracking risk gradually increases with time. The cracking risk exceeded 0.7 at 50 days. As a result, the change rule for cracking risk of concrete surface layer and core is not consistent. The initial phase should pay more attention to the maintenance of surface insulation, while the later phase should pay attention to the core cracking. Usually, engineers ignore the core cracking. This will have a greater impact on the concrete structure.

4.2 Pier Body with C35 Concrete

Figure 4 shows the cracking risk of pier body with C35 concrete. The results show that the surface cracking risk reaches the maximum value at 2–3 days. Even with strong insulation measures, the maximum cracking risk coefficient of the concrete surface is between 0.9 and 1.0, which is a high cracking risk. After the cracking risk reaches the maximum value, it decreases with the time. The cracking risk of the concrete core increases gradually after 20 days. The cracking risk is high as it exceeds 0.7 at 42 days and has approached 1.0 at 56 days. As the center temperature continues to decrease, the cracking risk coefficient remains at a high level. The cracking risk for C35 concrete is essentially the same as for C45. The maximum cracking risk factor differs due to the different volumes of concrete. In addition, lowering the concrete entry temperature increases the risk of surface cracking while decreasing the risk of core cracking.

Figure 5 shows the cracking of a test pier. The cracks are approximately 20–35 mm deep, 2.2–2.5 m long and 0.1–0.28 mm wide. It is located in the straight section part of the pier cross-section. And relatively few circular segments were found in the field.

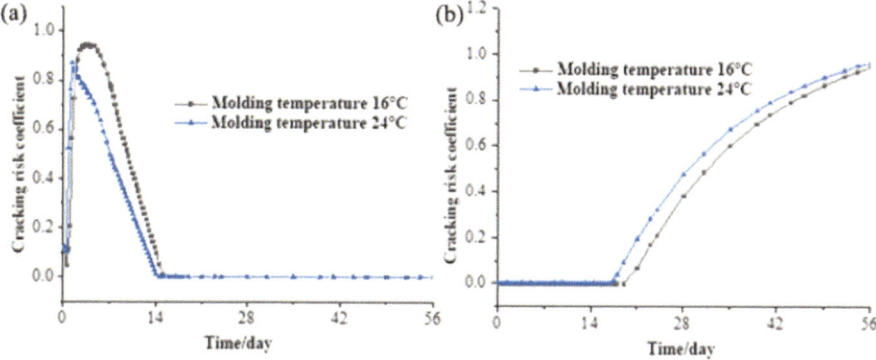

Fig. 4 Cracking risk of pier body with C35 concrete, **a** Surface, **b** Core

Fig. 5 Cracking of the actual pier body

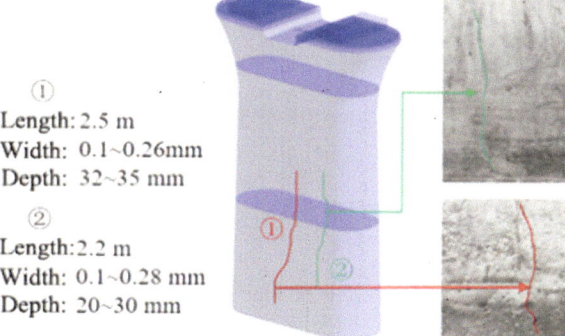

① Length: 2.5 m
Width: 0.1~0.26 mm
Depth: 32~35 mm

② Length: 2.2 m
Width: 0.1~0.28 mm
Depth: 20~30 mm

Therefore, the calculation results of the present cracking risk model are verified to be more reasonable with the actual measurement results.

5 Long-Term Cracking Analysis

Concrete cracking, development and stabilization is a long process. Environmental temperature, humidity, wind speed and other factors at high altitude not only show daily changes, but also reflect the pattern of change in the annual cycle. When evaluating concrete cracking, these factors need to be considered for a reasonable assessment of service risk. However, engineers often evaluate the cracking resistance of concrete by observing whether cracking occurs in the early stages. The long-term cracking risk of the concrete structure is ignored. However, this is precisely an important factor to be considered for concrete cracking control in ultra-high altitude environments. Therefore, this paper carries out a long-term cracking risk analysis of concrete based on the results of indoor tests. The effect of an anti-cracking product on concrete cracking control is further quantified. And long-term cracking and short-term cracking are defined. The long term counterpart is to experience at least one winter cycle of minimum temperatures. The early correspondence is within 60 days of the casting completion.

Based on the test results, a secondary calculation for the risk of concrete cracking at the bridge piers was carried out. The risk of long-term cracking was fully assessed. Considering that the concrete would be surface coated or wrapped when the mold was removed for curing. Insulation measures were taken to mitigate the effects of daily changes and sudden drops in temperature. Thus, the assessment of long-term cracking risk focused on the effect of annual variations in temperature. The risk of cracking was assessed during the hot season when the risk of cracking was high. The annual change pattern of air temperature is shown in the following formula:

$$T = T_0 + \frac{A}{2}\cos\left(\frac{\pi}{6}(t - t_0)\right) \qquad (4)$$

where T_0 is the average annual temperature, °C, taken as 8.6 °C; A is the annual variation of temperature, °C, taken as 26 °C; t_0 is the highest temperature in a year, °C, generally for the middle of July, taken as 6.5 °C.

5.1 Cracking Risk of the Pier Body with C45 Concrete

Figure 6 shows the long-term cracking risk of the pier body with C45 concrete. The results show that the maximum coefficient of early cracking risk for the baseline concrete core does not exceed 0.7. However, the long-term maximum cracking risk factor increases to 0.9 after considering the annual temperature change. With the addition of the crack resistance product, the long-term maximum cracking risk factor is 0.74, a reduction of about 15%. The risk of early cracking in the core decreases from 0.68 to 0.46, a reduction of about 32%. The maximum early cracking risk coefficient of surface concrete is 0.87. In contrast, the maximum early cracking risk with the addition of the anticracking agent is about 0.52, a reduction of about 40%. The cracking risk of the surface concrete decreases rapidly with time. The coefficient of cracking risk after 7 days for both the baseline concrete and the crack-resistant concrete is less than 0.7, and there is no longer a risk of cracking. The anti-cracking product can significantly reduce the risk of early and long-term cracking for concrete. Meanwhile, according to the cracking law of core and surface concrete, early cracking control of surface concrete is very important. The core should focus more on long-term cracking prevention and control.

Figure 7 shows the long-term cracking risk of the pier body with C35 concrete. The results show that the long-term maximum cracking risk of the baseline concrete core occurs near 350 days with a maximum value of 0.91. The long-term maximum cracking risk coefficient is 0.77 after the addition of the anti-cracking agent, a reduction of about 15%. With the addition of the anti-cracking product, the risk of early

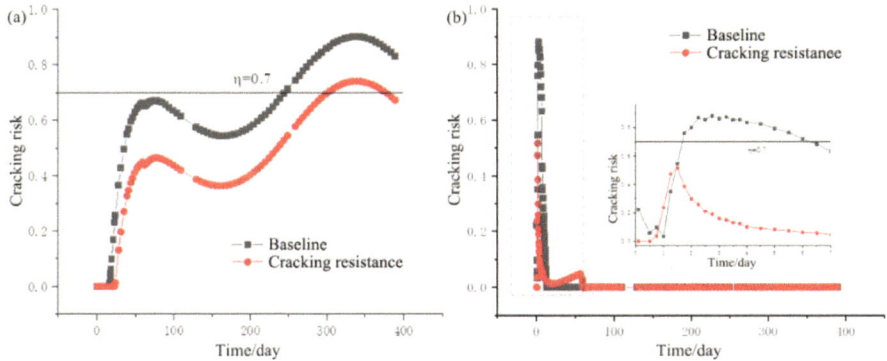

Fig. 6 Long-term cracking risk of pier body with C45 concrete, **a** Core, **b** Surface

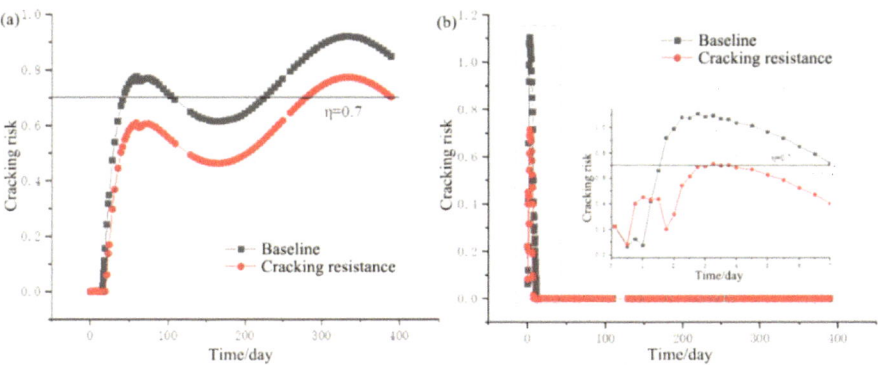

Fig. 7 Long-term cracking risk of pier body with C35 concrete, **a** Core, **b** Surface

cracking in the core decreases from 0.8 to 0.6, a 20% reduction. The early maximum cracking risk coefficient of surface concrete is 1.1. However, the maximum early cracking risk coefficient after addition of anti-cracking agent is 0.71, a 35% reduction. Similarly, the cracking risk of the surface concrete decreases rapidly with time. The baseline and crack-resistant concrete no longer has a cracking risk after 7 days.

5.2 Cracking Risk of the Pier Body with C35 Concrete

6 Conclusions

This paper took a key project in ultra-high altitude environment as an example. The extreme environment of concrete construction was studied in detail. Then, the bridge pier body and pier cap were evaluated for cracking risk in conjunction with a coupled hydration-temperature-moisture-constraint model. Crack control tests were carried out using a crack resistant product. And long-term cracking risk analysis was carried out. The main conclusions are as follows:

(1) The complex wind speed, air temperature, light, humidity and freeze–thaw environment in the ultra-high altitude environment presents a great threat to concrete curing and cracking. It is necessary to take reasonable crack control measures.
(2) The cracking risk of the pier surface layer increases first and then decreases. Generally, the maximum value occurs in about 2–4 days, and the maximum cracking risk coefficient ranges from 0.6 to 0.95. The cracking risk in the core increases progressively after 14 days. It is crucial to control cracking in the early stage of the pier body. And the core should pay more attention to the later cracking control.
(3) With the addition of HME®-V anti-cracking product, the expansion and deformation of concrete increases during the temperature rise phase. The volume shrinkage decreases in the temperature drop stage. A significant expansion deformation is produced, which compensates for the shrinkage deformation. At the same time, the risk of early and long-term cracking in the core and surface layer for concrete structures is significantly reduced. The crack resistance effect is remarkable.

Acknowledgements This research is supported by CSCEC Technology R&D Program Funding Projects (CSCEC-2021-S-1) and China Construction Science and Technology Innovation Platform Grant (CSCEC-PT-017).

References

1. Deng X, Zhang P, Wang R et al (2023) Frost resistance durability and damage model of fiber concrete in Tibet Plateau Area. Bulletin of the Chinese Ceramic Soc 42:3143–3153
2. He R, Wang T, Chen H et al (2020) Impact of Qinghai-Tibet Plateau's climate on strength and permeability of concrete. China J High Transport 33:29–41
3. Li Y, Wang Z, Xue C et al (2021) Influence of low air pressure on the performance of concrete in road engineering. China J High Transport 34:194–202
4. Nassif AY, Petrou MF (2013) Influence of cold weather during casting and curing on the stiffness and strength of concrete. Constr Build Mater 44:161–167
5. Wang S, Shui Z, Xuan D (2006) Investigation on surface cracking damage of concrete under big temperature difference. J Southeast Univ (Natural Science Edition)., 36:122–125
6. Emborg M, Bernander S (1994) Assessment of risk of thermal cracking in hardening concrete. J Struct Eng 120:2893–2912
7. Chen H, Wang T, He R et al (2020) Effect of complex climatic environment on pore structure and mechanical properties of concrete. J Chang'an University (Nat Sci Edn), 40:30–37
8. China Academy of Railway Sciences (2010) Code for durability design on concrete structure of railway: TB 10005–2010. China Railway Publishing Press, Beijing
9. Li H, Liu J, Wang Y (2015) Deformation and cracking modeling for early-age sidewall concrete based on the multi-field coupling mechanism. Constr Build Mater 88:84–93

Open Access This chapter is licensed under the terms of the Creative Commons Attribution 4.0 International License (http://creativecommons.org/licenses/by/4.0/), which permits use, sharing, adaptation, distribution and reproduction in any medium or format, as long as you give appropriate credit to the original author(s) and the source, provide a link to the Creative Commons license and indicate if changes were made.

The images or other third party material in this chapter are included in the chapter's Creative Commons license, unless indicated otherwise in a credit line to the material. If material is not included in the chapter's Creative Commons license and your intended use is not permitted by statutory regulation or exceeds the permitted use, you will need to obtain permission directly from the copyright holder.

The Mix Proportion Optimization Design of Coal Gangue Pervious Concrete

Junwu Xia, Chao Luo, and Enlai Xu

Abstract In order to improve the strength and permeability of coal gangue pervious concrete, an optimized mix design was conducted. An orthogonal experiment was employed to study the variations of compressive strength and permeability coefficient of coal gangue pervious concrete under the influence of aggregate particle size, water-cement ratio, designed porosity, and dosage of permeable admixture. After obtaining a relatively optimal mix proportion, further discussions were carried out by restricting the values of compressive strength and permeability coefficient to determine the appropriate range for the excess paste content ratio, total porosity, and effective porosity, resulting in the determination of the optimal mix design. Results indicated that the compressive strength reached its maximum at an aggregate particle size of 9.5–16 mm, with minimal impact on the permeability coefficient. As the water-cement ratio increased, the compressive strength gradually increased, while the permeability coefficient slightly decreased within the range of 0.25–0.29, and decreased by 60% within the range of 0.29–0.31. With the increase of designed porosity, the compressive strength gradually decreased, while the permeability performance gradually enhanced.

Keyword Coal gangue pervious concrete · Orthogonal experiment · Basic performance · Optimal mix design

---

J. Xia · C. Luo (✉)
State Key Laboratory for Intelligent Construction and Health Operation and Maintenance, China University of Mining and Technology, Xuzhou 221116, Jiangsu, China
e-mail: luochao355588@163.com

J. Xia
Jiangsu Collaborative Innovation Center of Building Energy-Saving and Construction Technology, Jiangsu Vocational Institute of Architectural Technology, Xuzhou 221116, Jiangsu, China

Jiangsu Design Institute of Geology for Mineral Resources (Testing Center, CNACG), Xuzhou 221006, China

E. Xu
School of Naval Architecture, Shandong Vocational University of Foreign Affairs, Weihai 264504, Shandong, China

1 Introduction

In recent decades, rapid urbanization and industrialization in China have led to extensive coverage of impermeable materials such as cement, concrete, and tiles on the ground surface, exacerbating the urban heat island effect. The use of porous materials represented by pervious concrete can effectively mitigate the urban heat island effect and achieve the recycling of urban water resources.

Pervious concrete, also known as porous concrete, thereby obtaining a higher porosity to achieve permeability and breathability [1]. In addition to its high permeability and breathability, the porous structure of pervious concrete also gives it excellent heat dissipation, sound absorption, and freeze–thaw resistance capabilities [2, 3]. Liv Haselbach from the University of Washington [4] monitored and compared the heating up and heat storage phenomena of pervious concrete pavement with traditional concrete pavement. The research results showed that the high surface area of pervious concrete enables efficient heat exchange, thereby reducing the impact of the urban heat island effect and potential heat shock caused by impervious surface runoff.

Most coal gangue is black or gray and relatively hard in texture. Chemical composition analysis reveals that its main components are SiO_2, Al_2O_3, Fe_2O_3, CaO. The mineral components mainly consist of clay minerals, quartz, calcite, pyrite, and carbonaceous matter [5]. To ensure permeability, fine aggregate is generally not added in pervious concrete. Based on this structural characteristic, most scholars choose to use the volumetric method for mix design [6–10].

2 Orthogonal Experimental Study on the Mix Ratio of Coal Gangue Permeable Concrete

2.1 Experimental Materials

In the orthogonal experiment designed in this study, the raw materials used for preparing coal gangue pervious concrete include: coal gangue coarse aggregate, cement, water, and permeable admixture. The coal gangue coarse aggregate is used in four particle size ranges: 4.75–9.5, 9.5–16, 16–19, and 19–26.5 mm. The cement used in the experiment is P.O. 52.5 cement, and the permeable admixture is supplied by Jiangsu Guangda Ecological Engineering Technology Co., Ltd. The water used in the experiment is ordinary tap water from Xuzhou City.

(a) Assembly process   (b) Device diagram

Fig. 1 Permeability coefficient test device diagram

2.2 Test Block Production and Testing Methods

The flowability of pervious concrete slurry should not be too high, hence the water-cement ratio is generally kept at a lower level, resulting in a relatively dry mixture. Considering relevant literature and practical experience, the mixing method adopted is the cement-wrapping method, while the forming method is the insertion and tamping method.

The compressive strength testing requirements follow the "Standard Test Method for Mechanical Properties of Concrete" GB/T 50081–2019, testing the compressive strength of cube specimens at specified ages. To ensure uniform pressure application, the lateral surface of the specimen is used as the loading surface. According to the requirements of GB/T 50081–2019, the loading rate is set at 0.3–0.5 MPa/s.

The permeability coefficient testing method follows the fixed water level method. In accordance with the permeability coefficient testing method outlined in Appendix A of "Technical Specification for Permeable Cement Concrete Pavement" CJJ/T135-2009, three measurements are taken for each set of specimens, and the arithmetic average of these three measurements is used as the experimental result. Data with an error exceeding 15% are discarded. The experimental setup and assembly process are illustrated in Fig. 1.

2.3 Experimental Design

This article explored four factors in the orthogonal experiment, including aggregate particle size with four levels: 4.75–9.5 mm, 9.5–16 mm, 16–19 mm, and 19–26.5 mm; water-cement ratio with four levels: 0.24, 0.27, 0.30, 0.33; designed porosity with four levels: 15, 18, 21, 24%; and permeable admixture dosage with four levels: 2.0, 2.4, 2.8, 3.2%. Experiment group design was conducted using the L16 (45) orthogonal table, with a blank column set, and three test blocks were cast for each group, resulting in a total of 48 test blocks.

Based on the orthogonal experimental group design, the quantities of raw materials for each group are calculated using the volumetric method, with an amplification factor of 120% to ensure sufficient mixing. The additional water quantity is obtained by multiplying the coarse aggregate quantity by its absorption rate. When the aggregate particle size is at levels 1–4, the additional water quantities are 104.78 ml, 103.72 ml, 91.45 ml, and 89.54 ml, respectively. The actual water quantity is the sum of the designed water quantity and the additional water quantity obtained from the mix proportion.

2.4 Test Results and Analysis

After curing 16 sets of test blocks in water for 28 days, when reaching the curing age, their compressive strength and permeability coefficient were determined according to the testing method in Sect. 1.2, and the test results were recorded as shown in Table 1.

Using compressive strength as the test indicator, the range analysis of the orthogonal experimental results was conducted, with the increase of aggregate particle size, the compressive strength first increases and then decreases, with a significant fluctuation. The water-cement ratio increased from 0.24 to 0.33, showing a trend of first increasing and then decreasing, reaching its maximum at 0.27. The designed porosity increased from 15 to 24%, resulting in a gradual decrease in compressive strength with a significant fluctuation, achieving the maximum value at 15%. The use of permeable agent at 2.0% led to the lowest compressive strength; when increased to 2.4% and beyond, the compressive strength significantly increased and remained within a relatively stable range. Comparing the range R values, the order of influence on compressive strength is determined as follows: aggregate particle size A >

Table 1 Results of orthogonal test

Group	Compressive strength (MPa)	Permeability coefficient (mm/s)	Group	Compressive strength (MPa)	Permeability coefficient (mm/s)
Z1	4.73	2.65	Z9	4.52	8.54
Z2	9.57	3.91	Z10	5.84	11.83
Z3	8.04	2.86	Z11	7.85	5.88
Z4	5.52	4.09	Z12	5.60	7.10
Z5	16.86	0.99	Z13	5.10	12.05
Z6	21.11	1.34	Z14	4.86	11.58
Z7	13.60	3.25	Z15	6.29	11.44
Z8	15.44	0.83	Z16	7.40	8.99

designed porosity C > water-cement ratio B > permeable agent usage D. The range R value for the blank column E is 1.89, all smaller than the considered influencing factors, indicating a small impact of experimental errors and a reasonable experimental setup. The optimal combination for compressive strength is determined as A2B2C1D4.

Similarly, using the permeability coefficient as the test indicator, the range analysis of the orthogonal experimental results was conducted, with the increase of aggregate particle size, the permeability coefficient first decreases and then increases. Under the influence of aggregate particle size, the overall fluctuation of the permeability coefficient is significant. The water-cement ratio increased from 0.24 to 0.33, showing a trend of first increasing and then decreasing in permeability coefficient, with a relatively stable overall change. The designed porosity increased from 15 to 24%, resulting in a gradual increase in permeability coefficient with a significant fluctuation, achieving the maximum value at 24% designed porosity. Under the influence of permeable agent usage, the permeability coefficient showed a trend of first decreasing and then increasing with minimal overall change, indicating a small impact of permeable agent usage on the permeability coefficient. When the usage of permeable agent reached 3.2%, the permeability coefficient was relatively high. Comparing the range R values, the order of influence on permeability coefficient is determined as follows: aggregate particle size A > designed porosity C > water-cement ratio B > permeable agent usage D. Through the range analysis, the optimal combination for permeability coefficient is determined as A4B2C4D4.

Considering both the compressive strength and permeability coefficient in line with the design requirements, the optimal mix proportion is determined to be A2B2C3D4, i.e., aggregate particle size of 9.5–16 mm, water-cement ratio of 0.27, designed porosity of 21%, and permeable agent usage of 3.2%.

3 Research on the Basic Properties of Coal Gangue Permeable Concrete

Based on the optimal mix proportion obtained from the orthogonal experiment in Chap. 1, single-factor experiments were conducted to investigate the variation patterns of the mechanical, workability, and permeability properties of coal gangue permeable concrete under the single-factor effects of aggregate particle size, water-cement ratio, and designed porosity. Subsequently, based on the basic performance analysis, the optimal state of coal gangue permeable concrete was determined, and further adjustments were made to optimize the test blocks under the optimal state.

Table 2 Single factor experiment design and material dosage

Group	Aggregate size (mm)	Water-cement ratio	Design porosity (%)	Coarse aggregate of coal gangue (g)	Cement (g)	Water (ml)	Permeable agent (g)
A1	4.75–9.5	0.27	20%	12,700.80	2991.41	807.68	95.73
A2	9.5–16	0.27	20%	13,829.76	4186.51	1130.36	133.97
A3	16–19	0.27	20%	13,547.52	2615.51	706.19	83.70
A4	19–26.5	0.27	20%	13,265.28	2197.95	593.45	70.33
B1	9.5–16	0.25	20%	13,829.76	4332.74	1083.18	138.65
B2	9.5–16	0.27	20%	13,829.76	4186.51	1130.36	133.97
B3	9.5–16	0.29	20%	13,829.76	4049.82	1174.45	129.59
B4	9.5–16	0.31	20%	13,829.76	3921.78	1215.75	125.50
C1	9.5–16	0.27	16%	13,829.76	4834.52	1305.32	154.70
C2	9.5–16	0.27	18%	13,829.76	4510.51	1217.84	144.34
C3	9.5–16	0.27	20%	13,829.76	4186.51	1130.36	133.97
C4	9.5–16	0.27	22%	13,829.76	3862.50	1042.87	123.60
C5	9.5–16	0.27	24%	13,829.76	3538.49	955.39	113.23

3.1 Experimental Design

Based on the research in Chap. 1, it was found that the dosage of permeable agent has minimal influence on both compressive strength and permeability coefficient. Therefore, it is not considered as a factor in this chapter. Using an aggregate particle size of 9.5–16 mm, water-cement ratio of 0.27, designed porosity of 20%, and permeable agent dosage of 3.2% as the reference mix proportion, single-factor experiments were designed. Considering the influence of aggregate particle size (A), water-cement ratio (B), and designed porosity (C), a total of 13 sets of mix proportions (4 + 4 + 5) were prepared, with 6 specimens cast for each set, totaling 78 specimens. The material quantities for each set were calculated based on the volume method, and the mix proportion designs and material quantities for each set are shown in Table 2.

3.2 Research on the Workability of Coal Gangue Permeable Concrete

The workability of permeable concrete is closely related to its permeability and mechanical properties. Based on the workability testing method mentioned in Sect. 2.1, the workability under the influence of each single factor was tested, and the variation of the excess paste ratio (β) calculated under each factor was plotted as

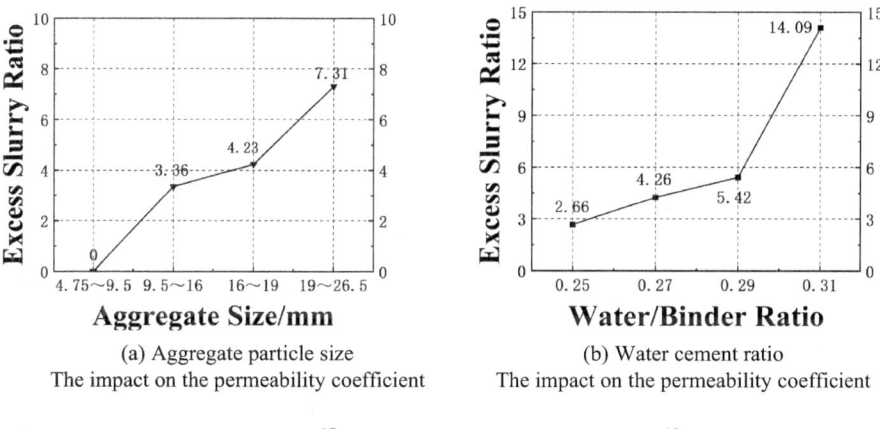

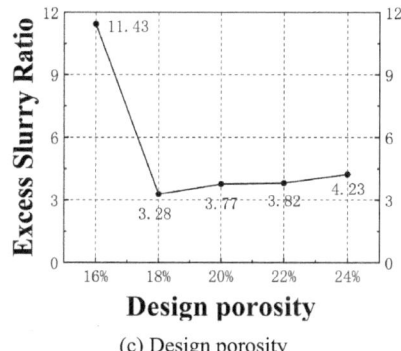

Fig. 2 Change of working performance under single factor

a line graph, as shown in Fig. 2. It can be observed from the graph that the excess paste ratio varies significantly under the influence of the mentioned factors.

In the study of workability, it was found that there is an optimal range for the excess paste ratio. Within this range, the flowability of the paste is moderate, and the thickness of paste coverage on the aggregate is moderate. Moderate flowability allows for good coverage on the surface of the aggregate without flowing into the lower voids or causing inadequate bonding, which could lead to reduced strength. A moderate thickness of paste coverage meets the strength requirements of the interfacial transition zone while ensuring that the voids are not blocked, meeting the requirements of permeability. Therefore, to determine the optimal range of the excess paste ratio, it is necessary to conduct a comprehensive analysis in combination with the mechanical and permeability properties.

3.3 Research on the Mechanical Properties of Coal Gangue Permeable Concrete

This section investigated the influence of single factors including aggregate particle size, water-cement ratio, and designed porosity on the compressive strength of coal gangue permeable concrete, using a single-variable control. The factors considered and the reference mix proportion were the same as in the previous section on workability research. The failure modes of the test blocks are shown in Figs. 3 and 4.

The study of the variation patterns in compressive strength of coal gangue permeable concrete under the influence of single factors revealed that the impact of aggregate particle size on compressive strength is very significant, with the optimal state achieved when the aggregate particle size is in the range of 9.5–16 mm. The influence

Fig. 3 A1 failure morphology

Fig. 4 A2 failure morphology

of water-cement ratio and designed porosity is comparatively smaller than that of aggregate particle size, and the variation trends at each stage are relatively uniform.

3.4 Study of Permeability of Coal Gangue Permeable Concrete

This section investigated the variation patterns of permeability coefficient and porosity of coal gangue permeable concrete under the single-factor influence of aggregate particle size, water-cement ratio, and designed porosity. The permeability coefficient was determined using the fixed water level method introduced in Sect. 1.1. Three test blocks were cast for each group, and each block was tested three times, with the arithmetic average taken as the final result. The variation of the permeability coefficient calculated under the influence of each factor was plotted as a line graph, as shown in Fig. 5.

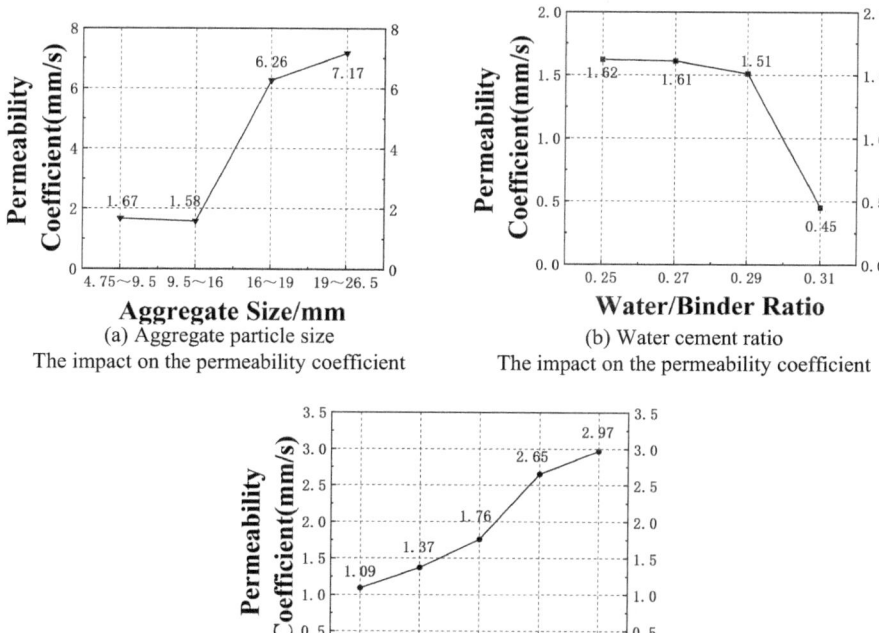

Fig. 5 Change of permeability coefficient under single factor

The analysis of the results revealed that aggregate particle size has a significant impact on the permeability coefficient, showing an overall trend of first decreasing and then increasing. The influence of water-cement ratio is relatively small, with the general trend indicating that a higher water-cement ratio leads to a smaller permeability coefficient. At water-cement ratios of 0.25 and 0.27, the permeability coefficient remains relatively unchanged, and it begins to decrease when the ratio is increased to 0.29. On the other hand, designed porosity has a significant impact on the permeability coefficient, with an increase in designed porosity leading to a gradual increase in the number of reserved voids and subsequently an increase in the permeability coefficient.

3.5 Determination of the Optimal State of Coal Gangue Permeable Concrete

To find the optimal combination point of compressive strength and permeability coefficient, the compressive strength and permeability coefficient obtained from all single-factor experiments were plotted as shown in Fig. 6. It can be observed from the graph that the performance of B3, C1, and C2 is relatively good. Through mutual comparison, it was found that the B3 group (aggregate particle size 9.5–16 mm, water-cement ratio 0.29, designed porosity 20%, permeable admixture dosage 3.2%) exhibited the highest compressive strength of 23.35 MPa and a permeability coefficient of 1.51 mm/s, making it the optimal mix proportion obtained in the current experiment.

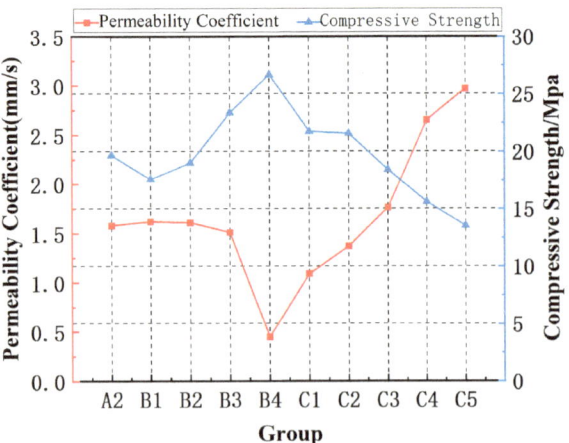

Fig. 6 Determination of the best binding point

4 Conclusion

Based on the above research, the following conclusions can be drawn:

1. With an increase in aggregate particle size, both compressive strength and permeability coefficient initially increase and then decrease. The water-cement ratio from 0.24 to 0.33 shows a similar trend of initially increasing and then decreasing for both properties. As designed porosity increases from 15 to 24%, compressive strength gradually decreases, while the permeability coefficient gradually increases. When the permeable admixture reaches 2.4%, the compressive strength reaches a relatively large value and remains stable, with little impact on the permeability coefficient.
2. According to the orthogonal experiment, aggregate particle size has a highly significant impact on both compressive strength and permeability coefficient, while designed porosity only has a highly significant impact on the permeability coefficient, and the water-cement ratio only has a significant impact on the permeability coefficient. The significance ranking of the influencing factors is as follows: aggregate particle size > designed porosity > water-cement ratio > permeable admixture dosage.
3. In the single-factor experiments, there is a correlation among the workability, mechanical properties, and permeability properties of coal gangue permeable concrete under the influence of aggregate particle size, water-cement ratio, and designed porosity. The optimal mix proportion obtained in the single-factor experiments is aggregate particle size 9.5–16 mm, water-cement ratio 0.29, designed porosity 20%, and permeable admixture dosage 3.2%. Considering workability, the excess paste ratio within the range of 3.3–7.4, total porosity within the range of 23.8–28%, and effective porosity within the range of 12–16.2%, both compressive strength and permeability coefficient of coal gangue permeable concrete can reach an optimal state.

References

1. Hu M, Zhang X, Siu Y et al (2018) Flood mitigation by permeable pavements in Chinese sponge city construction. Water 10(2):172
2. Wang WX, Xie XS, Xia GQ (1994) Performance and application of permeable concrete. China Building Materials Technology 1994(04):1–5
3. Kevern JT, Schaefer VR, Wang K (2009) Temperature behavior of pervious concrete systems. Transp Res Record: J Transp Res Board 2098(1):94–101
4. Haselbach L, Boyer M, Kevern JT et al (2011) Cyclic heat island impacts on traditional versus pervious concrete pavement systems. Transp Res Record: J Transp Res Board 2240(1):107–115
5. Gu QB (1997) The composition and comprehensive utilization of coal gangue. China's Mining Ind 05:14–16
6. Ling TQ, Cheng QQ, Qin X (2019) Research on the mix design and performance influencing factors of permeable concrete pavement. J Chongqing Jiaotong University (Nat Sci Edn) 38(3):38–43, 59

7. Chen Y, Wang D (2009) Research on the design method of permeable concrete mix proportion for road surface. J Build Mater 12(04):423–427
8. Cheng J, Yang Y, Chane WZ (2006) Research on the mix design of permeable concrete. Concrete 10:81–84
9. Zhang ZH, Wang QF, Yang J (2008) Research on factors influencing the strength and permeability of permeable concrete. Concrete 03:7–9
10. Wang Z, Qian JS, Zhang ZH (2008) Preliminary study on the design method of permeable concrete mix proportion. J Chongqing Jianzhu University 03:121–124

Open Access This chapter is licensed under the terms of the Creative Commons Attribution 4.0 International License (http://creativecommons.org/licenses/by/4.0/), which permits use, sharing, adaptation, distribution and reproduction in any medium or format, as long as you give appropriate credit to the original author(s) and the source, provide a link to the Creative Commons license and indicate if changes were made.

The images or other third party material in this chapter are included in the chapter's Creative Commons license, unless indicated otherwise in a credit line to the material. If material is not included in the chapter's Creative Commons license and your intended use is not permitted by statutory regulation or exceeds the permitted use, you will need to obtain permission directly from the copyright holder.

Research on Axial Tensile Mechanical Properties of Early-Strength High Ductility Cementitious Composites

Wenhong Duan, Jiaquan Yuan, Li Xiong, Weihong Jiang, Huimei Li, Lin Mou, Xiaohua Yang, Xiaomin Huang, Weibing Xu, and Kun Yang

Abstract Cement concrete is widely used in pavement, bridge decks and expansion joint anchorage zones. It is prone to cracking, potholes, spalling and other diseases during use. Cement-based repair materials generally have problems such as a long curing period, low bonding strength with existing concrete, and insufficient mechanical properties. Given this, this paper proposes an early-strength high-ductility cement composite material with the goal of short curing age, excellent tensile mechanical properties and relatively low price. The tensile mechanical properties of the composite material at different ages were tested, and the tensile constitutive model of the composite material was proposed. The results show that when the fly ash content is about 50 wt.%, the PVA fiber content is about 2%, and the sand-binder ratio is 0.36–0.50, the tensile properties of the early strength high ductility cement composite material at each age are better. Under the condition of optimum mix ratio, the tensile strength of 1, 3 and 28 d of early strength high ductility cement-based composites reached 4.58 MPa, 4.67 MPa and 4.64 MPa respectively, and the strain corresponding to 0.8 times of peak stress in softening section could reach 2%.

Keywords Cement-based repair materials · Curing age · PVA fiber · Early strength · High ductility

---

W. Duan · J. Yuan · L. Xiong · W. Jiang · H. Li · L. Mou · X. Yang
Dali Danan Expressway Co., Ltd., Dali, Yunnan, China

X. Huang (✉)
Kunming University of Science and Technology, Kunming, Yunnan, China
e-mail: Hjs.sy@163.com

W. Xu · K. Yang
Beijing University of Technology, Beijing, China

1 Introduction

Cement concrete has a wide range of applications in bridges and roads. It is prone to cracking, potholes, spalling and other diseases during use. The maintenance and maintenance of related diseases has become an important work in the daily operation and maintenance of roads and bridges. Traditional cement-based repair materials have problems such as a long curing period, low bond strength with existing concrete, insufficient tensile and compressive bearing capacity at an early age, and low ductility [1–5].

Given this, scholars at home and abroad have carried out a lot of research on new cement-based composite materials for road and bridge repair, and relevant improvement measures include the use of special cement, fiber addition, etc. Fu Tingyang pointed out that the 4 h compressive strength of early strength SAC can reach 40 MPa, and the flexural strength can reach 7 MPa. However, the flexural strength decreased after 28 days [6]. To overcome the shrinkage of flexural strength, Zhang et al.used ordinary Portland cement (OPC) to improve the early performance of SAC and studied the early compressive strength of concrete with different proportions of OPC and SAC. It was found that the addition of 10–20 wt.% OPC can make OPC and SAC compound (OPC-SAC) concrete quickly coagulate, and the setting time of OPC-SAC concrete is less than that of any single cement concrete, which provides a reference for the combined use of SAC and OPC [7, 8]. Relevant research results show that when the SAC content is high, the early compressive strength mainly comes from the hydration of SAC, and the hydration of OPC mainly occurs after the 3 d age [9]. When the SAC content is low, free water is consumed by SAC in the early stage, and OPC is covered by hydration products, which hinders the hydration of OPC, resulting in slow strength growth in the later stage [10]. OPC-SAC concrete not only has good early mechanical properties but also has better resistance to sulfuric acid and seawater erosion than OPC concrete [11, 12]. In addition, SAC hydration has the characteristics of micro-expansion. When the SAC content in OPC-SAC concrete is 40wt.%, the shrinkage of concrete is smaller than that of OPC concrete [13], which helps to improve the cooperation ability between repair materials and existing materials.

In recent years, the use of fibers to improve the mechanical properties of concrete has gradually become a hot spot [14, 15]. Xu et al. developed an ultra-high toughness cementitious composite (UHTCC) based on cement mortar and polyvinyl alcohol fiber. The results show that this material has good toughness [16]. Yan studied the influence of different design parameters on the tensile strength of plate dumbbell-shaped high ductility cement-based composite specimens, and pointed out that when the water-binder ratio and fiber volume ratio was low, the fiber only played a role in cracking resistance. The uniaxial tensile stress–strain curve of the specimen did not show strain hardening [17]. Ghosh developed a low-cost, high-early-strength cement-based material with self-compacting characteristics by using SAC to mix with other cement and adding fibers [18] but its ductility is poor. Doo-Yeol uses

Table 1 PVA fiber material parameters

Material	Density (g/cm^3)	Length (mm)	Diameter (mm)	Tensile strength (MPa)	Elastic modulus (GPa)
PVA fiber	1.30	12	0.04	1200	4.5

polyethylene fiber to reinforce SAC-OPC concrete and obtains early-strength high-ductility cement-based composites with 4 h, 28 d compressive and tensile strength of 38 MPa and 5 Mpa, respectively, and 60 and 9 MPa [19]. It should be pointed out that the price of polyethylene fiber used is 7 times that of PVA fiber [20], which is not conducive to the promotion and use. At present, the research results show that the research results of early strength and high ductility cement-based composites are still relatively few. It is difficult for the existing repair materials to have the characteristics of high early strength, small shrinkage deformation and high ductility at the same time.

Therefore, this paper first proposes an early strength high ductility cement composite material with the goal of short curing age, excellent tensile mechanical properties and relatively low price. Furthermore, the effects of PVA fiber content, fly ash content, sand-binder ratio and other parameters on the axial tensile mechanical properties of early-strength high-ductility cement composites at different curing ages were investigated through axial tensile performance tests. On this basis, the axial tensile constitutive model of this type of conforming material is given.

2 Generalization of Experiment

2.1 Raw Materials

Aiming at short curing age (1 d), excellent tensile mechanical properties (greater than 4 MPa) and relatively low price (SAC-OPC compound, domestic PVA fiber, etc.), 42.5 grade ordinary Portland cement, 42.5 grade fast hardening sulphoaluminate cement, polycarboxylate superplasticizer, polymer defoamer, quartz sand, first-grade fly ash and domestic PVA fiber were selected as the main matching materials. The mixing water is tap water, the particle size of quartz sand is 80–120 mesh, and the PVA fiber parameters are shown in Table 1. The preliminary coordination is shown in Table 2.

2.2 Specimen Molding and Maintenance

Using mechanical stirring, the weighed cement, quartz sand, fly ash, defoaming agent and water reducing agent were added to the mixer to stir slowly for 60 s, so that the

Table 2 The fit ratio of the test piece

Number	Number	PVA fiber (%)	FA/wt (%)	S/C	W/C	PCE/wt (%)	Defoamer content/wt (%)	Calcium formate content/wt (%)
1	PVA0	0	30	0.64	0.25	0.5	0.1	1.0
2	PVA0.5	0.5	30	0.64				
3	PVA1.0	1	30	0.64				
4	PVA1.5	1.5	30	0.64				
5	FA 30	2	30	0.64				
6	FA 40	2	40	0.64				
7	FA 50	2	50	0.64				
8	FA 60	2	60	0.64				
9	SA0.5	2	50	0.50				
10	SA0.36	2	50	0.36				

material was fully stirred, and then water was added to stir for 60 s. Finally, PVA fiber was dispersed into the mixer slowly and then stirred for 60 s after PVA fiber was added. After the vibration pouring of the specimen, in order to prevent the water from evaporating too fast, the specimen was covered with a plastic film; the specimens were placed at room temperature for 2 h and then demoulded. After demoulding, the specimens were cured at 20 ± 3 °C with relative humidity of $90 \pm 10\%$ to the specified age.

2.3 Test Method

According to the CECS 13-2009 'Standard test methods for fiber reinforced concrete', the test loading equipment adopts MTSLPS.305 electronic universal testing machine, and the test layout is shown in Fig. 1. The displacement control loading is adopted, and the speed is 0.02 mm/min. The tensile deformation is measured by an electronic extensometer, which is accurate to 0.001 mm, and the effective measurement area is 80 mm. The specimen is dumbbell-shaped, and the relevant dimensions are shown in Fig. 2.

Fig. 1 The failure mode of 2% PVA fiber content specimen

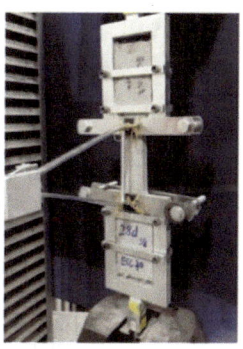

Fig. 2 Failure mode of specimens without PVA fiber

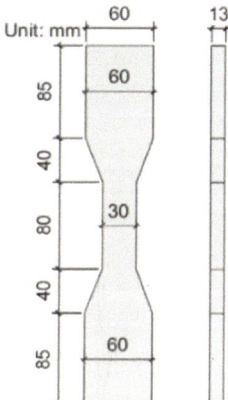

3 Experiment Results and Analyses

3.1 Breakdown Phenomenon

Figures 3 and 4 show the phenomenon of axial tensile failure of cement-based materials with PVA fiber and without PVA fiber.

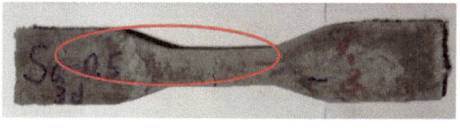 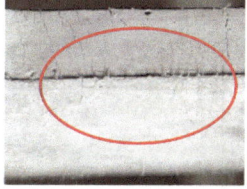

Fig. 3 Failure mode of the specimen with 2% PVA fiber content

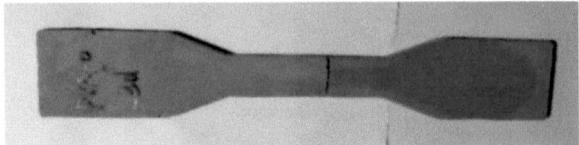

Fig. 4 Failure mode of the specimen without PVA fiber

It can be seen from Fig. 3 that the specimen with 2% PVA fiber content showed good ductility in the process of tensile failure. With the increase in load, when the load reached the initial crack load of the specimen, a very small crack appeared on the surface of the specimen. With the increase of load, the cracks appearing in the initial crack load gradually become wider and can be observed by the naked eye. As the load continues to increase, new transverse small cracks appear on the surface of the specimen. When the load reaches the ultimate load, the load begins to decline slowly, and the cracks continue to widen and connect. Then a main crack is formed, the bearing capacity of the specimen decreases rapidly, some micro cracks are closed, and the test is terminated. It can be seen from Fig. 4 that the cement mortar matrix specimen without PVA fiber shows obvious brittle failure characteristics when it is subjected to tensile failure. At the initial stage of loading, the tensile stress of the cement mortar matrix increases linearly with the increase of strain. When the load reaches the ultimate load of the specimen, a transverse crack suddenly appears on the surface of the specimen. The specimen is divided into two parts and destroyed, and the test is terminated.

3.2 Experiment Results

The direct tensile stress–strain curves of specimens at different ages are shown in Figs. 5, 6 and 7.

Based on Figs. 5, 6 and 7, it can be observed that the early-strength, high-ductility cement-based composite materials at 1 day, 3 days, and 28 days exhibit ductile failure characteristics. The influence of fiber content, fly ash content, and sand-cement ratio on the tensile stress–strain curves of specimens at different ages is significant. Generally, specimens using early-strength, high-ductility cement-based composite materials can be described in the tensile process through the following stages. First stage: Elastic stage, from the initiation of loading to the point of initial cracking, characterized by elastic deformation primarily borne by the cement mortar matrix. Second stage: Strain hardening stage, the curve segment between the initial cracking point and the peak point. In this stage, the tensile deformation of the specimen increases rapidly with the increase of tensile stress, showing a distinct strain hardening phenomenon. At this point, as PVA fibers start to play a role, new horizontal and elongated cracks continuously form on the specimen surface, and stress fluctuates with crack generation until reaching the peak stress. Third stage: Strain

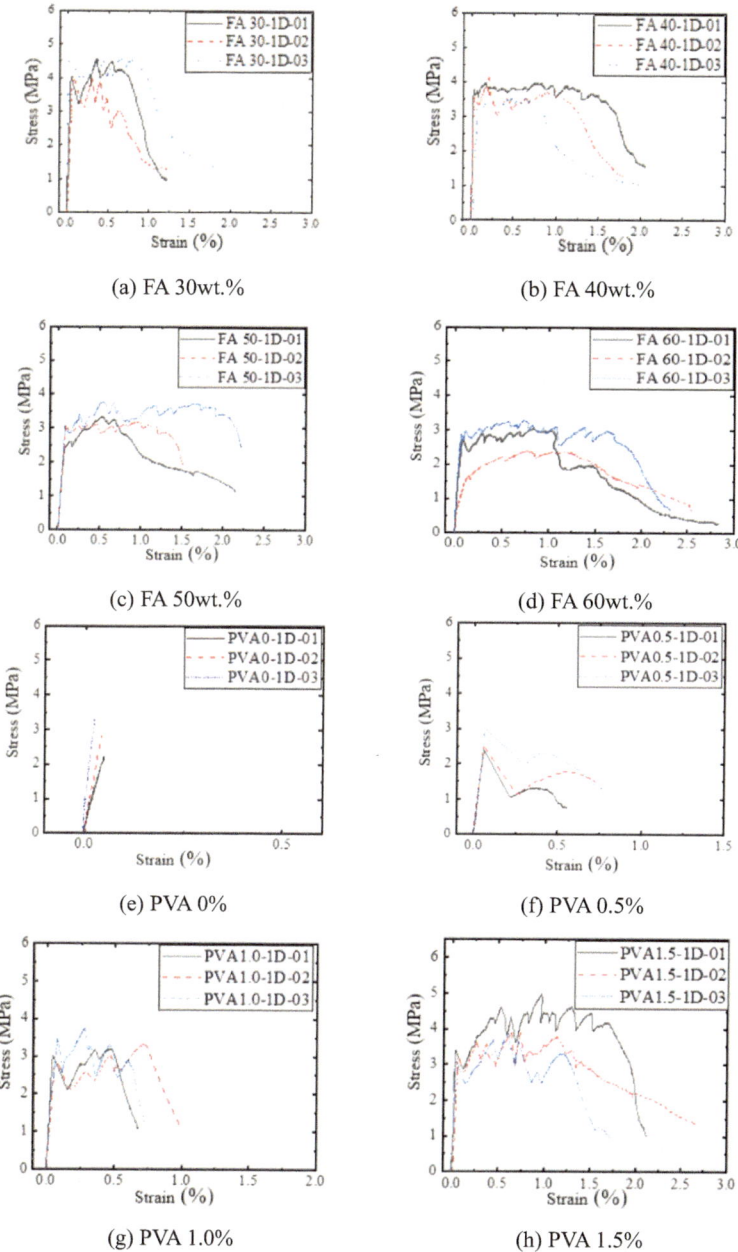

Fig. 5 Direct tensile stress–strain curve at 1 d age

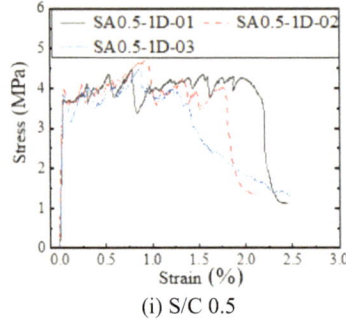
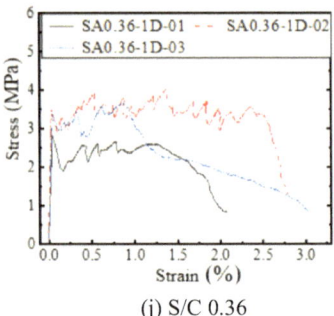

(i) S/C 0.5   (j) S/C 0.36

Fig. 5 (continued)

softening stage, the steady segment after the peak point. In this stage, stress is maintained at a stable level below the peak stress, and small cracks on the specimen surface gradually widen. Fourth stage: Failure stage, where a certain crack on the specimen surface gradually widens, PVA fibers rapidly fracture, and a main crack forms at a certain location within the gauge length of the specimen, leading to the ultimate failure of the specimen. The impact analysis of relevant parameters on the tensile mechanical properties is as follows.

The influence of fibers. Tensile mechanical performance indicators primarily include initial cracking strength, initial cracking displacement, peak strength, peak displacement, etc. Among them, initial cracking strength and initial cracking displacement are significantly influenced by initial defects. Figure 8 illustrates the impact of PVA content on the tensile mechanical properties of specimens at different ages.

(1) Initial Cracking Strength

As indicated in Fig. 8a, during the curing period of 1 day, the initial cracking tensile strength of the specimens shows a gradual increase with the addition of fiber content, ranging between 2.62 and 3.55 MPa. In the curing period of 3 days, the initial cracking strength initially increases and then decreases with the increase in fiber content, reaching values between 2.76 and 3.66 MPa. The maximum value is achieved when the fiber content is 1.5%. At the curing period of 28 days, the initial cracking strength of the specimens ranges between 2.06 MPa and 3.70 MPa, reaching its maximum value at a fiber content of 1.5%. Overall, the initial cracking strength of the specimens at each curing period exceeds 3 MPa when the fiber content is 1.5%.

(2) Peak Strength

As shown in Fig. 8b, during the curing period of 1 day, the specimens' peak strength exhibits a gradually increasing trend with the addition of fiber content, ranging between 2.62 and 3.96 MPa. At the curing period of 3 days, the peak strength initially increases and then decreases with the rise in fiber content, ranging between 2.76 and 4.41 MPa. The maximum value is achieved when the fiber content is 1.5%. During the curing period of 28 days, the peak strength of the specimens ranges

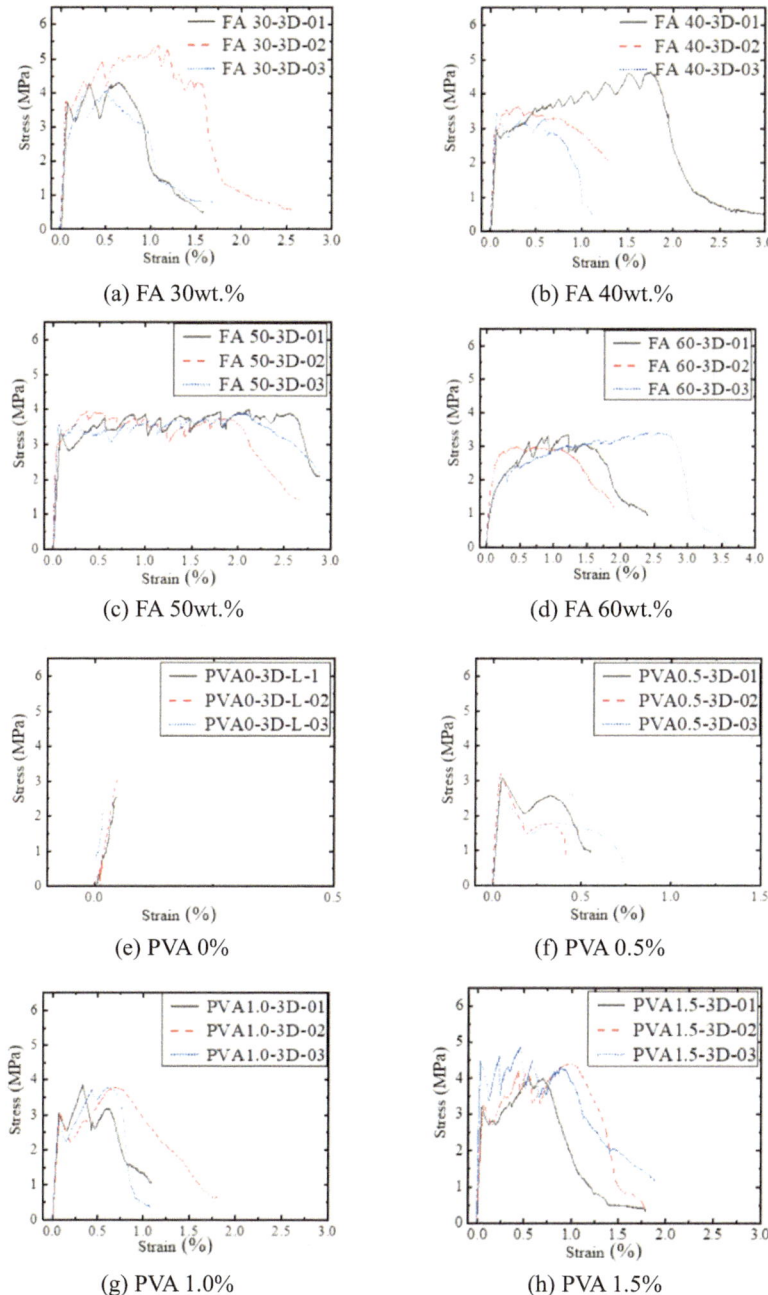

Fig. 6 Direct tensile stress–strain curve at 3 d age

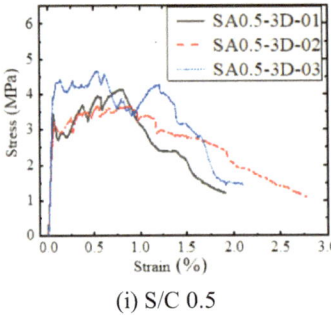
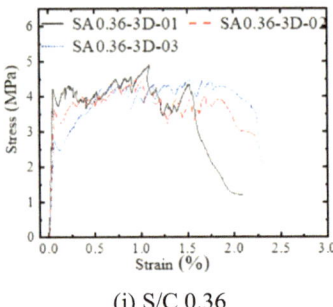

(i) S/C 0.5 (j) S/C 0.36

Fig. 6 (continued)

between 2.07 and 4.82 MPa. Overall, when the PVA fiber content is below 1%, the specimens do not exhibit the characteristics of strain hardening. Under this condition, the initial cracking strength is the same as the peak strength. When the fiber content is 1.5%, the peak strength of the specimens at each curing period exceeds 4 MPa.

(3) Initial Cracking Strain.

From Fig. 8c, it can be observed that due to the difficulty in accurately measuring the initial cracking strain, the influence of fiber incorporation on the pattern of initial cracking strain is not obvious. Generally, the addition of fibers enhances the initial cracking strain of specimens at 1-day and 3-day curing periods to varying degrees. However, fiber incorporation reduces the initial cracking strain of specimens at the 28-day curing period.

(4) Peak Strain

According to Fig. 8d, during the curing period of 1 day, the peak strain of the specimens shows an increasing trend with the addition of fiber content, initially rising and then decreasing. The maximum value is achieved when the PVA fiber content is 1.5%. When the fiber content is below 1%, the peak strain of each specimen is relatively small and close. Compared to specimens without fiber incorporation, the peak strain of specimens with fiber content of 1, 1.5, and 2% increased by 13.1 times, 16.0 times, and 9.8 times, respectively.

During the curing period of 3 days, the peak strain of the specimens increases with the addition of fiber content, reaching its maximum value when the PVA fiber content is 2%. Compared to specimens without fiber incorporation, the peak strain of specimens with fiber content of 1%, 1.5%, and 2% increased by 15.4 times, 20.3 times, and 28.7 times, respectively.

At the curing period of 28 days, compared to specimens without fiber incorporation, the peak strain of specimens with fiber content of 1%, 1.5%, and 2% increased by 5.3 times, 7.5 times, and 4.8 times, respectively. Overall, the peak strain of specimens at the 3-day curing period is most affected by fiber content, and the higher the fiber content, the greater the influence of the curing period on the peak strain of the specimens.

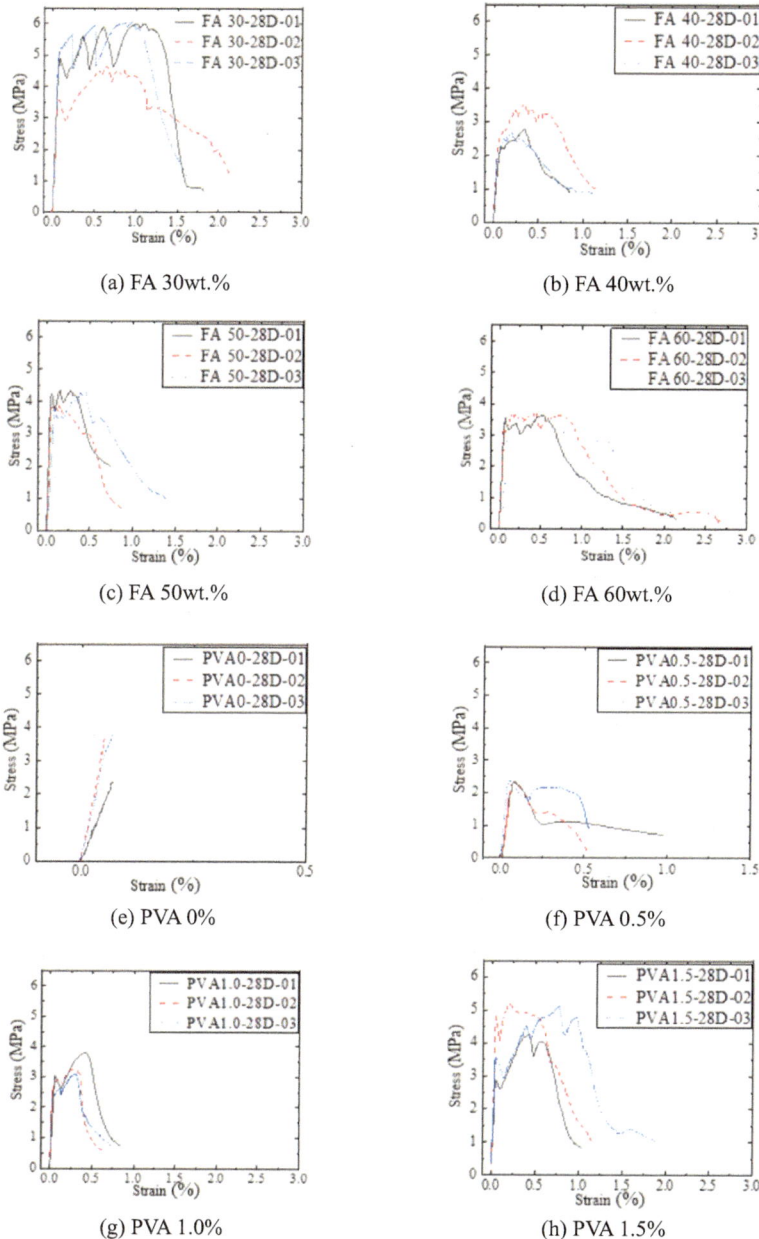

(a) FA 30wt.%
(b) FA 40wt.%
(c) FA 50wt.%
(d) FA 60wt.%
(e) PVA 0%
(f) PVA 0.5%
(g) PVA 1.0%
(h) PVA 1.5%

Fig. 7 Direct tensile stress–strain curve at 28 d age

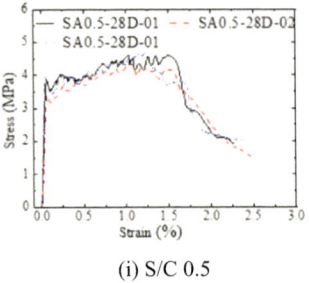
(i) S/C 0.5

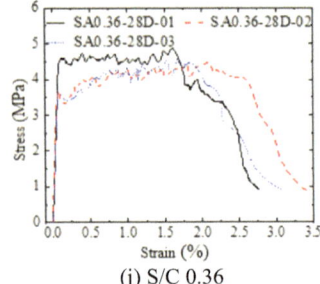
(j) S/C 0.36

Fig. 7 (continued)

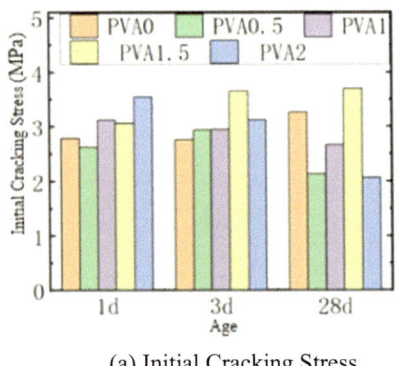

(a) Initial Cracking Stress

(b) Peak Stress

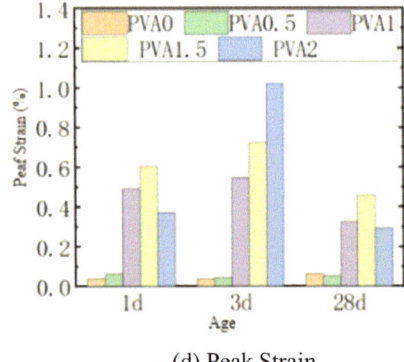

(c) Initial Cracking Strain

(d) Peak Strain

Fig. 8 Effect of PVA content

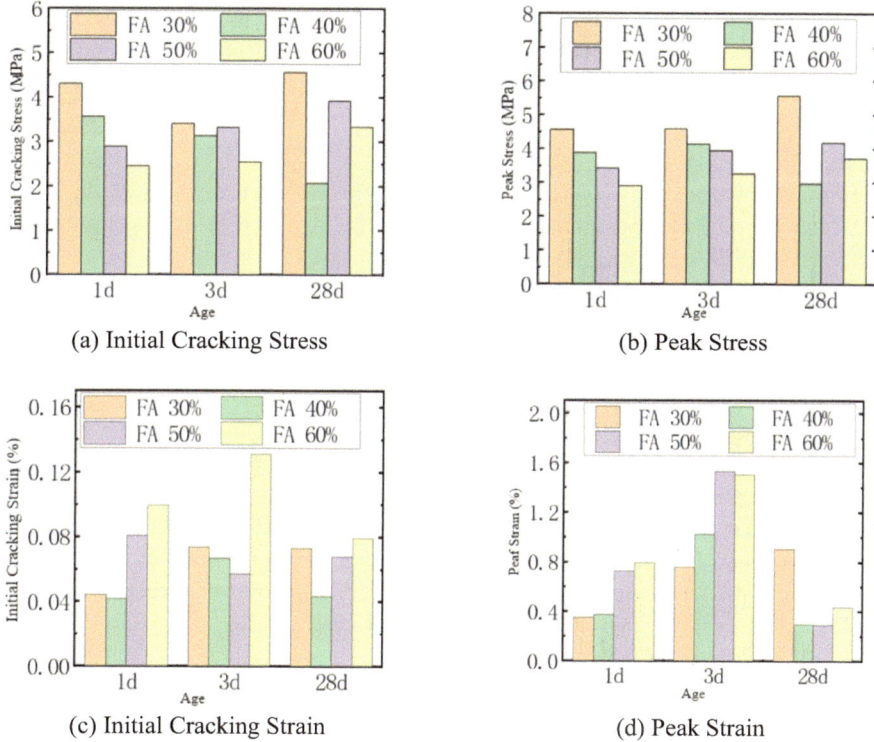

Fig. 9 Effect of FA content

The influence of fly ash. Figure 9 illustrates the impact of fly ash content on the tensile mechanical properties of specimens at different curing periods.

(1) Initial Cracking Strength

From Fig. 9a, it can be observed that during the 1-day curing period, the initial cracking strength of the specimens decreases with the increase in fly ash content. When the fly ash content increases from 30 to 60 wt.%, the initial cracking strength decreases by 43.2%. At the 3-day curing period, the difference in initial cracking strength is less than 8.2% for specimens with fly ash content between 30% and 50 wt.%, but a significant decrease (−23.3%) occurs when the fly ash content increases to 60 wt.%. At the 28-day curing period, the initial cracking strength of the specimens generally shows a decreasing trend with the increase in fly ash content. However, compared to the 3-day curing period, the specimens at the 28-day curing period exhibit a noticeable improvement in initial cracking strength.

(2) Peak Strength

As shown in Fig. 9b, the peak strength of specimens at different curing periods generally decreases with the increase in fly ash content. Additionally, the peak strength of specimens with different fly ash content tends to increase with the curing period. The decrease in peak strength ranges from -29.0% to -36.4% when the fly ash content increases from 30 wt.% to 60 wt.%. Over the curing period from 1 to 28 days, the peak strength of specimens with different fly ash content increases by + 21.1% to + 28.0%.

(3) Initial Cracking Strain

From Fig. 9c, it can be observed that during the 1-day curing period, the initial cracking strain of the specimens increases with the increase in fly ash content. At the 3-day and 28-day curing periods, except for specimens with 60 wt.% fly ash content, there is little difference in initial cracking strain among specimens. Overall, the change in initial cracking strain for specimens with different fly ash content is not significant with the increase in the curing period, mainly due to the substantial influence of initial defects on the initial cracking strain.

(4) Peak Strain

According to Fig. 9d, during the 1-day curing period, the peak strain of the specimens increases with the increase in fly ash content. When the fly ash content increases from 30 to 60 wt.%, the peak strain increases by 126.6%. At the 3-day curing period, the overall trend is an increase in peak strain with the increase in fly ash content. Compared to the 1-day curing period, the peak strain of specimens at the 3-day curing period increases by 90.2–174.0%. At the 28-day curing period, the change in peak strain is not clear, and the peak strain of specimens is generally lower than that at the 3-day curing period. Overall, the peak strain of specimens increases first and then decreases with the increase in curing period, and the peak strain of specimens at each curing period is significantly influenced by the fly ash content.

The influence of sand-cement ratio. Figure 10 illustrates the impact of the sand-cement ratio on the tensile mechanical properties of specimens at different curing periods.

(1) Initial Cracking Strength

From Fig. 10a, it can be observed that during the 1-day curing period, the initial cracking strength of the specimens initially increases and then decreases with the decrease in the sand-cement ratio, reaching its maximum value when the sand-cement ratio is 0.5. At the 3-day curing period, the initial cracking strength increases with the increase in the sand-cement ratio, with a growth rate of 1.1–5.7% for each 0.14 increase in the sand-cement ratio. However, at the 28-day curing period, the initial cracking strength initially decreases and then increases with the increase in the sand-cement ratio. Overall, the initial cracking strength of the specimens increases with the increase in the curing period.

(2) Peak Strength

As depicted in Fig. 10b, during the 1-day curing period, the peak strength of the specimens initially increases and then decreases with the decrease in the sand-cement ratio, reaching its maximum value when the sand-cement ratio is 0.5. At the 3-day curing period, the peak strength increases with the increase in the sand-cement ratio, with a growth rate of 5.5–11.5% for each 0.14 increase in the sand-cement ratio. However, at the 28-day curing period, the peak strength of the specimens increases

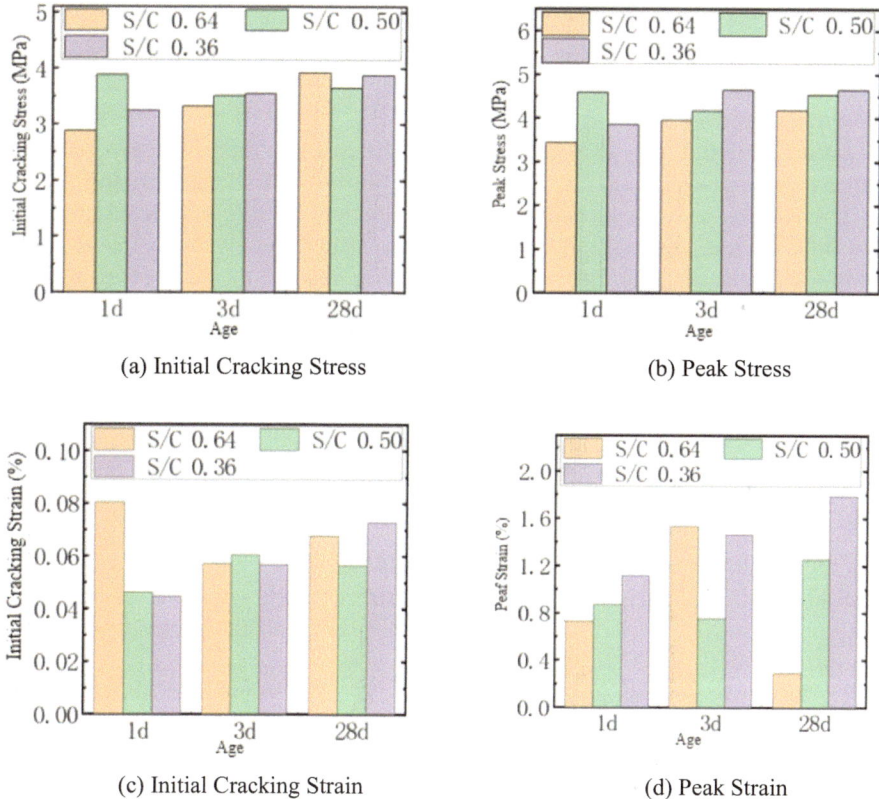

Fig. 10 Effect of S/C

with the decrease in the sand-cement ratio. Overall, the peak strength of the specimens increases with the increase in the curing period, and it is influenced by both the sand-cement ratio and the curing period.

(3) Initial Cracking Strain

From Fig. 10c, it can be observed that during the 1-day curing period, the initial cracking strain of the specimens decreases with the decrease in the sand-cement ratio. When the sand-cement ratio decreases from 0.64 to 0.5, the initial cracking strain decreases by −42.4%. At the 3-day curing period, the initial cracking strain initially increases and then decreases with the decrease in the sand-cement ratio, but the magnitude of the change in initial cracking strain for each specimen is relatively small (approximately 6.0%). At the 28-day curing period, the initial cracking strain of the specimens initially decreases and then increases with the decrease in the sand-cement ratio. Overall, the initial cracking strain is significantly influenced by the initial defects of the specimens, leading to a less clear pattern of influence by the curing period and sand-cement ratio.

(4) Peak Strain

According to Fig. 10d, during the 1-day curing period, the peak strain of the specimens increases with the decrease in the sand-cement ratio. For every 0.14 decrease in the sand-cement ratio, the peak strain increases by 20.1% to 28.3%. At the 3-day curing period, the peak strain initially decreases and then increases with the decrease in the sand-cement ratio. At the 28-day curing period, the peak strain of the specimens increases with the decrease in the sand-cement ratio, and there is a significant difference in peak strain among specimens with different sand-cement ratios. Overall, the peak strain of the specimens increases first and then decreases with the increase in the curing period. The peak strain of specimens at each curing period is significantly influenced by the sand-cement ratio, and both the sand-cement ratio and curing period play crucial roles in determining the peak strain of the specimens.

4 Conclusion

This study introduces an early-strength, high-ductility cement-based composite material and investigates the influence of fly ash content, sand-cement ratio, and PVA volume fraction on its tensile properties through axial tensile mechanical tests. The specific conclusions are as follows:

(1) The early-strength, high-ductility cement-based composite material at 1-day, 3-day, and 28-day curing periods exhibits ductile failure characteristics. Fiber content, fly ash content, and sand-cement ratio significantly impact the tensile stress–strain curves of specimens at different curing periods. The optimal mix ratio of the early-strength, high-ductility cement-based composite material shows four distinct stages of mechanical failure: elastic stage, strain hardening stage, strain softening stage, and failure stage.

(2) Increasing PVA fiber content significantly increases the peak strength and peak strain of specimens under axial tension. At a PVA volume fraction of 2%, the specimens exhibit distinct multiple-crack opening mode and strain hardening characteristics. With a fiber content of 1.5%, the initial cracking strength and peak strength of specimens at all curing periods exceed 3 MPa and 4 MPa, respectively. The peak strain of specimens at the 3-day curing period is most influenced by fiber content, and higher fiber content leads to a greater influence of the curing period on peak strain.

(3) Increasing fly ash content decreases the initial cracking strength and peak strength of specimens under axial tension. When the fly ash content is not less than 50 wt.%, the specimens show significant multiple-crack opening and strain-hardening phenomena. With a decrease in the sand-cement ratio, the peak strength of specimens increases. At a sand-cement ratio of 0.36, specimens at 1-day, 3-day, and 28-day curing periods all achieve $\varepsilon_{0.8u}$ of 2%, and the peak strength exceeds 4.0 MPa.

Acknowledgements This study was partly supported by the Yunjiaoke Jiaobian (Grant No. [2021] 91) and Beijing Natural Science Foundation (Grant No. 8232001). Their support is gratefully acknowledged.

References

1. Fan J (2015) Study on preparation and properties of ultra-early strength cement-based materials (Southeast University). (In Chinese)
2. Gao Z, Hang Y, Wang C (2009) Study on the preparation and mechanism of alkali-free concrete early. J Wuhan Univ Technol 31:81–83 (In Chinese)
3. Wang C, Zhang H, Luo Y and Zhou L (2015) Preparation and mechanism analysis of fast setting and fast hardening high strength concrete. J Huazhong Univ Sci Technol (Natural Science Edition), 89–93. (In Chinese)
4. Zhao M, Xin Y and Wang L (2011) Study on the compounding of inorganic salt early strength agent and polycarboxylate superplasticizer. Concrete, 91–94 (In Chinese)
5. Wang Y, Sun L, Liu S, Li S, Guan X, Luo S (2022) Development of a novel double-sulfate composite early strength agent to improve the hydration hardening properties of portland cement paste. Coatings 12:1485
6. Fu T, Guo B, Luo Y and Zhang J (2018) Research on the application of ultra-early strength sulphoaluminate cement concrete repair material. Concrete pp 140–144 (In Chinese)
7. Yang Q, Zhang X, Liu D, Zhang X, You Z (2018) Study on hydration properties and mechanism of silicate-sulphoaluminate composite cementitious system. Mater Guide 32(517–521):534 (In Chinese)
8. Zhang J, Li G, Ye W, Chang Y, Liu Q, Song Z (2018) Effects of ordinary Portland cement on the early properties and hydration of calcium sulfoaluminate cement. Const Build Mater 186:1144–1153
9. Trauchessec R, Mechling J-M, Lecomte A, Roux A, Le Rolland B (2015) Hydration of ordinary Portland cement and calcium sulfoaluminate cement blends. Cement Concr Compos 56:106–114
10. Zhang J, Ye C, Tan H, Liu X (2020) Potential application of Portland cement-sulfoaluminate cement system in precast concrete cured under ambient temperature. Constr Build Mater 251:118869
11. Bertola F, Gastaldi D, Irico S, Paul G, Canonico F (2020) Behavior of blends of CSA and Portland cements in high chloride environment. Constr Build Mater 262:120852
12. Cao R, Yang J, Li G, Liu F, Niu M, Wang W (2022) Resistance of the composite cementitious system of ordinary Portland/calcium sulfoaluminate cement to sulfuric acid attack. Constr Build Mater 329:127171
13. Kothari A, Tole I, Hedlund H, Ellison T, Cwirzen A (2022) Partial replacement of OPC with CSA cements–effects on hydration, fresh and hardened properties. Adv Cem Res 35:207–224
14. Kanda T, Li VC (1998) Interface property and apparent strength of high-strength hydrophilic fiber in cement matrix. J Mater Civ Eng 10:43598
15. Tetsushi K, CLV (1999) New micromechanics design theory for pseudostrain hardening cementitious composite. J Eng Mech 125:373–381
16. Xu S, Li H (2009) Study on direct tensile test of ultra-high toughness cementitious composites. Chin Civil Eng J 42:32–41 (In Chinese)
17. Zhu J, Li Z, Wang X, Li K, Liu W (2021) Uniaxial tensile constitutive relation model of engineering cement-based composites. J Basic Sci Eng 29:471–482 (In Chinese)
18. Li Y, Liu Z, Liang X (2013) Uniaxial tensile properties of high performance PVA fiber reinforced cementitious composites. Eng Mech 30:322–330 (In Chinese)

19. Ghosh D, Abd-Elssamd A, Ma ZJ (2021) Development of high-early-strength fiber-reinforced self-compacting concrete. Constr Build Mater 266:121051
20. Yoo D-Y, Oh T, Chun B (2021) Highly ductile ultra-rapid-hardening mortar containing oxidized polyethylene fibers. Constr Build Mater 277:122317

Open Access This chapter is licensed under the terms of the Creative Commons Attribution 4.0 International License (http://creativecommons.org/licenses/by/4.0/), which permits use, sharing, adaptation, distribution and reproduction in any medium or format, as long as you give appropriate credit to the original author(s) and the source, provide a link to the Creative Commons license and indicate if changes were made.

The images or other third party material in this chapter are included in the chapter's Creative Commons license, unless indicated otherwise in a credit line to the material. If material is not included in the chapter's Creative Commons license and your intended use is not permitted by statutory regulation or exceeds the permitted use, you will need to obtain permission directly from the copyright holder.

Development and Application of High Permeability and Low Shrinkage Synchronous Grouting Materials

Quanwei Liu, Zhijing Zhu, Weihao Li, Shoujie Ye, Rentai Liu, Mengjun Chen, and Linsheng Liu

Abstract In order to develop a synchronous grouting material with good comprehensive performance, this paper selected cement, fly ash, mineral powder and slag by orthogonal experimental design method to carry out the proportioning test research, which provided the most suitable slurry proportion for engineering construction. The effects of group proportioning on fluidity, compressive and flexural strength, impermeability and volumetric stability were investigated. The results show that the increase of mineral powder content improves the fluidity performance of the system; the slurry nodules with high cement dosage have higher mechanical properties and impermeability; the secondary hydration of fly ash plays a slower role, so it reduces the early strength of the material and increases the late strength; the slag reduces the impermeability and drying shrinkage of the nodules. The research focuses on the characteristics of synchronous grouting in a subway station construction project in Qingdao. It investigates the mechanical and engineering properties of a new type of material and applies it to the field of engineering control. To some extent, this material can replace cement and effectively prevent tunnel water leakage, reduce operation and maintenance costs, and extend the operation and maintenance cycle, showing significant potential for widespread application.

Keywords Shield method · High impermeability · Low shrinkage · Mortar · Synchronous grouting material · Proportioning design

Q. Liu · S. Ye · L. Liu
Qingdao Metro Line 6 Co., Ltd., Qingdao 266427, China

Z. Zhu · W. Li · R. Liu · M. Chen (✉)
Geotechnical and Structural Engineering Research Center, Shandong University, Jinan 250061, China
e-mail: mjun@sdu.edu.cn

© The Author(s) 2025
P. Xiang et al. (eds.), *Frontier Research on High Performance Concrete and Mechanical Properties*, Lecture Notes in Civil Engineering 518,
https://doi.org/10.1007/978-981-97-4090-1_36

1 Introduction

With the continuous expansion of urban areas, the available space on the surface is rapidly decreasing, leading many cities to turn to utilizing underground spaces [1, 2]. Shield tunneling, a construction method that utilizes shield machines to excavate tunnels below the ground surface, offers advantages such as safety, speed, and minimal environmental impact. Hence, it has been widely applied in subway tunnel construction [3–5]. After the assembly of shield tunnel segments, gaps occur between the inner wall of the shield and the outer wall of the segments as the shield advances. If these gaps are not promptly and effectively filled, significant ground settlement can occur, affecting surface buildings and underground pipelines [6–8]. In order to reduce the environmental impact during shield tunneling construction, it is necessary to fill the voids at the shield tail with grout material that exhibits good filling ability, flowability, and certain early and later stage strength. The solidified grout serves the purpose of filling the voids, providing a certain bearing capacity, and stabilizing the segment lining [9, 10]. During shield tunnel construction, the presence of loose or water-bearing strata can lead to water seepage issues. Synchronous grouting can address this by injecting stabilizing materials into the annular gap between the shield and the segments, reinforcing the surrounding strata and creating a sealed waterproof structure. This effectively prevents groundwater from entering the tunnel, ensuring its safe operation. The necessity of synchronous grouting in shield tunnel construction lies in its role in guaranteeing operational safety, enhancing stability, increasing water pressure resistance, and prolonging the lifespan and maintenance cycle of the tunnel. It plays a crucial role in safeguarding the quality and reliability of tunnel engineering [11].

Currently, most of the synchronous grouting materials are inert mortar, which suffer from issues such as easy layering and segregation, poor volume stability, low impermeability, and inadequate durability [12, 13]. Additionally, cement-based mortar consumes a substantial amount of cement, and the production process of cement requires significant energy, including the combustion of fossil fuels like coal, natural gas, and petroleum, resulting in the release of a large amount of carbon dioxide (CO_2). According to statistical data, the global CO_2 emissions in 2022 were estimated to be around 36.8 billion tons, with the cement industry contributing approximately 2.7 billion tons of carbon emissions, accounting for about 7.5% of the global energy-related carbon emissions [14]. This has significant implications for global greenhouse gas emissions and climate change. In the construction of shield tunnels, commonly used post-grouting materials can be broadly classified into two categories: single-fluid grout and dual-fluid grout [9]. The single-fluid inert grout is composed of fly ash, sand, bentonite, water, and additives, without the addition of cement or other cementitious materials. It exhibits a longer setting time, and both early and late strengths are relatively low [15]. The dual-fluid grout is a mixture of cement-based material in liquid A and sodium silicate in liquid B, characterized by a shorter setting time and higher early strength [16]. However, when the grout enters the physical gelation zone, it rapidly loses its fluidity within a short period. This often leads to suboptimal filling effects in the shield tail gap, subsequently affecting the control of soil settlement [15].

This paper investigates the formulation design of grouting materials suitable for shield tunnel construction in highly water-rich strata using ordinary Portland cement, mineral powder, fly ash, slag, and proprietary additives developed by Shandong University. The performance of the grouting materials is evaluated based on parameters such as fluidity, stone body strength, impermeability, and volume shrinkage. By utilizing industrial wastes such as mineral powder, fly ash, and slag as raw materials, an optimal ratio for high-performance, low-carbon synchronous grouting materials is obtained, contributing to resource utilization and waste reduction. Furthermore, field trials of the novel grouting materials were conducted at a shield tunnel construction project in a subway station in Qingdao. The results demonstrate excellent engineering application effectiveness with no tunnel water leakage issues, resulting in reduced operation and maintenance costs and extended operation and maintenance cycles. This research holds significant theoretical value and scientific significance in terms of ensuring operational safety of tunnels, promoting disaster prevention and reduction in the tunnel industry, and reducing energy consumption and carbon emissions.

2 Test Materials and Methods

2.1 Material Design Requirements

According to the "Technical Specifications for Synchronous Grouting Materials in Shield Tunneling", the slurry used for synchronous grouting should have a flowability of $\geq$ 160 mm, a water seepage rate of $\leq$ 3.5%, a strength of $\geq$ 2.5 MPa, and a stone formation rate of $\geq$ 95%.

2.2 Experimental Raw Materials

Cement: PO·42.5 cement produced by Shanshui Cement Plant was selected for this experiment, with a specific surface area of 304 m^2/kg and a compressive strength of 52.2 MPa at 28 days.

Mineral Powder: Grade S95 mineral powder with a calcium oxide content of 30–40% and a specific surface area of 400–500 m^2/kg.

Ground Granulated Blast Furnace Slag: Obtained from Qingdao Hengyuan Power Plant, with a specific surface area of 512 m^2/kg.

Fly Ash: Grade II fly ash obtained from Qingdao Hengyuan Power Plant, with a specific surface area of 413 m^2/kg.

Table 1 Chemical composition of each solid waste

Name of solid waste	SiO$_2$ (%)	Al$_2$O$_3$ (%)	CaO (%)	Fe$_2$O$_3$ (%)	SO$_3$ (%)	MgO (%)	K$_2$O (%)	Na$_2$O (%)	TiO$_2$ (%)
Fly ash	46.78	30.99	8.15	5.44	1.65	0.88	1.78	0.2	1.89
Slags	59.73	23.01	3.82	6.82	0.33	0.79	2.55	0.42	1.16
Mineral powder S95	34.31	11.79	41.62	0.43	2.85	6.38	0.37	0.11	0.94

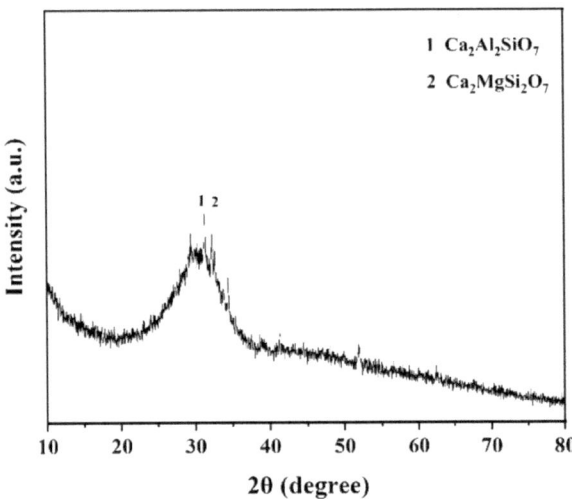

Fig. 1 XRD of S95 blast furnace slag powder

Composite Additives: Developed by the research team, these additives significantly improve the mechanical properties and impermeability of the stone formation and reduce the volume shrinkage of the stone formation.

Mixing Water: Deionized water. Prepared by reverse osmosis (RO) system, the ion content is less than 10 ppm.

The chemical composition of the raw materials is shown in Table 1 and the mineral composition is shown in Figs. 1 and 2.

2.3 Experimental Methods

The cement, mineral powder, ground granulated blast furnace slag, fly ash, and composite additives were mixed in certain proportions. The water-to-cement ratio was determined as 0.5 based on previous experience. After adding mixing water, the

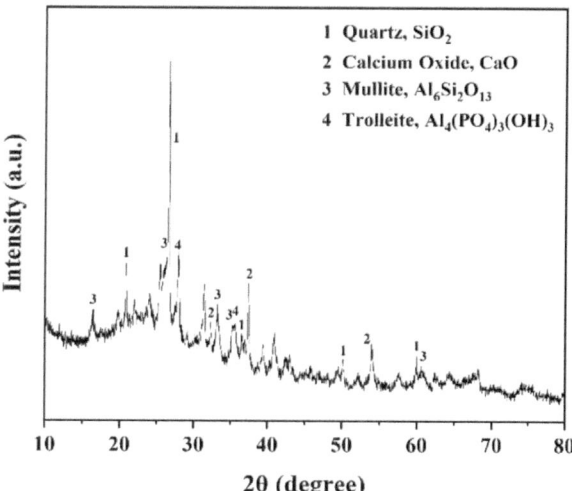

Fig. 2 XRD of coal ash

materials were uniformly mixed using a mixer to obtain a high-performance, low-carbon synchronous grouting material. Compressive strength tests, flexural strength tests, and shrinkage tests were conducted.

Compressive and Flexural Strength Tests: First, the specified amounts of the raw materials were weighed according to the component ratios. After adding water to the mixing pot, cement, mineral powder, ground granulated blast furnace slag, fly ash, and self-developed additives were added. The mixture was stirred at a low speed for 80 s, followed by high-speed stirring for 40 s. After mixing, the specimens were cast and cured. The curing temperature was 20 °C, and the relative humidity was 90%. After 24 h of demolding, the specimens were further cured. Compressive and flexural strength tests were conducted following the standard GB/T176176-2009 for cement strength testing. The testing instrument used was a universal pressure testing machine produced by Beijing Tuoxin Chengxin Development Co., Ltd.

Flowability Test: The flowability test of the slurry in this paper refers to the determination method for the flowability of cement paste in GB 8077-2000 "Test Method for Homogeneity of Concrete Admixtures". The diffusion diameter (mm) of the freshly prepared slurry on a glass plate is used as the representation. The conical mold used has dimensions of an upper mouth diameter of 36 mm, a lower mouth diameter of 60 mm, and a height of 60 mm. The glass plate has dimensions of 600 mm × 600 mm × 5 mm. The glass plate is placed on a horizontal surface or table. The glass plate and conical mold are wiped with a clean, damp cloth to ensure uniform wetting without water stains. The conical mold is placed at the center of the square glass plate. The prepared slurry is poured into the conical mold, and the surface is smoothed and any excess slurry on the glass plate is wiped off. The conical mold is lifted vertically while simultaneously starting the timing. The slurry is allowed to flow on the glass plate for 30 s. A ruler is used to measure the maximum diffusion diameter (mm) of

Table 2 Component ratios

Test no	Test ratio/%			
	Cement (%)	Mineral powder (%)	Fly ash (%)	Furnace slag (%)
1	5	5	15	75
2	5	10	25	60
3	5	15	35	45
4	10	5	25	60
5	10	10	35	45
6	10	15	15	60
7	15	5	35	45
8	15	10	25	50
9	15	15	15	55

the slurry in three uniformly intersecting directions on the glass plate. The average value of these measurements is taken as the flowability (mm) of the slurry.

Impermeability Test: The impermeability test involved using a truncated cone mold with an upper diameter of 70 mm, a lower diameter of 80 mm, and a height of 30 mm. After casting, the specimens were cured under standard conditions for 1 day. After completing the curing, the specimens were further cured until the testing age. The testing instrument used was the SS-25 Cement Soil Permeability Tester produced by Hebei Lisheng Instrument Co., Ltd.

Shrinkage Test: The dry shrinkage change rate of the grouting material was tested according to the standard JCT603-2004 "Test Method for Drying Shrinkage of Cement Mortar". Three stone formation specimens with dimensions of 25 mm × 25 mm × 280 mm were prepared for each group of grouting materials, with detachable copper probes embedded at both ends. After demolding, the initial length value and the length values at different ages of the specimens were measured using a length comparator..

Component Ratio Design. Orthogonal experimental design method was adopted to design the component ratios of the raw materials. The component ratios are shown in Table 2.

3 Material Performance Study

3.1 Slurry Fluidity

Fluidity is an important performance indicator of grouting materials, directly related to the difficulty of grouting construction and the injectability of grouting materials. If the fluidity of the slurry is too low, it cannot fully fill the fractures in the rock mass,

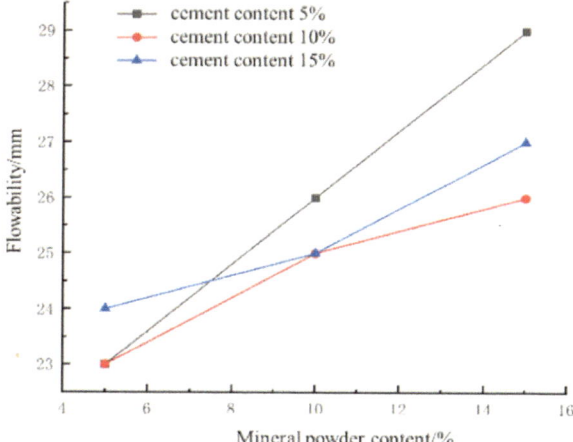

Fig. 3 Effect of group partition ratio on slurry fluidity

thereby failing to achieve the purpose of grouting and sealing [17]. The fluidity of the slurry at different component ratios is shown in Fig. 3.

From Fig. 3, it can be observed that as the content of mineral powder increases, the fluidity of the slurry gradually increases. Among them, the slurry with a cement content of 5% shows the largest variation in fluidity, almost exhibiting a linear increasing trend. The reason behind this is that cement particles have a wide range of particle size distribution and contain many voids between particles. When fine mineral powder is added, the mineral powder particles can fill the voids between cement particles, releasing some free water trapped by cement particles. This process lubricates the interaction between cement particles and promotes the improvement of the fluidity of the cement slurry [18]. Under the condition of a water-cement ratio of 0.5 and a fixed cement content, the dosage of ordinary mineral powder in the cementitious system determines the fluidity of the cement slurry. The inclusion of ordinary mineral powder can increase the free water content in the cement slurry, thereby increasing its fluidity [19]. For the cement slurries prepared with the component ratios of Group 1 and Group 4, their fluidity is below 24 cm, which does not meet the requirements for engineering construction.

3.2 Compressive and Flexural Strength of Stone Body

The strength of the stone body determines the effectiveness of filling and reinforcement between the tunnel lining and the rock mass. The higher the strength of the slurry stone body, the stronger its ability to resist water pressure and surrounding rock stress, resulting in better improvement of leakage for underground engineering [20]. The results of compressive and flexural strength tests on the slurry stone body at 3 days and 28 days are shown in Fig. 4.

Fig. 4 Physical and mechanical properties of serous stone body

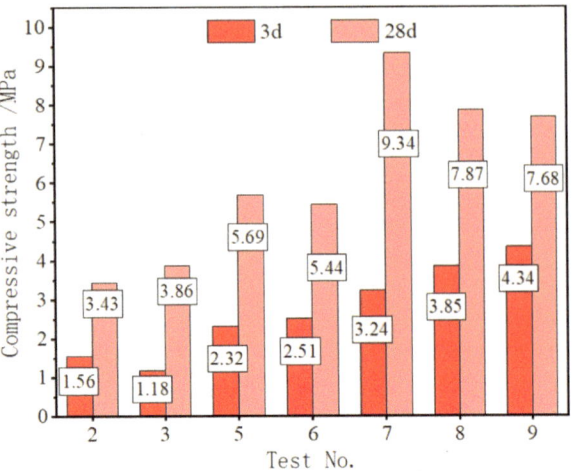
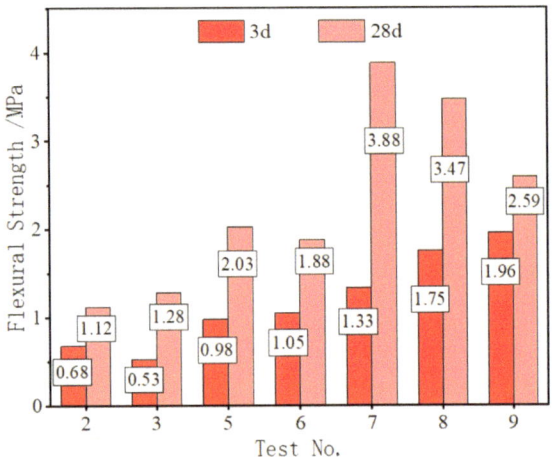

From Fig. 4, it can be observed that the cementitious materials with a higher cement content exhibit better mechanical properties in terms of both compressive and flexural strength. Ordinary fly ash does not possess "latent hydraulicity" but only volcanic ash reactivity, making it difficult to activate its activity [20]. Under the same cement content, the dosage of fly ash affects the early and later strength of the slurry stone body. A higher dosage of fly ash leads to lower early strength and higher later strength. When the cement content is 5% and the fly ash content is 35%, the flexural and compressive strengths at 3 days are the lowest. When the cement content is 15% and the fly ash content is 35%, the flexural and compressive strengths at 28 days are the highest, reaching 9.34 MPa for compressive strength and 3.88 MPa for flexural strength. Thus, it can be concluded that the inclusion of fly ash in the material system can fully utilize the reactivity of volcanic ash, effectively improve the later strength of the material. Considering the dilution effect of groundwater on the slurry, materials from test numbers 6–9 are selected for subsequent tests.

3.3 Permeability of Grout Stone

In practical engineering, the grout stone often faces the erosion of high-salinity seawater, which requires resistance against harmful substances such as chloride ions, sulfates, and carbonates from entering the interior of the grout stone[21]. The permeability of the grout stone is an important factor determining its durability, as it indirectly reflects the porosity and pore distribution of the grout stone and is closely related to its physical and mechanical properties. In order to further optimize the dosage of various minerals in the material system, typical sample ratios were selected for strength analysis, and the permeability characteristics of the samples after 28 days of hydration were tested. The test results are shown in Table 3.

From Table 3, it can be observed that under similar mineral content, the slurry stone body with a higher cement content can withstand higher water pressure, exhibit lower permeability coefficient, and possess stronger resistance to water infiltration. Under the same cement content, an increased dosage of fly ash and mineral powder leads to a stronger resistance against water permeability, while a higher dosage of slag weakens the water permeability. This is because fly ash has volcanic ash reactivity, reacts with the hydration products of cement, lowers the alkalinity of the slurry stone body, improves the pore structure, and increases the compactness of the stone body, thereby enhancing its water resistance. Mineral powder not only exhibits volcanic ash reactivity but also has certain cementitious properties, which can improve the water resistance of the stone body. Additionally, the fine particles of fly ash and mineral powder contribute to micro-filling effects, improving the micro-pore structure and further enhancing the water resistance of the stone body [22]. On the other hand, slag contains more flaky substances and impurities, which affect the pore structure of the stone body, subsequently reducing its water resistance. The material compositions of Test Numbers 6 and 7 in the system cannot meet the requirements for water resistance in actual engineering.

Table 3 Test result of impermeability after 28 days of hydration

Sample no	Sample thickness/mm	Osmotic pressure/MPa	Pervious time/s	Water permeability/mL	Permeability coefficient (m s^{-1})
6	23	0.178	301	22.0	2.43×10^{-8}
7	21	0.318	303	9.6	5.46×10^{-9}
8	22	0.332	305	8.3	4.71×10^{-9}
9	22	0.356	303	6.7	3.55×10^{-9}

3.4 Serum Stone Volume Shrinkage Test

The volume stability of serum stone is an important performance indicator for grouting materials, as it determines the reinforcement effect of the grouting material. Excessive volume changes of the stone can reduce the reinforcement strength and cause secondary water leakage. The volume stability of the serum stone ensures a tight bond between the grout and the rock mass, guaranteeing the effectiveness of grouting reinforcement. The 28-day volume stability is shown in Fig. 5.

From Fig. 5, it can be observed that the eighth group of test specimens had lower dry shrinkage rate, with a 28-day dry shrinkage rate of 12×10^{-4}. In contrast, the ninth group had a relatively higher shrinkage rate. This is because different mineral admixtures have varying effects on the later-stage drying shrinkage performance of the cementitious system. As the amount of fly ash and slag increases, the drying shrinkage of the stone decreases gradually. On the other hand, the addition of mineral powder increases the drying shrinkage of the stone. This is because fly ash and slag have latent activity and act as micro-aggregates in the cementitious system. When fly ash and slag replace cement in equal amounts, the hydration rate of the grout decreases, and the mineral admixtures act as micro-aggregates, causing coarsening of the pores and reducing the degree of hydration of the serum stone, thereby decreasing its shrinkage [23]. Although mineral powder also functions as a micro-aggregate, optimizing the pore structure of the grout and reducing a portion of the drying shrinkage, the increased addition of mineral powder mainly contributes to the drying shrinkage of the grout due to the higher reactivity and hydration degree of mineral powder compared to fly ash and slag. This accelerates the consumption of water and speeds up the internal drying process of the stone, promoting the growth of capillary negative pressure and increasing the area coefficient of action, thus exacerbating the drying shrinkage of the cementitious system.

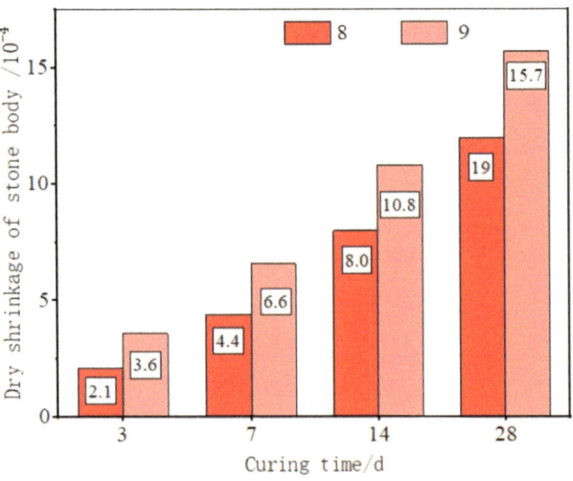

Fig. 5 Test results of dry shrinkage performance of samples at different ages

Based on the comprehensive analysis of various properties of the cementitious system, it is determined that the high-performance low-carbon grouting material consists of cement, mineral powder, fly ash, slag, and self-developed additives. The optimal dosage ratio for the grouting material is as follows: 15% cement, 10% mineral powder, 25% fly ash, 50% slag, and the dosage of self-developed additives should be controlled according to actual conditions.

4 Engineering Application

4.1 Project Overview

Xintun Metro Station on Line 6 of Qingdao Metro adopts shield tunneling construction method. The use of shield tunneling method inevitably causes disturbance to the surrounding strata, resulting in surface settlement and abundant groundwater. A grouting material with good impermeability is needed. Conventional cement-based mortar has poor water erosion resistance, prone to layering and segregation, poor uniformity, and low early strength, which cannot meet the requirements of tunnel engineering.

4.2 Engineering Application Results

In response to the requirements of synchronous grouting for shield tunneling, a green inorganic gelatinous grouting material was developed and field trials were conducted. The process included raw material acquisition, material production, material distribution, and on-site application. The engineering application process is shown in Fig. 6.

(a) Material production   (b) Slurry stirring   (c) synchronous grouting

Fig. 6 Pilot application

Through real-time control of parameters such as grouting pressure and grouting volume, the on-site grouting process achieved synchronous grouting during excavation, completing multiple continuous synchronized grouting rings. The grouting process did not encounter issues such as stratification or pipe blockage caused by the grout, demonstrating good operability. In this project, a total of 100 cubic meters of grout were used for field trials, covering 17 rings. Samples of the on-site grout were taken to evaluate its engineering application effectiveness. After 3 days, 28 days, and 180 days of completion of the grouting construction, in-situ core sampling was conducted at the lifting holes. The retrieved cores are relatively intact, as shown in Fig. 7, and performance tests were performed. Significant traces of gel filling in the grout are visible, indicating good integrity of the cores with a core recovery rate of 91%. The average compressive strength of the cores is 6.376 MPa. The results demonstrate that all parameters of the on-site grout meet or exceed the construction standards.

(a) Field drilling (b) drill hole sampling

(c) Coring sample

Fig. 7 In-situ core sampling

4.3 Research Limitation

Results Studies on the triaxial compression properties of grouting materials are scarce. When grouting materials are used in the backfilling of shield tunnel walls, it is required that they can quickly form a certain early strength. Under the complex stress environment behind shield tunnel walls, the triaxial compression performance of grouting materials should be studied, but the experimental conditions are difficult to simulate the triaxial stress state, and further research is needed. The impermeability and shrinkage of the material were tested, but the durability of the new synchronous grouting material was not involved in the study, and the scientific evaluation method for the durability of the grouting material was lacking, which also needed to be further studied in the future.

5 Conclusion

In response to the engineering requirements of synchronous grouting for shield tunneling, a high-performance low-carbon synchronous grouting material was developed and applied on-site, leading to the following conclusions:

(1) A high-performance low-carbon synchronous grouting material was prepared using industrial waste such as fly ash, mineral powder, and slag. The optimal ratio was obtained based on slurry flowability, stone strength, impermeability, and volume shrinkage: 15% cement, 10% mineral powder, 25% fly ash, 50% slag.
(2) With a constant water-binder ratio of 0.5 and a certain amount of cement, the higher the dosage of mineral powder in the cementitious system, the better the flowability of the grout. The higher the cement dosage, the better the mechanical properties of the grout stone. Fly ash has volcanic ash activity but is more difficult to activate. The early strength of the cementitious system with a higher dosage of fly ash is lower, but the later strength is higher.
(3) Grout stones with a higher cement dosage can withstand greater water pressure and have stronger impermeability. Under the same cement dosage, the higher the dosage of fly ash and mineral powder, the stronger the impermeability of the grout stone.
(4) As the dosage of fly ash and slag increases, the drying shrinkage of the grout stone decreases gradually, while the addition of mineral powder increases the drying shrinkage of the grout stone. The optimal ratio resulted in a 28-day shrinkage rate of the grout stone as low as 12×10^{-4}.

References

1. Feng Y (2011) The state council issued the 13th five-year plan for the development of modern comprehensive transportation system. Urban Mass Transit 20(03):67. Fang F (2011) Research on power load forecasting based on Improved BP neural network. Harb Inst Technol
2. Hong K (2015) Current situation and prospect of tunnel and underground engineering development in China. Tunnel Const 35(02):95–107
3. He C, Feng K, Fang Y (2015) Technical status and prospect of shield tunneling in subway tunnel construction. J Southwest Jiaotong Univ 50(01):97–109
4. Wang J, Zhang L, Feng X et al (2015) Experimental and applied research on alkali excited polymer double liquid grouting materials. Chin J Rock Mech Eng 34(S2):4418–4425
5. Gou C, Ye F, Zhang J et al (2013) Annular distribution model of synchronous grouting filling pressure in shield tunnel. Chinese J Geotech Eng 35(03):590–598
6. Yang Z, Chen H (2009) Analysis and optimal design of slurry ratio for synchronous grouting in shield tunnel. Tunnel Const 29(S2):29–32
7. Zhu D, Zhong X, Wang H (2016) Study on the influence of grouting slurry components on the permeability of shield tunnel. J Hebei Univ Eng (Natural Sci Edition) 33(01):27–30
8. Zheng W, Zou M, Wang Y (2019) Research progress of alkali excited gelling materials. J Build Struct 40(01):28–39
9. Ye F, Mao J, Ji M et al (2015) Research status and development trend of grouting behind shield tunnel wall. Tunnel Const 35(08):739–752
10. Wang M (2014) Technology status, existing problems and development ideas of shield tunneling and TBM tunneling in China. Tunnel Const 34(03):179–187
11. Chen J, Wang J, Zhen C et al (2018) Summary of experimental research on shield grouting materials. Fly Ash Comprehensive Utilization 01:75–80
12. Zhang S, Dai Z, Bai Y (2012) Study on the pressure dissipation of slurry for synchronous grouting in shield tunnel. China Railway Sci 33(03):40–48
13. Zhu X, Song P, Du F et al (2021) Preparation of high performance shield simultaneous grouting materials for water-rich sand layer and study on field grouting performance. New Build Mater 48(03):98–101
14. IEA (2022) CO^2 Emissions in 2022[R]
15. Zhang S, Dai Z, Bai Y (2015) Model test study on grouting pressure distribution in shield tunnel. China Rail Sci 36(5):43–53
16. Dou H (2014) Study on properties and microscopic properties of composite cement-based water glass double-liquid grouting materials. China University of Geosciences
17. Hu A, Xu H, Yang M (2005) Experimental study on new grouting materials. Chinese J Geotech Eng 02:210–213
18. Xu J, Wu K, Zhao D (2016) Development of geopolymer grouting materials for road repair and reinforcement. New Build Mater 43(03):26–28
19. Li Z, Li S, Zhang Q et al (2016) Simulation test and application study on grouting reinforcement of water-rich fractured rock mass. Chinese J Geotech Eng 38(12):2246–2253
20. Wang H, Zhang G, Ding Q et al (2007) Study on properties of alkali excitation and industrial waste grouting materials. J Build Mater 03:374–378
21. Tu P, Wang X (2011) Study on strength deterioration rule and service life of grouting materials for undersea tunnel. Hydrogeol Eng Geol 38(01):65–68+102
22. Xu J, Lin W, Xu K et al (2014) Research on fast hardening and high performance synchronous grouting materials for shield tunnel. Tunnel Construct 34(02):95–100
23. Cui X, Guo W, Wang L et al (2016) Study on the influence of mineral admixtures on the volume stability of mortar. Bulletin of the Chinese Ceramic Soc 35(06):1970–1975

Open Access This chapter is licensed under the terms of the Creative Commons Attribution 4.0 International License (http://creativecommons.org/licenses/by/4.0/), which permits use, sharing, adaptation, distribution and reproduction in any medium or format, as long as you give appropriate credit to the original author(s) and the source, provide a link to the Creative Commons license and indicate if changes were made.

The images or other third party material in this chapter are included in the chapter's Creative Commons license, unless indicated otherwise in a credit line to the material. If material is not included in the chapter's Creative Commons license and your intended use is not permitted by statutory regulation or exceeds the permitted use, you will need to obtain permission directly from the copyright holder.

Development and Field Application of Self-compacting and Highly Impermeable Backfill Materials

Peng Liu, Quanwei Liu, Shoujie Ye, Jia Yan, Rentai Liu, Mengjun Chen, Chao Zong, and Jinyan Jiang

Abstract Aiming at the characteristics of structural backfill engineering, combined with the backfill performance requirements of subway stations, and taking backfill material fluidity, crystalline body strength, seepage resistance and volumetric stability as the main assessment indexes, a self-compacting and high seepage-resistant backfill material has been researched and prepared by taking the bulk solid waste cinder slag and fly ash as the main raw materials, blast furnace slag powder as the dope, and cement as the curing component. Compared with traditional backfill, the material has self-leveling characteristics and good backfill uniformity. The 28-day compressive strength of the material can reach 10.2–21.3 MPa, which exceeds that of C10 concrete. The material has good seepage resistance, seepage pressure up to 0.7–0.8 MPa, good volume stability, 28 d shrinkage rate of 10^{-4} orders of magnitude. The research results were applied in the backfilling project of Xintun Station of Qingdao Metro Line 6, which showed excellent mobility, greatly saving construction cost and time, and the test results met the project requirements. This study takes into account the requirements of material performance and the proportion of solid waste mixing, and under the premise of meeting the engineering requirements of backfill materials, it can effectively absorb fly ash, slag and other bulk solid waste, and has achieved good economic and environmental benefits.

Keywords Self-compacting · Self-leveling · High impereability · Backfill material · Field application

---

P. Liu · J. Yan · R. Liu · M. Chen (✉)
Geotechnical and Structural Engineering Research Center, Shandong University, Jinan 250061, Shandong, China
e-mail: mjun@sdu.edu.cn

Q. Liu · S. Ye · C. Zong · J. Jiang
Qingdao Metro Line 6 Co., Ltd., Qingdao 266427, Shandong, China

1 Introduction

Bulk solid waste has attracted a lot of attention in recent years, the rapid development of China's economy and industrialization, thus generating huge amounts of solid waste, solid waste piled up into a mountain, threatening homeland security and ecological balance [1–3]. It is of great significance to carry out research on the recycling and green utilization of solid waste as a national dual-carbon strategic demand. Cement is the most commonly used cementitious material and occupies an important position in the national economy, but cement consumes a large amount of energy and emits a large amount of carbon dioxide during the production process, and it has defects such as long setting time, poor impermeability, low durability, etc., which is more and more unsuitable for the requirements of high-quality development, and the research and development of new green and low-carbon cementitious materials is an important topic that is of general concern to scholars at home and abroad. Bulk solid waste such as fly ash, granulated blast furnace slag, mineral powder, etc. is an ideal raw material for the synergistic preparation of cementitious materials due to its synergistic effect in chemical composition and mineral composition.

Fly ash and coal slag, as by-products of coal combustion with huge emissions and low cost, can be used for the preparation of bricks, blocks, ceramic granules, etc. [4]. More scholars at home and abroad have already studied their performance in preparing new cementitious materials. Nonetheless, fly ash and slag are still largely unutilized and stockpiled in huge quantities [5]. Therefore, their use in the research and development of cementitious materials can not only eliminate solid wastes to reduce the pressure on the environment, but also meet the engineering applications, which has a broad prospect.

China is in a critical period of large-scale development of infrastructure, major infrastructure projects have a huge demand for backfill materials. The traditional backfill materials are mainly Panax notoginseng ash soil and concrete [6], which has the problems of low strength, poor impermeability, complex and slow construction process, causing serious tunnel water leakage; concrete has the disadvantages of poor toughness, easy cracking, high carbon emission and high price, and it is difficult to promote large-scale [7, 8]. Therefore, it is urgent to study a new green backfill material to gradually replace the traditional backfill material to achieve safe, green and sustainable development. This study developed a backfill material with self-leveling, high impermeability, good volume stability and high economic benefit, which can solve the problems of water leakage of traditional materials, and then applied it in structural backfill engineering.

2 Dimensions of Specimens

2.1 Test Raw Materials

The cement used in the experiment is P.O 42.5 ordinary silicate cement produced by Shandong Shanshui Cement Group Co. The grade of blast furnace slag powder used in the experiment is S95, and the chemical composition as well as the composition of the mineral phase are shown in Table 1 and Fig. 1. From the results of XRF and XRD analysis, it can be seen that the main components of the ore powder are CaO, SiO_2 and Al_2O_3, and the main mineral phase compositions are Fuggerite ($Ca_2Al_2SiO_7$) and Akermanite ($Ca_2MgSi_2O_7$).

The fly ash used in the experiment was from Qingdao Hengyuan Thermal Power Co., Ltd. with a fineness (45 μm sieve residue) of 17.2%, a particle size D50 of 10.94 μm, a specific surface area of 4.70 m²/g, a water demand ratio of 104%, a strength-activity index of 98%. From the results of XRF and XRD analyses, it can

Table 1 Chemical composition of each solid waste

Name of solid waste	SiO_2 (%)	Al_2O_3 (%)	CaO (%)	Fe_2O_3 (%)	SO_3 (%)	MgO (%)	K_2O (%)	Na_2O (%)	TiO_2 (%)
Fly ash	46.78	30.99	8.15	5.44	1.65	0.88	1.78	0.2	1.89
Slags	59.73	23.01	3.82	6.82	0.33	0.79	2.55	0.42	1.16
Mineral powder S95	34.31	11.79	41.62	0.43	2.85	6.38	0.37	0.11	0.94

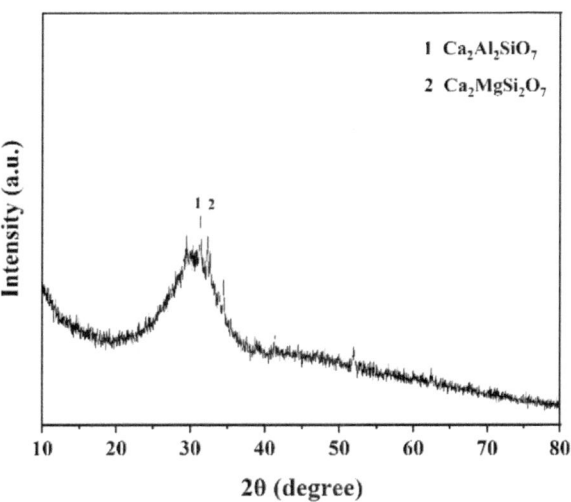

Fig. 1 XRD of S95 blast furnace slag powder

be seen that the main chemical compositions of fly ash are SiO_2 and Al_2O_3, and the main mineral phase compositions are Quartz (SiO_2), Calcium Oxide(CaO), Mullite ($Al_6Si_2O_{13}$), and Trolleite ($Al_2(PO_4)_3(OH)_3$). The mineral composition of fly ash is shown in Fig. 2.

The slag used in the experiment came from Qingdao Hengyuan Thermal Power Co., Ltd, with a particle size of 0.1–0.5 cm, the main chemical composition is SiO_2 and Al_2O_3, and the composition of the mineral phase is Quartz (SiO_2), Sanidine ($KAl_2Si_3O_8$) and Mullite ($Al_2(Al_{2.8}Si_{1.2})O_{9.6}$). The mineral composition is shown in Fig. 3.

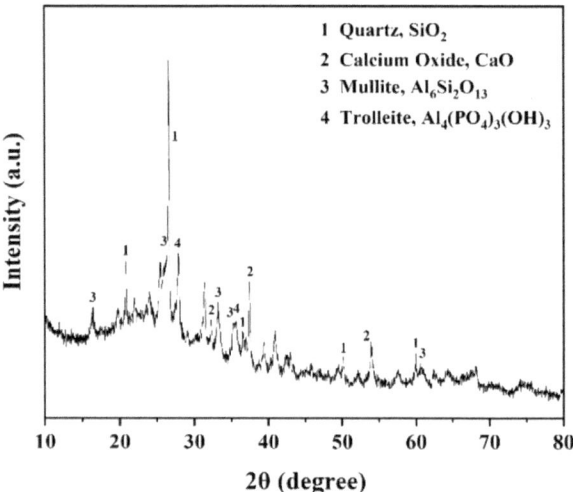

Fig. 2 XRD of coal ash

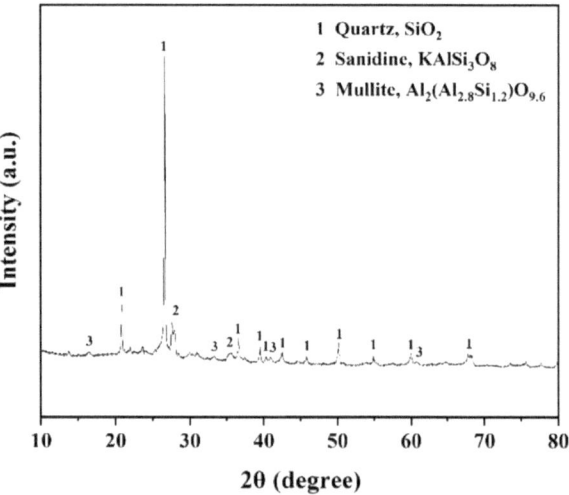

Fig. 3 XRD of cinder slag

The additives used in the experiment were self-formulated composite additives.

2.2 Test Method

The cinder, fly ash, cement, mineral powder as well as additives and water were mixed proportionally with a water–solid ratio of 0.46, and stirred well to obtain the backfill wet material. The mixed wet material was immediately tested for extensibility. The obtained backfill wet material was poured into the molds.

The material extension degree is tested according to GB/T 50080-2016 "Standard Test Methods for Properties of Ordinary Concrete Mixes": place the base plate on a solid horizontal surface and put the slump cylinder in the center of the base plate, and then step on the foot pedal on both sides of the slump, and the slump should be kept in a fixed position when loading materials; load the concrete mixture specimens into the slump cylinder evenly in three layers, and the slump cylinder is about one-third of the height of the cylinder after pounding. Each layer of concrete mix, the application of pounding rod from the edge to the center of the spiral uniformly inserted 25 times. After the top layer is inserted and pounded, take down the loading funnel, scrape away the excess concrete mixture and smooth it along the mouth of the cylinder; after clearing the concrete on the bottom plate of the cylinder, lift the slump cylinder vertically and smoothly, and the slump cylinder lifting process should be controlled in 3–7 s; when the concrete mixture is no longer spreading or the spreading duration has reached 50 s, a steel ruler should be used to measure the maximum diameter of the expanded surface of the concrete mixture and the diameter in a vertical direction with the maximum diameter, and the maximum diameter should be measured with a steel ruler. Maximum diameter is perpendicular to the diameter; when the difference between the two diameters is less than 50 mm, the arithmetic average should be taken as the expansion test results; when the difference between the two diameters is not less than 50 mm, should be re-sampled for further determination.

Impermeability according to JGJ/T 70-2009 "basic properties of building mortar test method standard" test: prepare the upper diameter of 70 mm, the lower diameter of 80, 30 mm high with the bottom of the truncated cone mold as a specimen molding mold. The prepared slurry is poured into the molding mold and vibration 30s, with a spatula to scrape the surface of the mold, will be filled with specimens of the mold into the standard maintenance box maintenance 1d, maintenance can be carried out after the completion of the specimen demolding, after the demolding of the specimen continues to be put into the standard maintenance box to the age of the test. Specimen maintenance can be loaded into the permeability test device to determine its permeability coefficient. When placing the test block, note that the bottom of the test block should be placed in the permeable stone, the upper part of the test block should be placed in the water filter paper, the test block around the application of butter or beeswax for sealing and waterproofing treatment. Permeability test device can be set through the monitor operating platform to adjust the parameters, water pressure from 0.1 MPa to start, constant pressure for 30 min to increase to 0.2 MPa, and then every 30 min to increase the water pressure by 0.1 MPa, through the measurement of a certain permeability pressure for a period of time out of the permeation of water to calculate the coefficient of permeability of the stone body.

Compressive strength in accordance with GB/T17671-1999 "cement sand strength test method (ISO method)" for testing. According to the experimental design to formulate new cement-based slurry and pour it into the treated molds, in turn, poured into the triple molds and ensure that the sealing of the molds, after the completion of the pouring of the slurry together with the molds into the standard conditions of the maintenance box under maintenance for 1 d. After the completion of the maintenance of the test blocks for demolding, the removal of the test block after the test blocks continue to be placed into the standard maintenance of the box to the age of the test. After maintenance to the age of the test block into the compression fixture in the compression and folding machine under the compression fixture through the computer control of the compression fixture column rate of decline for the compression test, until the test block is damaged, to get the test block of the stress with the displacement curve. The compressive strength of the test block is obtained by dividing the pressure on the test block by the area under pressure.

Shrinkage test was conducted in accordance with GB/T 29417-2012 Test Method for Dry Shrinkage and Cracking Properties of Cement Mortar and Concrete. Shrinkage was characterized by the rate of change in length of cementitious material specimens after undergoing age curing. Three hours before the test, the length gauge, standard rod and expansion and contraction limiting device were put into the dry shrinkage test chamber, the gauge was calibrated with the standard rod and the zero point of the micrometer was adjusted, the numbered side of the standard rod was facing up, and its direction and position were fixed; the probe of the gauge was wiped clean. Measure the initial length of the limiting expansion and contraction device. The method is to limit the expansion and contraction device with a numbered side up to measure the initial length, its direction and position will be fixed; limit the expansion and contraction device at both ends of the spherical probe and the measuring instrument's two flat probe

$$\varepsilon_t = \frac{L_t - L}{L_0} \times 100 \qquad (1)$$

where ε_t is the length change rate (%) at age t; L_t is the length measurement at age t (mm); L is the initial length measurement (mm); and L_0 is the baseline length value of the specimen (160 mm).

2.3 Group Distribution Ratio Design

The effects of three factors cement blast furnace slag powder mixing ratio, total cement mineral powder mixing ratio and slag fly ash mixing ratio on the performance of backfill materials were investigated by the controlled variable method respectively.

When studying the influence of cement blast furnace slag powder dosing ratio on the performance of the material law, the fixed coal slag content of 80%, fly ash dosing is 0, mineral powder dosing were 6, 8, 10, 12%, the rest of the ordinary silicate cement, water reducing agent dosing for the above 1% of the total mass of the dry material, the water-solids ratio of 0.46, the specific test allocation shown in Table 2.

To study the total mixing of cement and blast furnace slag powder on the material properties of the influence of the law, fixed the ratio of the two for 1:1, the two mixing were 8, 10, 12, 14%, fly ash mixing is 0, the rest of the coal slag, water reducing agent dosing for the above 1% of the total mass of the dry material, the water-solids ratio of 0.46, the specific test mixing shown in Table 3.

To study the effect of slag fly ash dosing ratio on the material properties of the law, fixed cement and blast furnace slag powder ratio of 1:1, the dosing of 12%, fly ash dosing were 0, 19%, 25%, 38% (the ratio of fly ash to coal furnace slag were 0:1, 1:3, 1:2, 1:1), the rest of the coal furnace slag, water reducing agent dosing for the above 1% of the total mass of the dry material, the water-solids ratio of 0.46, the specific The test mixes are shown in Table 4.

Table 2 Experimental proportioning design

Sample no	Experimental proportion (%)					
	Cement (%)	Mineral powder (%)	Fly ash	Furnace slag (%)	Extruder (%)	Water–solid ratio
A1	14	6	0	80	1	0.46
A2	12	8	0	80	1	0.46
A3	10	10	0	80	1	0.46
A4	8	12	0	80	1	0.46

Table 3 Experimental proportioning design

Sample no	Experimental proportion (%)					
	Cement (%)	Mineral powder (%)	Fly ash	Furnace slag (%)	Extruder (%)	Water
B1	8	8	0	84	1	0.46
B2	10	10	0	80	1	0.46
B3	12	12	0	76	1	0.46
B4	14	14	0	72	1	0.46

Table 4 Experimental proportioning design

Sample no	Experimental proportion (%)					
	Cement (%)	Mineral powder (%)	Fly ash (%)	Furnace slag (%)	Extruder (%)	Water
C1	12	12	0	76	1	0.46
C2	12	12	38	38	1	0.46
C3	12	12	25	51	1	0.46
C4	12	12	19	57	1	0.46

3 Specimens Results

3.1 Extension Degree of Material

In the process of subway station construction, there are often complex buildings above the roof slab and irregular pit space, which hinder the backfilling of the pit. Traditional backfill materials such as ready-mixed concrete and Sanqi gray soil need to be vibrated or compacted, which is complicated to construct and difficult to be applied to narrow spaces. The backfill material developed in this paper has the function of self-leveling, and the construction does not require vibration and compaction, which can greatly improve the construction efficiency. Therefore, the study of material fluidity performance is of great significance, and this paper characterizes the fluidity of the material in terms of the extension degree. The experimental results are shown in Fig. 4:

It can be seen from Fig. 4a that when the total cement and slag dosage is fixed, the extensibility of the freshly mixed backfill material increases with the increase in the dosage of mineral powder, which is due to the dense surface structure of the blast furnace slag microfine with poor hydrophilicity, which is conducive to the sliding of the material particles in the slurry, thus increasing the slurry fluidity [9, 10]. Figure 4b shows that the backfill slurry fluidity is enhanced with the increase in

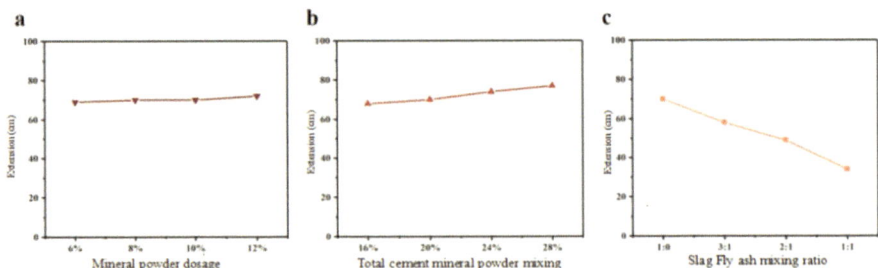

Fig. 4 Effect of various factors on the extensibility of freshly mixed backfill material: **a** Cement blast furnace slag powder mixing ratio, **b** Total cement mineral powder mixing ratio, and **c** Slag fly ash mixing ratio

the total dosage of cement and blast furnace slag powder because of the microporous filling effect of cement and mineral powder, and the dispersion effect is enhanced with the increase in dosage, and the degree of expansion is increased [11]. Figure 4c, on the other hand, illustrates that with the increase of fly ash doping, the material fluidity is significantly reduced, which is analyzed to be due to the surface effect of fly ash, i.e., large specific surface area, better hydrophilicity, adsorption of a large number of water molecules, and a large reduction in the free water of the slurry [12]. In order to meet the needs of engineering construction, the backfill material needs to have good fluidity, so the dosage of fly ash should not be too large [13].

3.2 Test of Compressive Strength of Material Caking Body

In order to protect the ground transportation, building safety, backfill material must have appropriate compressive strength. Three groups of experimental specimens were tested for compressive strength at the age of 3 d, 7 d and 28 d, and the test results are shown in Figs. 5, 6 and 7.

From Fig. 5, it can be seen that under the condition of constant total cement mineral powder dosage, with the increase of mineral powder dosage from 6 to 12%, the 3d, 7d, and 28d compressive strengths of the backfill material decreased from 6 MPa, 8.6 MPa, and 11.7 MPa to 4.9 MPa, 7.2 MPa, and 9.7 MPa, respectively, with the decreases of 18.3%, 16.3%, and 17.1%, respectively. Because the hydration activity of the cement is higher compared to the activity of the S95 mineral powder, the more the mineral powder is dosed, the less the cement is dosed, the lower the strength of the specimen, but the magnitude of the reduction is small, taking into account that the cost of cement is high compared to the cost of the mineral powder, under the circumstance of meeting the requirements of the intensity of the mineral

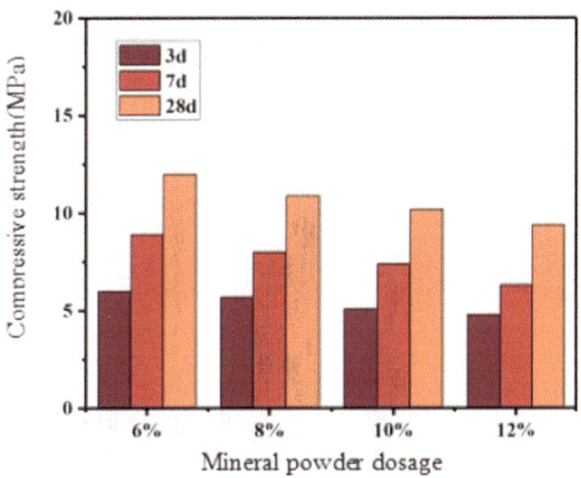

Fig. 5 Effect of cement mineral powder mixing ratio on compressive strength of materials

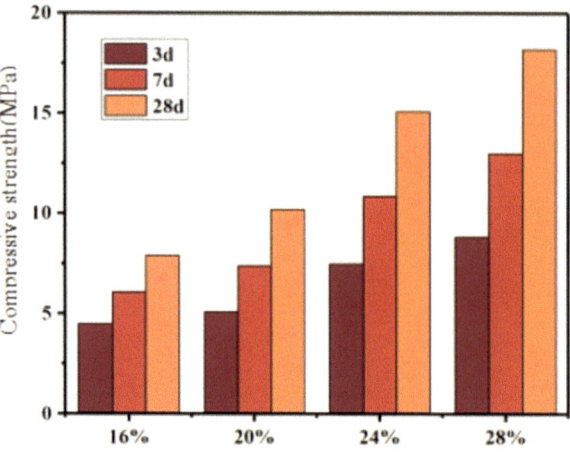

Fig. 6 Effect of total cement mineral powder mixing on compressive strength of materials

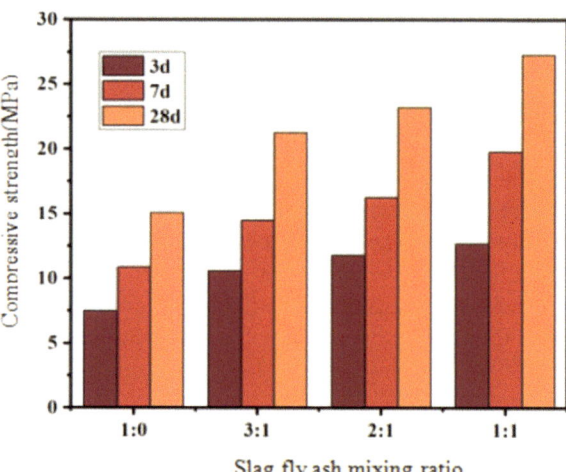

Fig. 7 Effect of slag fly ash dosing ratio on compressive strength of materials

powder dosage can be appropriately increased in order to improve the economic efficiency.

In order to further improve the densification of the material, more powdered admixtures need to be incorporated into the backfill material. Fly ash has the advantages of good water retention, high economic efficiency and potential gelling activity [14], so the effect of fly ash doping on the compressive strength of the material was further investigated. The data in Fig. 7 show that the strength of the specimens increased significantly after the incorporation of fly ash, The 3 d, 7 d, and 28 d compressive strengths of the backfill materials increased from 7.5 MPa, 10.8 MPa, and 15.1 MPa to 12.6 MPa, 20.3 MPa, and 27.5 MPa, respectively, for an increase in the mixing ratio of mineral powder fly ash from 0:1 to 1:1, with increases of 68%, 88%, and 82.1%, respectively. The reason for this is similar to the increase in strength with the increase in cement mineral powder doping in the previous section, fly ash also has a certain gelling activity, which is stimulated by the alkaline environment provided by cement hydration while filling the pores of slag particles to carry out vitreous depolymerization and remodeling to achieve hydration [15], thus increasing the compressive strength of the material. However, the data in Fig. 4c show that the fly ash dosage is too high, resulting in a sharp decrease in slurry extensibility, which is unable to meet the requirements of the material's operational properties, so the fly ash dosage can be appropriately reduced under the condition of meeting the strength requirements.

3.3 Testing of the Impermeability of the Material Nodular Body

Water-rich sand layer is widely distributed in coastal cities, and water seepage in subway stations is a common problem during their construction and operation. The station pit backfill material should have excellent impermeability performance, effectively blocking the seepage of groundwater and avoiding the occurrence of water seepage in the subway station. Combined with the results of the previous experimental data, the test blocks of backfill materials were optimized, and the test blocks with better material gradation, better operation performance and higher compressive strength were selected for the impermeability test, and the test results are shown in Table 5.

Analysis of the data in Table 5 shows that with the increase in the content of powder cementitious materials and admixtures, the impermeability pressure of the test block increases. The admixture of cement, mineral powder and fly ash fills the voids in the backfill material slurry, the material compactness increases, and the hydration products of the cementitious components further fill the pores, so that the impermeability of the material is further improved [16].

Table 5 Seepage pressure tests on specimen blocks of preferred materials

Sample No	Impermeability MPa
B3	0.6
B4	0.7
C4	0.8

Table 6 Shrinkage test of test pieces of preferred materials

Sample no	Shrinkage 10^{-4}		
	7d	14d	28d
B3	2.7	4.4	5.3
B4	2.8	4.8	5.6
C4	3.3	5.7	6.5

3.4 Shrinkage Test of Material Caking Body

Improper construction of underground engineering is prone to cause problems such as stratum deformation and surface collapse, which can lead to serious accidents and jeopardize social security. Shrinkage and collapse of backfill materials are one of the causes of the above problems, so the backfill materials of subway stations must have excellent volume stability. In this paper, some specimen blocks were selected for shrinkage testing, and the experimental results are shown in Table 6.

The data in Table 6 shows that the shrinkage of the material increases slightly with the increase in cement, mineral powder and fly ash content. Increase in cement mineral powder content results in early and rapid hydration of the material, water depletion in the pores of the specimen, decrease in the internal humidity of the pores, increase in self-drying and thus increase in shrinkage [17]. The specimens with large dosage of fly ash have more hydratable components compared to the specimens without fly ash, fly ash hydration consumes more water, and according to the capillary tension theory, the loss of water in the pore space leads to an increase in shrinkage [18]. However, in a comprehensive view, the shrinkage of the three groups of specimens was 10–4 orders of magnitude, and the shrinkage was very small, which met the requirements of the backfill project.

4 Engineering Applications

4.1 Project Overview

Qingdao Metro Line 6 Xintun subway station crosses the water-rich sand layer, which is rich in groundwater and has serious waterlogging. The construction site uses Sanqi gray soil and other waste rock fill for pit backfill. On the one hand, the backfill needs to be compacted, which makes the construction complicated, and it is more difficult and time-consuming for the construction of irregular and narrow space; on the other hand, the impermeability of Panax Grey Soil is poor, which can not effectively block groundwater and is not conducive to the waterproofing and leakage prevention of the station. Therefore, there is an urgent need for a backfill material that is easy to construct, self-leveling, impermeable and effective in blocking groundwater.

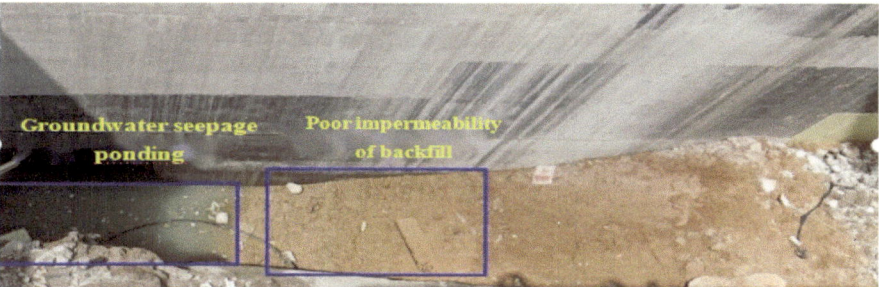

Fig. 8 Top view of backfilled pit

The pilot backfill area is located in the north area of Xintun subway station of Qingdao Metro Line 6, which is a north–south trench, and Fig. 8 shows the backfill pit.

4.2 Engineering Application Results

The construction site is backfilled by pumping with pump truck, the material realizes self-leveling, no need for vibration, and good dispersion resistance underwater. The backfill test total backfill 100 m^3, time-consuming 2.5 h, backfill is completed after 8 h that is to realize the final coagulation, to ensure that the subsequent construction, greatly saving the construction cost. During the test, on-site sampling and testing were carried out, and penetration testing was carried out after 16 h, and the on-site test data met the construction standards. The results of engineering applications are shown in Figs. 9, 10 and 11.

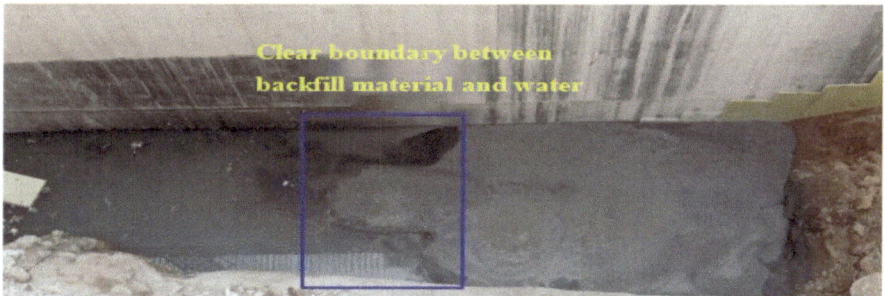

Fig. 9 Water-resistant dispersion of backfill material

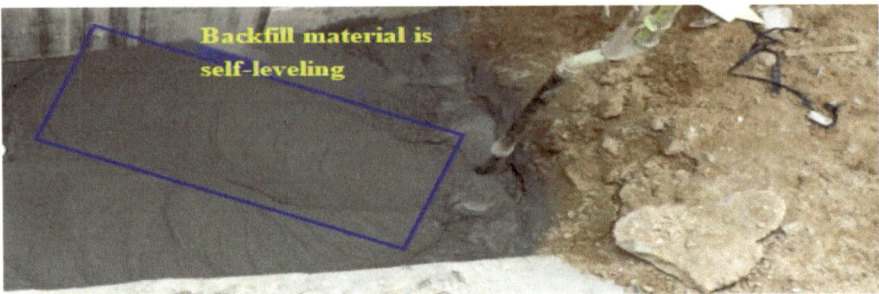

Fig. 10 Self-leveling of backfill material

Fig. 11 Site plan after 16 h of backfilling

5 Conclusion

Aiming at the demand of urban subway station construction project which is rich in groundwater, a self-compacting and highly impermeable backfill material was developed, and the effects of three factors, namely, cement blast furnace slag powder mixing ratio, total mixing amount of cement and mineral powder, and slag fly ash mixing ratio, on the flow performance and mechanical properties of backfill material as well as its impermeability performance, were investigated, and the following conclusions were obtained:

(1) The extensibility of freshly mixed backfill material increases with the increase of cement and mineral powder dosing; it decreases significantly with the increase of fly ash dosing.
(2) The compressive strength of the backfill material nodule body is significantly enhanced with the increase of fly ash dosage and total cement and mineral powder dosage, while when the total cement and mineral powder dosage is fixed and the mineral powder dosage is elevated, the compressive strength decreases slightly.
(3) The impermeability of the backfill material nodular body is enhanced with the increase of cement mineral powder as well as fly ash admixture.

(4) The fluidity of the material in the field test is excellent, which greatly saves the construction cost and time, and the test results meet the project requirements.

In summary, this study developed a self-compacting and highly impermeable backfill material using bulk solid waste such as fly ash and mineral powder, which has significantly better flow properties, mechanical properties and impermeability than traditional backfill materials, and can realize the substitution of traditional cement-based backfill materials, reducing energy consumption and carbon emissions. In the future, we will cooperate with Qingdao Metro Group to promote the use of this material, which is expected to be used 20,000 m^3/year.

References

1. Cai F, Xu H (2022) Current status and progress of solid waste disposal in China. Modern Salt Chem Ind 49(01):84–85
2. Li Y, Liu Y (2021) Research progress and trend of metallurgical solid waste bulk utilization technology in China. J Eng Sci 43(12):1713–1724
3. Yu C, Zhang L, Zheng D et al (2022) Research on solid waste-based mass polymers and their applications. Sci China Technol Sci 52(04):529–546
4. Zhang Z (2021) Analysis of the research contents and application prospects of curing agents based on coal-based solid wastes 3
5. Wang H (2021) Physicochemical characterization of coal-fired slag and feasibility study of its use in wastewater treatment. Energy and Energy Conservat 07:105–106
6. Yang W, Yan Y, Yang J (2022) Experimental study on the construction performance of pumpable backfill mixed with construction waste. J Shandong Agricul Univ (Nat Sci Edition) 53(01):131–137
7. Fan M (2009) Research on the basic properties and application of uncompacted backfill. Master's thesis, Beijing Institute of Technology
8. Xiong Z, Wang Z, Guo L et al (2021) Experimental study on permeability of backfill in pit fertilization trench. Sichuan Building Sci Res 47(06):52–57
9. Ding J, Hao Z, Hao J (2009) Study on the effect of slag powder on the compatibility and strength of concrete. Fujian Construct 12:16–17
10. Niu Y, Bi M (2019) On the main factors affecting the compatibility of fresh concrete. China Building Mater Technol 28(05):76–77
11. Yang Y, Shi S, Tong Z et al (2007) Preparation and strength characteristics of high-performance concrete with large admixture mixes. J Wuhan Univ Technol 11:21–24
12. Zhou X, Wang X (2003) Effect of fly ash on the workability of high performance concrete. Comprehensive Utilization of Fly Ash 02:27–30
13. Wang X, Shen X (2011) Experimental study on the strength of lightweight aggregate concrete with different dosages of fly ash. Silicate Bulletin 30(01):69–73
14. Wang X, Wang Y, Yang L et al (2013) High-performance fly ash concrete with high dosage of fly ash. Research on high-performance large-dose fly ash concrete. Silicate Bulletin 32(03):523–527
15. Li Z, Arjouhan Yan P (2010) Study on hydration degree of cement-fly ash composite cementitious materials. J Build Mater 13(05):584–588
16. Zheng Q, Xu X, Du Z et al (2011) Influence of mineral powder dosage on concrete properties. Concrete and Cement Prod 04:22–24
17. Liu J, Wang D, Song S et al (2008) Study on the properties and microstructure of high dosage mineral powder activated powder concrete. J Wuhan Univ Technol 11:54–57

18. Zhang X, Xiao R, Zhang X, Mao R, Lok J (2005) Analysis of the effect of fly ash dosage on concrete shrinkage and its mechanism. Concrete Cement Products 04:14–17

Open Access This chapter is licensed under the terms of the Creative Commons Attribution 4.0 International License (http://creativecommons.org/licenses/by/4.0/), which permits use, sharing, adaptation, distribution and reproduction in any medium or format, as long as you give appropriate credit to the original author(s) and the source, provide a link to the Creative Commons license and indicate if changes were made.

The images or other third party material in this chapter are included in the chapter's Creative Commons license, unless indicated otherwise in a credit line to the material. If material is not included in the chapter's Creative Commons license and your intended use is not permitted by statutory regulation or exceeds the permitted use, you will need to obtain permission directly from the copyright holder.

Preparation and Blast Responses of Basalt Fiber-Reinforced Polymer (BFRP) Bar Reinforced Shield Tunnelling Segments

Ruiyi Jiang, Jiang Feng, and Min Hou

Abstract BFRPs has the advantages of light weight, corrosion resistance and good durability compared with steels, which is expected to solve the above problems when applied to shield tunnelling segments. In this study, the preparation process of basalt fiber-reinforced polymer (BFRP) bar reinforced shield tunnelling segment (BSTS) is described in detail against the background of shield tunnelling segment actually used in engineering. In order to investigate the blast resistance performance of the BSTS under blast loading, the test process is simulated using LS-DYNA finite element analysis software. The results show that the main damage modes of BSTS under proximity blast loading can be categorized as cratering, cracking, spalling and rupture. With the decreasing Scaled distance, the top surface blast crater is expanding, and the concrete on the back surface is further spalling, showing more obvious local damage characteristics. Provide reference for the design of shield tunnel lining structure's anti-explosion performance.

Keywords BFRP · Shield tunnelling segment · Blast response · Numerical simulation

---

R. Jiang
Research Center of Lightweight Structures and Intelligent Manufacturing, State Key Laboratory of Mechanics and Control of Mechanical Structures, Nanjing University of Aeronautics and Astronautics, Nanjing 210016, China

J. Feng (✉)
Rocket Force University of Engineering, Xi'an 710025, China
e-mail: watermelon0724@163.com

M. Hou
College of Mechanics and Materials, Hohai University, Nanjing 210098, China

1 Introduction

With the continuous progress of science and technology, shield method as a safe and efficient construction technology is widely used in the field of underground engineering. Shield tunnels are involved in military and civilian transportation and have become the main target of terrorist attacks [1–3], so it is of great significance to carry out research on the explosion-proof performance of shield tunnels. Steel can corrode in certain specialized environments, resulting in damage to the engineered structure. Fiber-reinforced polymers (FRPs) are frequently used in protective engineering due to their excellent mechanical properties to address real-world needs [4].

Basalt Fiber Reinforced Polymer (BFRP) has the advantages of light weight, corrosion resistance and durability over steel reinforcement, and can be used as a substitute for steel reinforcement in special cases. Researchers have conducted numerous studies on BFRP reinforcement reinforced concrete. Feng et al. [5] compared the blast resistance of steel-reinforced concrete slabs and BFRP-reinforced concrete slabs, and the results showed that BFRP-reinforced concrete slabs have stronger blast resistance. Gao et al. [6] used sea sand seawater to cast BFRP reinforced concrete slabs to carry out proximity blast tests, and the BFRP reinforced concrete slabs showed excellent blast resistance and corrosion resistance. Li et al. [7] designed and prepared BFRP-reinforced autoclaved aerated concrete panels, and revealed their kinetic response laws and failure modes through proximity explosion experiments. Zhao et al. [8] designed and fabricated six reduced-size BFRP reinforcement-reinforced concrete arch members, and quantitatively analyzed the damage to the specimens by detecting the response displacements and accelerations of the arch members in the explosion test.

Currently, there are fewer studies on the damage modes and blast resistance of BSTS under different blast distances and blast loads. Based on the above background, this study takes the full-size shield tunnelling segment as the background, designs and products BSTS, and uses LS-DYNA finite element analysis software to simulate the experimental process of BSTS under the explosive load, to investigate its damage mechanism and damage characteristics, and to provide a reference for the design of shield tunnel lining structure's explosion resistance performance.

2 BFRP-Reinforced Shield Tunnelling Segments (BSTS)

2.1 BFRP Bars

Basalt fiber bundles are put into a mold and made into BFRP tendons of different diameters by pultrusion molding process. In order to ensure the bonding performance between the BFRP bars and concrete, the surfaces of the uncured BFRP bars are ribbed, and the finished products have an overall black color with a glassy sheen. For curved BFRP bars, it is necessary to place uncured BFRP bars into a bending mold

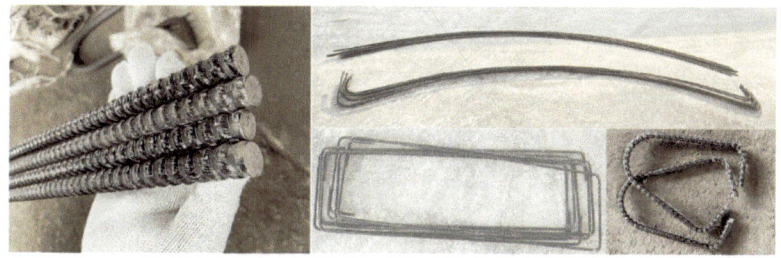

Fig. 1 BFRP bars

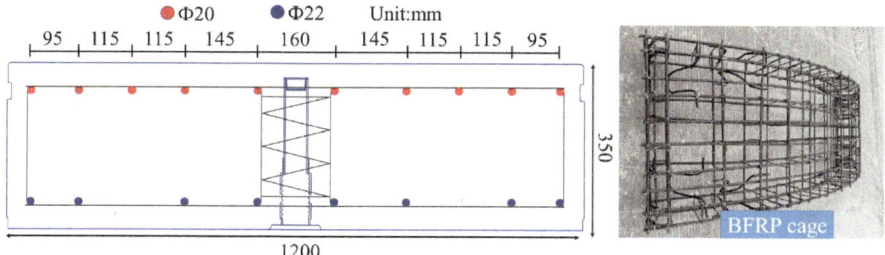

Fig. 2 Geometry of BSTS

in order to produce bars with different curvature shapes. The BFRP bars used in the test is shown in Fig. 1.

2.2 Specimen Size

The design and preparation of BSTS are carried out according to the Code for the Design of Concrete Structures (GB50010-2002), ACI 440.1R-2006 specification [9] and Technical Specification for the Application of Fiber-Reinforced Composite Materials in Construction Projects (GB50608-2010) [10]. The geometric feature of the BSTS is shown in Fig. 2. The shield tunnelling segment has an outer radius of 3100 mm, an inner radius of 2750 mm, a thickness of 350 mm, a width of 1200 mm, and a center angle of 67.5°. There are 10 longitudinal bars of 20 mm diameter on the outer arc and 8 longitudinal bars of 22 mm diameter on the inner arc.

2.3 Specimen Preparation

Figure 3 shows the preparation process of BSTS. Firstly, the BFRP cage needs to be fabricated, which is different from the welding process used for steel cage, where the

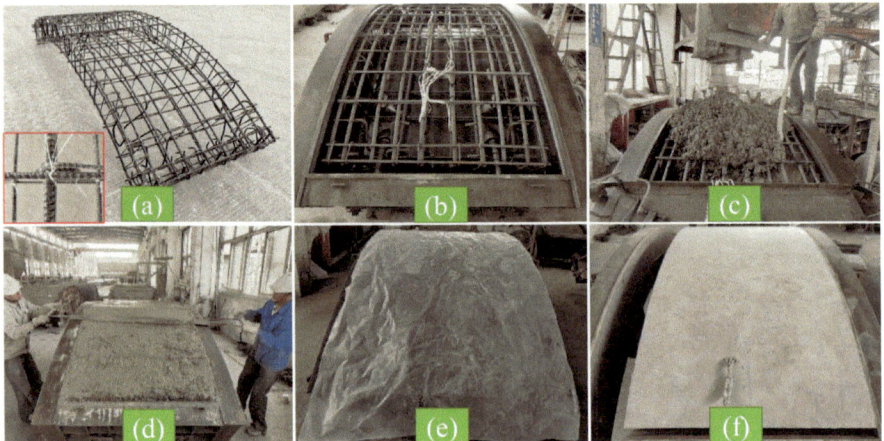

Fig. 3 Process of BSTS preparation

BFRP bars are secured to each other by wire ties, as shown in Fig. 3a. Subsequently, the BFRP cage is placed into the standard shield tunnelling segment mold which is uniformly coated with release agent beforehand, and cement mortar is injected after the position is determined. The strength class of concrete is designed as C50, so the ratio of cement: water: sand: stone is 123:52:233:342. The mortar inside the mold is vibrated with a vibrating rod to expel the air inside the mortar, and at the same time to make the gravel distribution more uniform. After grouting, the surface of the specimen is leveled with a scraper and cured with a film, demolded after 24 h, and finally maintained at room temperature for 28 days to meet the test standard.

3 Numerical Simulation

3.1 Material Model

The material parameters of concrete, air and TNT are shown in Table 1. Concrete is used *Mat_Concrete_Damage_Rel3, which is a model that requires only the input of the concrete unconfined compressive strength parameter, and the rest of the parameters are automatically generated. The maximum principal strains (MXEPS = 0.01) used in this study as the failure criterion, which has been proved to be reliable by many studies [11–13]. TNT explosives are described by the *MAT_HIGH_EXPLOSION_BURN material model, and the equation of state is defined by the keyword *EOS_JWL. The air is modeled using the *MAT_NULL material model and the equation of state is defined by the keyword *EOS_LINEAR_POLYNOMIAL.

BFRP is considered as a linear elastic material. The mechanical properties of different diameters of BFRP reinforcement used in this study are shown in Table 2.

Table 1 Concrete, air and TNT material parameters

Material	Parameter	Value	Material	Parameter	Value
Concrete	Density	2450 (kg/m^3)	TNT	Density	1630 (kg/m^3)
	Compression strength	49.5 (MPa)		E	6 (GJ/m^3)
	$\varepsilon_{failure}$	0.01		A	371.2 (GPa)
Air	Density	1.293 (kg/m^3)		B	3.747 (GPa)
	E	2.53 × 10^{-5} (GPa)		R_1	4.15
	C0, C1, C2, C3, C6	0		R_2	0.95
	C4, C5	0.4		P_{CJ}	21 (GPa)

Table 2 BFRP material parameters

Diameter (mm)	Density ρ (kg/m^3)	Elastic modulus E (GPa)	Tensile strength (MPa)
22	1780	58	769.72
20	1780	55	847.46

3.2 Model Description

The finite element model of the BSTS blast resistance test is shown in Fig. 4. The element type of reinforcement is BEAM161, and the element type of concrete, steel support and air, TNT is SOLID164. The steel support applies a fully fixed constraint and contact is provided between the steel support and BSTS. The bond between the BFRP and the concrete can be considered perfect. The Lagrangian elements are completely wrapped by the Eulerian element, and coupling is achieved through the keyword *CONSTRAINED_LAGRANGE_IN_SOLID. In order to prevent the blast shock wave bypass from affecting the simulation results, the no-reflection boundary condition is set on the outside of the air domain by the keyword *BOUNDARY_NON_REFLECTING.

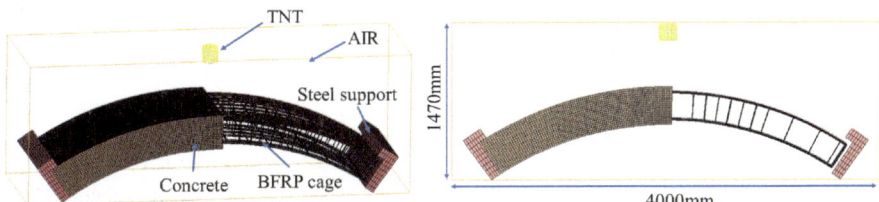

Fig. 4 Finite element model

4 Simulation Results

Figure 5 shows the top surface damage evolution of BSTS under 0.5 m blast distance and 2 kg TNT blast load. The shock wave from the explosion reaches the top of the surface at 0.159 ms and then expands around it in an approximate spherical shape. At 0.919 ms, the shock wave has been transmitted to the whole structure and the damage evolution has basically stopped. As can be seen from the figure, the top surface shows only tiny burst craters and virtually no damage overall.

Figure 6a shows the damage figure of BSTS under 0.5 m blast distance and 8 kg TNT blast load. When the shock wave generated by the explosion reaches the top surface of the BSTS, the concrete elements on the surface will be deleted due to exceeding the ultimate compressive strength, forming an approximately circular crater with a diameter of approximately 750 mm and a depth of up to 45 mm. The compression wave is reflected as a tensile wave after passing through the concrete specimen, causing damage to the back of the BSTS, creating a patch of spalled concrete that leads to the exposure of the bottom BFRP bars. Figure 6b shows the damage figure of BSTS under 0.3 m blast distance and 5 kg TNT blast load. The BSTS showed obvious localized damage characteristics, although the area of the top surface crater is smaller than the 8 kg–0.5 m case, the depth reached 73 mm, and the top BFRP reinforcement had been exposed, and the area and depth of the backside spalled area are also larger, with the crushing depth exceeding 1/2 of the thickness.

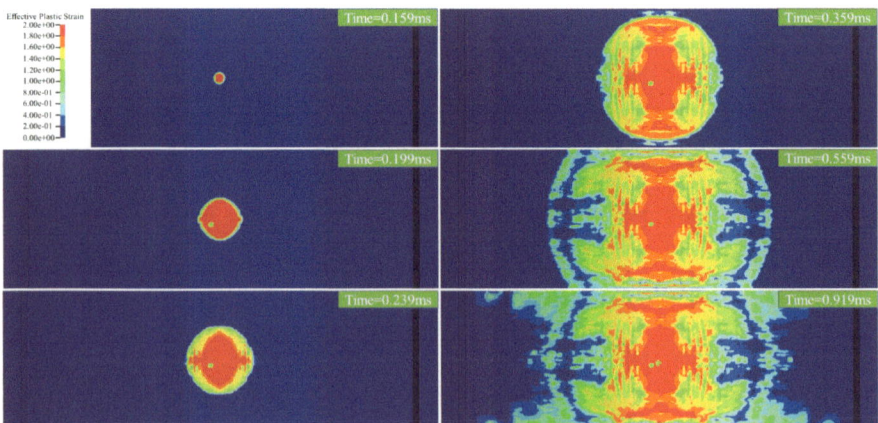

Fig. 5 Damage evolution of BSTS (2 kg–0.5 m)

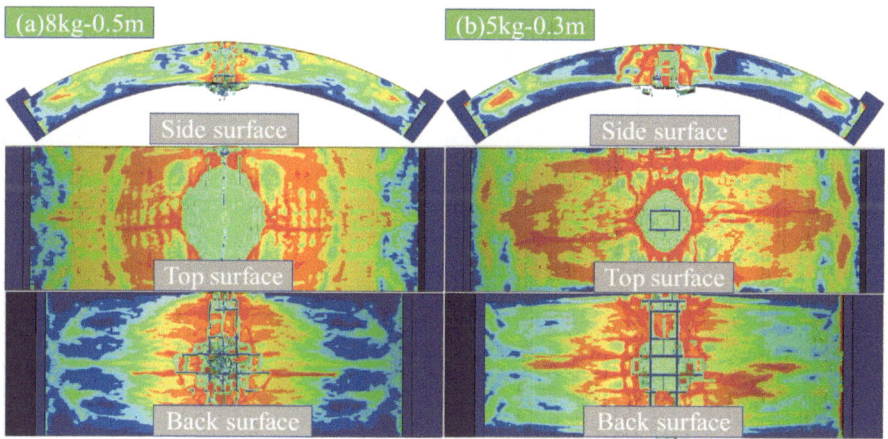

Fig. 6 Damage of BSTS **a** 8 kg–0.5 m, **b** 5 kg–0.3 m

5 Conclusion

In this paper, the preparation process of BSTS is described in detail, and the damage to BSTS under proximity blast loading is investigated using LS-DYNA software. The results show that the BSTS did not show obvious damage under 0.5 m blast distance and 2 kg TNT blast load, but when the mass of TNT is increased to 8 kg, there is a crater on the top surface and concrete spalling on the back surface.

As the scaled distance decreases, the top surface bursting crater continues to expand, and the backside concrete is further spalling. Under 0.3 m blast distance and 5 kg TNT blast load, the top and bottom of BSTS are exposed, and the total depth of crushing is more than 1/2 of the thickness, which showed more obvious localized damage characteristics.

References

1. Almustafa MK, Balomenos GP, Nehdi MLJES (2023) Data-driven reliability framework for qualitative damage states of reinforced concrete beams under blast loading. 294:116803. https://doi.org/10.1016/j.engstruct.2023.116803
2. Zhou L, Li X, Yan Q et al (2023) Blast test and probabilistic vulnerability assessment of a shallow buried RC tunnel considering uncertainty. 180:104717. https://doi.org/10.1016/j.ijimpeng.2023.104717
3. Li C, Aoude HJEFA (2023) Effect of retrofit type on the blast performance and failure mode of HSC beams retrofitted with UHPFRC. 152:107446. https://doi.org/10.1016/j.engfailanal.2023.107446
4. Wang G, Wei Y, Shen C et al (2023) Compression performance of FRP-steel composite tube-confined ultrahigh-performance concrete (UHPC) columns. 192: 111152. https://doi.org/10.1016/j.tws.2023.111152

5. Feng J, Zhou Y, Wang P et al (2017) Experimental research on blast-resistance of one-way concrete slabs reinforced by BFRP bars under close-in explosion. 150:550–561. https://doi.org/10.1016/j.engstruct.2017.07.074
6. Gao Y, Zhou Y, Zhou J et al (2020) Blast responses of one-way sea-sand seawater concrete slabs reinforced with BFRP bars. 232:117254. https://doi.org/10.1016/j.conbuildmat.2019.117254
7. Li S, He H, Yang D et al (2023) Blast responses of BFRP-bar reinforced AAC panels. 365:129655. https://doi.org/10.1016/j.conbuildmat.2022.129655
8. Zhao C, Tang Z, Wang P et al (2022) Blast responses of shallow-buried prefabricated modular concrete tunnels reinforced by BFRP-steel bars 7(2):184–198. https://doi.org/10.1016/j.undsp.2021.07.004
9. Busel JP, Shield CKJFRP (2015) 440.1R-15 guide for the design and construction of structural concrete reinforced with fiber-reinforced polymer (FRP) Bars. https://www.nssi.org.cn/nssi/front/88041353.html
10. GB50010-2002 (2010) Technical specification for application of fiber reinforced composite materials construction engineering. CNKI:SUN:JCJG.0.2003-08-000
11. Li J, Hao HJIJOIE (2014) Numerical study of concrete spall damage to blast loads. 68(jun.):41–55. https://doi.org/10.1016/j.ijimpeng.2014.02.001
12. Luccionia B, Aráoz G (2011) Erosion criteria for frictional materials under blast load. https://www.researchgate.net/publication/285482160_Erosion_criteria_for_frictional_materials_under_blast_load
13. Shu Y, Wang G, Lu W et al (2022) Damage characteristics and failure modes of concrete gravity dams subjected to penetration and explosion. 134:106030-. https://doi.org/10.1016/j.engfailanal.2022.106030

Open Access This chapter is licensed under the terms of the Creative Commons Attribution 4.0 International License (http://creativecommons.org/licenses/by/4.0/), which permits use, sharing, adaptation, distribution and reproduction in any medium or format, as long as you give appropriate credit to the original author(s) and the source, provide a link to the Creative Commons license and indicate if changes were made.

The images or other third party material in this chapter are included in the chapter's Creative Commons license, unless indicated otherwise in a credit line to the material. If material is not included in the chapter's Creative Commons license and your intended use is not permitted by statutory regulation or exceeds the permitted use, you will need to obtain permission directly from the copyright holder.

Seismic Behavior of Steel-Polypropylene Hybrid Fiber Reinforced Concrete Shear Wall

Luyang Zhang, Jitao Yao, and Yuting Tong

Abstract This study investigated the loading mechanism of SPFRC shear walls by conducting low-cycle repeated loading tests on two steel-polypropylene hybrid fiber reinforced concrete (SPFRC) shear walls and one reinforced concrete (RC) shear wall. Furthermore, this study analyzed the impact of fiber content and axial compression ratio on the failure mode, shear carrying capacity, ductility, and energy dissipation capacity of the shear walls. The test results show that hybrid fibers effectively restrain the development of cracks in shear walls, and improve the shear carrying capacity, deformation capacity, and energy dissipation capacity of the shear walls. Considering the contributions of hybrid fibers, horizontal and vertical distributed steel reinforcement, concrete diagonal struts, and concealed columns to shear carrying capacity, a calculation formula for the shear carrying capacity of SPFRC shear walls has been established based on the truss-diagonal brace mechanism. This formula has been validated using data from this study and relevant domestic literature. The average ratio of measured shear carrying capacity to calculated values is 1.01, with a standard deviation of 0.17, indicating a good agreement between them.

Keywords Steel-polypropylene hybrid fiber · Concealed column · Shear wall · Shear capacity · Seismic behavior

1 Introduction

Reinforced concrete shear walls play a crucial role in concrete structures by effectively resisting lateral shear forces. However, their application is constrained by drawbacks such as limited structural ductility and large cross-sectional dimensions. Steel-polypropylene hybrid fiber reinforced concrete presents several advantages,

---

L. Zhang · J. Yao (✉) · Y. Tong
School of Civil Engineering, Xi'an University of Architecture and Technology, Xi'an, Shaanxi 710055, China
e-mail: yaojitao1224@163.com

including superior ductility, high flexural and tensile strength, toughness, exceptional compressive and splitting tensile strength, as well as excellent durability [1, 2]. As a result, it can serve as a novel material solution to address the existing challenges of inadequate ductility, excessive cross-sectional dimensions, and self-weight issues in concrete shear walls. Currently, domestic and international scholars have conducted relevant research on the seismic performance of fiber-reinforced concrete shear walls. However, the majority of these studies primarily focused on the seismic performance of single-fiber concrete shear walls [3, 4], and limited information is available regarding the seismic performance of hybrid fiber concrete shear walls, particularly in the calculation of shear capacity under combined compression, bending, and shear effects. Based on the aforementioned analysis, this study focuses on SPFRC shear walls and conducts low-cycle repeated loading tests to analyze the effects of fiber dosage and axial compression ratio on shear capacity, ductility, and energy dissipation of the shear walls.

2 Experimental Research

2.1 Materials and Mixture Proportions

In the present research, the concrete mixture was designed for the concrete classes of C50 with target cube compressive strengths of 50 MPa. The concrete ratio for the shear wall specimens can be found in Table 1, while the main basic parameters of steel fibers and polypropylene fibers are presented in Table 2.

Table 1 Mixture proportions of concrete (kg/m^3)

Type	Cement	Water	Sand	Coarse aggregate	Water-reducing agent
NC	384	155	630	1100	5.76
SPFRC	384	155	630	1100	6.24

Table 2 Properties of fibers

Fiber type	Length (mm)	Diameter (μm)	Density (g/cm^3)	Ultimate tensile strength (MPa)	Elastic modulus (GPa)
Steel fibers	33	300	27.82	≥ 600	210
Polypropylene fibers	19	33	0.91	530	> 3.5

2.2 Details of Specimens

Three shear wall specimens were designed with fiber volume ratio and axial compression ratio as the main variable parameters. Among them, there were two SPFRC shear wall specimens and one RC shear wall specimen, all with a shear-span ratio of 1.1 and a scale ratio of 1:3. The cross-sectional dimensions of all specimens were b × h = 100 mm × 900 mm, and the vertical distance from the horizontal loading point to the top surface of the loading beam was 100 mm. Constraint boundary columns were provided at both ends of the cross-section. HRB400 steel bars were used for the reinforcement of the columns and the distribution reinforcement of the walls. The main parameters of the shear wall specimens can be found in Table 3, and the dimensions and reinforcement details of the shear wall specimens are shown in Fig. 1.

Table 3 Main parameters of shear wall specimens

Test specimens	V_{f1} (%)	V_{f2} (%)	λ	n
SWC-0.1	0	0	1.1	0.1
SWH-0.1	1.5	1.2	1.1	0.1
SWH-0.2	1.5	1.2	1.1	0.2

Note V_{f1} is the volume fraction of steel fibers; V_{f2} is the volume fraction of polypropylene fibers; λ is the shear-span ratio; n is the axial compression ratio

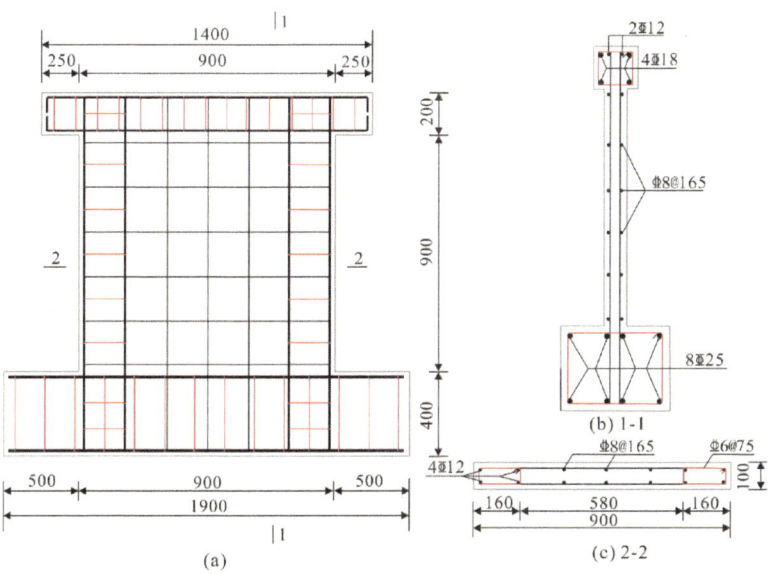

Fig. 1 Size and reinforcement of shear wall specimens

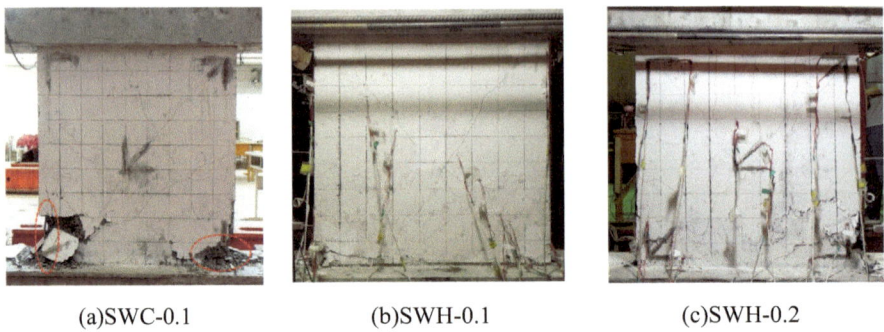

(a)SWC-0.1　　　　　　　(b)SWH-0.1　　　　　　　(c)SWH-0.2

Fig. 2 Failure pattern of specimen

3 Test Results and Discussion

3.1 Failure Modes

The failure modes of specimens SWC-0.1, SWH-0.1, and SWH-0.2 can be seen in Fig. 2. The SWH-0.1 and SWH-0.2 specimens with hybrid fibers show slower crack propagation due to the crack-arresting effect of the fibers. The presence of fibers in the cracks also helps distribute shear forces and enhances the bond between the fiber-reinforced concrete and the steel reinforcement, reducing the likelihood of debonding. Consequently, the load-carrying capacity of the fiber-reinforced shear wall specimens is improved. Compared to the SWC-0.1 specimen, the SWH-0.1 and SWH-0.2 specimens did not exhibit concrete spalling or outward bulging of steel reinforcement during failure. Additionally, these fiber-reinforced specimens had a higher number of cracks that were more evenly distributed. In terms of peak load, the SWH-0.1 specimen increased its load-carrying capacity by 44.1% compared to the SWC-0.1 specimen, while the SWH-0.2 specimen increased it by 46.2%.

3.2 Load–displacement Hysteretic Response

Figure 3 shows the hysteresis curves and skeleton curves of the specimens under horizontal load–displacement. From Fig. 3, it can be observed that all specimens were in an elastic state before the wall cracked, with the loading and unloading curves nearly coinciding and exhibiting linear behavior. The hysteresis loops were not significant, and there was no residual deformation. However, after the appearance of wall cracks, the loading and unloading curves gradually shifted towards the displacement axis, and the area of the hysteresis loop and the residual deformation after unloading increased with the number of cycles. Consequently, the load-bearing capacity increased while the unloading stiffness decreased. After reaching the peak load, the stiffness decreased significantly, and there was noticeable residual deformation, accompanied by a degradation of the load-bearing capacity. Comparative analysis of the hysteresis curves of the specimens reveals that the SWC-0.1 specimen and SWH-0.2 specimen exhibited a reverse "S" shape, while the SWH-0.1 specimen showed a bow-shaped hysteresis curve. The hysteresis curve of the SWH-0.1 specimen was more full, with a peak load approximately 1.4 times that of the SWC-0.1 specimen. It exhibited a higher initial stiffness and better energy dissipation capacity. Comparing the SWH-0.1 specimen and SWH-0.2 specimen, it can be observed that the residual deformation after unloading increased with the increase of the axial compression ratio, resulting in a higher stiffness and poorer energy dissipation capacity.

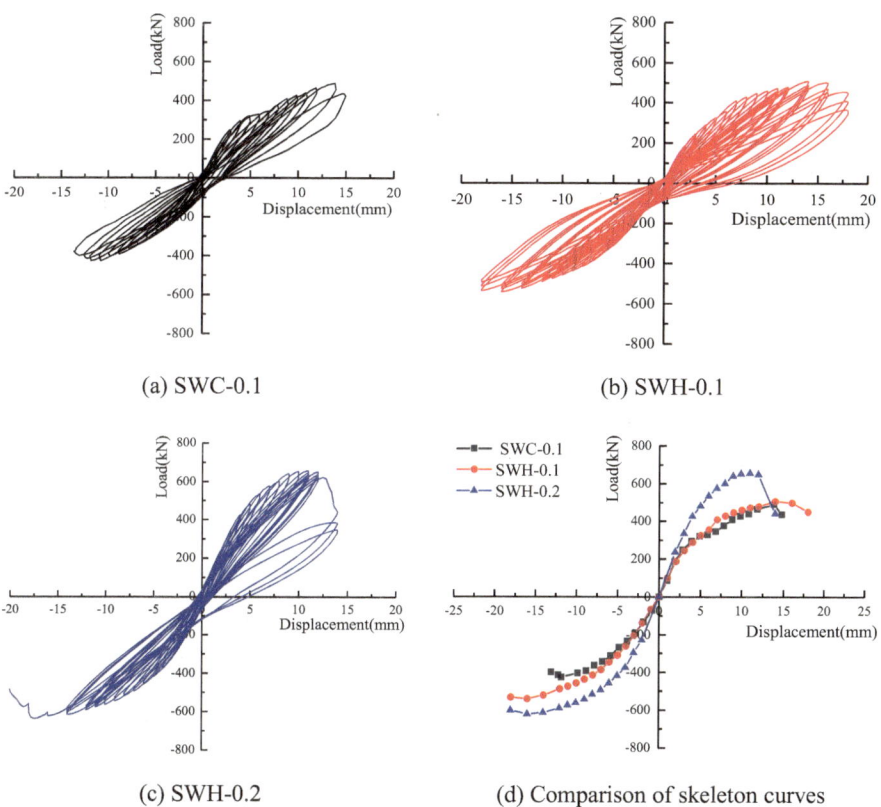

Fig. 3 Hysteretic curves of shear wall and skeleton curves of shear wall

3.3 Energy Dissipation Capacity

The energy dissipation capacity of the specimens reflects its ability to absorb seismic energy under earthquake action, typically measured by the area enclosed by the hysteresis curve and the equivalent viscous damping coefficient [5]. The calculation procedure for the equivalent viscous damping coefficient h_e of the specimens is illustrated in Fig. 4. and computed using Eq. (1).

$$h_e = \frac{1}{2\pi} \cdot \left(\frac{S_{ABC} + S_{ADC}}{S_{\triangle OBD} + S_{\triangle ODF}} \right) \quad (1)$$

The results of calculating the area enclosed by the hysteresis curve and the equivalent viscous damping coefficient for the specimens are presented in Table 4. It is evident from the table that specimen SWH-0.1 displays superior energy dissipation capacity. In comparison to specimen SWC-0.1, specimen SWH-0.1 shows a remarkable increase of 61.40% in energy dissipation capacity. However, specimen SWH-0.2 experiences a decrease of 45.11% relative to specimen SWH-0.1. These findings indicate that the inclusion of hybrid fibers enhances the energy dissipation capacity of the specimens, while an increase in the axial compression ratio reduces the overall energy dissipation capacity.

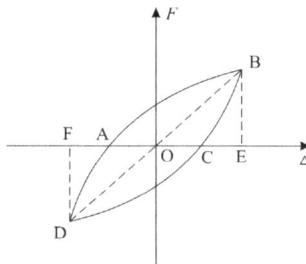

Fig. 4 Calculation of equivalent viscous damping coefficient

Table 4 Hysteresis curve area and equivalent viscous damping coefficient of the test piece

Test specimens	$S_{ABC} + S_{ADC}$ (mm²)	$S_{\triangle OBD} + S_{\triangle ODF}$ (mm²)	h_e
SWC-0.1	3983.664	5577.629	0.114
SWH-0.1	9235.661	7999.702	0.184
SWH-0.2	5051.534	8014.535	0.101

4 Calculation of Shear Bearing Capacity

The shear capacity of a hybrid fiber reinforced concrete shear wall with concealed columns is distributed between the wall structure and the concealed columns within the perimeter frame. The formula for calculating its shear capacity V can be expressed as:

$$V = V_w + V_e \tag{2}$$

where V_w is the shear capacity of the wall structure; V_e is the shear capacity of the concealed columns within the perimeter frame.

4.1 Shear Capacity of the Wall Structure

Under the combined effects of compression, bending, and shear, the force transmission mechanism and simplified shear capacity calculation model of the hybrid fiber-reinforced concrete shear wall structure are shown in Fig. 5.

Based on the force transmission mechanism of the shear wall and the simplified shear capacity calculation model, the shear capacity calculation formula for the hybrid fiber-reinforced concrete shear wall with concealed columns is as follows:

$$V_w = V_c + V_s + V_f \tag{3}$$

where V_c, V_s and V_f are the shear capacity of concrete, distributed steel reinforcement, and hybrid fibers, respectively.

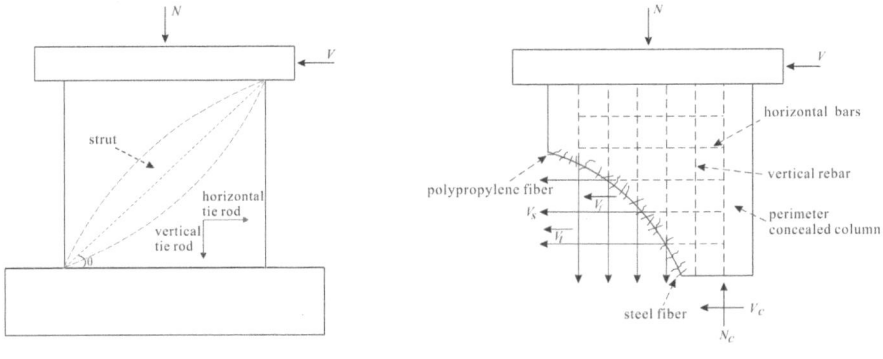

(a) shear transfer mechanism   (b) simplified calculation model

Fig. 5 Shear transfer mechanism and simplified calculation model

Shear capacity of concrete

According to the concrete strut mechanism, V_c can be represented as:

$$\begin{aligned} V_c &= \zeta f_c A_{str} \cos\theta \\ &= \zeta f_c a_w b \cos[\tan^{-1}(h/h_0)] \\ &= \zeta f_c (0.25 + 0.85N/f_c bh) hb \cos[\tan^{-1}(h/h_0)] \end{aligned} \quad (4)$$

where ζ is the concrete softening factor, which can be calculated as $\zeta \approx 3.35/\sqrt{f_c'} \leq 0.52$ [6]; f_c is the fiber reinforced concrete compressive strength; θ is the angle of inclination of the concrete strut; a_w is the height of the concrete strut; b is the width of the wall section; h is the wall height; h_0 is the effective height of the wall section; N is the vertical axial force acting on the wall.

Shear capacity of distributed steel reinforcement

The shear capacity of horizontally distributed steel reinforcement and vertically distributed steel reinforcement is calculated using the shear friction theory [7].

$$V_s = \frac{(H/h)^{4/3}}{1+(H/h)^{2/3}}\eta\omega_1\rho_{sh}f_{yh}bh + \frac{1}{1+(H/h)^{2/3}}\eta\omega_2\rho_s f_s bh \quad (5)$$

$$\omega_1 = \frac{1}{\sqrt{1+3(H/h)^{2/3}}} \quad (6)$$

$$\omega_2 = \frac{1}{\sqrt{3+(H/h)^{2/3}}} \quad (7)$$

where η is the shear friction coefficient, for ordinary concrete, η is taken as 0.85, and for fiber-reinforced concrete, η is taken as 1.0; ρ_{sh} and ρ_s are the horizontal and vertical distribution steel reinforcement ratios of the wall, respectively; f_{yh} and f_s are the yield strength of the horizontal and vertical distribution steel reinforcement of the wall, respectively; ω_1 and ω_2 are the stress coefficients for the horizontal and vertical distribution steel reinforcement, respectively.

Shear capacity of hybrid fibers

According to composite material theory, introducing the enhancement coefficients α and β, and considering the influence of fiber hybridization, the strength modification factor γ is introduced to account for the effect of fiber hybridization on the tensile strength of hybrid fiber-reinforced concrete f_{ft}, resulting in an empirical formula.

$$f_{ft} = f_t(1 + \alpha\lambda_1 + \beta\lambda_2 + \gamma\lambda_1\lambda_2) \quad (8)$$

where f_t is the tensile strength of the ordinary concrete; λ_1 is the characteristic value of steel fibers, which can be calculated as $\lambda_1 = V_{f1}(l_{f1}/d_{f1})$, V_{f1} is the steel fiber volume fraction, and l_{f1}/d_{f1} is the aspect ratio of steel fibers; λ_2 is the characteristic value of polypropylene fibers, which can be calculated as $\lambda_2 = V_{f2}(l_{f2}/d_{f2})$, V_{f2} is the polypropylene fiber volume fraction, and l_{f2}/d_{f2} is the aspect ratio of polypropylene fibers; γ is the fiber hybridization effect enhancement factor, which is calculated to be 0.96; The values of α and β are obtained through experimental data from this study and reference [8] and are fitted to $\alpha = 0.077$ and $\beta = 0.096$.

The shear-carrying capacity of hybrid fiber can be expressed as: $V_f = V_c \cdot \alpha_1 f_{ft}/f_t$, α_1 is the enhancement coefficient of hybrid fibers on the shear carrying capacity of shear walls, which is calculated to be 0.3 [9].

4.2 Shear Carrying Capacity of Hidden Columns

The shear carrying capacity of hidden columns, calculated using the formula for shear carrying capacity of hybrid fiber concrete columns, is as follows [10]:

$$V_e = 0.42\sqrt{f_c} \cdot \sqrt{1 + \frac{n\sqrt{f_c}}{0.42}} \cdot b_e(0.9h_e - a_s) + \frac{A_s f_{ys}(0.9h_e - a_s)}{s} \quad (9)$$

where b_e is the cross-sectional width of the hidden column; h_e is the cross-sectional height of the hidden column.

4.3 Comparisons Between Experimental and Calculated Values

Formula (2) has been used to compute the shear carrying capacity of 13 shear walls, as referenced in this study and literature [11]. The comparison between these calculated values and the experimental values is shown in Fig. 6. It is clear that the average ratio of the measured shear force values to the calculated values for these 13 shear walls is 1.01, with a standard deviation and coefficient of variation of 0.17. This indicates that the calculated results based on the proposed formula for shear walls made of hybrid fiber concrete in this paper align well with the experimental values, demonstrating the feasibility of considering the contribution of hidden columns to shear carrying capacity.

Fig. 6 The comparison between calculated values and experimental values

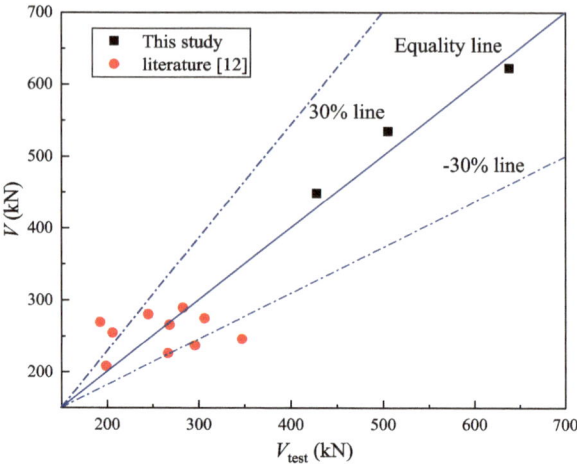

5 Conclusions

1. Compared to ordinary concrete specimens, hybrid fiber reinforced concrete specimens exhibit delayed crack initiation, a higher number of cracks at failure, smaller maximum crack width, and no occurrence of significant concrete spalling.
2. The addition of hybrid fibers effectively suppresses the development of cracks in shear walls, thereby improving their shear carrying capacity, deformation capacity, and energy dissipation capacity. In terms of peak load, ductility, and energy dissipation capacity, the hybrid fiber-reinforced concrete specimens with an axial compression ratio of 0.1 demonstrate respective increases of 44.1, 13.07 and 61.40% compared to ordinary concrete shear wall specimens.
3. The calculation results obtained from the established shear capacity calculation formula align well with experimental values. This indicates that it can effectively reflect the stress mechanism of shear walls and predict the shear capacity of shear walls with concealed columns in hybrid fiber-reinforced concrete.

Acknowledgements The author gratefully acknowledges the funding supported by the National Natural Science Foundation of China (Program No. 51968070).

References

1. Sukontasukkul P, Pongsopha P, Chindaprasirt P et al (2018) Flexural performance and toughness of hybrid steel and polypropylene fibre reinforced geopolymer. Constr Build Mater 161:37–44
2. Afroughsabet V, Ozbakkaloglu T (2015) Mechanical and durability properties of high-strength concrete containing steel and polypropylene fibers. Constr Build Mater 94:73–82
3. Seo MS, Kim HS, Truong GT et al (2017) Seismic behaviors of thin slender structural walls reinforced with amorphous metallic fibers. Eng Struct 152:102–115

4. Zhao J, Dun H (2014) A restoring force model for steel fiber reinforced concrete shear walls. Eng Struct 75:469–476
5. Tong XL, Fang Z, Luo XI (2016) Experimental study on seismic behavior of reactive powder concrete shear walls. J Build Struct 37(1):21–30
6. Hwang SJ, Lee HJ (2002) Strength prediction for discontinuity regions by softened strut-and-tie model. J Struct Eng 128(12):1519–1526
7. Tang XR, Jiang YS, Ding DJ (1993) Application of the theory of softened truss to low-rise steel fiber high strength concrete shear walls. J Build Struct 14(2):2–11
8. Mei GD, Xu LH, Lu WM et al (2013) Effect of fiber content on axial tensile properties of hybrid fiber reinforced concrete. Eng J Wuhan Univ 46(6):752–758
9. An YJ, Zhao GF, Huang CK (1993) Study on methods for calculating bearing capacity of SFRC elements reinforced with steel bars. China Civ Eng J 26(1):38–46
10. Xu LH, Huang Y, Wei CM et al (2014) Experimental tests of seismic bearing capacity of steel-polypropylene hybrid fiber reinforced concrete columns. J Build Struct 35(8):95–103
11. Xia GZ, Xia DT (2008) Research on behavior of hybrid fiber reinforced high-performance concrete shear walls. J Huazhong Univ Sci Technol (Urban Sci Ed) 4(4):103–106

Open Access This chapter is licensed under the terms of the Creative Commons Attribution 4.0 International License (http://creativecommons.org/licenses/by/4.0/), which permits use, sharing, adaptation, distribution and reproduction in any medium or format, as long as you give appropriate credit to the original author(s) and the source, provide a link to the Creative Commons license and indicate if changes were made.

The images or other third party material in this chapter are included in the chapter's Creative Commons license, unless indicated otherwise in a credit line to the material. If material is not included in the chapter's Creative Commons license and your intended use is not permitted by statutory regulation or exceeds the permitted use, you will need to obtain permission directly from the copyright holder.

Investigation of Dynamic Response of Concrete Slab Under Air Blast Loading

Qindong Lin, Chun Feng, Yundan Gan, Jianfei Yuan, and Ying Yang

Abstract The dynamic response of concrete components under blast loading is of great military and social importance. Based on the continuum-discontinuum element method and fluid–structure coupling algorithm, the dynamic response of concrete slab is simulated and analyzed. First, a full-time numerical simulation is conducted. Then, the displacement and crack characteristics of concrete slab are analyzed quantitatively. The result indicates that the displacement is distributed in a circular pattern, and the maximum value is located at the center of concrete slab. The crack ratio increases gradually with the growth of time, while the growth rate is not a constant value. The fracture type of concrete slab includes tensile fracture and shear fracture, the interface at the center mainly undergoes shear fracture, and a small number of interface on the outside of cracked region suffers tensile fracture.

Keywords Dynamic response · Concrete slab · Air blast · CDEM

1 Introduction

As an important component material of engineering structure, concrete is widely used in civil and military engineering and other fields. Today, local conflicts, terrorist attacks and gas explosions are frequent worldwide, and the damage to concrete structures caused by explosive loading has resulted in serious loss of life and property. The investigation of the dynamic response of concrete components under blast loading is of great military and social importance [1–4].

Based on experimental study and theoretical analysis, scholars analyzed on the mechanical response of concrete and generalized the change laws. Combining with SEM and FTIR, Sun et al. [5] obtained the macroscopic failure mode of structure

---

Q. Lin · Y. Gan (✉) · J. Yuan · Y. Yang
Xi'an Modern Chemistry Research Institute, Xi'an 710065, Shaanxi, China
e-mail: ganyundan@163.com

C. Feng
Institute of Mechanics, Chinese Academy of Sciences, Beijing 100190, China

and the microscopic damage behavior of coating under the 10 kg TNT. Through 11 independent explosion tests, Wang et al. [6] analyzed the influence of different PODZ coating thickness on the anti-explosion performance, and observed the reinforced concrete slab failure modes and damage characteristics. Yang et al. [7] conducted the contact explosion experiments on corrugated steel reinforced concrete slabs, and obtained the typical damage characteristics of components. Yu et al. [8] compared the damage characteristic of three concrete slabs under different charges, and determined the range of critical perforation charges. Yue et al. [9] derived the equations of material resistance to explosions in both the infinite and the semi-infinite medium, and obtained the critical depth.

With the rapid development of computer technology and numerical algorithm, many scholars studied the mechanical response of concrete components under explosive loading based on numerical simulation. Sun et al. [10] esflished a meso-structure model of plain/reinforced concrete slabs with stochastic aggregate method, and simulated the response of reinforced concrete slabs by LS-DYNA software. By introducing the zero thickness cohesive element to simulate the dynamic breakup process of concrete structure, Wu et al. [11] studied the dynamic failure process of reinforced concrete under explosive loading. Yin et al. [12] established a new compression dynamic increase factor (CDIF) mode, and simulated the dynamic response of SFRC plate. Fatima et al. [13] investigated the response and extent of damage in reinforced concrete framed building under varying blast load pulse shapes.

Currently, scholars mainly study the dynamic response of concrete components under blast loading based on the fluid–structure coupling algorithm, and the structure is simulated by finite element method or finite difference method. Due to the deficiency of numerical algorithm in simulating crack, the initiation and expansion of crack are mainly characterized by deleting elements, which violates the conservation of mass. Based on the continuum-discontinuum element method and fluid–structure coupling algorithm, the dynamic response of concrete slab under blast loading is simulated and analyzed.

2 Numerical Simulation

2.1 Numerical Model

The continuum-discontinuum element method (CDEM) is introduced into the numerical simulation. The numerical model of concrete slab and TNT is plotted in Fig. 1. The horizontal length of concrete slab is 150 cm, the horizontal width is 150 cm, and the vertical height is 9 cm. The explosive TNT is a sphere with a radius of 7.6 cm, and a vertical distance of 50 cm from the concrete slab. To simulate the propagation process of shock wave, an air domain is created, and the concrete slab and TNT are inside the air domain. To better realize the fluid–structure interaction simulation of

Fig. 1 Numerical model of concrete slab and TNT

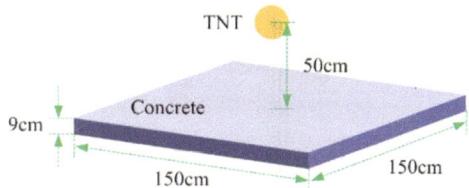

Table 1 Mechanical parameters of concrete

Material	Density (kg/m³)	Elastic modulus (GPa)	Cohesive strength (MPa)	Tensile strength (MPa)
Concrete	2320	15	10	5

TNT, air and concrete slab, the air domain and concrete slab are divided by orthogonal mesh. In order to simulate concrete fracture process more accurately, a bilinear cohesive fracture model is used as the constitutive model of concrete interface, and the mechanical parameters of concrete are listed in Table 1.

2.2 Numerical Results

Based on the CDEM numerical algorithm and the fluid–structure coupling algorithm, the propagation process of shock wave, the interaction process between shock wave and concrete slab, and the dynamic fracture process of concrete slab are accurately simulated. Based on the full-time numerical result, the displacement and fracture characteristics of concrete slab are investigated.

Displacement characteristic

The displacement nephograms of concrete slab under the blast shock wave are plotted in Fig. 2. It is observed that the displacement is distributed in a circular pattern, and the maximum value is located at the center of concrete slab, which is because it is the closest to the explosive and is the first to bear the shock wave. With the increase of distance from the center of concrete slab, the displacement value gradually decreases. Influenced by the boundary condition and the shape of concrete slab, the displacement in the central region is circularly distributed, but the displacement near the boundary is no longer circularly distributed and gradually changes to a rectangular distribution. Due to the shock wave, the displacement of concrete slab gradually increases and keeps moving downward, the crack of interface and separation between the concrete elements occur during this process. The element at the center of concrete slab is the first to separate, and the region of separation between the concrete elements is continuously extended to the surrounding area.

To accurately study the displacement spatial–temporal characteristic of concrete slab, the displacement curves at the midline corresponding to different times are

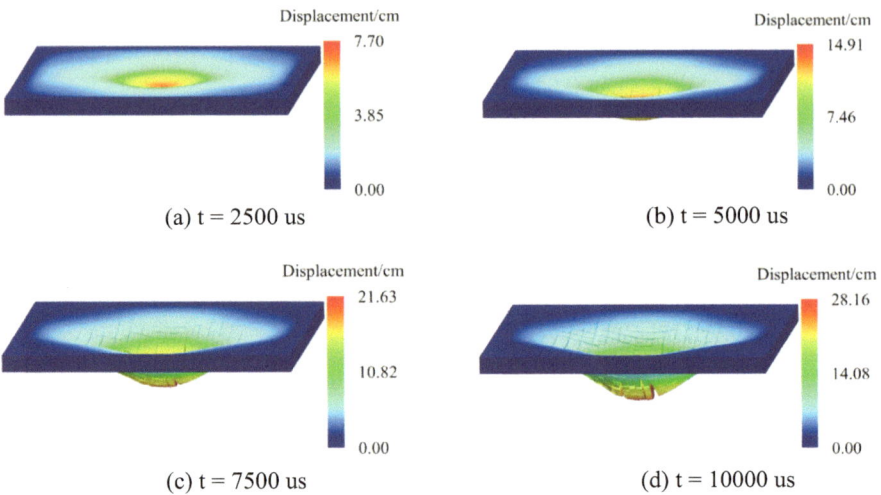

Fig. 2 Displacement nephograms of concrete slab

plotted in Fig. 3. It is observed that the change trend of displacement curves at different times is basically the same, the displacement first increases and then decreases with the increase of distance. The maximum displacement is achieved at the distance $d = 0.75$ m, which is located exactly at the center of concrete slab. The change trend of whole displacement curve is basically symmetrical with $d = 0.75$ m, and this law is consistent with the spatial law that the displacement is circularly distributed in Fig. 2. For the same position, the displacement value gradually increases with the growth of time, which indicates that the concrete slab keeps moving downward. According to the difference in displacement at different times, it is concluded that the difference in displacement at the same time interval decreases as time increases, which is due to the decreasing velocity of movement caused by the decreasing pressure of shock wave.

Crack characteristic

Due to the inconsistency of deformation between the concrete elements, the concrete interface cracks. To investigate the change trend of fracture degree of concrete slab, a dimensionless index, crack ratio α, is introduced, and the time-history curve is plotted in Fig. 4. It is observed that the crack ratio α increases gradually with the growth of time, and there is an inflection point (point A) on the curve. In the OA range, the crack ratio α increases sharply; In the AB range, the crack ratio α increases slowly, and the growth rate decays with the growth of time. There are two main factors that cause the different growth trend in the OA range and AB range. First, the overpressure of shock wave is large at the initial moment, which leads to rapid growth of interface stress, a large number of interfaces satisfy the fracture criterion and crack. Subsequently, as the overpressure of shock wave gradually decreases, the growth rate of interface stress decays, which leads to the number of newly cracked interfaces decreasing.

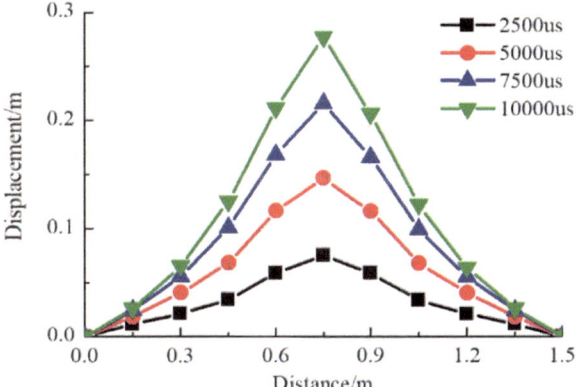

Fig. 3 Displacement curves at the midline

Second, due to the downward movement of concrete slab, the distance from the explosive increases continuously, which further weakens the destruction of shock wave on the concrete slab.

Although the crack ratio α can accurately describe the development trend of cracked interface, it can not reflect the fracture type and spatial distribution of the cracked interface. The interface fracture nephograms of concrete slab are plotted in Fig. 5, it is observed that the fracture type includes tensile fracture and shear fracture, and the interface mainly suffers shear fracture. At $t = 2500$ us, a large number of interfaces in the center region of concrete slab crack, most of them are shear fracture, while the interface at the outside of cracked region and the boundary of concrete slab mainly undergoes tensile fracture. With the increase of time, the cracked region expands continuously to the outside, and most of the newly cracked interface suffers shear fracture. There is still some tensile cracked interface on the outside of cracked region, and the interface at the boundary of concrete slab mainly undergoes shear fracture. By comparing the fracture nephograms for $t = 5000$ us, $t = 7500$ us and

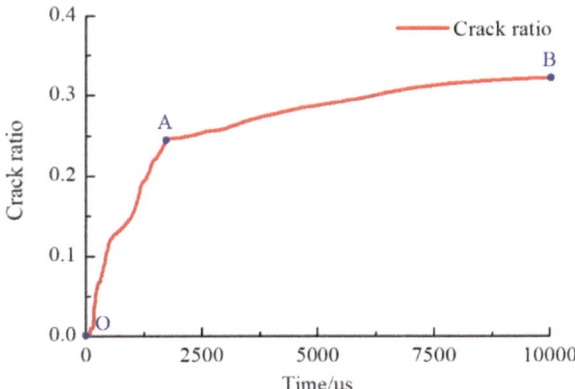

Fig. 4 Time-history curve of crack ratio α

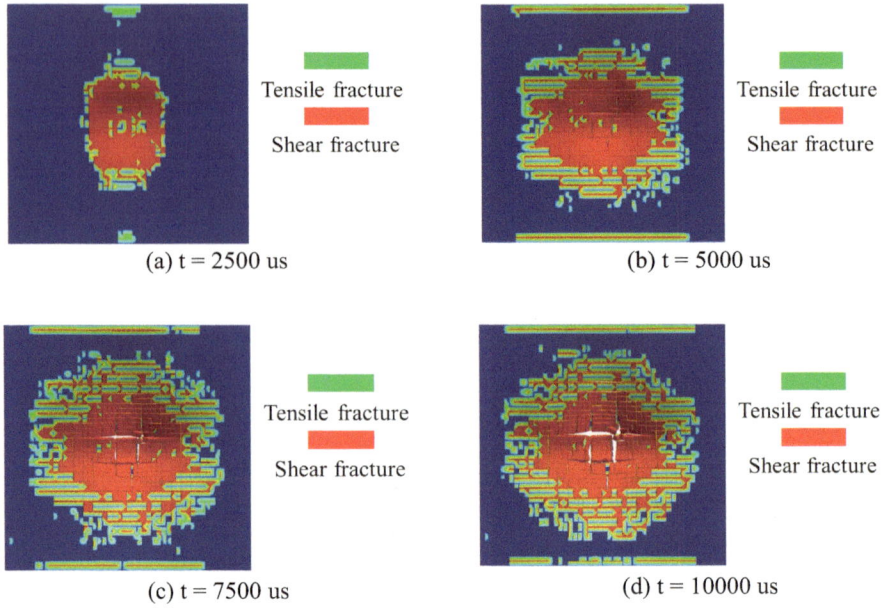

Fig. 5 Interface fracture nephograms of concrete slab

$t = 10{,}000$ us, it is concluded that the growth region of cracked interface is small, which is consistent with the change trend of time-history curve of crack ratio α.

3 Conclusions

Based on the continuum-discontinuum element method and fluid–structure coupling algorithm, the dynamic response of concrete slab under blast loading is simulated and analyzed accurately. The following conclusions can be drawn:

(1) The displacement is distributed in a circular pattern, and the maximum value is located at the center of concrete slab. Influenced by the boundary condition and the shape of concrete slab, the displacement near the boundary is no longer circularly distributed and gradually changes to a rectangular distribution.
(2) The crack ratio α increases gradually with the growth of time, while the growth rate is not a constant value, which decays with the increase of time. The interface at the center cracks first, and most of the cracked interface suffers shear fracture. There is still some tensile cracked interface on the outside of cracked region, and the interface at the boundary of concrete slab mainly undergoes shear fracture.

References

1. Peyman S, Eskandari A (2023) Analytical and numerical study of concrete slabs reinforced by steel rebars and perforated steel plates under blast loading. Results Eng 19:101319
2. Mudragada R, Bhargava P (2023) Effect of masonry infill on the response of reinforced concrete frames subject to in-plane blast loading. Structures 57:105317
3. Almustafa MK, Balomenos GP, Nehdi ML (2023) Data-driven reliability framework for qualitative damage states of reinforced concrete beams under blast loading. Eng Struct 294:116803
4. Fan Y, Chen L, Xiang HB, Fang Q, Han FY (2023) Lead spall velocity of fragments of ultra-high-performance concrete slabs under partially embedded cylindrical charge-induced explosion. Defence Technol 23:50–59
5. Sun PF, Lu P, Huang WB, Zhang R, Fang ZQ, Sang YJ (2021) Research of explosion resistance of sprayed anti-blast polyurea reinforced concrete slab. Mater Rep 35:642–648
6. Wang W, Yang JC, Wang JH, Gao WL, Wang X (2020) Experimental research on anti-contact explosion of POZD coated square reinforced concrete slab. Explosion Shock Waves 40:14–23
7. Yang CF, Yan JB, Liu Y, Lü ZJ, Huang FL (2022) Damage characteristics of corrugated steel concrete slab under contact explosion load. Trans Beijing Inst Technol 42:453–462
8. Yu X, Zhou BK, Hu F, Zhang Y, Xu XY, Fan CF, Zhang W, Jiang HW, Liu PQ (2021) Experimental research on the characteristics of BFRP grid-reinforced concrete slabs under contact explosion. Protective Eng 43:11–16
9. Yue SL, Wang MY, Zhang N, Qiu YY, Wang DR (2016) A method for calculating critical spalling and perforating thicknesses of concrete slabs subjected to contact explosion. Explosion Shock Waves 36:472–482
10. Sun JC, Chen XW, Deng YJ, Yao Y (2019) Dynamic response of mesoscopic plain/reinforced concrete slabs under blast loading. Explosion Shock Waves 39:32–42
11. Wu ZJ, Zhang PL, Liu QS, Li WF, Jiang WZ (2018) Dynamic failure analysis of reinforced concrete slab based on cohesive element under explosive load. Eng Mech 35:79–90
12. Yin HW, Jiang K, Zhang L, Wang L, Wang CL (2020) K&C model of steel fiber reinforced concrete plate under impact and blast load. Chin J High Pressure Phys 34:134–144
13. Fatima A, Sangi AJ, Mohammad AF, Joohi M (2023) Global response of reinforced concrete framed building under varying blast load pulse shapes. Structures 50:482–493

Open Access This chapter is licensed under the terms of the Creative Commons Attribution 4.0 International License (http://creativecommons.org/licenses/by/4.0/), which permits use, sharing, adaptation, distribution and reproduction in any medium or format, as long as you give appropriate credit to the original author(s) and the source, provide a link to the Creative Commons license and indicate if changes were made.

The images or other third party material in this chapter are included in the chapter's Creative Commons license, unless indicated otherwise in a credit line to the material. If material is not included in the chapter's Creative Commons license and your intended use is not permitted by statutory regulation or exceeds the permitted use, you will need to obtain permission directly from the copyright holder.

Research on Cumulative Damage of Quasi-Static Reinforced Concrete Short Columns with Low Cycle Fatigue

Hongyu Zhou, Juxin Guo, Qi Tang, and Haoda Wang

Abstract Columns are the main load-bearing elements in many structures, but with the advancement of service time, damage to reinforced concrete columns before reaching their design service life occurs from time to time. In this experiment, the cumulative damage of reinforced concrete short columns is investigated by using the research method of proposed static structural test. Through the low-week fatigue proposed static test of reinforced concrete short columns, the changes in the process are analyzed, and the changes in physical quantities such as cracks, strains, and deflections are recorded to study the shear damage characteristics of fatigue of reinforced concrete short columns, the damage mechanism, and the mechanical properties of the members such as the load carrying capacity after the damage.

Keywords Reinforced concrete short columns · Cumulative damage · Low-cycle fatigue

1 Introduction

Columns are the main load-bearing members in many structures, but with the advancement of service time, the damage of reinforced concrete columns before reaching the design service life occurs from time to time. Therefore, in order to utilize the mechanical properties of reinforced concrete structures more reasonably and effectively, some scholars at home and abroad have studied the cumulative damage performance of reinforced concrete structures, such as the study of the mechanism of cumulative damage affected by different design parameters of reinforced concrete

H. Zhou (✉) · J. Guo · H. Wang
Faculty of Architecture, Civil and Transportation Engineering, Beijing University of Technology, Beijing, China
e-mail: ZHYktztgyx@163.com

Q. Tang
Nanfaxin Township, Shunyi District, Beijing, China

members, the study of the cumulative damage mechanism of reinforced concrete members with different destructive forms of bending and shear damages, and so on.

Alliche et al. conducted a three-point bending test [1] using two ratios of unreinforced concrete beams and investigated the evolution of damage by measuring strain records. The results show that increasing the stress ratio helps to improve the fatigue life under the same loading. More fatigue studies on fiber-reinforced concrete members have been conducted [2–8], pointing out that the bond strength between the reinforcement and the concrete is a key factor affecting the fatigue life of the structure. Zhu et al. [5] investigated the fatigue modeling of concrete beams filled with fiber-reinforced polymers. Since fatigue loading reduces the stiffness and strength of structural members, the post-fatigue residual strength was used to express the fatigue strength of the members after any given time or residual static strength after any number of fatigue cycles Katakalos et al. [6] investigated the fatigue performance of reinforced concrete beams and conducted fatigue tests using 16 reinforced concrete beams. The result was that the flexural stiffness of reinforced concrete beams did not decrease by the increase in the number of cycles throughout the fatigue tests, but the maximum deflection increased with the increase in the number of cycles. Zhang [9] mainly investigated the fatigue degradation effect of the number of loading cycles and loading amplitude on reinforced concrete columns, and concluded that the ultimate deflection of columns affected by the number of loading cycles and loading amplitude was significantly reduced compared to the ultimate deflection of columns under monotonous loading. Zhu [10] conducted a study on the cumulative damage and horizontal capacity of low-cycle fatigue of bending-shear damaged reinforced concrete (RC) piers, loaded 23 square bending-shear damaged piers, and investigated the low-cycle fatigue performance of the piers under variable-amplitude and equal-amplitude reciprocating load with different number of cycles. The test results show that: under the action of low circumferential cyclic loading, the cumulative hysteretic dissipation energy of the specimen decreases with the increase of displacement amplitude, and the number of cycles of specimen destruction also decreases with the increase of displacement amplitude; it is also found that the specimen's strength and stiffness degradation is serious when it is close to destruction, and at the same time, the residual deformation of the specimen increases with the increase of the number of cycles of loading.

Previous research on the cumulative damage of reinforced concrete members mainly focused on the performance of beam members, exploring the effects of the number of cyclic loading on the mechanical properties of different forms of hoop reinforcement, different stress tendons, and different cross-section forms. When studying the mechanical property decline caused by cumulative damage of reinforced concrete short columns, the study also focuses on exploring the component strength, stiffness and other macro indexes, without quantitative analysis of the test results. Therefore, this paper is based on the lack of concrete structure shear fatigue research, for axial loading and transverse loading mixed action, reinforced concrete column shear damage behavior of the research is relatively weak, for the repetitive load under the influence of the mechanism of the understanding is not yet comprehensive, the industry's common norms proposed by the design method has not been involved in the relevant parameters. Monotonic pushover and low-week fatigue proposed static tests of reinforced concrete columns were carried out to study the fatigue shear damage characteristics of reinforced concrete short columns, damage mechanisms, and mechanical properties such as bearing capacity of members after damage by analyzing the changes in physical quantities such as cracks, strains, and deflections in the process of cumulative damage of reinforced concrete short columns. Based on the effect of cumulative damage of reinforced concrete short columns, data calculation reference is provided for practical engineering design and safety assessment.

2 RC Columns Proposed Static Cyclic Loading Damage Test

2.1 Test Piece Design and Fabrication

This test takes the design and the actual situation similar to the state of force, the four (notated as YJ-1~4) dimensions of 300 mm × 300 mm × 546 mm concrete strength of C30, longitudinal reinforcement selection of HRB400, hoop reinforcement selection of HPB300 reinforced concrete columns for the loading test, the specific dimensions and reinforcing bars are shown in Fig. 1. The damage-free specimens are loaded by static loading to simulate damage. The test was conducted by proposed static cyclic loading, and the specimens were loaded according to the specified loading mode after preloading.

2.2 Test Material Properties

Before the test on the components, the use of mechanical testing machine, the reinforcement, concrete material mechanical properties of the actual measurement, the material test data of steel bar is shown in Table 1. For the determination of the basic properties of concrete materials, the use of reinforced concrete components poured with exactly the same material for the production of specimens, a total of 24 specimens were produced by the same conservation conditions as the reinforced concrete test columns were maintained and then measured, the individual data After processing, the axial compressive strength of the concrete cube was 27.66 MPa, and the split tensile strength was 2.25 MPa.

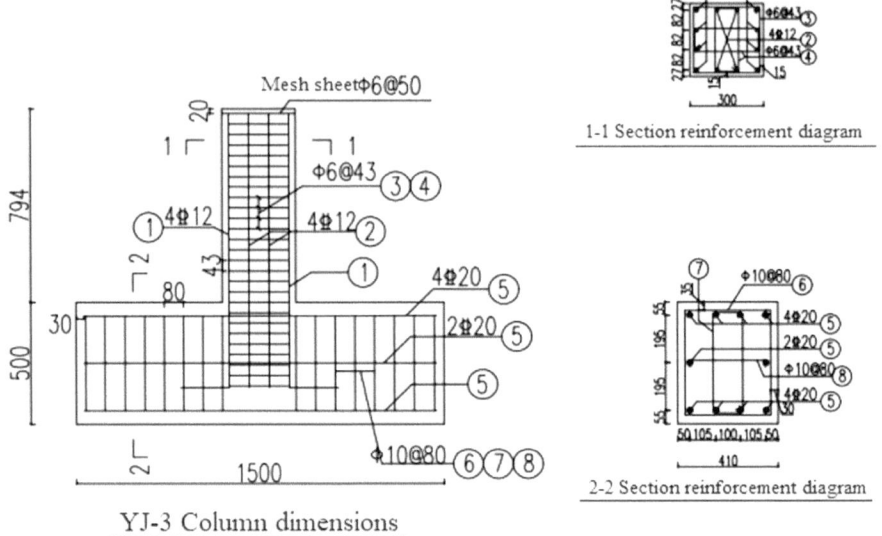

Fig. 1 Specimen size and reinforcement diagrams

Table 1 Mechanical properties of reinforcing steel materials

Serial number	Rebar diameter (mm)	Reinforcing steel grade	Yielding strength (MPa)	Tensile strength (MPa)
1	6	HPB300	486.67	669.42
2	10	HPB300	453.33	593.43
3	12	HRB400	471.67	705.35
4	20	HRB400	433.67	695.91

2.3 Test Loading and Acquisition Program

The specimen is loaded by a 200-ton proposed static testing machine, and the test is completed by applying axial and horizontal loads and constraints to the specimen according to the set monotonic loading regime, and the loading equipment is shown in Fig. 2. Since the column root boundary constraint condition is simulated by the reinforced concrete base, the base needs to be fixed with the ground by I-beam and bolts to prevent the base from shifting and rotating during the test. At the same time, the ground level is checked and adjusted to prevent damage to the column base caused by stress concentration.

The location and number of rebar strain gauges are shown in Fig. 3. After the concrete casting and curing is completed, white paint is brushed on the surface in order to observe the cracks, and the concrete surface is polished after the paint dries,

Fig. 2 Test piece installation site plan

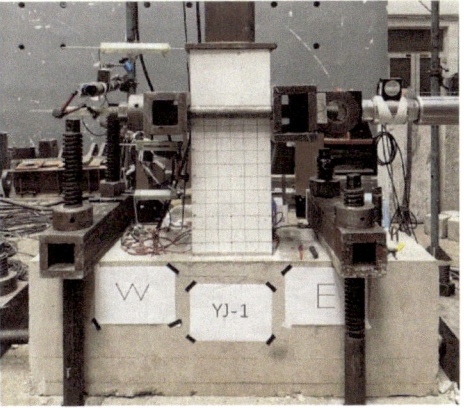

and the concrete strain gauges are pasted; 100mm concrete strain gauges are used, and the strain gauges are pasted in the range of 50–150 mm in the height of the three side columns in addition to the observing surface, and the position of its patch is as shown in Fig. 4.

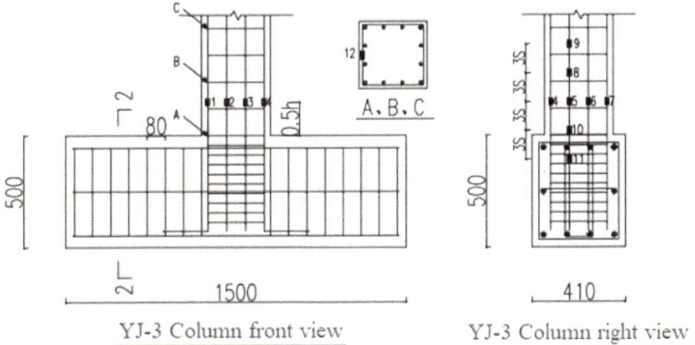

Fig. 3 Schematic diagram of the distribution location of reinforcement strain gauges

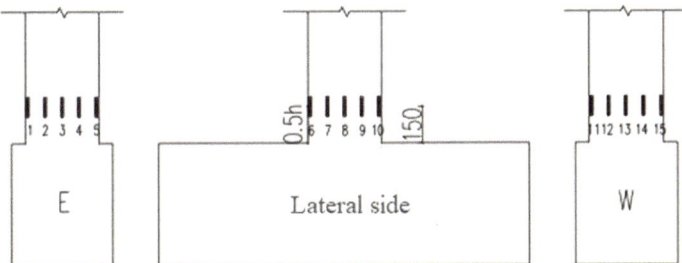

Fig. 4 Concrete strain gauge distribution location diagram

When the material experiment is completed and the specimen maintenance is completed, the specimen will be lifted to the loading position of the experimental machine, and the specimen will be fixed after adjusting the position and calibrating the level. Then the strain gage pre-posted to the design position of the reserved wires connected to the collector, loaded displacement meter and access to the collector; and then turn on the power supply, check whether the sensors are intact, ready to start the experiment.

The test first applied axial load, to be stabilized after the beginning of the application of horizontal load, the formal loading of the specimen before the preload, set the preload load value of 76 kN. For monotonic test loading selection of the first half of the load-controlled loading, the loading rate of 5 kN/s; loaded to 342 KN after switching to the displacement-controlled loading, the loading rate of 0.02 mm / s, until the destruction of the specimen. For the damage test choose equal amplitude cyclic loading system, as shown in Fig. 5, cyclic loading test process first apply thrust (i.e., to the W side of the load for forward loading), followed by the application of tensile force (i.e., to the E side of the load for the reverse loading); forward loading a time, reverse loading once for the completion of a cyclic loading, the cycle of 30 times after the end of the cumulative damage loaded to start monotonous push overloading until the destruction of the specimen. Cyclic loading with displacement control loading, specimen YJ-2~4 displacement rate of 0.4%, 0.7%, 0.8%, respectively, the calculated displacement of the top of the column were 2.18 mm, 3.82 mm, 4.37 mm.

The results were all in accordance with the design stresses for compression shear damage. Without damage members in the loading process with the increase in horizontal displacement cracks increase and development, specimen damage eventually showed tensile surface to the compression surface 45° damage, a large number of concrete peeling, damage to the surface of the rebar is convex, detached from the

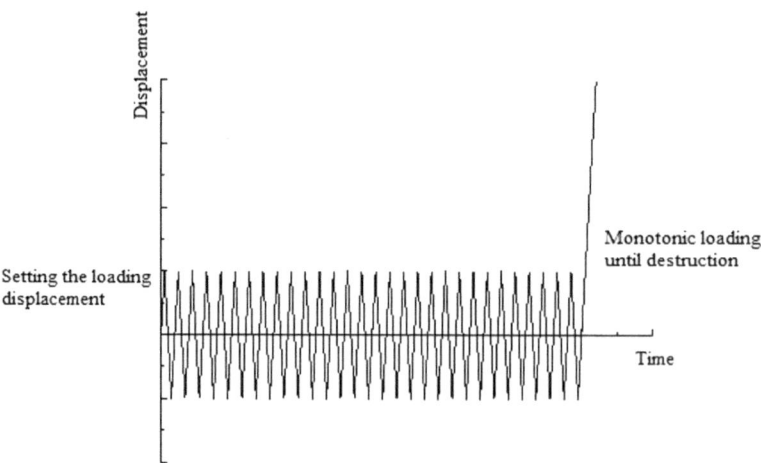

Fig. 5 Cumulative damage test loading regime

concrete and the hoop constraints protruding. Damaged members in the application of cyclic loading process in the loading surface and near the loading surface of the observation surface of the lateral side of the alternating emergence and development of cracks, in the ten cycle loading crack development is basically stable, no longer change; cyclic loading is over, close to the foot of the column within the scope of the formation of horizontal cracks and vertical cracks.

3 Cumulative Damage Experimental Results and Characterization

Organize and analyze the cumulative damage test results, by the specimen YJ-2, YJ-3, YJ-4 cumulative damage process displacement-load curve, such as Fig. 6.

The damage process curve of specimen YJ-2 has a smooth and full date-palm shape. This is due to the fact that the reinforcement reached the maximum strain value in the first cycle of repeated loading. However, in the subsequent cycles of loading, the strain of the reinforcement instead decreases slowly when the preset displacement loading is performed again due to the concrete cracking and the decrease in the stiffness of the member.

The damage process curves of specimens YJ-3 and YJ-4 were shuttle-shaped. Especially the curve of specimen YJ-4 is more narrow. This is because in the early stage of cyclic loading, the stressed reinforcement of the specimen is constantly subjected to alternating tensile and compressive stresses that are approximately equal to the yield load, resulting in a continuous superposition of specimen damage. At the same time, under the action of alternating tensile and compressive stresses, the bond between the reinforcement and the concrete gradually failed, and the concrete damage was obvious, forming bond splitting cracks. As a result, the vertical microcracks of specimens YJ-3 and YJ-4 develop significantly, and the damage process curve shows obvious "pinch shrinkage phenomenon". Since the loading displacement of

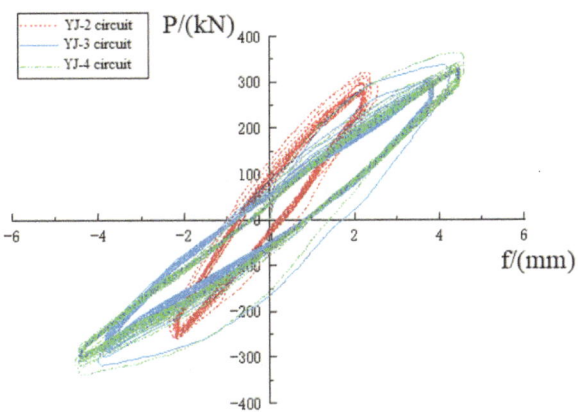

Fig. 6 Comparison of displacement-load curves of specimens YJ-2, YJ-3 and YJ-4 during

specimen YJ-4 is larger, the damage caused is more, and the hysteresis curve is more elongated. Yi et al. [11] pointed out that the displacement of the member is composed of horizontal displacement formed by bond slip, shear and bending deformation of the reinforcement and concrete. Meanwhile, it is due to the formation and development of split crack bond that the protective layer of concrete is no longer restrained, and with the alternation of tension and compression, the cumulative damage to the concrete accumulates internally, showing a gradual increase in the width of the cracks in the specimen, which is also the reason for the spalling and detachment of concrete in the footing of columns of specimens YJ-3 and YJ-4. The study of Goodnight et al. [12] showed that the load history can influence the reinforcement buckling and has an effect on the cumulative strain within the longitudinal and transverse reinforcement of reinforced concrete columns, with larger concrete compressive strains leading to inelastic strains in the transverse reinforcement, which results in the loss of the concrete's effect on preventing buckling of the longitudinal reinforcement.

For the horizontal loads of specimens YJ-2, YJ-3 and YJ-4, the magnitude of the horizontal loads at the time of reaching the loading amplitude was not equal during each cycle of loading. Especially in the first 5 cycles, the horizontal loads showed obvious degradation. The rate of horizontal load degradation was not consistent between hysteresis loops, with the largest degradation between the 1st and 2nd loop, followed by a gradual decrease. In the subsequent loops, the total amount of horizontal force degradation was smaller than that of the first 5 weeks, which indicated that the damage inside the member gradually accumulated and developed and gradually converged to a steady state.

4 Specimen Performance After Damage

The monotonic pushover test results of specimens YJ-1, YJ-2, YJ-3, and YJ-4 are organized and analyzed to obtain the monotonic pushover displacement-load curve comparison Fig. 7. It can be seen from the comparison graphs that the ultimate load capacity of the specimen decreases after the damage caused by cyclic loading. Since the specimen is brittle damage, the bearing capacity decreases rapidly after reaching the peak load, so the peak load point is taken as the ultimate displacement point of the member, and the yield load and yield displacement of each specimen are obtained by using the energy equivalence method, which is expressed in Table 2.

The test results show that the short reinforced concrete columns under compression and lateral horizontal force are damaged suddenly with a sharp decrease in bearing capacity and brittle damage, which is in line with the design expectation. During the cyclic loading process, the damage volume energy dissipation gradually decreases with the increase of loading times, but the rate of decrease gradually slows down and stabilizes after 10 cycles of loading. The energy dissipation in the first loading cycle is the largest, and the damage volume energy dissipation gradually decreases with the increase of loading times. Larger displacement loading reduces the energy dissipation capacity of the specimen.

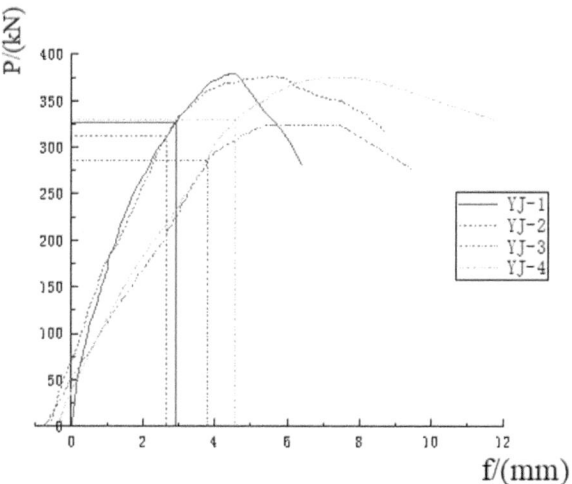

Fig. 7 Comparison of monotonic pushover displacement-load curves

Table 2 Test main result parameters

Specimen number	Displacement corresponding to peak load (mm)	Peak load (kN)	Yield displacement (mm)	Ultimate load (kN)	Limit displacement (mm)	Ductility factor	Yield load (Py/kN)
YJ-1	4.47	379.05	2.82	321.23	6.40	2.27	320.89
YJ-2	5.63	376.58	2.66	320.09	8.56	3.31	312.41
YJ-3	5.50	324.05	3.63	275.44	7.48	1.97	284.97
YJ-4	7.43	373.99	4.38	317.89	8.49	1.94	329.55

The ultimate load capacity of the specimen decreases after cyclic loading. For the member with displacement within the range of yield displacement, the tensile and compressive strains of reinforcement are small, there is no obvious bond damage between reinforcement and concrete, cracks do not produce bond splitting cracks along the longitudinal reinforcement, and the damage curves are in the form of rounded and full date-palm shape, and the ductility is slightly increased. For the member whose displacement exceeds the yield displacement, the plastic deformation leads to the gradual failure of the bond between the reinforcement and the concrete, and the concrete damage is obvious, forming bond splitting cracks and vertical microcracks, and the damage curve is in the shape of a pike, with the deterioration of the ductility.

5   Conclusion

1. If the displacement rate during fatigue loading is small and does not exceed the yield displacement of the member, then there is no obvious bond damage between the reinforcement and concrete, cracks do not sprout along the longitudinal reinforcement, and there is no damage to the outer protective layer of concrete, and the hysteresis curve is in the form of rounded and smooth jujube shape.
2. If the loading displacement exceeds the yield displacement, the reinforcement will be subjected to alternating yield load tension and pressure, specimen damage superposition, bond failure gradually, the formation of bond splitting cracks, hysteresis curve has a "pinch shrinkage phenomenon". With the increase of plastic deformation, the steel package Singh effect makes the component bending stiffness decrease, and then loaded again when the strain in the steel bar decreases, the stress decreases, the bond is gradually stabilized, the hysteresis curve in the subsequent cycle is relatively stable.
3. After cyclic loading, the ultimate bearing capacity of the specimen decreases. Displacement within the yield displacement of the member, the ductility coefficient increases, the ductility is improved; displacement rate exceeds the yield displacement of the undamaged state of the member, as the displacement rate increases, the ductility coefficient decreases, the ductility deteriorates.

References

1. Alliche A, Frangois D (1992) Damage of concrete in fatigue. J Eng Mech 118(11)
2. El-Ragaby A, El-Salakawy E, Benmokrane B (2007) Fatigue life evaluation of concrete bridge deck slabs reinforced with glass FRP composite bars. J Compos Constr 11(3):258–268
3. Shan B, Xiao Y, Guo Y (2006) Residual performance of FRP-retrofitted RC columns after being subjected to cyclic loading damage. J Compos Constr 10(4):304–312
4. Kim YJ, Heffernan PJ (2008) Fatigue behavior of externally strengthened concrete beams with fiber-reinforced polymers: state of the art. J Compos Constr 12(3):246–256
5. Zhu Z, Ahmad I, Mirmiran A (2009) Fatigue modeling of concrete-filled fiber-reinforced polymer tubes. J Compos Constr 13(6):582–590
6. Katakalos K, Papakonstantinou CG (2009) Fatigue of reinforced concrete beams strengthened with steel-reinforced inorganic polymers. J Compos Constr 13(2):103–112
7. Parvez A, Foster SJ (2015) Fatigue behavior of steel-fiber-reinforced concrete beams. J Struct Eng 141(4)
8. Chen C, Cheng L (2016) Fatigue behavior and prediction of NSM CFRP-strengthened reinforced concrete beams. J Compos Constr 20(5)
9. Zhang Q (2014) Study on seismic performance and residual deformation of RC columns considering shear action. Doctoral dissertation, Dalian University of Technology, p 212
10. Zhu J (2020) Study on cumulative damage and horizontal bearing capacity of RC pier columns damaged by bending and shear with low peripheral fatigue. Doctoral dissertation, Dalian University of Technology, p 171

11. Yi W, Zhou Y, Hwang H, Cheng Z, Hu X (1984) Cyclic loading test for circular reinforced concrete columns subjected to near-fault ground motion. Soil Dyn Earthq Eng 2018(112):8–17
12. Goodnight JC, Kowalsky MJ, Nau JM (2013) Effect of load history on performance limit states of circular bridge columns. J Bridg Eng 18(12):1383–1396

Open Access This chapter is licensed under the terms of the Creative Commons Attribution 4.0 International License (http://creativecommons.org/licenses/by/4.0/), which permits use, sharing, adaptation, distribution and reproduction in any medium or format, as long as you give appropriate credit to the original author(s) and the source, provide a link to the Creative Commons license and indicate if changes were made.

The images or other third party material in this chapter are included in the chapter's Creative Commons license, unless indicated otherwise in a credit line to the material. If material is not included in the chapter's Creative Commons license and your intended use is not permitted by statutory regulation or exceeds the permitted use, you will need to obtain permission directly from the copyright holder.

Study on the Damage Evolution Law of Railway High Pier of New Replaceable Components Under Near-Fault Ground Motion

Xudong Zhang, Xiushen Xia, and Heng Zhang

Abstract In order to improve the energy dissipation effect of the connection components of steel truss joints and optimize the arrangement of the components, a new replaceable component railway pier structure was proposed, referring to the transmission tower structure. The new pier consists of four pier columns and the steel truss connection system between the columns. Based on the OpenSees platform, a dynamic analysis model of the new railway pier was established. 56 seismic records were input, and incremental dynamic analysis (IDA) was carried out to investigate the damage evolution law of the connection components and the influence of component yielding on the stress of the rebar in the middle and bottom of the pier. The results show that the steel truss connection system yields in a row and consumes energy in a graded manner. The yielding order of the second and third rows of the steel truss connection system is always in the forefront. The second and third rows are the main energy-consuming rows, followed by the fourth, fifth, and sixth rows, and finally the seventh, eighth, and first rows, with the first and eighth rows being less likely to yield. The outer chord members in each row yield first, followed by the inner chord members, and the vertical and diagonal members are less likely to yield. The yielding of the pier columns occurs after the yielding of the chord members of the steel truss. The yielding of the outer chord members in the steel truss connection system has a certain inhibitory effect on the stress increase of the rebar in the pier columns and at the bottom of the pier, and the inhibitory effect on the bottom of the pier is better than that in the middle.

Keywords Railway pier · Replaceable component · Seismic damage · Incremental dynamic analysis

X. Zhang · X. Xia (✉) · H. Zhang
Faculty of Civil Engineering, Lanzhou Jiaotong University, Lanzhou, Gansu, China
e-mail: xiaxiushen@mail.lzjtu.cn

X. Zhang
e-mail: 11210226@stu.lzjtu.edu.cn

1 Introduction

The traditional ductile design concept prioritizes the protection of life as the first goal. Based on this design concept, it can ensure that bridge structures will not collapse under strong earthquake actions. However, if bridge structures suffer severe damage after an earthquake, they will be difficult to repair, and the bearing capacity of the damaged areas is difficult to predict [1–3]. In addition, based on the ductile design, bridge structures with large residual displacements under strong earthquake actions need to be rebuilt [4–6].

To improve the seismic performance of bridge piers, seismic isolation piers with recoverable functions have received extensive attention from researchers [7, 8]. Due to the influence of higher mode effects, rocking and self-centering piers may still have the possibility of plastic hinge zones in the middle region of the piers, which brings difficulties to seismic design [9, 10]. Based on the concept of easy repair or replaceability after an earthquake [11–14], a replaceable component railway pier structure was proposed in literature [15–18]. The damage law of replaceable components was studied, the influence of cross braces and diagonal chord members on the internal forces of the pier columns was investigated, and the effect of yielding of replaceable components in the high pier structure on the internal forces of the key sections of the pier columns was examined. The seismic performance of the replaceable component railway pier structure was evaluated based on a probabilistic approach. The replaceable components in the above-mentioned replaceable component railway pier structure are all arranged in a parallel chord steel truss layout. Due to the relatively small horizontal displacement between the pier columns under earthquakes, most of the components cannot yield, and the energy dissipation of the yielding diagonal members is less efficient and not conducive to the full utilization of materials.

Based on the above research, in order to improve the energy dissipation effect of the replaceable components and optimize the arrangement of the components, this paper proposes a new replaceable component railway pier structure, referring to the transmission tower structure. Compared to traditional structures, this type of structure can achieve graded energy consumption, predictable damage, simple and clear stress, high strength, easy adjustment of stiffness, high structural ductility and plasticity, strong energy consumption, high structural reliability, simple construction period, low maintenance cost, easy repair or replacement after earthquakes, and economical and durable. A finite element model of the structure is established, and 56 seismic records are input along the transverse direction for incremental dynamic analysis (IDA). The damage evolution law of the replaceable components and pier columns is investigated, and the influence of component yielding on the stress of the rebar at the bottom and in the middle of the pier columns is examined. This study can provide reference for the seismic design of railway piers.

2 New Type of Replaceable Component Railway Pier Structure and Finite Element Model Establishment

2.1 New Type of Replaceable Component Railway Bridge Pier Structure Form

Based on a railway prestressed concrete bridge with a span of $(78 + 2 \times 136 + 78$ m) as the engineering background, a continuous rigid structure scheme is adopted for the span arrangement. Pier 5 of the bridge is a column-plate type pier with a height of 105 m, as shown in Fig. 1.

The new type of replaceable component railway pier is designed based on the principle of equivalent vibration period, and is designed based on the original pier. Its structure mainly includes four pier columns and the steel truss connection system between each pier column. The main structure of the pier consists of four pier columns located at the four corners of the cross-section, made of reinforced concrete material. The steel truss connection system can fix the four pier columns as a whole, providing initial stiffness to the structure, and is a replaceable component. Each pier column has the same square section, with a top section size of 300 cm × 300 cm and a bottom section size of 550 cm × 550 cm. The outer dimensions of the pier top and bottom are 1000 cm × 1000 cm and 1900 cm × 2800 cm (length × width), respectively. The cross-sectional shape of the pier body and the outer dimensions in the longitudinal and transverse directions change according to a parabolic curve. The longitudinal reinforcement strength grade of the pier column is HRB400, and the pier column concrete is C50.

Referring to the structural arrangement of the steel truss connection system members between the pier columns of transmission towers, the structure of the new type of replaceable component railway pier is shown in Fig. 2, and the cross-sections of the steel truss connection system and each member are shown in Fig. 3. The steel material used for the chord members is Q345, and the steel material used for the web members and vertical members is Q235. The new type of pier is equipped with

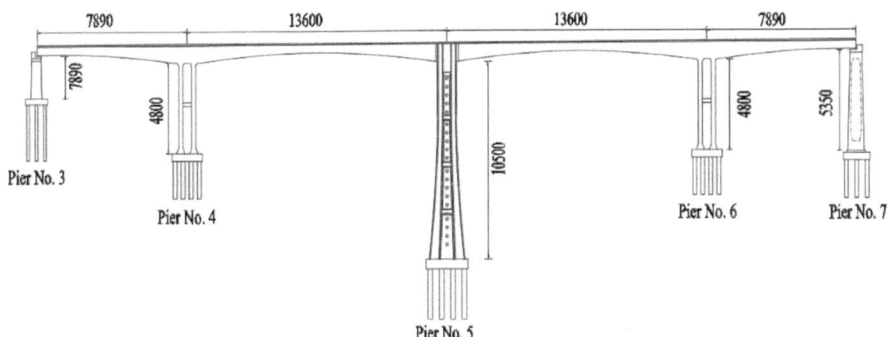

Fig. 1 Bridge façade layout (unit: cm)

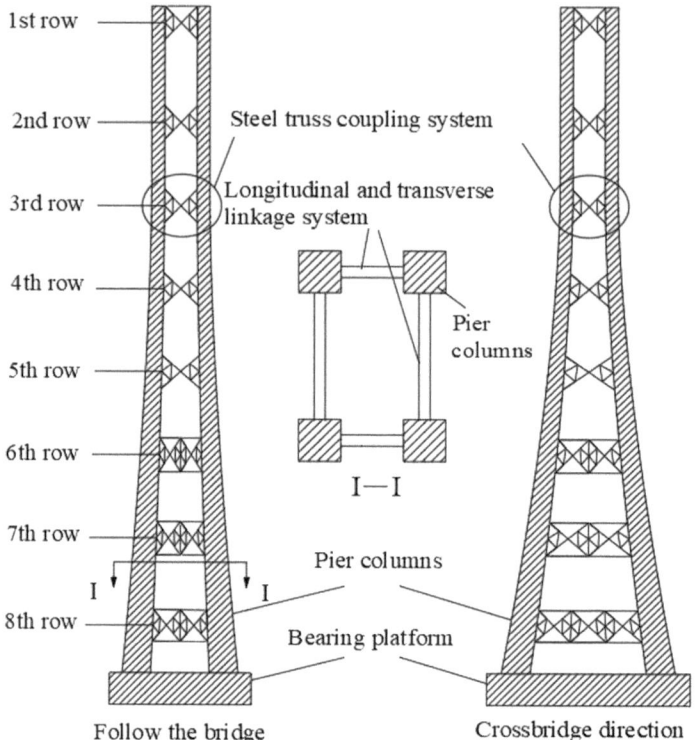

Fig. 2 Schematic diagram of the new high pier structure

eight rows of steel truss connection systems along the pier body from top to bottom. The upper half of the pier body is a straight section, and the lower half is a curved section. Except for the spacing between the first and second rows, which is 7.5 m, the spacing between the other rows is 10 m. The spacing between each row in the curved section is uniformly 7.5 m. The steel truss connection systems are numbered from the first row to the eighth row from top to bottom.

2.2 Establishment of Finite Element Model

Referring to Ref. [19], a single pier model is created by concentrating half of the mass of adjacent beams on each side of the pier top as shown in Fig. 1. An OpenSees platform is used to create a dynamic analysis model for the new type of railway pier, and the four pier columns and the steel truss connection system are simulated using fiber beam–column elements [20]. The concrete material is modeled using the concrete01 constitutive model, and the steel reinforcement is modeled using the steel01 constitutive model. The pier model has a total of 488 nodes, and the four pier

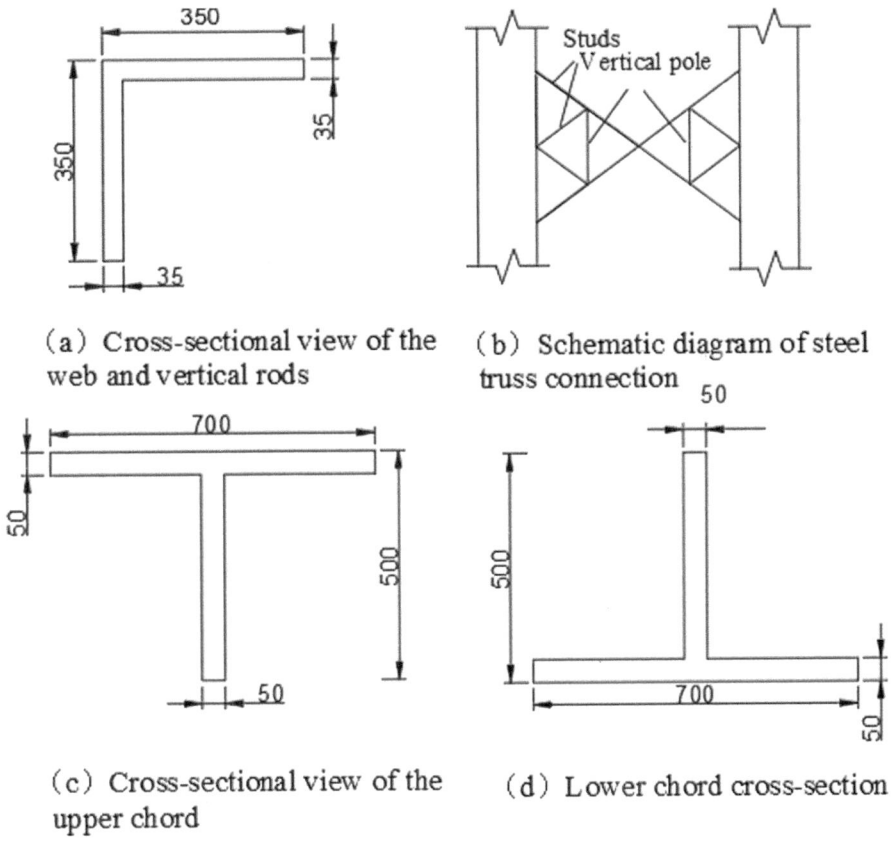

Fig. 3 Steel truss connection system and section (unit: mm)

columns and the steel truss connection system have a total of 936 elements. The pier top is connected by 4 rigid arm elements, resulting in a total of 940 elements in the structure. The pier bottom is fixed to the foundation.

3 Selection and Scaling of Seismic Ground Motions

Seven near-fault ground motions with epicentral distances between 2 and 20 km are selected from the database of the Pacific Earthquake Engineering Research Center. The 7 ground motions are scaled based on the principle of yielding of each row of members, resulting in a total of 56 ground motions. All ground motions are input along the transverse direction of the bridge, and the information of the ground motions is listed in Table 1.

Table 1 Basic information on ground motion

Serial number	Numbering	Earthquake shaking	Magnitude	Epicenter distance (km)	PGA (g)	PGV (cm·s^{-1})
1	1197	Chi-Chi Taiwan	7.6	3.12	0.760	85.755
2	1547	Chi-Chi Taiwan	7.6	14.91	0.135	36.691
3	1605	Duzce Turkey	7.1	6.58	0.515	84.234
4	1158	Kocaeli Turkey	7.5	15.37	0.364	55.661
5	1148	Kocaeli Turkey	7.5	13.49	0.134	40.067
6	1176	Kocaeli Turkey	7.5	4.83	0.321	71.889
7	881	Landers	7.3	17.36	0.164	22.219

4 Incremental Dynamic Analysis Results

In order to study the damage mechanism of the web members and vertical members of the steel truss joint system and their influence on the seismic response of the pier column, several key position members were selected from each row of trusses as the research objects, as shown in Fig. 4. Under the excitation of 56 near-fault ground motions, the outer web members of each row of trusses yielded first, followed by the inner web members, while the vertical members and chord members did not yield. The evolution of damage for the outer web members of each row of steel truss joint systems is listed in Table 2, with the row yielding first being the most unfavorable, and the yielding order is from left to right.

From Table 2, it can be seen that as the seismic intensity gradually increases, each row of steel truss joint systems will gradually go through the yielding stage, until all eight rows of steel truss joint systems reach the yielding stage.

Fig. 4 Schematic position of key members

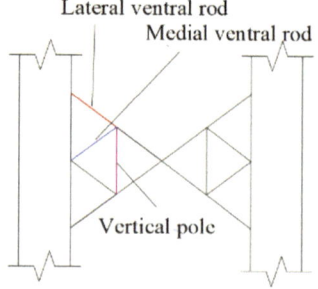

Table 2 Evolution of external abdominal rod damage of steel truss connection

Serial number	Numbering	Earthquake shaking	Yield sequence and its corresponding ground motion PGA
1	1197	Chi-Chi Taiwan	Six (0.315 g), Five (0.330 g), Four (0.417 g), Seven (0.427 g), Three (0.524 g), Two (0.560 g), Eight (0.750 g), One (2.55 g)
2	1547	Chi-Chi Taiwan	Three (0.106 g), Two (0.122 g), Four (0.128 g), Five (0.204 g), Six (0.308 g), Seven (0.391 g), One (0.577 g), Eight (0.601 g)
3	1605	Duzce Turkey	Three (0.238 g), Six (0.284 g), Two (0.293 g), Four (0.311 g), Five (0.341 g), Seven (0.369 g), Eight (0.717 g), One (0.1.258 g)
4	1158	Kocaeli Turkey	Three (0.250 g), Two (0.280 g), Four (0.310 g), Five (0.322 g), Six (0.336 g), Seven (0.536 g), Eight (0.820 g), One (1.754 g)
5	1148	Kocaeli Turkey	Three (0.248 g), Two (0.318 g), Four (0.338 g), Five (0.520 g), Six (0.576 g), Seven (0.773 g), Eight (1.08 g), One (1.16 g)
6	1176	Kocaeli Turkey	Three (0.230 g), Two (0.244 g), Four (0.250 g), Five (0.284 g), Six (0.320 g), Seven (0.50 g), One (0.850 g), Eight (0.972 g)
7	881	Landers	Two (0.238 g), Three (0.249 g), Four (0.274 g), Five (0.336 g), Six (0.390 g), Seven (0.560 g), One (0.927 g), Eight (0.960 g)

From Table 2, it can be seen that under the excitation of wave 1148, when PGA = 0.248 g, the outer web members of the third row of steel truss joint systems start to yield, but at this time, the nonlinearity of the hysteresis curve is not obvious due to the recent yielding of the members. The stress–strain curves of the outer web members of the third row of steel truss joint systems under the excitations of PGA = 0.338 g and PGA = 1.16 g are given (Figs. 5 and 6). From Fig. 5, it can be seen that when PGA = 0.338 g, the curve shows obvious yielding, indicating that the members have undergone significant yielding. From Fig. 6, it can be seen that when PGA = 1.16 g, the hysteresis curve gradually widens and remains at the same height. Figures 7 and 8 respectively show the stress and strain time history curves of the outer web members of the third row steel truss connection system under 1148 excitation. It can be seen from Fig. 7 that under wave 1448 excitation, all three curves appear as platforms, and as the peak ground acceleration increases, the platform of the time history curve gradually widens. From Fig. 8, it can be seen that under wave 1448 excitation, the peak values of all three curves exceed the yield strain of Q235 steel bars, and as the peak ground acceleration increases, The higher the peak value of the curve, the higher it is. Taking a comprehensive look at Figs. 5, 6, 7 and 8 indicating that with the gradual increase of seismic intensity, the yielding degree of the members becomes more obvious and the energy dissipation capacity becomes stronger.

Fig. 5 Stress–strain relationship of lateral ventral of lateral rod at PGA = 0.338 g

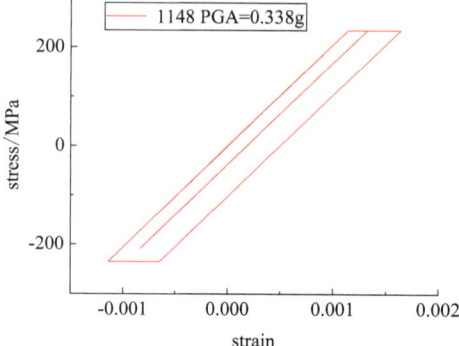

Fig. 6 Stress–strain relationship ventral rod at PGA = 1.16 g

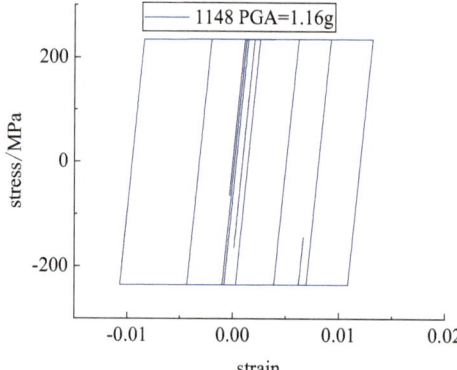

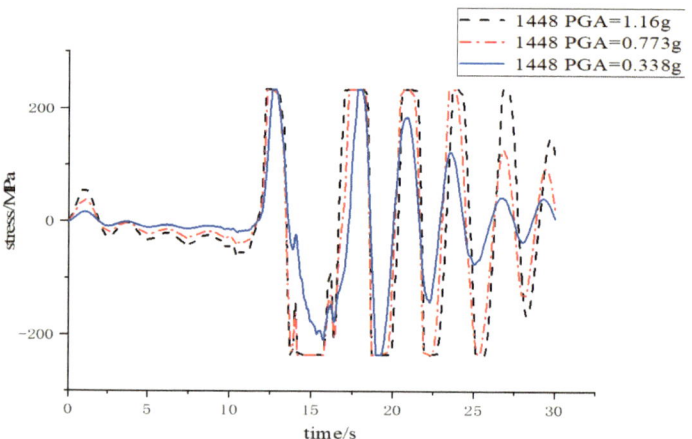

Fig. 7 Stress time history curves of different PGA outer web members

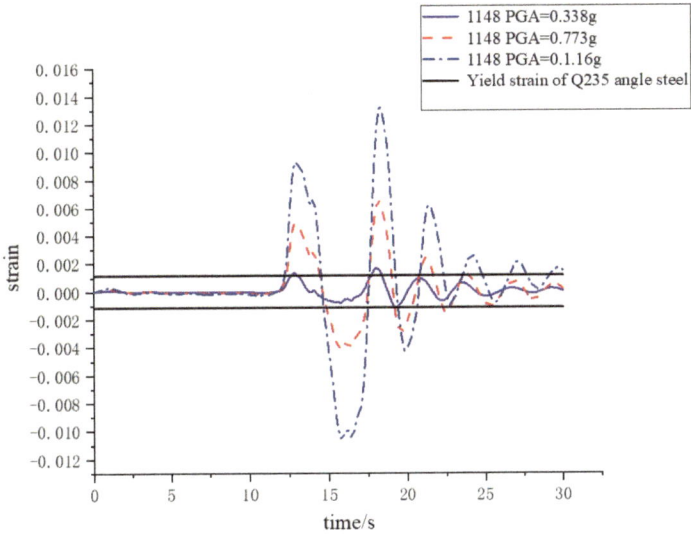

Fig. 8 Strain time history curves of different PGA outer web members

From Table 2, it can be seen that the second, third, and sixth rows of outer web members of the steel truss joint systems always rank at the top as the seismic intensity gradually increases, especially the second and third rows as the main energy dissipation rows, followed by the fourth, fifth, and sixth rows, and finally the seventh, eighth, and first rows, where the first and eighth rows are less likely to yield and dissipate little seismic energy.

Table 3 shows the yielding conditions of the most unfavorable vertical members and inner web members under different wave excitations.

Figures 9 and 10 show the stress and strain time history curves of the third row of vertical bars under wave 1148 excitation. It can be seen from Fig. 9 that under wave 1448 excitation, as the peak ground acceleration increases, the three curves

Table 3 Yield of the most unfavorable row shaft and inner web

Serial number	Numbering	Earthquake shaking	Vertical pole		Internal ventur	
			PGA (g)	σ (MPa)	PGA (g)	σ (MPa)
1	1197	Chi-Chi Taiwan	2.55	73	2.55	235
2	1547	Chi-Chi Taiwan	0.601	100	0.577	235
3	1605	Duzce Turkey	1.258	89	1.258	235
4	1158	Kocaeli Turkey	1.754	95	1.754	235
5	1148	Kocaeli Turkey	1.16	98	1.08	235
6	1176	Kocaeli Turkey	0.972	93	0.85	235
7	881	Landers	0.96	86	0.96	215

never show a plateau and the peak curve never reaches 235 MPa. From Fig. 10, it can be seen that under wave 1448 excitation, as the peak ground acceleration increases, the peak values of the three curves never exceed the yield strain of Q235 steel bars, Figs. 9 and 10 indicate that as the intensity of seismic motion increases, the vertical pole does not yield and does not participate in energy dissipation. The remaining 6 seismic motion time history curves are consistent with the patterns in Figs. 9 and 10. Due to space limitations, only one seismic motion data is cited. From Table 3, it can be seen that the stress of the vertical members is less than 235 MPa under all wave excitations, indicating that the vertical members are in an elastic state and do not participate in dissipating seismic energy. They are only structural members designed to reduce the calculated length of the outer web members and prevent compression buckling. Figures 11 and 12 show the stress and strain time history curves of the third row inner web member under wave 1148 excitation. It can be seen from Fig. 11 that the stress time history curve only appears as the peak acceleration of the ground motion increases under wave 1448 excitation when PGA = 1.08 g. From Fig. 12, it can be seen that the peak value of the time history curve under wave 1448 excitation only exceeds the yield strain of Q235 steel bars when PGA = 1.08 g, Figs. 11 and 12 indicate that as the seismic intensity increases, the inner web member only yields when the seismic intensity is high, and participates in less energy consumption. The remaining six seismic time history curves are consistent with the patterns in Figs. 11 and 12, but due to space limitations, only one seismic data is cited. The inner web members in the first six ground motions in Table 3 reach a stress of 235 MPa and yield when the PGA reaches the maximum amplitude or approaches the maximum amplitude (see Table 3). Under the seventh ground motion in Table 3, the stress is less than 235 MPa at the PGA of the maximum amplitude, indicating that the inner web members start to dissipate seismic energy when the PGA is very large.

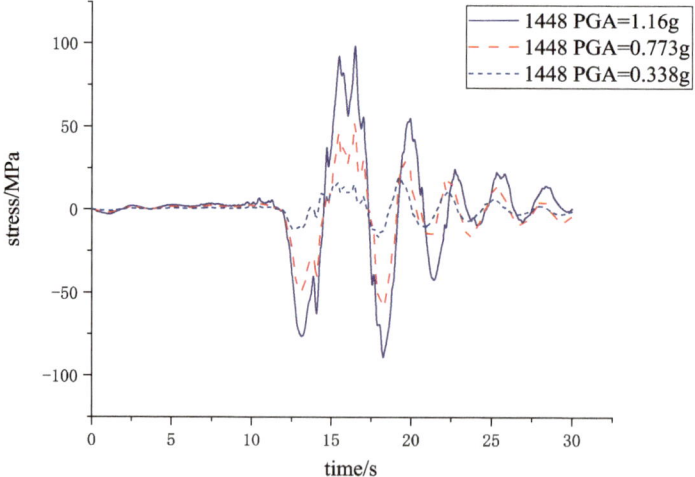

Fig. 9 Vertical bar stress time history curve

Study on the Damage Evolution Law of Railway High Pier of New … 523

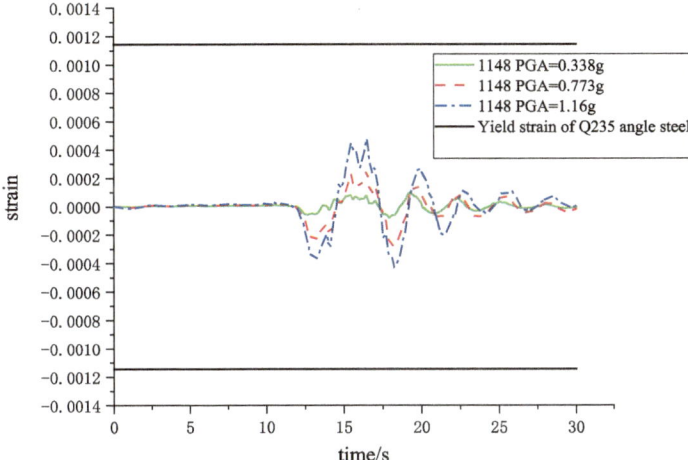

Fig. 10 Vertical rod strain time history curve

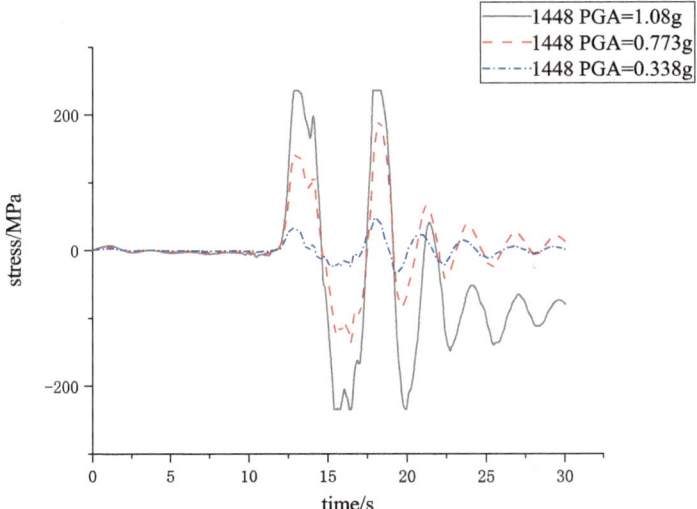

Fig. 11 Stress time history curve of inner web member

Figures 13 and 14 show the stress time history curves of the steel bars at the bottom and middle of the pier under 1148 wave excitation. It can be seen from Figs. 13 and 14 that as PGA gradually increases, the peak value of the curve also increases. The time history curve does not show a platform and the peak value does not reach 400 MPa. Figures 15 and 16 show the strain time history curves of the steel bars at the bottom and middle of the pier under 1148 wave excitation. It can be seen from Figs. 15 and 16 that the peak value of the time history curve has never reached

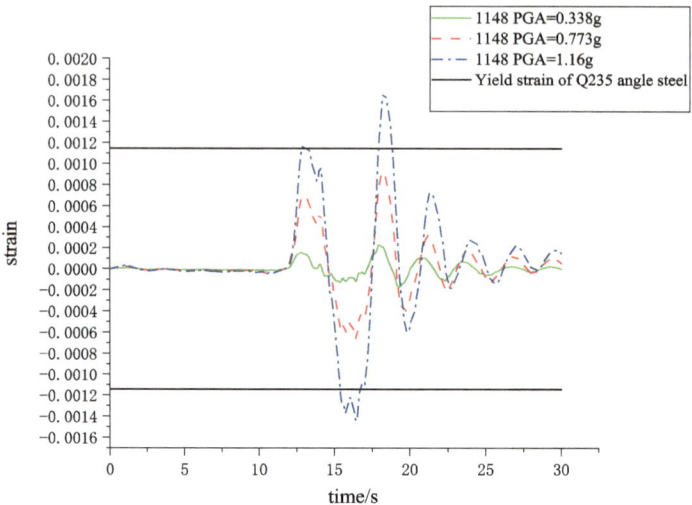

Fig. 12 Strain time history curve of inner web member

the yield strain of HRB400 steel bars, Taking a comprehensive look at Figs. 13, 14, 15, and 16, it is shown that during the process of seismic enhancement, the steel bars at the bottom and middle of the pier did not yield. Table 2 shows that when PGA reached 1.16 g, all the outer web members of the steel truss had already entered a yielding state. This indicates that the yielding of the members occurred before the pier column yielded, consuming energy before the pier column, which played a role in protecting the pier column. The other six seismic time history curves are basically consistent with the laws in Figs. 13, 14, 15, and 16, Due to space limitations, only one seismic data example is provided. Figures 17 and 18 show the variations of steel stress at a distance of 52.5 m (mid-span) from the bottom of the pier under seven near-fault ground motions and the corresponding variations of steel stress at the bottom of the pier. From Fig. 17, it can be seen that for waves 1158 and 1197, which have relatively obvious variations, the upward trend of the curve gradually slows down and the slope of the curve decreases as the PGA gradually increases. This indicates that the yielding of the steel truss joint system row by row can slow down the increase of stress at the mid-span, dissipating a portion of the seismic energy and achieving a damping effect. From Fig. 18, it can be seen that for waves 1197, 1176, 1605, 881, and 1158, which have relatively obvious variations, the upward trend of the curve gradually slows down, and the deceleration magnitude is greater than that of the mid-span curve. This indicates that the damping effect brought by the yielding of the steel truss joint system row by row is more pronounced at the bottom of the pier, dissipating more seismic energy and improving the seismic performance of the new tall pier. Moreover, from Fig. 17, it can be seen that at the PGA of the maximum amplitude of waves 1197, 1605, and 1158 (see Table 2), the steel stress at a distance of 52.5 m from the bottom of the pier reaches 400 MPa and yields, while in other cases.

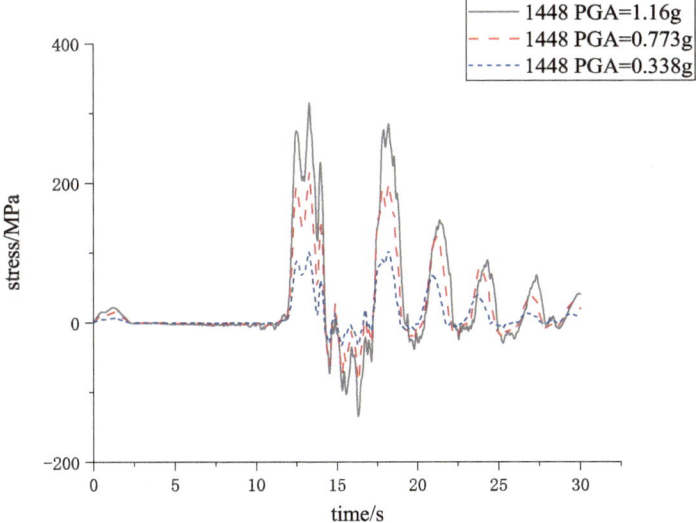

Fig. 13 Stress time history curve of pier reinforcement bottom reinforcement

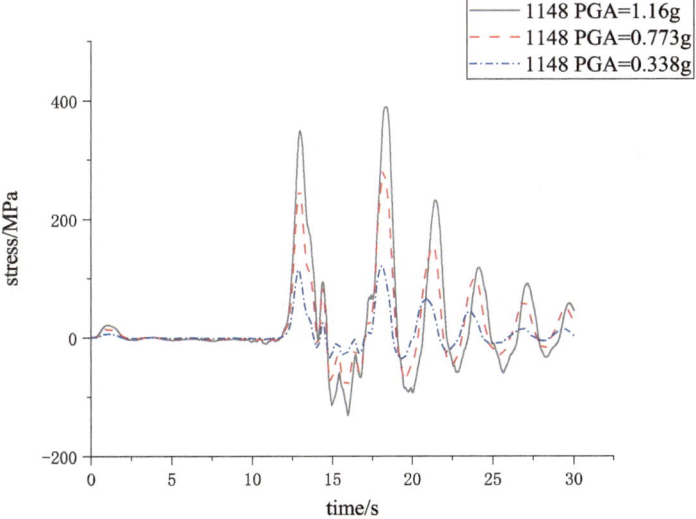

Fig. 14 Stress time history curve of in piers

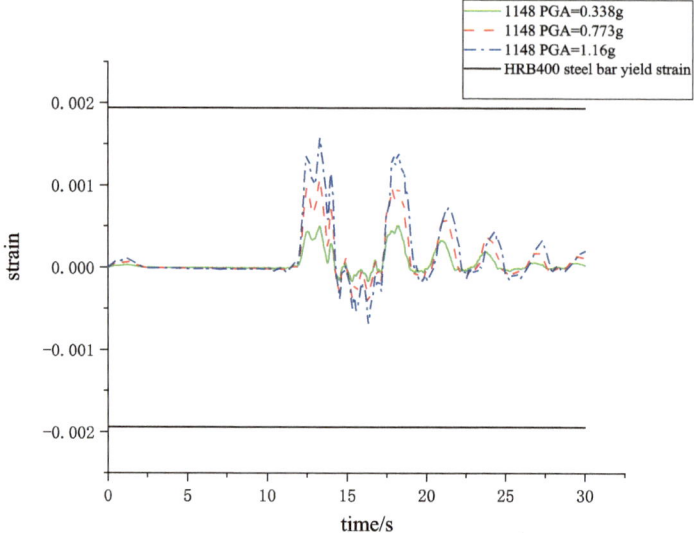

Fig. 15 Strain time history curve of pier bottom reinforcement

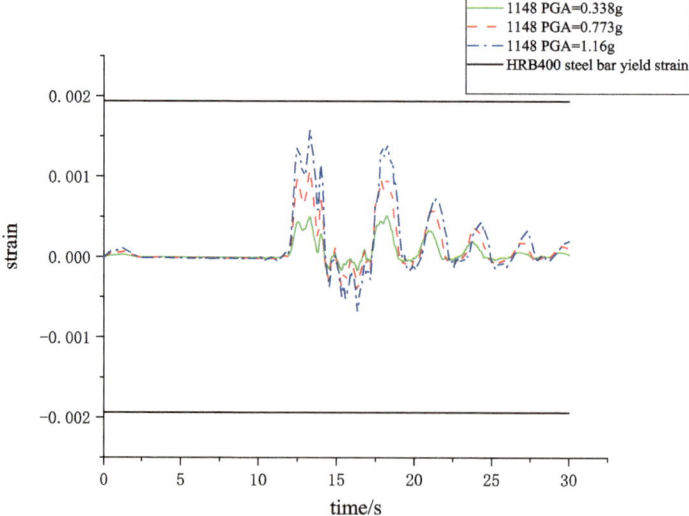

Fig. 16 Strain time history curve of steel bars in the pier

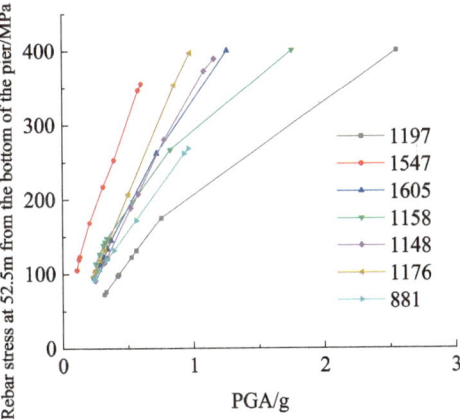

Fig. 17 Tensile stress of rebar at 52.5 m from the bottom of the pie

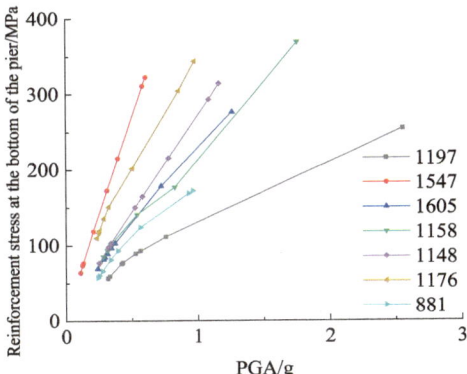

Fig. 18 Tensile stress of steel bar at the bottom of pier

5 Conclusion

(1) A new type of replaceable component railway high pier structure has been proposed. Incremental dynamic analysis shows that as the seismic intensity increases, the outer web members of the steel truss joint systems between piers yield and dissipate energy in a graded manner, and the yielding degree of the outer web members increases with the increase of seismic intensity.

(2) The second, third, and sixth rows of outer web members of the steel truss joint systems always rank at the top as the seismic intensity gradually increases. Among them, the second and third rows have the highest frequency of yielding, dissipate more seismic energy, and serve as the main energy dissipation members. The fourth, fifth, and sixth rows rank next, and the seventh, eighth, and first rows yield the least and dissipate little seismic energy. The first and eighth rows are less likely to yield.

(3) The vertical members and inner web members of the steel truss joint systems are less likely to yield and have limited energy dissipation capacity. They are designed to reduce the compressive calculation length of the outer web members and avoid buckling of the compression members.

(4) The yielding and energy dissipation of the diagonal web members in the steel truss joint systems help reduce the stress in the piers and bottom reinforcement. The seismic isolation effect at the bottom of the piers is better than that in the middle. The yielding of the piers occurs after the yielding of the diagonal web members of the steel truss.

References

1. Bassam A, Iranmanesh A, Ansari FA (2011) Simple quantitative approach for post earthquake damage assessment of flexure dominant reinforced concrete bridges. J Eng Struct 33(12):3218–3225
2. Chen X, Li JZ, Liu XX (2017) Seismic performance of tall piers influenced by higher-mode effects of piers. J Tongji Univ (Nat Sci) 45(02):159–166
3. Lu H, Li JZ (2013) Analysis of seismic performance characteristics of bridge with high piers under strong earthquake motion. China Earthq Eng J 35(04):858–865
4. Han Q, Dong HH, Wang LH et al (2021) Review of seismic resilient bridge structures with replaceable members. China J Highw Transp 34(09):15–230
5. Yie AJ, Guan ZG (2017) Bridge seismic resistance. People's Communications Press, Beijing
6. Kawashima K, Macrae GA, Hoshikuma JI et al (1998) Residual displacement response spectrum. J Struct Eng 124(5):23–530
7. Zhou Y, Wu H, Gu AQ (2019) Earthquake engineering: From earthquake resistance, energy dissipation, and isola-teion, to resilence. Eng Mech 36(06):1–12
8. Li JZ, Guan ZG (2017) Research progress on bridge seismic design: target from seismic alleviation to post-earthquake structural resilience. China J Highw Transp 30(12):1–9+59
9. Du Q, Xia XS, Chen XC et al (2019) Study of higher-mode effects on self-centering piers under near-field and far-field earthquakes. J. Bridge Constr 49(06):60–65
10. Du Q, Xia XS, Chen XC et al (2019) Research on high-order modal contribution of railway self-reducing high piers. J Vib Eng 32(06):1003–1010
11. Liu XL, Chen Y, Jiang HJ (2013) Research progress in new replaceable coupling beams. J Earthq Eng Eng Vib 33(01):8–15
12. Liu XL, Chen C (2014) Research progress in structural systems with replaceable members. J Earthq Eng Eng Vib 34(01):27–36
13. Liu XL, Wu DY, Zhou Y (2019) State-of-the-art of earthquake resilient structures. J Build Struct 40(02):1–15
14. Zhou Y, Gu AQ (2019) Displacement-based seismic design of self-centering shear walls under four-level seismic fortifications. J Build Struct 40(03):118–126
15. Xia XS, Zhang XY, Wang JB (2021) Shaking table test of a novel railway bridge pier with replaceable components. J. Eng Struct 232:1–9
16. Mang L (2019) Study on seismic behavior of new high-rise railway piers. Lanzhou Jiaotong University, Lanzhou
17. Huang YB (2021) Research on the seismic damage evolution law of new type high pier. Lanzhou Jiaotong University, Lanzhou
18. Gao ZL, Xia XS, Huang YB et al (2022) Fragility analysis of railway high pier with replaceable components under near fault ground motions. J Railway Sci Eng 19(09):2682–2690

19. Chen X, Dario DD, Li CX (2023) Seismic resilient design of rocking tall bridge piers using inerter-based systems. Eng Struct 281
20. Du K, Sun JJ, Xu WX (2012) The division of element, section and fiber in fiber model. J Earthq Eng Eng Vibr 32(05):39–46

Open Access This chapter is licensed under the terms of the Creative Commons Attribution 4.0 International License (http://creativecommons.org/licenses/by/4.0/), which permits use, sharing, adaptation, distribution and reproduction in any medium or format, as long as you give appropriate credit to the original author(s) and the source, provide a link to the Creative Commons license and indicate if changes were made.

The images or other third party material in this chapter are included in the chapter's Creative Commons license, unless indicated otherwise in a credit line to the material. If material is not included in the chapter's Creative Commons license and your intended use is not permitted by statutory regulation or exceeds the permitted use, you will need to obtain permission directly from the copyright holder.

Damage Assessment of RPC Strengthened RC Columns Subjected to Blast Loading

Yu Fu, Siyuan Qiu, Zhifu Yu, Juan Su, and Xiaomeng Hou

Abstract In our country, there are 60 billion square meters of existing leasable area, reactive powder concrete (RPC) has the characteristics of high toughness and high strength-to-weight ratio. RPC strengthened concrete (RC) forms RPC-RC structure, it is an important way for sustainable development of civil engineering to improve the anti-explosion ability of building structure while strengthening and reforming the structure. The dynamic response and damage assessment methods of RPC-RC columns subjected to blast loading are not clear. In order to solve this problem, the dynamic response of RPC-RC column under blast loading is simulated by using LS-DYNA finite element software, the Pressure-Impulse curve of RPC-RC column is established by using simplified numerical method, and the parameters of Pressrue-Impulse curve are analyzed. Finally, based on the results of parameter analysis, the formulas for calculating the overpressure asymptote and impulse asymptote are obtained, and the damage assessment method for RPC-RC columns is established.

Keyword RPC reinforcement · Blast resistance · Numerical simulation · Damage assessment

1 Introduction

There are about 60 billion square meters of existing leasable area in our country, and the safety of existing building structures decreases with the increase of years. RPC with high strength, durability and toughness has been applied more and more in the reinforcement of existing buildings. Zheng Wenzhong [1] used RPC as a template to form RPC and RC composite structure, and carried out research on

---

Y. Fu · J. Su
Offshore Oil Engineering Co., Ltd, Tianjin 300461, China

S. Qiu · Z. Yu · X. Hou (✉)
Key Lab of Structures Dynamic Behavior and Control of Ministry of Education, Harbin Institute of Technology, Harbin 150090, China
e-mail: houxiaomeng_hit@126.com

bending and shear properties and design methods, and applied them to prefabricated building. Professor Deng Zongcai [2], professor Bu Liangtao [3] and Talayeh [4] have studied RC members strengthened by RPC under bending, compression and shear respectively. The research and application show the advantages and prospects of RPC in strengthening and improving the structural life. Professor Sun Wei, Professor Fang Qin, Professor Wu Hao, Professor Ju Yang and others [5, 6] have studied the dynamic behavior and failure mode of RPC structural members under explosive loading, when the strain rate ranges from 10 to 1000 s^{-1}, the strength of RPC increases with the increase of strain rate. Sometimes the explosion accident may lead to the damage of the reinforced structure. The dynamic response and damage assessment methods of RPC reinforced concrete columns are not clear. The application of P-I curve to the damage assessment of building structures can be traced back to the 1940s and 1950s, the damage degree of structural members under the possible explosion can be predicted. At present, many scholars have used P-I curve method to evaluate the damage of RC columns, beams, slabs and block walls. Cormie [7] and others divided the P-I curve into three parts according to the ratio of the duration of explosive action and the inherent period of the component. Hou [8] established the P-I curve of RPC unidirectional plate based on SDOF method. The P-I curve evaluation method of RPC reinforced concrete columns has not been studied. In this paper, based on the failure criterion of residual axial compression capacity, the P-I curves of RPC-RC columns are established, and the parameters of the P-I curves are analyzed.

2 Establishing P-I Curve by Numerical Simulation

The overpressure P and impulse I of P-I curve established by numerical simulation are two basic parameters of explosion action. The coordinates (P, I) corresponding to a point in P-I curve represent an explosion working condition. The damage degree of column under explosion is measured by damage index. Three critical damage grades ($D = 0.2, D = 0.5, D = 0.8$) were used to divide the damage grade of the column into four grades: mild, moderate, severe and collapse. It needs a lot of trial calculation to draw P-I curve by numerical simulation, and then the points near three critical damage grades are selected to fit the curve, the steps of drawing the curve are as follows: (1) select a suitable set of overpressure and impulse for initial calculation, first keep the overpressure value unchanged, and continuously adjust the duration of explosion to change the impulse value, (2) to reduce the overpressure value and keep the same, and to adjust the impulse value until the damage degree of the column reaches the predetermined value again; (3) to repeat (1)–(2) steps of the work, a large number of data points are obtained and fitted to draw a complete P-I curve.

3 The P-I Curves of RPC-RC Columns Are Established by Using the Simplified Numerical Method

To analyze the P-I curves of RPC-RC columns. The parameters are RPC thickness and RPC strength in RPC reinforced layer. The finite element model of RPC reinforced concrete column is shown in Fig. 1. The parameters of each column are shown in Table 1. The effects of parameters on P-I curve, overpressure asymptote and impulse asymptote are studied, and the formulas for calculating overpressure asymptote and impulse asymptote are established based on the analysis results.

3.1 RPC Thickness

In order to study the influence of the thickness of RPC reinforced layer on the P-I curve, overpressure asymptote and impulse asymptote of RPC-RC column, the thickness of RPC reinforced layer is 20 mm, 30 mm and 40 mm respectively, the values of the remaining parameters of the column are the same as those of the ordinary RPC-RC column in Table 1. The P-I curves of RPC-RC columns with three different reinforced layer thicknesses are established by using a simplified numerical method, as shown in Fig. 2, the P-I curves of RPC-RC columns move to the right. The overpressure asymptotes and impulse asymptotes of the three damage grades of the P-I curves increase with the increase of the thickness of the reinforced layer. This is because the overpressure asymptote corresponds to the quasi-static load, the bending failure of the column occurs mainly, and the impulse asymptote corresponds to the impulse load, when the shear failure of the column occurs, therefore, it can be concluded that the flexural capacity and shear capacity of columns increase with the increase of RPC thickness, which is related to the increase of positive moment

Fig. 1 Finite element model of RPC-RC column

Table 1 RPC-RC column parameters

RPC thickness t (mm)	RPC strength f_{RPC} (MPa)	RPC longitudinal reinforcement ratio ρ_{s0} (%)	Strength of concrete f_c (MPa)	Width of column section b (mm)	Height of column section h (mm)	Column height H (mm)
20	145	0.5	30	300	300	3300

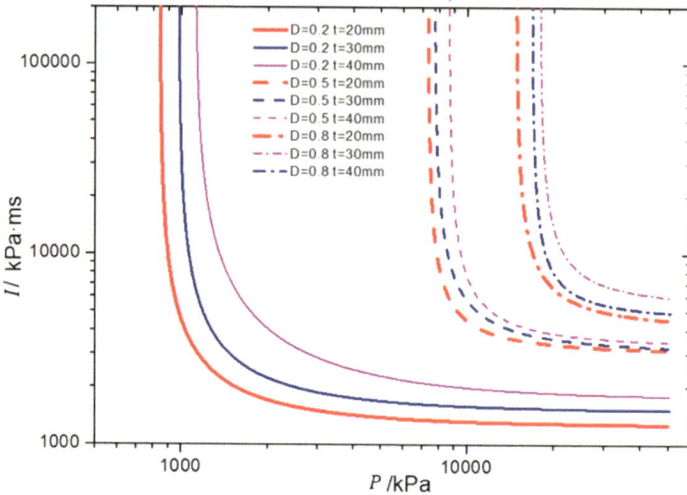

Fig. 2 The P-I curves of RPC-RC columns with three different thickness of RPC strengthening layer

of inertia of the section of columns, therefore, when using RPC strengthening, the thick RPC strengthening layer should be given priority.

3.2 RPC Strength

Considering the influence of the strength of RPC reinforced layer on the P-I curve of RPC-RC column, the overpressure asymptote P_0 and the impulse asymptote I_0, the strength of RPC reinforced layer is 145 MPa, 175 MPa and 215 MPa respectively, the values of the remaining parameters are the same as those of the RPC-RC column in Table 1. Based on the simplified numerical method, the P-I curves of three different strength RPC-RC columns are also established. Figure 3 shows that the P-I curve moves slightly to the right with the increase of RPC strength, and the P_0 and I_0 corresponding to the three critical damage grades also increase with the increase of RPC strength, but the increase is not significant, from the above results, it can be concluded that when the thickness of RPC strengthening layer is small, the effect of increasing the strength of RPC strengthening layer on reducing the damage degree of columns is small, and when the thickness of RPC strengthening layer is large, the strength of RPC strengthened layer has an obvious effect on the damage degree of columns. Therefore, the RPC strengthening design should ensure that the RPC has a greater thickness in order to give full play to the ultra-high strength of RPC.

Damage Assessment of RPC Strengthened RC Columns Subjected …

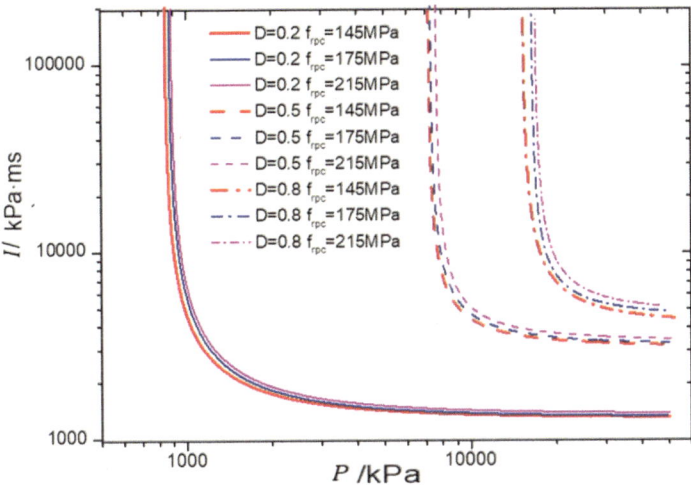

Fig. 3 P-I curves of RPC-RC columns with three different RPC strengths

4 Establishment of the Calculation Formula

Based on the parametric analysis of the P-I curves of RPC-RC columns in Chap. 3, the empirical formulas of the overpressure asymptote, the impulse asymptote and the nine parameters of RPC-RC column under three critical damage levels are established by using the multivariate nonlinear regression method, as shown in Formulas (1)–(6).

$$P_0(0.2) = (6.561 f_c/p_0 + 5.556 h/l_0 - 0.558 b/l_0 - 1.104 H/l_0 \\ + 26587.312 \rho + 17621.817 \rho_s + \alpha_1) p_1 \tag{1}$$

$$I_0(0.2) = (0.706 f_c/p_0 + 3.111 h/l_0 - 0.115 b/l_0 - 0.369 H/l_0 \\ + 3807.634 \rho + 6827.972 \rho_s + \alpha_2) I_1 \tag{2}$$

$$P_0(0.5) = (26.831 f_c/p_0 + 14.983 h/l_0 - 1.077 b/l_0 - 3.377 H/l_0 \\ + 75117.158 \rho + 31493.027 \rho_s + \alpha_3) p_1 \tag{3}$$

$$I_0(0.5) = (11.748 f_c/p_0 + 25.335 h/l_0 - 0.657 b/l_0 - 4.485 H/l_0 \\ + 30796.443 \rho + 62634.248 \rho_s + \alpha_4) I_1 \tag{4}$$

$$P_0(0.8) = (0.364 f_c/p_0 + 25.424 h/l_0 - 1.449 b/l_0 - 4.456 H/l_0 \\ + 69585.615 \rho + 1484.431 \rho_s + \alpha_5) p_1 \tag{5}$$

$$I_0(0.8) = (14.745f_c/p_0 + 29.834h/l_0 - 1.091b/l_0 - 16.906H/l_0 \\ + 54859.693\rho + 255489.232\rho_s + \alpha_6)I_1 \tag{6}$$

The α_1–α_6 are 0 for unreinforced RC columns, and the formulas for $\alpha1$–$\alpha6$ in RC columns strengthened with RPC are as followed.

$$\alpha_1 = \exp(0.002f_{rpc}/p_0 + 0.025t/l_0 + 0.067\rho_{s0} + 5.527)$$

$$\alpha_2 = \exp(0.001f_{rpc}/p_0 + 0.033t/l_0 + 0.085\rho_{s0} + 4.238)$$

$$\alpha_3 = \exp(-0.346f_{rpc}/p_0 + 1.268t/l_0 + 7.346\rho_{s0} - 42.327)$$

$$\alpha_4 = \exp(0.003f_{rpc}/p_0 + 0.031t/l_0 + 0.114\rho_{s0} + 5.681)$$

$$\alpha_5 = \exp(0.004f_{rpc}/p_0 + 0.043t/l_0 + 0.205\rho_{s0} + 3.969)$$

$$\alpha_6 = \exp(0.002f_{rpc}/p_0 + 0.011t/l_0 + 0.116\rho_{s0} + 8.234)$$

In the formula, P_0 is the overpressure asymptote, unit KPa; I_0 is the impulse asymptote, unit KPa·ms; f_{RPC} is the strength of RPC strengthened layer, unit MPa; t is the thickness of RPC strengthened layer, unit mm; ρ_{S0} is the ratio of longitudinal reinforcement of RPC strengthened layer; f_c for compressive strength of concrete, unit MPa; h is column height, unit mm; b is section width, unit mm; d is section height, unit mm; ρ is longitudinal reinforcement ratio; ρ_S is stirrup ratio, and p_0 is 1 Mpa, p_1 is 1 Kpa, l_0 is 1 mm, I_1 is KPa·ms. The formula is only derived from numerical simulation and has not been tested.

5 Conclusion

The P-I curves of RC column and RPC-RC column are established based on the numerical simulation method. The damage degree of RC column strengthened by RPC is obviously reduced. In order to study the influence of the parameters of RPC strengthened layer (the thickness of RPC strengthened layer and the strength of RPC strengthened layer) on the P-I curve of RPC-rc column, the simplified numerical method is used to establish the P-I curve of RPC-rc column and the parametric analysis is carried out, the conclusions are as follows: (1) When the thickness of RPC is between 20 and 30 mm, the overpressure asymptote P_0 increases about 25% and the impulse asymptote I_0 increases about 20%. With the increase of the thickness, and when the thickness of RPC is 40 mm, the strength of RPC increases from 145 to 215 MPa, P_0 and I_0 increases about 25%. (2) Based on the results of parametric analysis, the calculation formulas of overpressure asymptote and impulse asymptote are presented.

Acknowledgements This paper is one of the stage results of the national natural science foundation project "Dynamic Performance and blast resistance design method of RPC reinforced concrete frame" (52078169).

References

1. Zheng W, Lv X, Wang Y (2016) Flexural behaviour of concrete composits slabs with precast ribbed reactive powder concrete bottom panels. J Comput Theor Nanosci 13(3):1831–1839
2. Zongcai D, Wuchen Z (2015) Flexural behavior of RC beams strengthened with hybrid fiber reinforced RPC. J Harbin Eng Univ 36(09):1199–1205
3. Tao PL, Zhu J, Tao Jian J, Hou Q (2014) Experimental study on high strength concrete columns strengthened with reactive powder concrete under small eccentric loading. Architecture 44(11):14–19
4. Talayeh N, Eugen B (2013) Experimental investigation on reinforced ultra-high-performance fiber-Reinforced concrete composite beams subjected to combined bending and shear. ACI Struct J 110(2):251–261
5. Lai J, Sun W (2009) Dynamic behaviour and visco-elastic damage model of ultrahigh performance cementitious composite. Cem Concr Res 39(11):1044–1051
6. Li X, Fang Q, Kong X, Wu H (2018) Inertial effect correction of strain rate effect curve of concrete-like materials in numerical simulation. Eng Mech 35(12):46–53
7. Cormie D, Mays G, Smith P (2009) Blast effects on buildings
8. Hou X, Cao S, Rong Q, Zheng W et al (2018) A P-I diagram approach for predicting failure modes of RPC one-way slabs subjected to blast loading. Int J Impact Eng 120:171–184

Open Access This chapter is licensed under the terms of the Creative Commons Attribution 4.0 International License (http://creativecommons.org/licenses/by/4.0/), which permits use, sharing, adaptation, distribution and reproduction in any medium or format, as long as you give appropriate credit to the original author(s) and the source, provide a link to the Creative Commons license and indicate if changes were made.

The images or other third party material in this chapter are included in the chapter's Creative Commons license, unless indicated otherwise in a credit line to the material. If material is not included in the chapter's Creative Commons license and your intended use is not permitted by statutory regulation or exceeds the permitted use, you will need to obtain permission directly from the copyright holder.